METHODS IN MOLECULAR BIOLOGY

Series Editor
John M. Walker
School of Life and Medical Sciences
University of Hertfordshire
Hatfield, Hertfordshire, AL10 9AB, UK

For further volumes:
http://www.springer.com/series/7651

Biological Networks and Pathway Analysis

Edited by

Tatiana V. Tatarinova

Keck School of Medicine
University of Southern California
Los Angeles, CA, USA

Yuri Nikolsky

Prosapia Genetics
Solana Beach, CA, USA

Humana Press

Editors
Tatiana V. Tatarinova
Keck School of Medicine
University of Southern California
Los Angeles, CA, USA

Yuri Nikolsky
Prosapia Genetics
Solana Beach, CA, USA

ISSN 1064-3745 ISSN 1940-6029 (electronic)
Methods in Molecular Biology
ISBN 978-1-4939-7025-4 ISBN 978-1-4939-7027-8 (eBook)
DOI 10.1007/978-1-4939-7027-8

Library of Congress Control Number: 2017937364

Printed on acid-free paper

This Humana Press imprint is published by Springer Nature
The registered company is Springer Science+Business Media LLC
The registered company address is: 233 Spring Street, New York, NY 10013, U.S.A.

Preface

Google queries for *systems biology* and *pathway analysis* fetch over 9 million and 14 million entries, respectively. These numbers speak volumes about the utility and popularity of systems data analysis in modern bioscience. These days, any gene expression or SNP-analyzing manuscript would feature a chapter on pathways, ontology enrichment, and/or biological networks. The application of systems biology approaches now spreads widely from basic research and preclinical drug discovery to translational research and personalized healthcare.

Systems biology "focuses on the systematic study of complex interactions in biological systems, thus using a new perspective (integration instead of reduction) to study them" (Wikipedia). From a practical standpoint, it translates as an integration of accumulated biological knowledge in a computer-readable format, followed by a creation of tools for the analysis of biological and chemical experimental data. Starting in the 1970s, biochemistry was the first field codified into databases such as BRENDA, EMP/MPW, and, later, KEGG. Over the years, a regulation and signaling components were added to biochemistry in the form of protein interaction databases such as HPRD and BIND. On top of that, comprehensive ontologies of cellular processes and protein functions were developed and integrated, the best known of which is Gene Ontology (GO).

Functional analysis is inseparable from high-throughput, or "omics"-driven experimental biology, which has been rapidly evolving since the late 1990s. At that time, the "genome-wide," noisy assays with thousands of data points were nearly illegible for a majority of wet lab researchers, in part, due to the "diaper stage" of development for the statistical tools which only helped to reduce data complexity, but largely failed to aid in understanding of the underlying biology. Gradually, bioinformaticians and wet lab biologists found efficient ways of communicating. As a result, wet lab biologists acquired the skill of using existing databases of pathways and processes for mapping and prioritization of experimental data (enrichment analysis). Later, biological networks were added to analysis toolboxes, borrowing from years of research in graph theory and physics.

Recent technological advances and scalability in next-generation sequencing (NGS) and other genomics technologies enable production of biological "big data" at unprecedented tera- and petabyte scales. Efficient mining of these vast and complex datasets for the needs of biomedical research critically depends on an integration of the clinical and omics information, sophisticated analytical tools, and taking into account prior knowledge about genotype-phenotype relationships and protein functionality. Experimental "omics" data has been accumulated in publicly available and private databases for over 20 years.

Analytical tools are described in hundreds of computational biology and bioinformatics publications and scattered across code repositories and commercial bioinformatics suites. Information about protein functionality is structured and accumulated in computer-readable format in several curated databases on protein-protein interactions, pathways, and network modules. Such curated content is then used for analysis of "omics" datasets, by means of ontology enrichment, interactome density analysis, pathways activation analysis, network modeling, and other approaches. In this book, we collected cutting-edge material on the latest methods and studies on "data-driven" and "knowledge-based" analysis from the internationally recognized leaders in this field.

This book represents a compilation of methods of functional analysis and their applications, written by experts from academy, governmental research organizations, pharmaceutical industry, and bioinformatics laboratories. It begins with the modeling of protein-protein interactions (PPIs) and protein-nucleic acids interactions as these are the building blocks of protein functionality and the most essential tools for functional analysis of large experimental datasets. There are several ways to extract protein interactions. Information on many of them is scattered in hundreds of thousands of experimental articles and can be extracted in both human- and machine-readable form. We have two chapters devoted to extracting PPIs from literature and an experimental one, using a modified yeast two-hybrid assays. In the former approach, a team from GeneGo (now acquired by Thomson Reuters) developed a sophisticated approach of structured manual annotations to assemble a comprehensive database of over 1 million experimentally proven interactions of different types.

In the second chapter, the authors present high-throughput, quantitative, yeast two-hybrid screening approach coupled with the NGS approach. This strategy allows identification of interacting proteins that are preferentially associated with a bait of interest and helps eliminate nonspecific interacting proteins.

As an example of the large-scaledata-driven network approach, we included a chapter on co-expression modules in cancer datasets. The analysis of differentially expressed gene sets (in a form of functionally related genes or pathways) in a form of either RNA-Seq or microarray experiments has been a method of choice for extracting the strongest signals from "omics" data. The authors combined an experimental approach of extracting co-expression modules from cancer expression datasets via meta-analysis with calculation of promoter motifs. Analysis of gene co-expression networks is a powerful "data-driven" tool, invaluable for understanding cancer biology and mechanisms of tumor development.

The most common and intuitive approach to functional analysis of "omics" datasets is ontology enrichment. Essentially, it consists of labeling each gene, protein, and RNA species on the experimental list with a certain functional category (cellular process, pathway, network module etc.), followed by grouping them according to the "collective" labels. The motivation behind using gene sets instead of individual genes is twofold. First, this approach incorporates pre-existing biological knowledge into the analysis and facilitates the interpretation of experimental results. Second, it employs a statistical hypotheses testing framework.

In this book, we include a comprehensive review of the Gene Set Analysis (GSA) approaches for testing differential expression of gene sets and several GSA approaches for testing statistical hypotheses beyond differential expression that allow to extract additional biological information from the data. Gene sets frequently can be analyzed as pathways. A novel algorithm OncoFinder evaluates the activation of molecular pathways on the basis of gene/protein expression data in the objects of interest. OncoFinder enables performing both quantitative and qualitative analysis of the intracellular molecular pathways. Another approach enables causal analysis of multidimensional "omics" dataset using an "upstream analysis" strategy which combines TRANSFAC database with analysis of the upstream signal transduction pathways that control the activity of these TFs. This analysis highlighted a substantial heterogeneity of specific TF-DNA binding sites in terms of their observed relative binding avidity and correlations between avidity for specific TF-DNA binding sites with the levels of mRNA transcription at the proximal gene target. Combined gene expression/promoter sequence analysis has been applied to extract novel insight from cancer biology.

Another novel method, weighted SNP correlation network analysis (WSCNA), can be used to identify SNP networks from GWAS data, create network-specific polygenic scores, examine network topology to identify hub SNPs, and gain biological insights into complex traits. An automatic annotation system (Association Rule Mining Annotator for Pathways) utilizes rule mining techniques to predict metabolic pathways across a wide range of prokaryotes. This system can be used to enhance the quality of automatically generated annotations as well as annotating proteins with unknown function.

The increasing amount and variety of data in biosciences call for innovative methods of visualization, scientific verification, and pathway analysis. sbv IMPROVER is a platform that uses crowdsourcing and verification to create biological networks with easy public access. Currently, it contains 120 networks built in Biological Expression Language to interpret data from PubMed articles with high-quality verification available for free on the CBN database. Another solution is an integrated computational platform Lynx—aweb-based database and knowledge extraction engine, which provides its users with advanced search capabilities and an access to a variety of algorithms for enrichment analysis and network-based gene prioritization. User-friendly web services and interfaces connect its users both to the Lynx integrated knowledge base (LynxKB) and integrated analytical tools.

MetaCore and Key Pathway Advisor constitute an integrated platform for functional data analysis. This platform enables analysis of sequencing data, annotation of gene variants, gene expression, proteomics, and other high-throughput (OMICs) data, which is routinely challenging because of its biological complexity and high level of technical and biological noise. We present techniques and concepts used to represent complex biomedical networks. The BioXM Knowledge Management Environment (BioMax AG, Germany) is an example of how a domain such as oncology is represented and how this representation is utilized for research. We also discuss the ArrayTrack (National Center for Toxicology Research, FDA) that is also used in the routine review of genomic data submitted to the FDA. ArrayTrack stores a full range of information related to DNA microarrays and clinical and nonclinical studies as well as the digested data derived from proteomics and metabolomics experiments.

Recent advances in genome sequencing and "omics" technologies are opening new opportunities for improving diagnosis and treatment of human diseases. The precision medicine initiative in particular aims at developing individualized treatment options that take into account individual variability in genes and environment of each person. Systems biology approaches that group genes, transcripts, and proteins into functionally meaningful networks will play a crucial role in the future of personalized medicine. By that, systems biology enables comparisons of healthy and disease-affected tissues and organs from the same individual, as well as these between healthy and disease-afflicted individuals. However, the field faces a multitude of challenges ranging from data integration to statistical and combinatorial issues in data analyses. Here, we collected computational approaches developed to tackle challenges in network analyses. Successful application of systems biology approach to psychiatric diseases opens the application part of our book. Another chapter is using an example of Alzheimer's disease to identify and analyze the candidate gene lists, and divide them up into different tiers of evidence consistency established by enrichment analysis across sub-datasets collected within the same experiment and across different experiments and platforms. Ingenuity Pathway Assistant tool was used to expand these gene lists and interpret the outputs.

One chapter is devoted to a different kind of networks, the connectome of brain cells affected in mental diseases. It has been long recognized that schizophrenia, unlike certain other mental disorders, appears to be delocalized, i.e., difficult to attribute to a dysfunction of a few specific brain areas, and may be better understood as a disruption of brain's emergent network properties. The authors focused on topological properties of functional brain networks obtained from fMRI data, in order to demonstrate that some of those properties can be used as discriminative features of schizophrenia in multivariate predictive setting.

We have also included a chapter on in-depth clinical analysis of a particular pathway, with pleiotropic effects on key cellular functions. Wnt (Wingless-related integration site) is one of the key signaling pathways in eukaryotes, which orchestrates self-renewal programs in normal somatic stem cells as well as in cancer stem cells. Aberrant Wnt signaling is associated with a wide variety of malignancies and diseases. Although our understanding has increased tremendously over the past decade, therapeutic targeting of the dysregulated Wnt pathway remains a challenge and the effect of Wnt-targeted compounds poorly predictable. The chapter revised recent preclinical and clinical therapeutic approaches to target the Wnt pathway.

Functional data analysis is evolving quickly as a discipline. Novel network algorithms and software tools are published almost weekly, and the scope of applications expands with every new DNA, RNA, or protein assay hitting the market. Therefore, we could not and had no intention to pack as many tools as possible into this volume. Instead, we tried to focus on the established methods and software packages we see in the marketplace every day and provide readers with a broad understanding of issues and applications of this fascinating new field.

Los Angeles, CA, USA *Tatiana V. Tatarinova*
Solana Beach, CA, USA *Yuri Nikolsky*

Contents

Contributors

ALEXANDER M. ALIPER • *Drug Research and Design Department, Pathway Pharmaceuticals, Wan Chai, Hong Kong, Hong Kong SAR; Department of Personalized Medicine, First Oncology Research and Advisory Center, Moscow, Russia; Laboratory of Bioinformatics, D. Rogachev Federal Research Center of Pediatric Hematology, Oncology and Immunology, Moscow, Russia*

ANCHA BARANOVA • *The Center of the Study of Chronic Metabolic and Rare Diseases, School of Systems Biology, George Mason University, Fairfax, VA, USA; Russian Centre for Medical Genetics RAMN, Moscow, Russia; Moscow Institute of Physics and Technology, Dolgoprudny, Russia; Atlas Biomed Group, Moscow, Russia*

NIKOLAY M. BORISOV • *Department of Personalized Medicine, First Oncology Research and Advisory Center, Moscow, Russia; Laboratory of Bioinformatics, D. Rogachev Federal Research Center of Pediatric Hematology, Oncology and Immunology, Moscow, Russia; National Research Centre "Kurchatov Institute", Centre for Convergence of Nano-,Bio-, Information and Cognitive Sciences and Technologies, Moscow, Russia*

IMANE BOUDELLIOUA • *Computational Bioscience Research Center (CBRC), Computer, Electrical and Mathematical Sciences and Engineering Division (CEMSE), King Abdullah University of Science and Technology (KAUST), Thuwal, Saudi Arabia*

SERGEY BRUSKIN • *Laboratory of Functional Genomics, Vavilov Institute of General Genetics RAS, Moscow, Russia; Moscow Institute of Physics and Technology, Dolgoprudny, Russia*

ANTON A. BUZDIN • *Drug Research and Design Department, Pathway Pharmaceuticals, Wan Chai, Hong Kong, Hong Kong SAR; Department of Personalized Medicine, First Oncology Research and Advisory Center, Moscow, Russia; Laboratory of Bioinformatics, D. Rogachev Federal Research Center of Pediatric Hematology, Oncology and Immunology, Moscow, Russia; Group for Genomic Regulation of Cell Signaling Systems, Shemyakin-Ovchinnikov Institute of Bioorganic Chemistry, Moscow, Russia; National Research Centre "Kurchatov Institute", Centre for Convergence of Nano-,Bio-, Information and Cognitive Sciences and Technologies, Moscow, Russia*

GUILLERMO A. CECCHI • *IBM T.J. Watson Research Center, Yorktown Heights, NY, USA*

EVGENY CHEKALIN • *Laboratory of Functional Genomics, Vavilov Institute of General Genetics RAS, Moscow, Russia*

MINJUN CHEN • *Division of Systems Toxicology, National Center for Toxicological Research (NCTR), FDA, Jefferson, AR, USA*

ROSER COROMINAS • *Department of Psychiatry, University of California San Diego, La Jolla, CA, USA; Genetics Unit, Universitat Pompeu Fabra, Hospital del Mar Research Institute (IMIM), Barcelona, Spain; Centro de Investigación Biomédica en Red de Enfermedades Raras (CIBERER), Instituto de Salud Carlos III, Barcelona, Spain*

MARK D'SOUZA • *Department of Human Genetics, University of Chicago, Chicago, IL, USA*

DARRELL DESVEAUX • *Department of Cell and Systems Biology and Centre for the Analysis of Genome Evolution and Function, University of Toronto, Toronto, ON, Canada*

INNA DUBCHAK • *Genomics Division, Lawrence Berkeley National Laboratory, Berkeley, CA, USA; Department of Energy Joint Genome Institute, Walnut Creek, CA, USA*

ALEXEY DUBOVENKO • *Clarivate Analytics, Thomson Reuters, Philadelphia, PA, USA*

YANN DUCHARTRE • *Division of Hematology and Oncology, Department of Pediatrics and Pathology, Children's Hospital Los Angeles, Keck School of Medicine, University of Southern California, Los Angeles, CA, USA*

FRANK EMMERT-STREIB • *Computational Medicine and Statistical Learning Laboratory, Tampere University of Technology, Tampere, Finland*

HONG FANG • *Z-Tech Corporation, An ICF International Company, Jefferson, AR, USA*

T. CONRAD GILLIAM • *Department of Human Genetics and Computation Institute, University of Chicago, Chicago, IL, USA*

GALINA GLAZKO • *Department of Biomedical Informatics, University of Arkansas for Medical Sciences, Little Rock, AR, USA*

YUNCHEN GONG • *Department of Cell and Systems Biology and Centre for the Analysis of Genome Evolution and Function, University of Toronto, Toronto, ON, Canada*

ANNA GURYANOVA • *Global Health Governance Journal, Seton Hall University, South Orange, NJ, USA*

SVETLANA GURYANOVA • *Laboratory of Peptide Chemistry, Shemyakin-Ovchinnikov Institute of Bioorganic Chemistry (IBCH RAS), Moscow, Russia*

DAVID S. GUTTMAN • *Department of Cell and Systems Biology and Centre for the Analysis of Genome Evolution and Function, University of Toronto, Toronto, ON, Canada*

STEPHEN C. HARRIS • *Division of Systems Toxicology, National Center for Toxicological Research (NCTR), FDA, Jefferson, AR, USA*

SOMAYE HASHEMIFAR • *Toyota Technological Institute at Chicago, Chicago, IL, USA*

KLAUS HEUMANN • *Biomax Informatics AG, Planegg, Germany*

ROBERT HOEHNDORF • *Computational Bioscience Research Center (CBRC), Computer, Electrical and Mathematical Sciences and Engineering Division (CEMSE), King Abdullah University of Science and Technology (KAUST), Thuwal, Saudi Arabia*

STEVE HORVATH • *Departments of Human Genetics and Biostatistics, University of California, Los Angeles, CA, USA*

LILIA M. IAKOUCHEVA • *Department of Psychiatry, University of California San Diego, La Jolla, CA, USA*

ALEXANDER IVLIEV • *Thomson Reuters, Boston, MA, USA*

MICHAEL KAHN • *Department of Biochemistry and Molecular Biology, Keck School of Medicine, as well as Norris Comprehensive Cancer Research Center, University of Southern California, Los Angeles, CA, USA*

ALEXANDER E. KEL • *Institute of Chemical Biology and Fundamental Medicine, SBRAN, Novosibirsk, Russia; Biosoft.ru, Ltd, Novosibirsk, Russia; geneXplain GmbH, Wolfenbüttel, Germany*

YONG-MI KIM • *Division of Hematology and Oncology, Department of Pediatrics and Pathology, Children's Hospital Los Angeles, Keck School of Medicine, University of Southern California, Los Angeles, CA, USA*

MICHAEL B. KORZINKIN • *Department of Personalized Medicine, First Oncology Research and Advisory Center, Moscow, Russia; Laboratory of Bioinformatics, D. Rogachev Federal Research Center of Pediatric Hematology, Oncology and Immunology, Moscow, Russia*

NATALIA B. KUZMINA • *Laboratory of Systems Biology, A.I. Burnazyan Federal Medical Biophysical Center, Moscow, Russia*

VLADIMIR A. KUZNETSOV • *Bioinformatics Institute, Agency of Science, Technology and Research, Singapore, Singapore; School of Computer Engineering, Nanyang Technological University, Singapore, Singapore*

PETER LANGFELDER • *Department of Human Genetics, University of California, Los Angeles, CA, USA*

MORGAN E. LEVINE • *Department of Human Genetics and Semel Institute for Neuroscience and Human Behavior, University of California, Los Angeles, CA, USA*

JENNIFER D. LEWIS • *Plant Gene Expression Center, United States Department of Agriculture, Albany, CA, USA; Department of Plant and Microbial Biology, University of California Berkeley, Berkeley, CA, USA*

GUAN NING LIN • *Department of Psychiatry, University of California San Diego, La Jolla, CA, USA*

SASCHA LOSKO • *Biomax Informatics AG, Planegg, Germany*

NATALIA MALTSEV • *Department of Human Genetics and Computation Institute, University of Chicago, Chicago, IL, USA*

MARIA J. MARTIN • *European Molecular Biology Laboratory, European Bioinformatics Institute (EMBL-EBI), Cambridge, UK*

ANATOLY MAYBURD • *The Center of the Study of Chronic Metabolic and Rare Diseases, School of Systems Biology, College of Science, George Mason University, Fairfax, VA, USA*

ROHINI MEHTA • *The Center of the Study of Chronic Metabolic and Rare Diseases, School of Systems Biology, George Mason University Fairfax, Fairfax, VA, USA*

HYUN-JUN NAM • *Department of Psychiatry, University of California San Diego, La Jolla, CA, USA*

TATIANA NIKOLSKAYA • *RosGenDiagnostika, Moscow, Russia*

YURI NIKOLSKY • *Prosapia Genetics, Solana Beach, CA, USA; School of Systems Biology, George Mason University, Fairfax, VA, USA*

ROGER PERKINS • *Z-Tech Corporation, An ICF International Company, Jefferson, AR, USA*

VLADIMIR PRASSOLOV • *Engelhardt Institute of Molecular Biology, Russian Academy of Sciences, Moscow, Russia*

FENG QIAN • *Z-Tech Corporation, An ICF International Company, Jefferson, AR, USA*

YASIR RAHMATALLAH • *Department of Biomedical Informatics, University of Arkansas for Medical Sciences, Little Rock, AR, USA*

EUGENE RAKHMATULIN • *Clarivate Analytics, Thomson Reuters, Philadelphia, PA, USA*

IRINA RISH • *IBM T.J. Watson Research Center, Yorktown Heights, NY, USA*

RABIE SAIDI • *European Molecular Biology Laboratory, European Bioinformatics Institute (EMBL-EBI), Cambridge, UK*

LEMING SHI • *Division of Systems Toxicology, National Center for Toxicological Research (NCTR), FDA, Jefferson, AR, USA*

PHILIP YU SMIRNOV • *Laboratory of Systems Biology, A.I. Burnazyan Federal Medical Biophysical Center, Moscow, Russia*

VICTOR SOLOVYEV • *Softberry Inc., Mount Kisco, NY, USA*

ZHENJIANG SU • *Division of Systems Toxicology, National Center for Toxicological Research (NCTR), FDA, Jefferson, AR, USA*

DINANATH SULAKHE • *Department of Human Genetics and Computation Institute, University of Chicago, Chicago, IL, USA*

TATIANA V. TATARINOVA • *Spatial Sciences Institute, University of Southern California, Los Angeles, CA, USA; Center for Personalized Medicine, Children's Hospital Los Angeles, Los Angeles, CA, USA; A.A. Kharkevich Institute for Information Transmission Problems RAS, Moscow, Russia*

ANDREW TAYLOR • *Department of Human Genetics, University of Chicago, Chicago, IL, USA*

WEIDA TONG • *Division of Systems Toxicology, National Center for Toxicological Research (NCTR), FDA, Jefferson, AR, USA*

MARTIN TRISKA • *Spatial Sciences Institute, University of Southern California, Los Angeles, CA, USA*

JORGE URRESTI • *Department of Psychiatry, University of California San Diego, La Jolla, CA, USA*

LARISA S. VENKOVA • *Department of Personalized Medicine, First Oncology Research and Advisory Center, Moscow, Russia; Laboratory of Bioinformatics, D. Rogachev Federal Research Center of Pediatric Hematology, Oncology and Immunology, Moscow, Russia*

SHENG WANG • *Department of Human Genetics, University of Chicago, Chicago, IL, USA; Toyota Technological Institute at Chicago, Chicago, IL, USA*

BING XIE • *Department of Human Genetics, University of Chicago, Chicago, IL, USA; Department of Computer Science, Illinois Institute of Technology, Chicago, IL, USA*

ALEXANDER A. ZENIN • *Laboratory of Systems Biology, A.I. Burnazyan Federal Medical Biophysical Center, Moscow, Russia*

ALEX A. ZHAVORONKOV • *Drug Research and Design Department, Pathway Pharmaceuticals, Wan Chai, Hong Kong, Hong Kong SAR; Department of Personalized Medicine, First Oncology Research and Advisory Center, Moscow, Russia; Laboratory of Bioinformatics, D. Rogachev Federal Research Center of Pediatric Hematology, Oncology and Immunology, Moscow, Russia*

ALENA ZOLOTARENKO • *Laboratory of Functional Genomics, Vavilov Institute of General Genetics RAS, Moscow, Russia*

BORIS ZYBAILOV • *Department of Biochemistry and Molecular Biology, University of Arkansas for Medical Sciences, Little Rock, AR, USA*

Chapter 1

A Practical Guide to Quantitative Interactor Screening with Next-Generation Sequencing (QIS-Seq)

Yunchen Gong, Darrell Desveaux, David S. Guttman, and Jennifer D. Lewis

Abstract

Yeast two-hybrid screens are a powerful approach to identify protein-protein interactions; however, they are typically limited in the number of interactions identified, and lack quantitative values to ascribe confidence scores to the interactions that are obtained. We have developed a high-throughput, quantitative, yeast two-hybrid screening approach coupled with next-generation sequencing. This strategy allows the identification of interacting proteins that are preferentially associated with a bait of interest, and helps eliminate nonspecific interacting proteins. The method is high-throughput, allowing many more baits to be tested and many more candidate interacting proteins to be identified. Quantitative data allows the interactors to be ascribed confidence scores based on their enrichment with particular baits, and can identify both common and rare interacting proteins.

Key words Yeast two-hybrid screen, Next-generation sequencing, High-throughput screening, Quantitative

1 Introduction

Protein-protein interactions are an essential aspect of cellular function and signaling. First developed in 1989 by Fields and Song [1], yeast two-hybrid screening has become a key tool in the identification of protein-protein interactions. In the yeast two-hybrid system, the bait protein is fused to the DNA-binding domain while the prey protein is fused to the transcriptional activation domain (or vice versa). When the bait and prey proteins interact, the DNA-binding domain and activation domain are brought into close proximity, which allows activation of reporter genes. The original nuclear-based eukaryotic yeast two-hybrid system of Fields and Song uses the galactose (Gal4) transcriptional activation and DNA-binding domains from yeast [1]. A slightly modified version, the LexA system, employs prokaryotic-binding partners, the B42 acid blob activation domain, and LexA

Tatiana V. Tatarinova and Yuri Nikolsky (eds.), *Biological Networks and Pathway Analysis*, Methods in Molecular Biology, vol. 1613, DOI 10.1007/978-1-4939-7027-8_1, © Springer Science+Business Media LLC 2017

DNA-binding domain from *E. coli*, and is also nuclear-based [2]. The split-ubiquitin yeast two-hybrid system is a membrane-based screen, and uses the two halves of ubiquitin and the VP16 activation domain from herpesvirus [3]. Reporter genes can include genes for amino acid biosynthesis to enable rescue of auxotrophic yeast strains, colorimetric reporters like beta-galactosidase, or antibiotic resistance to aureobasidin A.

These different yeast two-hybrid systems have been used to query the interaction of a bait with a specific prey protein or set of proteins (binary screens), or with proteins encoded by a cDNA library prepared from a particular tissue or treatment (library screens). cDNA libraries can be prepared by random hexamer priming that does not yield full-length clones, poly T priming that may lack 5′ end of the gene, or by selecting mRNAs with 5′ caps and priming with poly T oligos for full-length cDNAs. *See* [4] for a detailed review on cDNA libraries. cDNA libraries can be normalized to remove highly abundant clones; however, normalized libraries tend to be much more expensive than non-normalized libraries. Screens can occur by mating, where haploid bait-containing or prey-containing yeast strains are mated to form a diploid strain, or by transformation where yeast is transformed with the plasmid DNA encoding the bait and prey constructs. In either case, interacting proteins are identified through the activation of reporter genes. The flexibility of yeast two-hybrid screens is one of its strengths, allowing researchers to investigate specific interactions between known proteins, and to identify unknown binding partners with a protein of interest. Library screens are of particular interest to many researchers, as they allow an unbiased assessment of potential binding partners. However, nonspecific interactions may arise from highly abundant cDNAs in non-normalized cDNA libraries, and "sticky" interactions of intrinsically promiscuous proteins may obscure true binding partners. A major bottleneck in yeast two-hybrid screening involves identifying the prey vector cDNAs from yeast colonies that express the reporter genes for interaction. This is typically done by extracting the plasmids from each yeast colony, followed by Sanger sequencing of the cDNA to identify the gene. As this is a laborious process, the most common interacting proteins are typically identified while more rare interactors may be missed, and the data is not quantitative.

We developed a high-throughput yeast two-hybrid screen that employs next-generation sequencing (QIS-Seq) to overcome some of these obstacles [5]. QIS-Seq allows quantitative identification of interacting proteins, and high-throughput screening of experimental versus control bait proteins (i.e., luciferase). Screening of colonies carrying putative interactors is carried out in a typical yeast two-hybrid fashion by selection for prototrophy and/or expression of marker genes. Colonies carrying putative interactors are harvested en masse and the plasmids are extracted. We use next-generation sequencing to identify cDNAs cloned into the prey plasmid for each screen with a bait of interest or negative control, and clones present in the original cDNA library. Sequencing reads can then be normalized and compared among the bait, negative control, and cDNA library at each locus. This quantitative method allows the calculation of an enrichment score for each interacting prey to identify cDNAs that are specifically enriched for a bait of interest compared to the negative control. It also allows nonspecific and likely false positives to be excluded by identifying interacting proteins found in all or many baits or the negative control, as well as cDNAs that are equivalently abundant in both the library and the experimental sample. Since sequencing of clones is not limiting, QIS-Seq identifies both rare and common interacting proteins, and may reduce the need for normalized libraries. We provide protocols for a split-ubiquitin membrane-based yeast two-hybrid screen by transformation of competent yeast cells [5]. Nevertheless, this approach may be applied to any type of yeast two-hybrid screen, or used in a mating approach. QIS-Seq is most effectively used for organisms where genome information is available as the reads are mapped to specific loci; however, in organisms with incomplete genomes, the sequences could be analyzed to identify particular domains that interact with the bait.

2 Materials

2.1 Yeast Media

1. YPAD (yeast extract-peptone-adenine-dextrose) Medium: 1% Yeast extract, 2% Bacto peptone, 2% D-+-glucose >99.5%, 2% Bactoagar, 0.004% Adenine sulfate. *See* Table 1 for details (*see* **Note 1**).
2. SC (synthetic complete) Medium: 0.062% drop-out (DO) supplement –His/–Leu/–Trp, 2% D-+-glucose >99.5%, 0.17% yeast nitrogen base without amino acids or ammonium sulfate, 0.5% ammonium sulfate, 2% Bactoagar (*see* **Note 2**). *See* Table 2 for specific media combinations required for a split-ubiquitin yeast two-hybrid screen.

Table 1
YPAD medium

Ingredient	Broth	Plates
Yeast extract	6 g	6 g
Peptone	12 g	12 g
Glucose	12 g	12 g
Adenine sulfate	40 mg	40 mg
H_2O	to 600 mL	to 600 mL
Bactoagar	n/a	12 g

Table 2
Synthetic defined (SD) medium

Ingredient	–His/–Leu/–Trp broth	–His/–Leu/–Trp plates	–Leu broth	–Leu plates	–Leu/–Trp broth	–Leu/–Trp plates
DO–His/–Leu/–Trp	0.372 g	0.372 g	0.372 g	0.372 g	0.372 g	0.372 g
Nitrogen base	1.02 g	1.02 g	1.02 g	1.02 g	1.02 g	1.02 g
Ammonium sulfate	3 g	3 g	3 g	3 g	3 g	3 g
Glucose	12 g	12 g	12 g	12 g	12 g	12 g
H_2O	to 600 mL	to 600 mL	to 600 mL	to 600 mL	to 600 mL	to 600 mL
Bactoagar	n/a	12 g	n/a	12 g	n/a	12 g
Histidine 10 mg/mL	n/a	n/a	1.2 mL	1.2 mL	1.2 mL	1.2 mL
Leucine 10 mg/mL	n/a	n/a	n/a	n/a	n/a	n/a
Tryptophan 10 mg/mL	n/a	n/a	1.2 mL	1.2 mL	n/a	n/a
2 M 3-AT	n/a	Must be optimized	n/a	n/a	n/a	n/a
Purpose:	n/a	Selection of interacting clones	Growth of pBT3-N or pTLB-1 strains	Growth of pBT3-N or pTLB-1 strains	Growth of pBT3-N and pPR3-N/C strains	Growth of pBT3-N and pPR3-N/C strains

Ingredients below dotted line are added in sterile hood after autoclaving and cooling media to 50 °C. In place of DO–His/–Leu/–Trp, DO–Leu/–Trp or DO–Leu may be used. With either DO supplement, please check the bottle for the number of grams per liter. If DO–Leu/–Trp or DO–Leu is used, this would not require supplementation of the media with additional amino acids after autoclaving

3. 10 mg/mL histidine: Prepared in ultrapure water, filter sterilized, stored at 4 °C (*see* **Note 3**).
4. 10 mg/mL leucine: Prepared in ultrapure water, filter sterilized, stored at 4 °C (*see* **Note 3**).
5. 10 mg/mL tryptophan: Prepared in ultrapure water, filter sterilized, stored at 4 °C (*see* **Note 3**).
6. 2 mg/mL uracil: Prepared in ultrapure water, filter sterilized, stored at 4 °C (*see* **Note 3**).
7. 2 M 3-aminotriazole (3-AT): Prepared in ultrapure water, filter sterilized, stored at 4 °C (*see* **Note 4**).
8. 90 mm and 150 mm sterile disposable petri dishes.

2.2 Yeast Competent Cells

1. *Saccharomyces cerevisiae* strain AP-4.
2. 50% PEG 3350: Prepared in ultrapure water, filter sterilized, and aliquoted (*see* **Note 5**).
3. 1 M Lithium acetate: Prepared in ultrapure water, autoclaved and aliquoted (*see* **Note 6**).
4. Sterile 4.5 mm glass beads (Zymo Research).
5. 10 mg/mL sheared salmon sperm DNA.
6. Purified bait and prey plasmids.
7. Sterile toothpicks.
8. Sterile 250 mL and 2 L Erlenmeyer flasks.
9. Sterile glass or plastic pipets.
10. Sterile ultrapure H_2O.
11. Sterile 2 mL Eppendorf tubes.
12. Sterile 250 mL centrifuge bottles.

2.3 Isolation of Plasmids from Yeast

1. 0.1 M sodium phosphate buffer pH 7.4/1.2 M sorbitol: Prepared with 100 mL of 0.1 M sodium phosphate buffer pH 7.4 using 7.74 mL of 1 M NaH_2PO_4 and 2.26 mL of 1 M Na_2HPO_4. Add 1.2 M sorbitol. Bring up to 100 mL volume with ultrapure H_2O and filter sterilize.
2. 0.1 M sodium phosphate buffer pH 7.4: Prepared from 100 mL of 0.1 M sodium phosphate buffer pH 7.4 using 7.74 mL of 1 M NaH_2PO_4, 2.26 mL of 1 M Na_2HPO_4 and sterile ultrapure H_2O.
3. Lyticase (25KU): Dissolved in 50 μL of 0.1 M sodium phosphate buffer pH 7.4 + 137.5 μL ultrapure H_2O + 312.5 μL 80% glycerol, freshly made and used immediately.
4. 10 mg/mL RNAseA.
5. Sterile 80% glycerol.

6. Qiagen spin miniprep kit (*see* **Note 7**).
7. Sterile 1.5 mL Eppendorf tubes.
8. Cell spreader.
9. Sterile 50 mL centrifuge tubes rated for high speed.

2.4 cDNA Amplification

1. Platinum Taq High Fidelity polymerase and buffer.
2. dNTPs.
3. Vector-specific forward and reverse primers.
4. Qiagen PCR purification kit.

2.5 Equipment

1. Stir plates and magnetic stir bars.
2. Biosafety cabinet (class II type A2) or laminar flow hood.
3. Ultra-low freezer.
4. 42 °C water bath.
5. Benchtop microcentrifuge.
6. 37 °C incubator.
7. 28 °C shaker-incubator.
8. 28 °C incubator.
9. Spectrophotometer.
10. High-speed refrigerated floor centrifuge (i.e., Beckman Coulter) with a JA-20 rotor for 50 mL tubes, and a JA-14 rotor for 250 mL bottles.
11. Thermocycler.
12. Horizontal gel electrophoresis system, agarose gels and power supply.

3 Methods

3.1 Comments on Working with Yeast

Maintain sterile technique throughout. Perform manipulations in sterile hood. Yeast are fragile, particularly after transformation, so make sure you pipet gently. If Eppendorf tube lids are not sufficiently opened, use a sterile pipet tip or toothpick to push the lid further open, instead of nonsterile gloves.

3.2 Testing a New Bait to Determine Suitability for Yeast Two-Hybrid Screening

1. Promoter strength and the protein of interest can impact a bait protein's suitability for a yeast two-hybrid screen. Our system uses the CYC1 weak promoter or the TEF1 strong promoter. We also advise constructing different versions of your bait protein, based on the annotation of different domains in the protein, as some baits can lead to autoactivation of the reporter genes (*see* **Note 8**). For our split-ubiquitin yeast two-hybrid screen, we test our bait constructs in pBT3-N-HA-K8-CAAX

(weak CYC1 promoter) or pTLB-1-HA-K8-CAAX (strong TEF1 promoter). Both these vectors have been modified with a polybasic region (K8) and a prenylation signal (CAAX) for strong membrane association [5–7].

2. If using a histidine reporter for interactions, 3-AT is used to titrate the sensitivity of the HIS3 reporter as it can be leaky. To determine the optimal concentration of 3-AT, we co-transform the bait of interest with pFur4-NubI, pFur4-NubG, or pPR3-N/pPR3-C. Fur4 is a yeast transmembrane protein that is used as a negative control [8]. NubI contains the N-terminus of ubiquitin with a wild-type isoleucine at residue 13, and interacts readily with the C-terminus of ubiquitin (Cub), regardless of the bait or prey fusion. NubG has a glycine at residue 13 which prevents interaction with Cub, unless NubG and Cub are brought into close proximity by the interaction between the bait and prey [3]. Therefore, you can use pFur4-NubI as a positive control for interaction and pFur4-NubG as a negative control for interaction. The empty library vector (pPR3-N where Cub is at the N-terminus of the fusion, or pPR3-C where Cub is at the C-terminus of the fusion) is used as an additional negative control. From one co-transformation, plate 100 μL of cells on SD–His/–Leu/–Trp + 5, 10, 20, 30 mM 3-AT and 5 μL of cells on SD–Leu/–Trp to determine the transformation efficiency. This range should work if the bait construct is under the weak promoter (pBT3-N backbone). If the bait construct is under the strong promoter (pTLB-1 backbone), try a higher range of 3-AT (up to 100 mM). Each bait protein needs to be independently tested to determine whether it autoactivates the reporter genes. Co-transformations are always less efficient than single transformations. However, co-transformations will still produce enough colonies to optimize the 3-AT concentration.

3. Once you have determined the general range of 3-AT that appears effective, the concentration of 3-AT needs to be further optimized using 2.5 mM intervals in the range of 3-AT concentration where there are lots of colonies with the bait and pFur4-NubI and few or no colonies with the bait and pFur4-NubG. For instance, if SD–His/–Leu/–Trp with 5 mM 3-AT looked promising, test a new co transformation with the same constructs as in **step 2** on SD–His/–Leu/–Trp containing 0.5 mM, 2.5 mM, 5 mM, and 7.5 mM 3-AT. If there are few colonies from the co-transformations in **step 2**, you can try 0.5–2.5 mM 3-AT or alternatively test the bait under the strong promoter (pTLB-1-HA-K8-CAAX).

4. You may want to include a bait protein with known interacting proteins to use as a positive control for the screen. However, it is more common to work with baits that have no known

interacting proteins. In this case, it is critical to titrate the system as described above to ensure that your bait protein shows some level of specific interaction.

5. We advise including a bait protein to be used as a negative control for interaction. We used luciferase for this purpose, as it was not expected to interact with Arabidopsis proteins. The negative control will identify proteins that are intrinsically sticky, allowing the exclusion of these candidate interacting proteins from the data set.

3.3 Preparing and Transforming Competent Cells for a Bait Strain or Test Library Screening

1. Streak AP-4 yeast strain from glycerol stocks onto YPAD plates. Grow at 28 °C for 5 days. It is best to streak a new plate from the glycerol stock each week.
2. Pick four to five colonies and resuspend in 0.5 mL YPAD in a sterile 1.5 mL microfuge tube with a sterile toothpick. Yeast cells are quite sticky, so make sure the colonies are thoroughly resuspended.
3. Transfer to 250 mL flask with 50 mL YPAD (*see* **Note 9**). Incubate at 250 rpm 28 °C overnight (16–20 h).
4. Measure the OD_{600} of a ½ or ¼ dilution of the overnight culture. The OD_{600} of the undiluted culture must be >1, and should be a minimum of 4.5. If the OD_{600} is too low, there is a problem with the yeast and a new culture should be set up.
5. Subculture cells to obtain a culture that is actively growing. To do this, inoculate 300 mL YPAD in a sterile 2 L flask with sufficient overnight culture so that the OD_{600} is 0.1. Calculate the amount to add based on the initial OD_{600} of the culture. Incubate at 250 rpm 28 °C for 3–4 h (*see* **Note 9**).
6. Harvest the culture when the OD_{600} reaches 0.6. Use a sterile long glass or plastic 1 mL pipette to take aliquots so that the inside of the flask remains sterile, as a Pipettor will not be sterile along its length.
7. Harvest the culture at 1000 × *g* for 5 min at room temperature in sterile 250 mL bottles.
8. Resuspend the pellet in 20 mL sterile water. Use a new bottle of sterile water to prevent contamination. Use a 10 mL sterile glass pipet and pipet gently. Transfer to a sterile 50 mL centrifuge tube. Centrifuge again at 1000 × *g* for 5 min at room temperature.
9. Resuspend the pellet in 1.5 mL sterile water. Use a 10 mL sterile glass pipet and pipet gently.
10. Aliquot 100 μL of competent cells to microfuge tubes for the controls, and 200 μL of competent cells to tubes for the library transformation. Better transformation efficiencies are obtained with multiple small-scale library transformations.

11. Mix in plasmids to aliquots of yeast.
    (a) For the controls: use 500 ng of each plasmid. We would test the bait with pPR3-N for the transformation efficiency, with pFur4-NubI as a positive control, and with pFur4-NubG as a negative control.
    (b) For a test library transformation: use 1 μg of each plasmid. When performing large-scale library transformations, it is better to transform the library into a strain that already carries the bait plasmid (*see* Subheading 3.4).
12. Flick tubes to make sure yeast have not settled. Add 300 μL of PEG/LiAc to controls or 600 μL of PEG/LiAc to library transformations. Mix into yeast while adding so that yeast do not aggregate. To prepare 1X PEG/LiAc for one reaction, mix 240 μL 50% PEG, 36 μL 1 M LiAc, 10 μL 10 mg/mL boiled and cooled salmon sperm DNA, and 14 μL water. Scale up as needed (*see* **Note 10**).
13. Heat shock for 45 min at 42 °C. Flick tubes every 10 min.
14. Pulse spin, wash with 500 μL water. Twirl toothpick in Eppendorf tube to resuspend yeast. Make sure to pipet gently!
15. Pulse spin, wash with 250 μL water if plating on 90 mm plates. If you are optimizing the 3-AT concentration, wash with 500 μL water. Wash with 400–500 μL water if plating on 150 mm plates.
16. Library transformations should be plated on 150 mm SD–His/−Leu/−Trp + appropriate 3-AT.
    (a) Controls (pPR3-N, pFur4-NubI, pFur4-NubG) should be plated onto 90 mm SD–His/−Leu/−Trp + appropriate 3-AT.
    (b) The transformation efficiency (pPR3-N) plates should have 10 μL and 100 μL of bait + empty prey vector on SD–Leu/−Trp (*see* **Note 11**).
    (c) The water control should be plated onto 90 mm SD–Leu/−Trp.
    (d) Use sterile glass beads for even plating.
17. Incubate at 28 °C for 1 week.
18. *See* Subheading 3.2 to evaluate your results.

3.4 Preparation and Single Transformation of Competent Cells

This should be used for higher efficiency transformations needed for large-scale library screening (*see* **Note 12**).

1. Streak AP-4 carrying bait construct from glycerol stock onto SD–Leu plates. Grow at 28 °C for 7 days. It is best to streak a new plate from the glycerol stock each week.

2. Pick four to five colonies and resuspend in 0.5 mL SD–Leu with a toothpick. Yeast cells are sticky, so make sure the colonies are thoroughly resuspended.
3. Transfer to 250 mL flask with 50 mL SD–Leu. Incubate at 250 rpm 28 °C for 24 h.
4. Take OD_{600} of the overnight culture (must be >1). Prepare at least a ½ dilution to get an accurate reading.
5. Inoculate competent cell culture with a small aliquot of the overnight culture, in 300 mL SD–Leu in a sterile 2 L flask. You may want to start with 250 μL, but may have to optimize the volume depending on how well your bait strain grows. Incubate at 250 rpm 28 °C overnight (12–15 h). Determine the OD_{600} early in the morning so that the culture does not grow beyond $OD_{600} = 0.6$ (*see* **Note 9**).
6. Harvest culture when the $OD_{600} = 0.6$. Use a sterile long glass 1 mL pipette to take aliquots so that the inside of the flask remains sterile as a Pipettor will not be sterile along its length.
7. Harvest culture 1000 × *g* 5 min RT in sterile 250 mL bottles.
8. Resuspend pellet in 20 mL sterile water. Use a 10 mL glass pipet and pipet gently. Centrifuge at 1000 × *g* 5 min at room temperature.
9. Resuspend pellet in 1.5 mL water. Use a 10 mL glass pipet and pipet gently.
10. Aliquot 200 μL yeast to tubes for transformations. Better efficiency is achieved with multiple tubes of 200 μL, rather than with fewer large volume transformations.
11. Mix in plasmids to aliquots of yeast. For controls or library transformation, use 500 ng of plasmid. For the controls, we would test the bait with pPR3-N for the transformation efficiency, with pFur4-NubI as a positive control, and with pFur4-NubG as a negative control.
12. Flick tubes to make sure yeast have not settled. Add 600 μL of PEG/LiAc to transformations. Mix into yeast while adding so that yeast do not aggregate. Vortex 5 s, setting 6. To prepare 1X PEG/LiAc for one reaction, mix 480 μL 50% PEG, 72 μL 1 M LiAc, 20 μL 10 mg/mL boiled and cooled salmon sperm DNA, and 28 μL water. Scale up volumes as needed (*see* **Note 10**).
13. Heat shock 45 min 42 °C. Flick tubes every 10 min.
14. Pulse spin, wash with 500 μL water. Twirl toothpick in tube to resuspend yeast.
15. Pulse spin, wash with 250 μL water if plating on 90 mm plates. Wash with 400–500 μL water if plating on 150 mm plates.
16. Library transformations should be plated on two 150 mm plates of SD–His/–Leu/–Trp + appropriate 3-AT.

(a) Controls (pPR3-N, pFur4-NubI, pFur4-NubG) should be plated onto 90 mm SD–His/−Leu/−Trp + appropriate 3-AT.

(b) The transformation efficiency (pPR3-N) plates should be plated with 5 μL and 50 μL of bait + empty prey vector on SD–Leu/−Trp (*see* **Note 11**).

(c) The water control should be plated onto 90 mm SD–Leu/−Trp.

(d) Use sterile glass beads for even plating.

17. Incubate at 28 °C for 1 week.
18. Pick colonies and restreak on SD–His/−Leu/−Trp to create master plates. If your colonies are a range of sizes, choose the larger and medium size colonies (as compared to your pFur4-NubG negative control).

3.5 Isolation of Plasmids from Yeast

Plasmid DNA is extracted en masse to create a pool of prey interactors for each bait screen (*see* **Note 13**).

1. Restreak colonies containing putative interacting proteins from the SD–His/−Leu/−Trp master plate on SD–Trp to preferentially retain prey plasmid. Grow at 28 °C for 3–4 days.
2. Repeat #1.
3. Streak onto SD–Trp 150 mm plates. For each colony, make two streaks with a sterile flat toothpick about 1 cm long. Grow at 28 °C for 3–4 days.
4. Harvest yeast off plates with a cell spreader into SD–Trp. For a specific bait, pool all yeast together. Yeast cells lyse most efficiently when fresh. The following volumes are for a pellet of ~5 g.
5. Centrifuge at 1000 × *g* for 5 min at room temperature. Pipet off supernatant.
6. Wash pellet in 0.1 M sodium phosphate buffer pH 7.4/1.2 M sorbitol. Centrifuge at 1000 × *g* for 5 min at room temperature.
7. Resuspend pellet in 7.12 mL 0.1 M sodium phosphate buffer pH 7.4/1.2 M sorbitol +500 μL lyticase +50 μL 10 mg/mL RNAseA.
8. Incubate at 37 °C overnight. Keep tube upright so that it does not leak.
9. Transfer to a 50 mL centrifuge tube. Add 12.5 mL (use equivalent volume to whatever volume your pellet is at after resuspending in **step 7**) of 0.2 N NaOH+1% SDS (Qiagen miniprep solution P2). Invert four to six times. Incubate at room temperature for 15 min. You can also incubate at 65 °C for 15 min if the cells do not look lysed.

10. Add 17.5 mL of chilled Qiagen miniprep buffer N3. If your volume from **step 9** is different, then use a similar proportion of N3 to the volumes listed here. Invert four to six times. Place on ice for 20 min.
11. Using a large centrifuge (i.e., Beckman Coulter floor centrifuge with a JA-20 rotor) and 50 mL polypropylene centrifuge tubes rated for high speed, centrifuge your sample at 14,000 rpm (24,000 g) for 30 min at 4 °C. Pipet the supernatant into a new tube to help prevent the particulate from clogging the spin columns in **step 13**.
12. Centrifuge the sample again at 14,000 rpm (24,000 g) for 15 min at 4 °C. Pipet the supernatant into new tube to help prevent the particulate from clogging the spin columns in **step 13**.
13. Use multiple Qiagen spin columns (8–10+) to purify plasmid DNA. Repeatedly load supernatant onto columns but stop before columns clog. Continue loading supernatant onto new columns if initial columns clog.
14. Add 0.5 mL Qiagen buffer PB, wait 5 min. Spin at 14,000 rpm (maximum speed) for 1 min in a microcentrifuge.
15. Add 0.75 mL Qiagen buffer PE. Spin at 14,000 rpm (maximum speed) for 1 min in a microcentrifuge.
16. Spin at 14,000 rpm (maximum speed) for 1 min in a microcentrifuge to remove residual PE buffer.
17. Add 50 μL Qiagen buffer EB to the center of the column. Let it sit for 1 min. Spin at 14,000 rpm (maximum speed) for 1 min in a microcentrifuge.
18. Add 35 μL buffer EB to the center of the column. Let it sit for 1 min. Spin at 14,000 rpm (maximum speed) for 1 min in a microcentrifuge.
19. Combine all eluates for a single bait screen.
20. Run a 1% gel to check for the integrity of your plasmid. Quantitate the concentration of plasmid DNA using a spectrophotometer or Nanodrop. You should have ~100–150 ng/μL of DNA in ~800 μL of TE. The plasmids may run with a bit of a smear as the sizes may vary.

3.6 Amplify cDNAs from Prey Plasmids

In this step, you will conduct low cycle amplification of the cDNAs from your prey plasmids to generate sufficient DNA for an Illumina run. Since the primers sit down in the prey vector close to the cDNA, most of the sequence will correspond to the genome from which your mRNA was extracted, rather than vector sequence. Most bait proteins are likely to interact with a number of different prey proteins, resulting in prey plasmids containing different cDNAs of varied size. Low cycle amplification of the cDNAs from the prey plasmids (recovered from yeast colonies on interaction plates) should result in representation of all prey cDNAs that encode proteins interacting with the bait of interest. We also carry out low cycle amplification of the cDNAs present in the cDNA library, to determine which genes from the genome are represented. Make a 100 ng/μL stock of DNA to use as the template in PCR.

1. Use a high-fidelity proofreading polymerase (i.e., Fermentas High-fidelity mix K0191 5 U/μL).
2. Do a test PCR to make sure that you are amplifying a range of different sized cDNAs.
3. Set up multiple 25 μL reactions with 2.5 μL 10× buffer + 2 μL 2.5 mM dNTPs + 1 μL 10 μM of each forward and reverse vector-specific primers + 2 μL 100 ng/μL template + 0.5 μL enzyme mix + water to 25 μL.
4. PCR conditions: preheat lid; 94 °C 3 min 1×; 94 °C 30 s, 57 °C 30 s, 72 °C 3 min 15×; 72 °C 5 min, 15 °C hold. The annealing temperature may vary depending on your primers. The extension time will depend on the average and maximum size of the cDNAs in your library. It is preferable to do multiple (10) separate reactions so that the amplified clones are independent from one reaction to the next. Our vector-specific primers (5′pPR3Nbp438 and 3′pPR3Nbp649) are shown below.

5′pPR3Nbp438	CGTTAAGTCGAAAATTCAAGACAAGGAAGGAAT
3′pPR3Nbp649	GCGTGACATAACTAATTACATGACTCGAGGTCGA

5. Pool your reactions after PCR. Run 5 μL on gel. The amplified products should run with a bit of a smear, but it is likely there will be some distinct bands. If the average size of your cDNAs is 1 kb, the visible range of PCR products will be about 500–1500 bp.
6. Purify the amplified products on a Qiagen PCR purification column. Expected yields are 200–500 ng/μL.

3.7 Library Generation and Next-Generation Sequencing (NGS)

To identify the cDNAs in the prey plasmids from each screen with your bait or luciferase, you must first construct an Illumina library for each different screen. We also construct an Illumina library for the cDNA library to identify all of the genes that are present in the cDNA library. Reads from the bait screen (or luciferase screen) refer to the prey cDNAs whose proteins interact with the bait of interest.

1. Commercial kits for Illumina library generation are readily available (i.e., Bioo Scientific). Alternatively, many core facilities provide library generation and quality control services prior to an Illumina run.

2. Next-generation sequencing services (i.e., Illumina) are available through the core facilities at many institutions.
3. Obtain your next-generation sequence reads from the core facility in standard FASTQ format (*see* **Note 14**).

3.8 Align Reads to Genome

The sequence reads from the bait screen, the luciferase screen, or the cDNA library itself can be mapped to a genomic reference using software that produces an output file in SAM/BAM formats, which are standard in NGS (*see* **Note 15**). NGS mapping programs typically run on a command line in a UNIX environment, and employ third-party software. This software is under constant revision, so specific commands may change as the programs are revised. We mapped Illumina reads to Arabidopsis gene models downloaded from NCBI. The following steps describe the use of the popular read alignment tool BWA version 0.7.12 (http://bio-bwa.sourceforge.net/) for this purpose (*see* **Note 16**).

1. To make reference index, run this command line (*see* **Note 17**):

```
bwa index -a bwtsw prey_seq.fasta
```

2. To align the reads (paired-end reads in this example) to the reference sequences and output results in sai format: (*see* **Note 18**).

```
bwa aln prey_seq.fasta reads1.fastq > alignment1.sai
bwa aln prey_seq.fasta reads2.fastq > alignment2.sai
```

3. To generate a paired-end alignment in sam format:

```
bwa sampe prey_seq.fasta alignment1.sai alignment2.sai
reads1.fastq reads2.fastq > alignment.sam
```

4. To convert sam format to bam format and sort the bam format file using samtools version 1.1 (http://samtools.sourceforge.net/) (*see* **Note 19**):

```
samtools view -b -S alignment.sam > alignment.bam
samtools sort alignment.bam alignment_sort.bam
```

3.9 Count Reads for Each Gene

Using the mapping data, you can determine the number of mapped reads for each gene, and calculate the coverage of each gene for each sample (i.e., reads from the bait screen) or all samples combined (using the reads from the bait screen, the luciferase screen, and the cDNA library) (*see* **Note 20**).

1. If you use the BWA read alignment tool and create bam files as described in Subheading 3.8, you can use htseq-count (http://www-huber.embl.de/users/anders/HTSeq/doc/count.html) or bedtools version 2.15.0 (using multicov command) (http://bedtools.readthedocs.org/en/latest/content/tools/multicov.

html) to count the reads mapped to each gene. Below is the example for bedtools (*see* **Notes 21** and **22**)

```
bedtools multicov -bams alignment_sort.bam -bed reference.
fasta.bed > alignment.coverage
sort alignment.coverage > alignment.coverage.sort
cut -f1,13 alignment.coverage.sort > alignment.coverage.
sort.cut
```

2. To determine the combined coverage for all samples, you must combine the data for each sample. Each sample has one alignment.coverage.sort.cut file. To combine the coverage data into one file, use the following command line (*see* **Note 23**).

```
join *.cut > combined_coverage.txt
```

3.10 Normalize Reads for Each Gene

Each sample will contain a different number of reads, depending on the input DNA and the number of reads that pass the quality filter. The number of reads must be normalized for all samples. Read numbers per gene are normalized as cpm (count per million) or rpkm (reads per kilobase per million) within each sample for comparison among the samples (*see* **Note 24**). To calculate these values, the R version 3.1.1 EdgeR package version 3.12.0 can be used (http://bioconductor.org/packages/release/bioc/html/edgeR.html). To use EdgeR one should execute the following commands in R environment:

```
X <- read.delim("combined_coverage.txt",header=FALSE,row.
names=1,sep=" ")
cpm_value <- cpm(X)
write.table(cpm_value,"combined.coverage.cpm")
genelength<-read.delim("prey_seq.fasta.length.sort",head-
er=TRUE,row.names=1)
rpkm_value <- rpkm(X, log=FALSE,gene.length=genelength
$Length)
write.table(rpkm_value,"combined.coverage.rpkm")
```

3.11 Determine Enrichment of Putative Interacting Proteins with a Bait of Interest

The enrichment of a specific interactor with a bait of interest can be determined by calculating the number of reads obtained from the bait screen or luciferase screen (as a negative control) and normalizing against the abundance of reads from the luciferase screen, the bait screen, and the cDNA library as below. By including the read counts for the genes present in the cDNA library, this helps to account for highly abundant genes. This calculation uses the normalized reads (rpkm) from each sample (*see* Subheading 3.10).

$$\text{Percentage Enrichment} = [(\text{reads from bait screen} - \text{reads from luciferase screen}) / (\text{reads from bait screen} + \text{reads from luciferase screen} + \text{cDNA library reads})] * 100$$

This calculation can be easily done in an Excel spread sheet. To import the data into the spreadsheet to do the enrichment calculation, open the combined.coverage.rpkm file with a text editor, copy the contents, open a spreadsheet file, and paste the contents into it. Note the use of a space as the separator (*see* **Note 25**).

3.12 Calculate the Percentage of the Mapped Length

This calculation enables you to determine the portion of the gene that is represented in the cDNA library and your sample. The percentage of the mapped length is calculated using the length of mapped regions and the theoretical length of the gene model. The mapped regions of all aligned reads are collected, assembled and the overlapping regions are identified to produce a list of continuously mapped region(s). The length of these regions is summed up and divided by the length of the gene model, to provide the percentage of the mapped length. The bedtools program with R commands can be used to do this as shown below.

1. Determine depth for each gene and each location (*see* **Note 26**):

```
bedtools genomecov -d -ibam alignment_sort.bam -g prey_-
seq.fasta > alignment_sort.bam.genomecov
```

2. The percentage of the mapped length is calculated in R:

```
data<-read.delim("alignment_sort.bam.genomecov",head-
er=FALSE)
data_no_zero<-data[data$V3>0,]
table(data_no_zero$V1)/table(data$V1)
```

3.13 Select Candidate Interacting Proteins to Test in Downstream Assays

Based on the enrichment of specific interacting proteins for your bait and the coverage of these loci, you can generate a list of putative interacting proteins to test in downstream assays (*see* **Note 27**). This could include but is not limited to the following assays:

1. Comparing a wild-type line to a knockout line of the putative interacting protein for your phenotype of interest.
2. Analyzing the expression pattern or localization of the gene or protein to determine if this is consistent with the putative role of the interacting protein.
3. Testing for direct interactions between your bait and a specific interacting protein candidate in vitro or in the organism of interest.
4. Testing for expression changes of the mRNA for the interacting protein under an appropriate stimulus.

4 Notes

1. YPAD rich medium is used for routine growth of yeast strains. Adenine sulfate is added to the media to reduce the reversion of the *ade2* mutation to *ADE2*. Liquid media can be prepared in larger volumes, and aliquoted into 1 L Pyrex bottles or other suitable autoclavable bottles. If YPAD will be aliquoted, put the appropriate amount of agar for the volume of YPAD into a 1 L bottle. As the glucose is autoclaved with the other components, the autoclaving time should be kept short to prevent caramelization. For 1 L of media, autoclave for 20 min, and for each additional L, add 5 min of autoclaving time. If you have more than 3 L of media, autoclave it in separate batches. Determine the volume of media by the total volume in the autoclave, not by the volume in each bottle. It is also better to use shallow trays (2.75″ high; i.e., Thermo Scientific Nalgene #6902-3000) so that the heat from the autoclave adequately circulates around the bottles. For media with agar, swirl bottles after removing the media from the autoclave to disperse agar. 600 mL of media is sufficient to pour 1 sleeve of 20 plates. Media should be used within 3 months of preparation.
2. Yeast strains used in genetic screens are commonly auxotrophic for tryptophan (*TRP1*), histidine (*HIS3*), uracil (*URA3*), leucine (*LEU2*), and/or adenine (*ADE2*). SC selective media is used for the growth of specific yeast strains carrying plasmids conferring prototrophy for these amino acids. It is prepared by mixing yeast nitrogen base with amino acid mixtures that exclude specific amino acids (drop-out [DO] supplement). Depending on the drop-out media we are making, we use DO supplement –His/−Leu/−Trp, DO supplement –Leu/−Trp, or DO supplement –Trp. The DO supplement –His/−Leu/−Trp can be used to make any combination of single, double, or triple dropout for –His/−Leu/−Trp, with the necessary amino acids added after autoclaving (*see* **Note 3**). Alternatively, a DO supplement lacking only one amino acid could be used and would not require supplementation with additional amino acids. To achieve high-efficiency transformation, media should be prepared with a pH of 5.6 using 1 M NaOH, before adding Bactoagar. Media should be used within 3 months of preparation.
3. For the uracil stock (and any other amino acids that are not dissolving well), prepare by stirring on a stir plate while you add 1 M NaOH dropwise to help dissolve amino acids, to a maximum concentration of 0.1 M NaOH. Make sure to handle the stocks in a sterile manner so that these stocks can be used for each batch of media you have to make.

4. Make sure to handle in a sterile manner so that these stocks can be used for each batch of media you have to make. Handle carefully using appropriate personal protective equipment and environmental controls as 3-AT is toxic. Stock solutions should be disposed in accordance with environmental regulations.
5. As PEG is highly insoluble, begin dissolving PEG in 40% of the final volume of ultrapure H_2O by placing the solution in a 50 °C water bath. Transfer the solution between the stir plate and water bath as the PEG starts to dissolve. Once dissolved, bring up to the final volume and filter sterilize. Make sure to handle the PEG in a sterile manner so that these stocks can be used for each batch of media you have to make. Aliquot into smaller tubes for experiments so that sterility is maintained. Make sure that caps fit tightly and wrap caps in parafilm, so that PEG does not inadvertently become more concentrated due to evaporation of the water.
6. Make sure to handle in a sterile manner so that these stocks can be used for each batch of media you have to make. Aliquot into smaller tubes for experiments so that sterility is maintained.
7. Qiagen spin mini kit includes buffers P2, N3, PB, PE, and EB.
8. Domain prediction can also be done using the Phyre2 protein homology server (http://www.sbg.bio.ic.ac.uk/phyre2/html/page.cgi?id=index) [9].
9. It is important that the yeast are sufficiently aerated. Use a flask that is five times larger than the liquid volume of media.
10. The salmon sperm DNA should be boiled in a heat block for 10 min, quick chilled, then added to 1× PEG/LiAc. Prepare the amount of salmon sperm DNA you require for one experiment. The PEG/LiAc mixture should be made right before use.
11. To calculate the transformation efficiency, use the following equation.

$$\begin{aligned}&\text{Transformation efficiency (cfu}/\mu g)\\&\quad = (\text{number of colonies} \times \text{total suspension volume})\\&\quad /(\text{volume of transformation plated } (\mu L)\\&\quad \times \text{amount of DNA used } (\mu g))\end{aligned}$$

12. Transformation efficiencies will be significantly higher when transforming one plasmid rather than two.
13. We usually streak many different colonies from the interaction plates for a specific bait screen, and pool all of these colonies together. Alternatively, you can pool colonies for one set of transformations, and extract the plasmids from this smaller pool. This would allow you to compare the interacting proteins obtained from several independent sets of transformations.

14. A FASTQ file is the sequence read file with quality information for the error probability of each base pair in the sequence.
15. A Binary Alignment Map (BAM) file is a compressed Sequence Alignment Map (SAM) file. SAM format specifications can be found at this site: https://samtools.github.io/hts-specs/SAMv1.pdf
16. The alignment is carried out separately for the prey cDNAs identified after screening each different bait.
17. The dash followed by a letter ("–a" in our example) is a switch that controls the program execution. –a bwtsw directs BWA to use a specific algorithm, as implemented in BWT-SW, to construct the BWT index, and works well for large genomes, particularly if there are many prey sequences.
18. The ">" command directs the output of the program to a new file.
19. –b directs the output to a BAM format. –S was required in previous samtools versions, but newer versions of samtools will automatically detect the correct format.
20. Different samples may have different representations of specific cDNAs. By combining the reads for all of the prey cDNAs from your screens and the library, you can determine the quality of your library and how much coverage you have for each gene.
21. The multicov command uses the indexed and sorted BAM files to determine the number of alignments for an interval, which is a gene in our analysis. The -bams switch is followed by (multiple) bam files; -bed is followed by a bed file that includes the following columns: chrom (the name of the chromosome), chromStart (the starting position on the chromosome), chromEnd (the ending position on the chromosome), name (name of the BED line), score, strand, thickStart, thickEnd, itemRgb, blockCount, blockSizes, blockStarts. The first four columns are necessary for this analysis. For more information about bed format, *see* https://genome.ucsc.edu/FAQ/FAQformat.html#format1.
22. The sort and cut programs are version 8.4.
23. The join program is version 8.4.
24. The cpm results are written in combined.coverage.cpm file and rpkm results are written in combined.coverage.rpkm file.
25. We have found that our approach can identify loci that are rare in the library (1.9 rpm) and enriched more than 500 times for a specific bait (~1000 rpm) [5].
26. The genomecov command allows you to determine the coverage of sequences in the genome. –d determines the coverage per base on each chromosome. –ibam indicates that a BAM file

is used and that the coverage is grouped by chromosome. –g indicates the genome reference sequence fasta file.

27. We exclude candidate interacting proteins whose cDNAs are 100 nt or less, as they are unlikely to form a large enough protein domain for interaction.

Acknowledgments

We would like to thank Maël Baudin, Jana A. Hassan, and Vasanth Singan for critically reviewing the manuscript. This work was supported by Natural Sciences and Engineering Research Council of Canada awards to D.S.G. and D.D; a Canada Research Chair in Plant-Microbe Systems Biology (D.D.) or Comparative Genomics (D.S.G.); the Centre for the Analysis of Genome Evolution and Function (D.D. and D.S.G.); United States Department of Agriculture Agricultural Research Service 5335-21000-040-00D (J.D.L).

References

1. Fields S, Song OK (1989) A novel genetic system to detect protein-protein interactions. Nature 340:245–246
2. Gyuris J, Golemis E, Chertkov H, Brent R (1993) CDI1, a human G1-phase and S-phase protein phosphatase that associates with CDK2. Cell 75:791–803
3. Johnsson N, Varshavsky A (1994) Split ubiquitin as a sensor of protein interactions in vivo. Proc Natl Acad Sci U S A 91:10340–10344
4. Harbers M (2008) The current status of cDNA cloning. Genomics 91:232–242
5. Lewis JD, Wan JR, Ford R, Gong Y, Fung P, Wang P, Desveaux D, Guttman DS (2012) Quantitative interactor screening with next-generation sequencing (QIS-Seq) identifies *Arabidopsis thaliana* MLO2 as a target of the *Pseudomonas syringae* type III effector HopZ2. BMC Genomics 13:8
6. Hancock JF, Paterson H, Marshall CJ (1990) A polybasic domain or palmitoylation is required in addition to the CAAX motif to localize p21ras to the plasma membrane. Cell 63:133–139
7. Yeung T, Gilbert GE, Shi J, Silvius J, Kapus A, Grinstein S (2008) Membrane phosphatidylserine regulates surface charge and protein localization. Science 319:210–213
8. Iyer K, Burkle L, Auerback D, Thaminy S, Dinkel M, Engels K, Stagljar I (2005) Utilizing the split-ubiquitin membrane yeast two-hybrid system to identify protein-protein interactions of integral membrane proteins. Sci STKE 2005:I3
9. Kelley LA, Mezulis S, Yates CM, Wass MN, Sternberg MJE (2015) The Phyre2 web portal for protein modeling, prediction and analysis. Nat Protoc 10:845–858

Chapter 2

sbv IMPROVER: Modern Approach to Systems Biology

Svetlana Guryanova and Anna Guryanova

Abstract

The increasing amount and variety of data in biosciences call for innovative methods of visualization, scientific verification, and pathway analysis. Novel approaches to biological networks and research quality control are important because of their role in development of new products, improvement, and acceleration of existing health policies and research for novel ways of solving scientific challenges. One such approach is sbv IMPROVER. It is a platform that uses crowdsourcing and verification to create biological networks with easy public access. It contains 120 networks built in Biological Expression Language (BEL) to interpret data from PubMed articles with high-quality verification available for free on the CBN database. Computable, human-readable biological networks with a structured syntax are a powerful way of representing biological information generated from high-density data. This article presents sbv IMPROVER, a crowd-verification approach for the visualization and expansion of biological networks.

Key words Systems Biology, Network Model, Signaling Pathway, Crowdsourcing, Crowd Verification, sbv IMPROVER, Biological Expression Language (BEL)

1 Introduction

Over the last few decades, there was a surge in biomedical sciences that has resulted in increasing amount of diversified data. For instance, in 2014, MEDLINE counted over 21 million citations from 5647 indexed journals [1]. That is more than a 5% increase in the amount of citations from academic journals from the previous year, and more than a 100% increase from 2000. This increasing number of peer-reviewed publications in biomedical sciences creates several challenges.

First, visualization, which helps scientists in understanding biological pathways and uncovering important properties of the underlying processes. There are different pathway databases, the most popular being: KEGG, Reactome, PID, BioCyc, Cyclone, RegulonDB, WikiPathways, Pathway Commons, Pathway Assist, and NetPath. The KEGG database, initiated in 1995 by Minoru Kanehisa, is one of the first databases of biological signaling pathways freely available to the general scientific community [2].

Tatiana V. Tatarinova and Yuri Nikolsky (eds.), *Biological Networks and Pathway Analysis*, Methods in Molecular Biology, vol. 1613, DOI 10.1007/978-1-4939-7027-8_2, © Springer Science+Business Media LLC 2017

Next, verification, which aims to extract the maximum value out of verified data, is another challenge. In the wake of high-profile controversies, scientists are facing up problems with replication [3, 4]. There is growing alarm about results that cannot be reproduced. Strict guidelines to improve the reproducibility of experiments are a welcome move [5]. Verification helps to avoid inaccurate conclusions and determine the right algorithms and models. Advancements in research process verification can contribute to challenges made to existing theories. More evidence can either prove existing theories or reveal different interpretations and flaws within it. The quality of scientific predictions has become more dependent on the samples of systems that are modeled, measured, and analyzed. When only a small minority of results is tested, it raises concerns over the legitimacy of the results and the entire set of predictions. Therefore, scientific developments and diversification of data now require a community approach for its scientific verification. Community feedback is the basis of crowdsourcing, which highlights a new trend in science and technology: people working together to innovate and create extraordinary data and to find new solutions for extant challenges. Community approaches are seen as an attempt to reach consensus in the sciences. Some see progress in science as a social process dominated by the scientific community at a particular moment in time. Therefore, it can be a reflection of the paradigm of "what is right," as adopted by scientific society.

Among the most exemplary projects that utilize crowdsourcing as a data analysis tool in biosciences is the sbv IMPROVER Challenge, also known as the System Biology Verification project. IMPROVER is an abbreviation for **I**ndustrial **M**ethodology for **PRO**cess **VE**rification in **R**esearch [6]. The sbv IMPROVER project is a collaborative effort that includes scientists from IBM Research (Yorktown Heights, NY) and Philip Morris International (PMI), Research & Development (Neuchâtel, Switzerland). The goal of the project is to develop a more transparent and robust process for assessing complex scientific data in systems biology (the study of biological organisms, viewed holistically as integrated and interacting networks of genes, proteins, and biochemical reactions). This approach has implications for a wide variety of industries including pharmaceuticals, biotechnology, nutrition, and environmental safety—essentially any area that requires a more meaningful scientific analysis of Big Data [7]. Systems biology verification and industrial methodology for process verification in research are the basis of sbv IMPROVER. Researchers at IBM and Philip Morris International R&D (PMI; Neuchâtel, Switzerland) have been collaborating on a vision for quality assurance in systems biology research. The goal of collaboration is to assure the validity of complex scientific results in the area of systems biology, and recognize the power of communities to assess methodological aspects of scientific research. Although industry shares many of the same needs for validation as academia, a methodology for verifying research is needed in the industrial setting that recognizes both speed and protection of proprietary data constraints, as well as the importance of market considerations and consumer protection. sbv IMPROVER has further advanced crowdsourcing and implemented crowd-verification; a strategy scientists use to verify networks [8]. It shows what is possible to create by combining science, technology, and organized human and social capital. Researchers who are participants in the challenge compete for grants and opportunities to present their data at the sbv IMPROVER Symposium, an international symposium that features the work of scientists from Belgium, France, Germany, India, Italy, Japan, Luxemburg, Malaysia, Poland, Russia, Spain, Switzerland, UK, and the US [9].

The collection of networks that resulted is freely available to the scientific community in a centralized web-based repository: The Causal Biological Network database. It is composed of over 120 manually controlled and well-annotated biological network models. It can be accessed at http://causalbionet.com. The website uses a MongoDB tool that allows users to search for genes, proteins, biological processes, small molecules, and keywords in the network descriptions. This systematic approach allows users to retrieve biological networks of interest. The content of networks can be searched and visualized. Nodes and edges can be filtered with all supporting evidence. The information on the resource is linked to the original articles in PubMed. Moreover, networks can be downloaded for further visualization and evaluation [10].

2 Materials

Peer-reviewed scientific articles from PubMed constituted the majority of the project's resources. They were used to analyze investigations on the topic, to combine the data, and to determine the most effective methods for their visualization and verification.

3 Methods

There are different tools and methods for pathway analysis that help determine the pathways in comprehensive biological networks. Among the most important tools for pathway analysis are GEPAT, PAGE, CPath, and EASE, as well as Cytoscape, ONDEX HTML, and Pathview. Some of these tools and methods also require the use of the biological pathway exchange languages, such as SBML, Kappa, BioPAX, and BEL.

3.1 Network Language

The networks at the sbv IMPROVER project were built using the Biological Expression Language (BEL), which is an open-source language (http://www.openbel.org/) that can represent scientific findings from life sciences in a computable form. BEL was designed to represent research by capturing causal and correlative relationships in context, where the context can include information about the biological and experimental system in which the relationships were observed, as well as the supporting publication citations. The structure of a BEL node, which includes the biological entity, the namespace, or database to standardize the nomenclature of the entity, and the function that describes the type of entity (protein, chemical, biological process, family, complex, etc.), shows the definition of the prefixes for BEL namespaces and functions that appear in the networks.

BEL statements contain three components: a subject, a predicate, and an object, representing discrete scientific findings and their relevant contextual information as qualitative causal relationships. Subjects and objects are visualized as nodes in the biological networks. Predicates are statements that connect two nodes (i.e., network edges), maintain the computability of networks, and are supported by evidence from the scientific literature. All semantic triples are in a defined ontology, for example, HGNC (www.genenames.org), SwissProt (www.uniprot.org), EntrezGene (www.ncbi.nlm.nih.gov/gene), Rat Genome Database (www.rgd.mcw.edu), or ChEBI (www.ebi.ac.uk/chebi). BEL provides the means to describe biological interactions qualitatively, but not to quantify the magnitude or rate of these interactions. This limitation is by design, as quantitative information has significant variability and is not consistently reported in the literature. BEL-based models not only represent all molecular species but also preserve the directionality of interactions [11].

3.2 IMPROVER Methodology

sbv IMPROVER is an open database for the scientific community: https://bionet.sbvimprover.com/

The crowd-verification of biological network models is performed through the following steps [12]:

1. Develop a high-performance platform for the crowd-verification of biological network models and import created biological network models onto the platform.
2. Start the crowd-verification phase by making the platform accessible to the research community, with associated incentives to stimulate online verification of nodes and edges supported by scientific findings.
3. Interpret the results after a predetermined period to identify questionable edges (e.g., edges that did not obtain a consensus from the community).

4. Organize a "jamboree" session where community members that contributed significantly to the online verification can meet recognized experts and analyze scientific evidence for the questionable edges identified in the previous step. Publish the verified and extended networks.
5. Assess the resulting networks and determine to what extent the biological mechanisms were further expanded, revised, or invalidated. Disseminate the networks for public use.

sbv IMPROVER is a robust methodology that verifies systems biology approaches using double-blind performance assessments and applies the wisdom of crowds to solve scientific challenges. The sbv IMPROVER Network Verification Challenge (NVC) asks participants to verify, modify, or create edges in selected biological network models. Its aim is to build consensus around which parts of the networks are accurate, incorrect, or incomplete.

IMPROVER building blocks need to accommodate a priori unknown input–output functions. The development of appropriate scoring metrics is a key element for the verification methodology that helps identify the strength or weakness of a building block when precise knowledge of an input–output relationship is not possible. The verification can be done internally by members of a research group, or externally by crowdsourcing to interested community members. IMPROVER is, therefore, a mix of internal/non-public and external/public assessment tests or challenges.

Biological network models are a representation of known biology within defined contextual boundaries (e.g., species, tissue, and disease). Networks consist of nodes (e.g., DNA, RNA, proteins, etc.) and edges, where edges are causal or correlative relationships between the nodes. For instance, the protein MDM2 negatively regulates the activity of the protein p53. MDM2 and p53 are the nodes, and "negatively regulates" is the edge. NVC participants are requested to verify this kind of relationship on the basis of peer-reviewed scientific literature.

The NVC website visualizes available networks, enabling participants to scrutinize relationships, and make submissions that will either extend the network or verify existing parts of the network.

Each new edge that is created must respect the network's contextual boundary conditions and be submitted with a supporting peer-reviewed academic article. New nodes can only be created as part of creating an edge. Participants can capture new edges using the Biological Expression Language (BEL).

Verification of the network includes the following:

- Supplementation of existing evidence to provide further support for an existing edge.
- Confirmation or rejection of evidence for edges, based on whether the provided reference supports the edge and whether an evidence form has been filled accurately.

When submitting additional evidence or voting on edges, participants ought to fill in the evidence and complete the vote form as completely and accurately as possible. This helps others to understand the rationale for submissions in the network and helps in the creation of the consensus-building process.

The outcome of the online verification process is the combination of submissions by different participants. Based on this, each edge can have four possible states by the end of the challenge:

- Verified: there is at least one verified piece of evidence associated with the edge. A piece of evidence is verified if the overwhelming majority of participants approved rather than rejected the evidence.
- Ambiguous: participants are divided on whether a piece of evidence supports the edge (less than 80% of participants approve or reject the edge).
- Rejected: all evidence that has been suggested in favor of an edge has been rejected by the overwhelming majority of participants during the course of the challenge.
- Not verified: the evidence for an edge did not receive sufficient submissions from participants to be considered verified.

Selection of edges that attracted a lot of attention and controversy from challenge participants is reviewed and discussed at the "jamboree." This face-to-face meeting takes place after the online verification process is completed.

4 Notes

Worldwide explosions of data generation in biomedical sciences have confronted a scientific community with a necessity for creating innovation in data visualization and high-throughput data verification.

BEL was adopted as the structured language to represent the network models in the sbv IMPROVER Network Verification Challenge (NVC). It enables the visualization of causal and correlative relationships between biological nodes and edges in computable and human-readable statements.

Biological Network Models in the sbv IMPROVER Network Verification Challenge (NVC) are verified by participants. The networks are split into five tracks: cell stress, cell fate, cell proliferation, immune response, and tissue response. The evidence is primarily based on human biology non-diseased respiratory tissue biology augmented with chronic obstructive respiratory disease biology.

Structure of the network models in the Network Verification Challenge includes nodes, edges, and context.

Nodes that are a wide range of biological entities are represented as nodes in the network models. They include proteins, DNA variants, noncoding RNA, phenotypic or clinical observations, chemicals, lipids, methylation states, and other modifications (e.g., phosphorylation). Existing nodes were identified using biological databases, such as SwissProt (www.uniprot.org), EntrezGene (www.ncbi.nlm.nih.gov/gene), Rat Genome Database (www.rgd.mcw.edu), and ChEBI (www.ebi.ac.uk/chebi).

Edge: the causal or correlative nature of relationships between nodes is represented as an edge. This allows the biological intent of the network model to be easily digested by a scientist. An example of a relationship, or edge, is TGF Beta 1 *increases* SMAD1.

Context: each edge is constructed within precisely defined contextual boundaries and based on a literature reference to justify the edge's existence. The context of an edge may include species, tissue, cell, and disease.

The nodes and edges in a network model are captured in BEL, a computable language designed for network biology.

The networks, as implemented on the NVC website, are dynamic. They can be modified as new knowledge becomes available and current edges and pieces of evidence are verified by the community.

The network models selected for the NVC were derived from CausalBioNet network models and represent important biological processes implicated in human lung physiology and specific processes related to COPD.

Non-disease networks include the following: cell proliferation, cellular stress, cell fate, pulmonary inflammation, tissue repair, and angiogenesis.

Chronic obstructive pulmonary disease (COPD) networks are: B-cell Activation and T-cell Recruitment and Activation subnetworks to represent immune processes and their role in COPD, Extracellular matrix (ECM) Degradation and Efferocytosis subnetworks were constructed by heavily modifying healthy models to specifically represent COPD-relevant mechanisms.

Networks are available for download upon registration on the sbv IMPROVER website (https://bionet.sbvimprover.com/) and are of great use to both academic and industry users in promoting future research in this area of great therapeutic importance.

Therefore, crowdsourcing efforts that take advantage of new trends in social networking have flourished. These initiatives match discipline specific problems with problem solvers who are motivated by different incentives to compete and show that their solution is the best.

Challenge-based approaches create metrics for the comparison of possible solutions to those challenges designed to verify building blocks. The effectiveness of one methodology can promote community acceptance of the best performing methodology and can then be used as a reference standard. sbv IMPROVER offers a complement and enhancement to the peer-review process in which the results of a submitted paper are measured against benchmarks in a double-blind, challenge-assisted peer-review process. The sbv IMPROVER approach can be applied to a variety of fields where the output of research projects is fed as input into other projects, as is the case in industrial research and development, and where verification of the individual projects or building blocks is elusive, as it is in the case of systems biology.

This approach allows for the application of network pharmacology and systems biology beyond toxicological assessment and can be applied in areas such as drug development, consumer product testing, and environmental impact analysis [13, 14].

The sbv IMPROVER approach differs from other scientific crowdsourcing approaches in that it focuses on the verification of processes in industrial contexts in addition to basic scientific questions.

Web-based graphical interfaces allow for visualization of causal and correlative biological relationships represented using crowdsourcing principles. It enables participants to communally annotate these relationships based on evidence. Gamification principles are incorporated to further engage domain experts throughout the biological sciences to gather robust peer-reviewed information from which relationships can be identified and verified.

The resulting network models represent the current status of biological knowledge within the defined boundaries, in this case, for processes relating to human lung disease. These models are amenable to computational analysis. For some period following the conclusion of the challenge, the published models will remain available for continuous use and expansion by the scientific community.

Collaborative competition has the unique ability to facilitate analysis of high-throughput data and to become an elevator to solutions. Such approaches to research allow for the organization and processing of information in a trustworthy and effective way.

References

1. Medline/Pubmed resources Detailed Indexing Statistics: 1965–2014. http://www.nlm.nih.gov/bsd/index_stats_comp.html. Accessed 24 Feb 2016
2. Kanehisa M, Goto S (2000) KEGG: kyoto encyclopedia of genes and genomes. Nucleic Acids Res 28:27–30
3. Yong E (2012) Replication studies: bad copy. Nature 485:298–300. doi:10.1038/485298a
4. Ioannidis JP, Allison DB, Ball CA et al (2009) Repeatability of published microarray gene expression analyses. Nat Genet 41:149–155
5. Repetitive flaws (2016) Nature 529:256. http://www.nature.com/news/repetitive-flaws-1.19192
6. Meyer P, Alexopoulos LG, Bonk T et al (2011) Verification of systems biology research in the age of collaborative competition. Nat Biotechnol 29(9):811–815. doi:10.1038/nbt.1968
7. Peitsch M C (2013) sbv IMPROVER: species translation challenge open to the scientific community for submissions. American Laboratory, http://www.americanlaboratory.com/913-Technical-Articles/138841-sbv-IMPROVER-Species-Translation-Challenge-

Open-to-the-Scientific-Community-for-Submissions/. Accessed 24 Feb 2016

8. Meyer P, Hoeng J, Rice JJ et al (2012) Industrial methodology for process verification in research (IMPROVER): toward systems biology verification. Bioinformatics 28(9):1193–1201. doi:10.1093/bioinformatics/bts116
9. Boue S, Fields B, Hoeng J et al (2015) Enhancement of COPD biological networks using a web-based collaboration interface. F1000Res 4:32. doi:10.12688/f1000research.5984.1
10. Boue S, Talikka M, Westra JW et al (2015) Causal biological network database: a comprehensive platform of causal biological network models focused on the pulmonary and vascular systems. Database (Oxford) 2015:bav030. doi:10.1093/database/bav030
11. Younesia E, Hofmann-Apitius M (2013) Biomarker-guided translation of brain imaging into disease pathway models. Sci Rep 3:3375. doi:10.1038/srep03375
12. Ansari S, Binder J, Boue S et al (2013) On crowd-verification of biological networks. Bioinform Biol Insights 7:307–325. doi:10.4137/BBI.S12932
13. Hoeng J, Deehan R, Pratt D et al (2012) A network-based approach to quantifying the impact of biologically active substances. Drug Discov Today 17(9–10):413–418
14. Sewer A, Hoeng J, Deehan R et al (2014) Systems biology approaches for compound testing. In: Hoffmann RD, Gohier A, Pospisil P (eds) Data mining in drug discovery, 1st edn. Wiley-VCH Verlag GmbH & Co. KGaA, Weinheim, Germany

Chapter 3

Mathematical Justification of Expression-Based Pathway Activation Scoring (PAS)

Alexander M. Aliper, Michael B. Korzinkin, Natalia B. Kuzmina, Alexander A. Zenin, Larisa S. Venkova, Philip Yu. Smirnov, Alex A. Zhavoronkov, Anton A. Buzdin, and Nikolay M. Borisov

Abstract

Although modeling of activation kinetics for various cell signaling pathways has reached a high grade of sophistication and thoroughness, most such kinetic models still remain of rather limited practical value for biomedicine. Nevertheless, recent advancements have been made in application of signaling pathway science for real needs of prescription of the most effective drugs for individual patients. The methods for such prescription evaluate the degree of pathological changes in the signaling machinery based on two types of data: first, on the results of high-throughput gene expression profiling, and second, on the molecular pathway graphs that reflect interactions between the pathway members. For example, our algorithm OncoFinder evaluates the activation of molecular pathways on the basis of gene/protein expression data in the objects of the interest.

Yet, the question of assessment of the relative importance for each gene product in a molecular pathway remains unclear unless one call for the methods of parameter sensitivity/stiffness analysis in the interactomic kinetic models of signaling pathway activation in terms of total concentrations of each gene product.

Here we show two principal points:

1. First, the importance coefficients for each gene in pathways that were obtained using the extremely time- and labor-consuming stiffness analysis of full-scaled kinetic models generally differ from much easier-to-calculate expression-based pathway activation score (PAS) not more than by 30%, so the concept of PAS is kinetically justified.
2. Second, the use of pathway-based approach instead of distinct gene analysis, due to the law of large numbers, allows restoring the correlation between the similar samples that were examined using different transcriptome investigation techniques.

Key words Systems biology, Mitogenic cell signaling, Protein-protein interaction, Parameter sensitivity/stiffness analysis, RNA microarray analysis

Tatiana V. Tatarinova and Yuri Nikolsky (eds.), *Biological Networks and Pathway Analysis*, Methods in Molecular Biology, vol. 1613, DOI 10.1007/978-1-4939-7027-8_3, © Springer Science+Business Media LLC 2017

1 Introduction

1.1 Methods for the Analysis of Intracellular Pathway Activation

Numerous molecular pathways that determine the mitogenic fate (proliferation with the risk of cancer development and progression, differentiation, necrosis, apoptosis, etc.) of a cell have been in the focus of research interest in a couple of previous decades [1–3]. These pathways that are mostly initiated by various receptor tyrosine kinases (RTK) involve a plethora of types of proteins, which, in turn, possess multiple binding sites/domains. As a result, pure experimental research methods may not always give an exhaustive answer to the question of molecular etiology for a certain cancer case, since it may require detailed measurements of interactions between several dozens of proteins that transfer the mitogenic signal. On the other hand, valuable information on the details of protein-protein interaction may be obtained using in silico analysis of chemical kinetics for signal transduction [4]. Such studies can be useful to understand what protein activity should be either downregulated or enhanced for the prevention of carcinogenesis or tumor suppression and destruction.

Complex kinetic models for activation of cell signaling pathways that integrate the systems of ordinary differential equations (ODE) for the concentrations of multiple chemical species that change their configurations during the process of signal propagation have been developed and thoroughly investigated at least about last 15 years [5–11]. Considerable difficulties that had arisen during such investigations were more or less successfully resolved. First, there is essential combinatorial complexity for emerging plethora of chemical species, which can be overcome using the universal software packages for rule-based description of highly branched signaling networks [12–18] and domain-oriented methods for combinatorial complexity reduction [19–22].

Even more terrifying problem is related to multiple unknown parameters such as dissociation/Michaelis constants and total concentrations of certain signaling proteins in the ODE systems. The general approach to fitting these parameters that may describe the experimental (e.g., Western blotting) data in their best way was formulated in the series of works [9–11]. This approach comprises that the researcher constructs a kinetic model of protein-protein interaction, performs the ODE integration for concentrations of various chemical species (i.e., protein complexes in certain states), adjusts the unknown kinetic constants to fit the experimental data of activation kinetics, and, finally, makes predictions on the different details and conditions of signal propagation. To learn the hints and clues what parameters should be tuned for the most effective parameter fitting, one may use the methods for parameter sensitivity [23] and/or sloppiness/stiffness analysis [24].

1.2 Possible Use of Kinetic (Interactomic) Models for Gene Expression-Based Pathway Activation Analysis

Despite the achievements specified above, these kinetics models of highly branched networks remained almost totally in the domain of "pure fundamental" science that do not deal with any practical medical application. Nevertheless, since signal transduction at every stage depends on the concentrations of the interacting gene products, numerous approaches have been proposed for the conversion of the information on abundances of mRNA/proteins for all genes/gene products into the values that correspond to overall activation/inhibition of intracellular pathways. The most advanced methods take into account the pathway topology (e.g., TAPPA [25], topology-based score [26], Pathway-Express [27], and signaling pathway impact analysis, SPIA [28]).

Similarly, we have recently proposed OncoFinder [29–37], a systems bioinformatics tool for the analysis of changes in intracellular molecular pathways (e.g., signaling, metabolic, and cytoskeleton pathways). As an input data set, OncoFinder operates with the results of various "omics" profiles obtained for the biosamples under investigation, e.g., taken from the patients and from the healthy donors. These profiles may be transcriptomic (e.g., obtained with ether microarray hybridization or next-generation sequencing), proteomic, epigenomic, etc. The data of full mRNA/protein abundancies are integrated by OncoFinder into the assessment values for activation of different cellular pathways (signalome).

In all these methods, the relative expression levels for each gene can be found, respectively, by comparison of expression levels in an individual case sample and the average level for the corresponding normal sample or set of samples. These data on gene/protein/miRNA, etc., expression levels are accumulated by the software into the signalome-based entities, and for each molecular pathway the individual measure for pathological perturbations is evaluated. One important issue, however, remained unresolved until recently: how the assessment function for *PAS* can take into account the relative importance of different genes and gene products for the whole process of the pathway activation/inhibition? Several clues, however, were provided in our recent publication [30], where we applied, for the assessment of the relative importance of distinct gene products, the concepts of parameter sensitivity and stiffness/sloppiness that may be analyzed using the interactomic kinetic models for signaling pathway activation.

Here, we formulate and demonstrate on the example of the ERK activation upon EGF stimulation, the general approach to development, fitting according to the experimental data, as well as gene product importance analysis for the robust kinetic models of signaling pathways activation that involves the "low-level" (mass action law) description of most protein-protein acts, yet is useful for the practical problems like analysis of pathological perturbations in signaling pathways for a given cancer patient. Moreover, we demonstrate that the OncoFinder-based pathway approach restores correlations between the similar samples that were examined using the different methods of mRNA investigation (e.g., microarray hybridization versus next-generation sequencing), even when these correlations were poor at the level of individual genes.

2 Materials and Methods

2.1 Western Blotting Measurements of the HEK293 Cells

As an experimental source for the development of the protein-protein interaction kinetic model for the EGFR signaling pathway, we have taken the results of Western blotting investigation of activated protein abundances upon the EGF stimulation of the HEK293 cells [10]. For these procedures, there were used the antibodies for anti-phospho-EGFR (Y1173), anti-Src (GD11), anti-phospho-Shc (Y317), anti-phospho-GAB1 (Y627), anti-phospho-MEK (S217/S221), anti-phospho-ERK1/2 (T202/185 and Y204/187), anti-phospho-AKT1 (S473), general anti-phosphotyrosine (pY20), as well as anti-GRB2 (C-23), anti-GAB1 (H-198), anti-PI3K-p85, anti-Ras, anti-glyceraldehyde-3-phosphate dehydrogenase (GAPDH) (6C5), and anti-a-tubulin (DM1A) as a housekeeping gene. Chemiluminescence signals from immunoreactive bands were detected on a KODAK Image Station 440CF.

2.2 Rule-Based Modeling Software

Since our goal was to distinguish different scenarios of protein-protein interaction using computational methods, we decided to build as detailed model as possible. That is why the model was developed using the software BioNetGen 2 [12], which, along with StochSim [13, 14], Kappa [15, 16], Moleculizer [17], and *Lillte b* [18], describes highly branched kinetic networks using the rule-based approach. This approach means that all possible chemical complexes (species) that emerge during the signal propagation are generated algorithmically according to the user-specified reactions rules that describe certain events on certain sites on certain protein molecules. Along with the entire graph of chemical transformations of molecules and their complexes, the rule-based systems biology software builds the corresponding system of ordinary differential equations (ODE) for concentration of each species. Like most code packages for rule-based description of molecule interactions, BioNetGen 2 allows ODE integration using both deterministic Runge-Kutta and stochastic Monte Carlo [38] method.

2.3 OncoFinder Algorithm for Processing of Transcriptomic/Proteomic Data

We processed the transcriptomic/proteomic data from the human tissue samples under investigation to establish pathway activation strength (PAS) profiles corresponding to intracellular signaling pathways. The formula for PAS calculation accounts for gene expression data and for information on the protein interactions in a pathway, namely, individual protein activator or repressor roles in a pathway [30]; for pathway p, $PAS_p = \sum_n ARR_{np} \cdot \log(CNR_n)$. The relative role of a gene product in signal transduction is reflected by a discrete flag *activator/repressor role* (ARR), which equals 1 for an activator gene product, −1 for a repressor, and shows intermediate values −0.5; 0.5 and 0 for the gene products that have repressor, activator, or unknown roles, respectively. The CNR_n value (*case-to-normal ratio)* is the ratio of the expression level of a gene n in the sample under investigation to the average expression level in the reference sampling. The positive value of PAS indicates activation of a signaling pathway, and the negative value stands for its repression. The analysis included 271 intracellular signaling pathways.

2.4 Datasets for Studying the Correlations Between the Same Samples Examined using Different Transcriptome Investigation Methods

To study the effects, which are introduced by the examination of the same samples at different transcriptome investigation platforms, and to check if the signalome-based approach instead of the gene-based approach can increase the correlations between these samples, we compared different gene expression datasets generated using both next-generation sequencing (NGS) and microarray hybridization.

Gene expression data were downloaded from the Gene Expression Omnibus (GEO) repository of transcriptomic information [39]. The overview of materials, methods, and results of these cross-investigation datasets is shown in Table 1.

Table 1
Transcriptomic data deposited in the GEO database that were used for the current study

Dataset ID	Origin	Case samples versus control samples	Experimental platforms	# of samples
GSE36244	HepG2 cells	Treated vs. untreated with benzopyrene	Transcriptome at Affymetrix Human Genome U133 Plus 2.0 arrays and transcriptome et Illumina Genome Analyzer sequencer	4
GSE41588	HT-29 cells	Treated vs. untreated with 5-aza-deoxy cytidine	Transcriptome at Affymetrix Human Genome U133 Plus 2.0 arrays and transcriptome at Illumina Genome Analyzer sequencer	6
GSE37765	Lung adeno-carcinoma	Tumor samples vs. matched samples of normal tissue	Transcriptome at Agilent 1M CNV arrays and transcriptome Illumina Genome Analyzer sequencer	6

3 Results

3.1 Building Large-Scale, Low-Level Models of Mitogenic Cell Signaling

Among signaling pathway kinetic models that take into account the combinatorial complexity, there were descriptions of highly branched networks that lead to the activation of one of the most important mitogenic effectors, extracellular regulatory kinase (ERK), as well as the serine/threonine kinase AKT that is particularly important for anti-apoptotic response, upon the stimulation of the HEK293 cells with epidermal growth factor (EGF) [9, 11] and with insulin [10].

We have shown experimentally that in HEK293 cell line, insulin does not significantly activate ERK, although it considerably amplifies the ERK response upon EGF stimulation [9, 10]. Using mathematical modeling, we hypothesized the essential role of the adapter protein Grb2-associated binder 1 (GAB1) [10] that plays the key role in signal propagation upstream of the small GTPase Ras [9, 10], whereas the major mechanism of GAB1-dependent signal amplification is recruitment of GAB1 (and other adapter proteins) to the plasma membrane via phosphatidylinositol-3,4,5-triphosphate (PIP_3) [9, 10]. This hypothesis was confirmed in a series of experiments including Western blotting, chemical inhibition of key components in signaling pathways, and short interfering RNA (siRNA)-based depletion of important proteins in signal transduction [9, 10].

Nevertheless, some details of GAB1 interaction with its partners remain unclear. This protein molecule is known to possess multiple docking sites that specifically bind numerous partners such as phosphatidylinositol-3 kinase (PI3K) [40], GTPase activation protein RasGAP [41], tyrosine phosphatase SHP2 [9], tyrosine kinases of the Src family [42], etc. Moreover, one of major GAB1 partners, growth factor receptor binder 2 (Grb2), has been reported to bind to GAB1 in many ways. First, the association of Grb2 and GAB1 can be performed via phosphorylated tyrosine residues of GAB1 and SH2 domain of Grb2 [43, 44]. Second, the binding may be performed via the C-terminal Src homology 3 (SH3) domain of Grb2 and the proline-rich domain (PRD) of GAB1 [45, 46]. Our previous in silico modeling studies have not favored any of these possible scenarios due to the limited description of combinatorial complexity in these models [9, 10]. To obtain several hints and insights on the details of protein-protein interaction within the mitogenic signaling network, we concentrated on building and investigating a full-scale combinatorial complex network model.

3.2 Protein-Protein Interactions in the Current Model

We have developed a highly branched model for activation of mitogenic (Ras/ERK) and survival (AKT) targets upon EGF stimulation. Our model is a further step in computational research of mitogenic signaling pathways. Our previous results showed the ability to explain most of measured data and to make experimentally verifiable predictions on the details of cell signaling propagation [9, 10]. However, they both were "manually" developed: all species and reactions in the network were specified by the model developers.

Although the network structure of the current model differs significantly from our previous models, most proteins and their principal relationships are quite similar (Fig. 1) [9, 10]. Signal propagation starts with the activation of the cell-surface receptor [5]. Upon EGF binding, the receptor dimerizes and undergoes transphosphorylation at tyrosine residues in the cytoplasmic tail [5]. These residues can bind Shc, Grb2 (followed by binding of the guanine exchange factor SOS), p85 subunit of PI3K and GTPase RasGAP [5–7]. All the species that contain phosphorylated EGFR may be endocytosed and degraded, releasing the binding partners of EGFR. Primary activation of the tyrosine kinases of the Src family (Src) is also implemented by EGFR [8]. The overall picture of the first stages of the pathway activation is shown in Fig. 2.

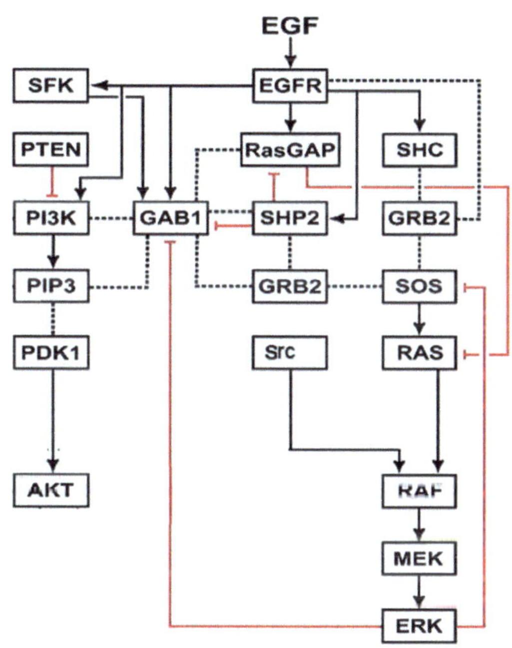

Fig. 1 Flow chart of signal propagation through the EGFR signaling network, pretty similar to our previously published models [10]. *Solid lines* with *arrows* show the activation or tyrosine phosphorylation of proteins and lipids. *Dotted lines* represent direct protein-protein and protein-lipid interactions. *Red lines* with blunt ends show inhibition

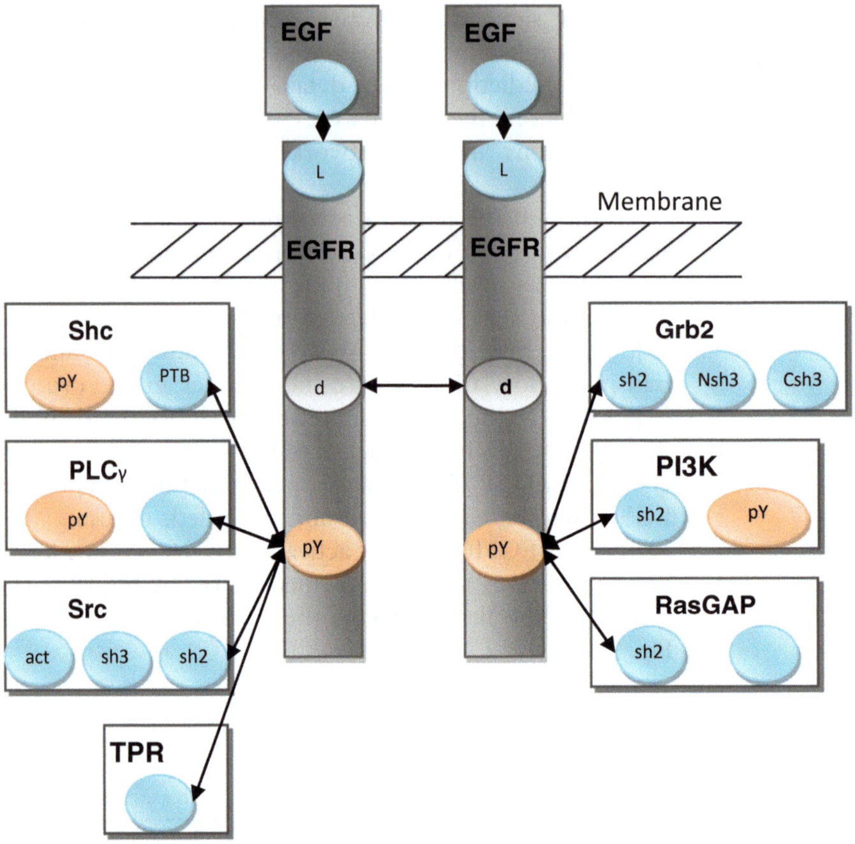

Fig. 2 Epidermal growth factor receptor (EGFR) and its partners in our kinetic model: the ligand (EGF), adapter proteins Shc and Grb2, phospholipase C_γ (PLC_γ), phosphatidylinositol 3-kinase (PI3K), GTPase RasGAP, tyrosine kinase Src, and unspecified tyrosine phosphatase (TPR). Protein molecules are represented as *rectangles*, their structural and functional subunits—as *blue ovals*, phosphorylated tyrosine residues that bind specific partners—as *pink ovals*

The major role in the amplification of the Ras/MEK/ERK signaling is played by the adapter protein GAB1. GAB1 is recruited to the plasma membrane through PIP_3 via the plekstrin homology (PH) domain [47]. When bound to the membrane, either via PIP_3 or via Grb2-(Shc)-EGFR, GAB1 may be phosphorylated by EGFR or SFK, which is followed by binding Grb2 [43, 44], PI3K [40], RasGAP [41]—for the sake of simplicity, we assume competitive binding of these partners to GAB1) or SHP2 [40]. When bound to GAB1, SHP2 exhibits phosphatase activity against the phosphorylation sites on GAB1, as well as on EGFR [40].

Likewise, Grb2 also possesses scaffolding properties. We took into account in [9] that it has two SH3 domains (the N-terminal domain specifically binds SOS, while the C-terminal one binds GAB1) [45, 46]. In addition, the SH2 domain at Grb2 binds phosphotyrosine residues, both on EGFR and GAB1 [43, 44].

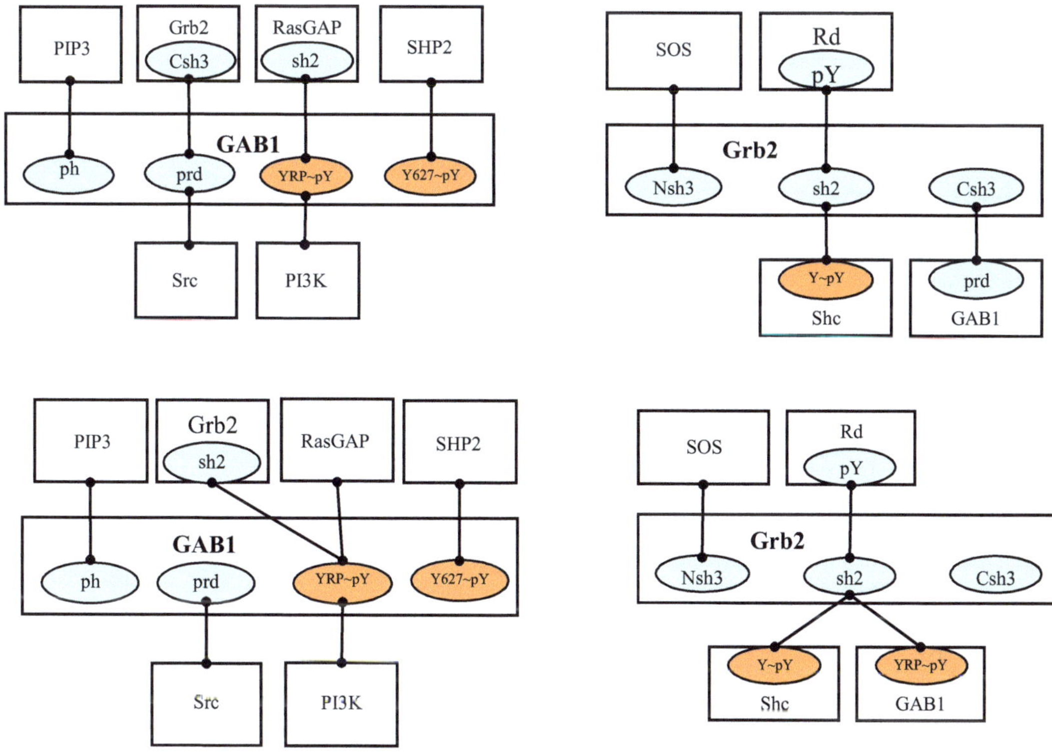

Fig. 3 Domain/site structure and binding partners of two major scaffold proteins, GAB1 and Grb2, according to the variants A (*upper row*) and B (*lower row*) of the signaling network model. *Rectangles* represent protein or lipid molecules, *blue ovals*—protein domains, *orange ovals*—tyrosine residues

To make distinction between the two modes of GAB1-Grb2 binding, we created two variants of signaling models that are symbolically called A and B (*see* Fig. 3). The only difference between protein-protein interaction in models A and B is the sites/domains for Grb2-GAB1 binding. Whereas the model A assumes that these proteins bind each other via the proline-rich domain (PRD) of GAB1 and C-terminal SH3 domain of Grb2, in the model B these proteins associate via one of the numerous tyrosine residues at GAB1 and SH2 domain at Grb2.

PIP_3 (which is produced by the membrane-recruited PI3K) recruits to the plasma membrane adapter protein GAB1 and a serine/threonine kinase PDK1. Membrane-recruited PDK1 causes Akt phosphorylation at Thr308 residue [48, 49].

Membrane-recruited SOS produces transformation of Ras-GDP into Ras-GTP complex [9, 50]; the contrary process is catalyzed by membrane recruited RasGAP. Ras-GTP causes primary activation of Raf protein [51], however, for full activation of Raf, active Src is needed [52]—*see* Fig. 4. Active Raf causes MEK activation by dual phosphorylation of MEK activation loop [53]. Active MEK causes ordered phosphorylation of ERK [7].

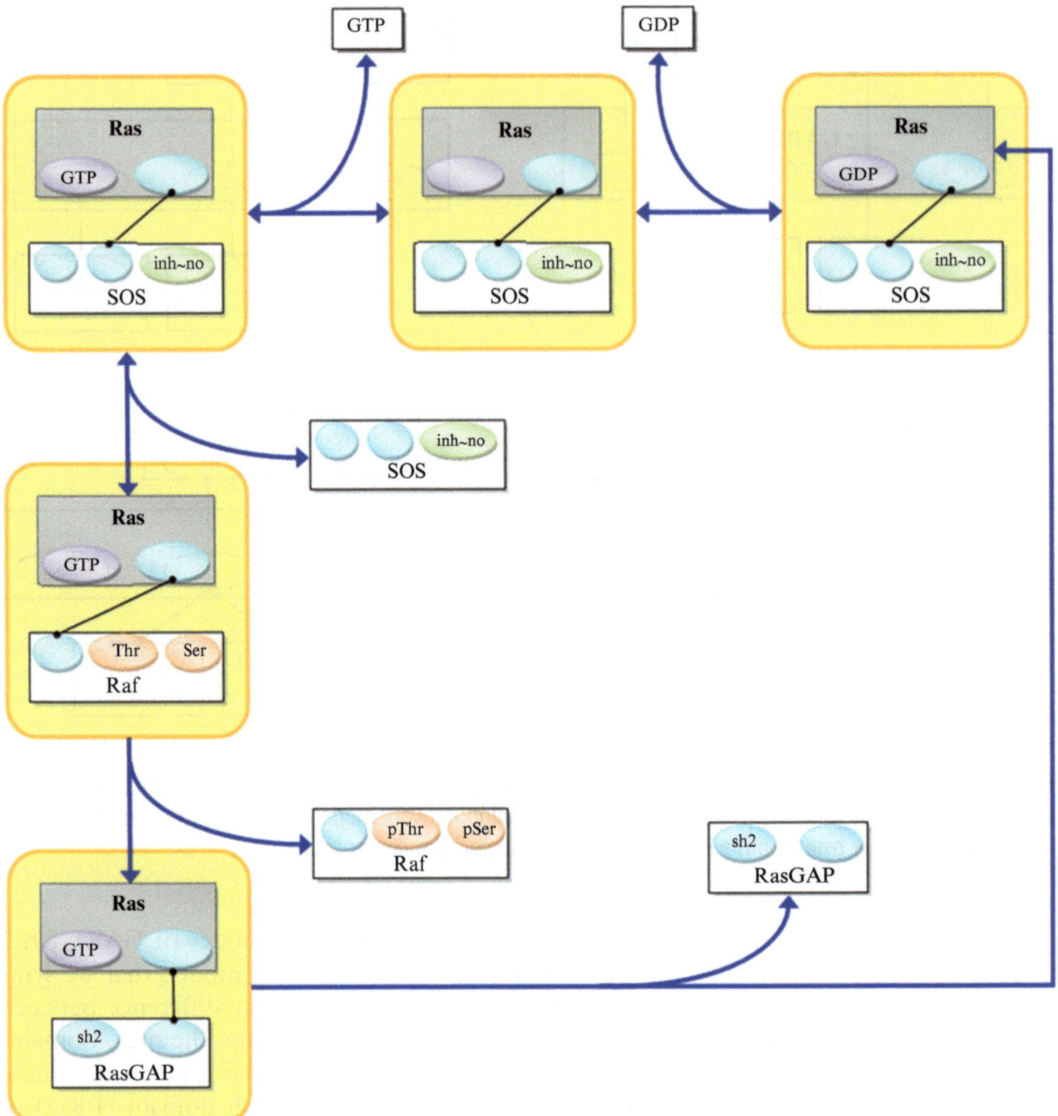

Fig. 4 Activation on Ras and Raf oncogenes in our kinetic model

As a serine/threonine kinase, active ERK may impose negative feedbacks via phosphorylation of serine/theronine inhibitory sites at SOS and GAB1 [54–56].

3.3 Model Size in Comparison with Previous Results

Our first large-scale model of signaling networks [9] assumed independent binding of multiple partners to the scaffolding proteins such as GAB1 and Grb2. Since it was constructed manually, we needed certain model reduction methods [19–22] that replaced highly branched networks describing transitions between the states of a scaffold protein with more compact pathways involving several virtual ("macroscopic") proteins that possess fewer sites than a real scaffold. Contrary, for the sake of simplicity, our model of coupled

Table 2
Overview of our signaling network models

	Reference (Kiyatkin et al. [9])	Reference (Borisov et al. [10])	Current work
Initiating ligands	EGF	EGF + insulin	EGF
Number of species	~200	78	2022 (model A) 665 (model B)
Number of reactions	~500	111	12148 (model A) 5733 (model B)
Combinatorial complexity	Independent binding with "manual" model construction	Competitive binding with "manual" model construction	Independent binding with automated model construction

insulin-EGF signaling [10] employs the assumption of competitive binding of all partners to the scaffold proteins. This assumption reduces the number of chemical species and reactions, however, it makes impossible to do any predictions on the details of protein-proteins interaction (identification of binding sites/domains etc.).

In contrast to our previous work, the current model exploits the ability of BioNetGen 2 to generate all possible complexes and reactions that may arise during the signal propagation according to the user-specified reaction rules. In our example, BioNetGen 2 generated more than 2000 chemical species and 12,000 reactions for the model A and more than 650 species and 5500 reactions for the model B, which makes our model one of the biggest in current systems biology of mitogenesis. Table 2 presents the comparison of size and overall topology of signaling network models developed during the past decade.

Despite rather large model size, total computation time (including signaling network generation, equilibration of species concentration prior to the stimulation followed by calculation of signaling kinetics after adding EGF) was 15.8 min for the variant A and only 1.43 min for the variant B at a personal computer with CPU frequency of 1.82 GHz and RAM capacity of 1 Gb.

3.4 Fitting Kinetic Parameters and Model Predictions

Methods for network signaling model handling and verification that we applied here were similar to previously used ones [9, 10]. During the "training"/"fitting" process, it is important not to exceed the boundaries for kinetics parameters that are imposed by experimental hints for the similar processes. In addition, any reaction should not be faster than it is prohibited by the diffusion limit.

To make the model more "robust," after the completion of parameter fitting, the researcher can make experimentally verifiable predictions, which include, for example, some model

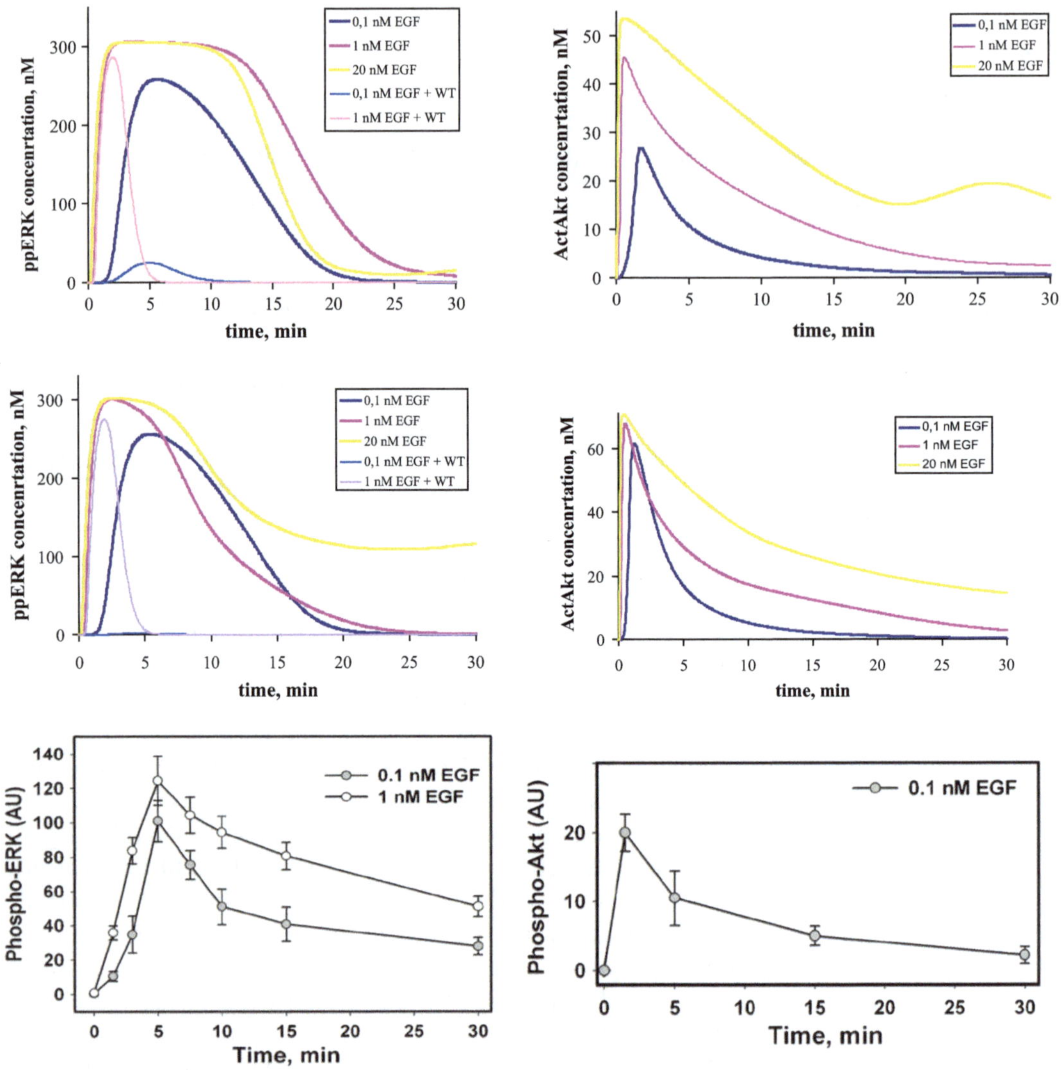

Fig. 5 Model "training/fitting" according to the experimental data on ERK (*left column*) and Akt (*right column*) activation. *Upper row*: results of model A fitting for different EGF doses and application of wortmannin (WT), a PI3K inhibitor. *Middle row*: the same results for model B fitting. *Lower row*: experimental western blotting data, which we obtained previously on HEK293 [10]

perturbations, such as inhibition of certain enzymes or increases/decrease of certain protein abundances via protein overexpression or, perhaps, siRNA-assisted depletion.

Figure 5 shows the results of our model "training" on the example of ERK (left column) and Akt (right column) activation patterns. Models A (upper row) and B (middle row) were "trained" for three values of EGF dose (20, 1, and 0.2 nM), as well as for action of PI3K inhibitor wortmannin (WT), according to the experimental data (Western blotting) that were published previously for the HEK293 culture cells—lower row, taken from [10].

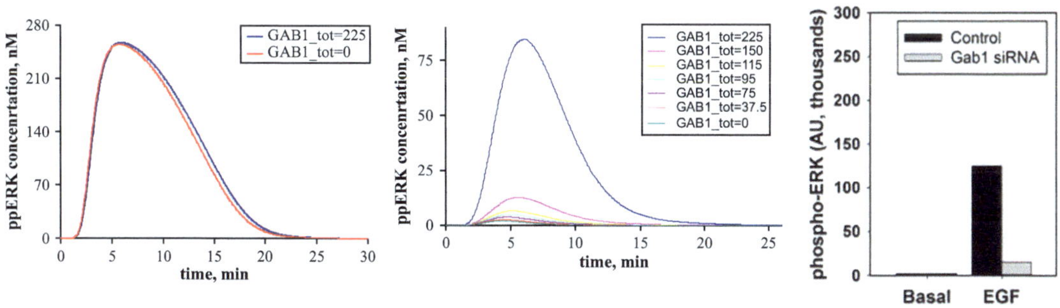

Fig. 6 Model predictions on effects of siRNA-induced GAB1 depletion for EGF concentration of 0.1 nM. *Left panel*: model A. *Middle panel*: model B. *Right panel*: experimental (western blotting) validation [10] for the influence of siRNA-induced GAB1 depletion on ERK activation at 1.5 min after EGF stimulation of HEK293 cells

Although the activation curve shapes for models A and B were not always strictly similar, it was impossible to make clear preferences for the variant A or B at this "training"/"fitting" step. The same conclusion can be made for the GAB1, Ras, and MEK activation curves (data not shown).

Previously, we demonstrated the crucial role of GAB1 as an enhancer of mitogenic signaling upon EGF simulation [9, 10]. When GAB1 is recruited to the plasma membrane via PIP_3, it may bind PI3K, which produces more PIP3, thus increasing the concentration of the membrane-recruited GAB1 and closing the positive feedback loop. The effects of feedback loop disruption by GAB1 depletion, which we computationally predicted in our earlier models, were experimentally verified using the siRNA method [9, 10].

Interestingly, our calculations (*see* Fig. 6) show that only model B (middle panel) was capable of reproducing experimental results (right panel) for the influence of GAB1 depletion on the ERK response to EGF stimulation of HEK293 cells. Contrarily, the model A (left panel) did not show the decrease of ERK signal even for the total removal of GAB1 for the cell. This surprising effect is caused by the specific "sequestration" of Grb2 by GAB1 in the model A. If a large GAB1-containing complex binds Grb2-EGFR or Grb2-Shc-EGFR complex via the C-terminal SH3 domain of Grb2, the resulting reaction product may exceed the critical number of protein molecules in the complex (we assumed that any complex cannot contain more than five molecules), thus preventing SOS binding to Grb2 via the N-terminal SH3 domain. However, in the model B, GAB1 and EGFR cannot bind to Grb2 simultaneously (*see* the lower right panel in Fig. 3), so that GAB1 cannot "sequester" the membrane-recruited (via EGFR) Grb2 from the pool of molecules that are capable of recruiting SOS to the membrane.

4 Discussion

4.1 Difficulties with Experimental Validation of Computational Findings

Although our results showed feasibility and traceability of large-scale combinatorially complex signaling network models that are automatically generated using the rule-based software for systems biology, our finding that model B may be more adequate than model A, needs further verification.

It should be notified that the experimental validation, which may favor scenario A or B, may introduce several experimental errors or uncertainties. Direct measurements may lead to triple immunoblotting (detection of certain phosphotyrosine residue in the simultaneous Grb2-GAB1 precipitate), which has very low registration efficiency, and, consequently, large relative errors.

Experiments with mutant proteins (for example, substitution of GAB1 tyrosine residue that binds SH2 domain for Grb2 with phenylalanine) inevitably involve model laboratory animals, and the use of animals may introduce extra uncertainties, e.g., due to the human-murine differences in the genome and proteome.

4.2 Potential Usefulness of Kinetic Models for Signaling Pathways

Although current versions of OncoFinder [30] software packages are based on the assumption of equal relative importance of all gene products, both signal activators and signal inhibitors, this hypothesis may seem rather artificial. As far as we previously mentioned, at least two ways for the determination of relative importance of genes/proteins may be suggested. The former operates with the concept of sensitivity of the ODE system on the free parameters [23], which is generally applied to kinetic constants (such as the dissociation constant, the Michaelis-Menten constants, etc.), but also may be used (exactly as here) for the total concentrations of certain proteins in the kinetic model of a pathway, as follows, $w_j^{(1)} = \lim_{t \to \infty} \frac{1}{T} \int_0^T \left| \frac{\partial \ln [EFF(t)]}{\partial \ln C_j^{tot}} \right| dt$. Here, $[EFF(t)]$ is the time-dependent concentration for the active form of definitive pathway effector, and C_j^{tot} is the total concentration for the protein j in the kinetic pathway model.

The latter way to calculate the importance function for the genes/proteins in a pathway is related to the stiffness/sloppiness analysis [24] for the effector activation upon total protein concentrations. According to such an approach we interrogated the Hesse matrix, $H_{ij} = \frac{\partial^2}{\partial C_i^{tot} \partial C_j^{tot}} \sum_k ([EFF(\mathbf{C}^{tot}, t_k)] - \frac{[EFF]_k^{\exp})^2}{\sigma_k^2}$, where $\mathbf{C}^{tot}$ is the vector of total concentrations for every protein type in the pathway model, $[EFF(\mathbf{C}^{tot}, t_k)]$ is the concentration for the active form of the definitive effector calculated at the time point t_k, $[EFF]_k^{\exp}$ is the experimentally measured (using, e.g., the Western blotting technique) concentration at the same time, and σ_k is the experimental error for this measurement. The sloppiness/stiffness analysis searches for the eigenvalues, λ_m, and eigenvectors, ξ_m, for the Hesse matrix, $\mathbf{H}\xi_m = \lambda_m \cdot \xi_m$. The higher is the absolute value of λ_n, the "stiffer" is the direction within the n-dimensional space of $\mathbf{C}^{tot}$ (where n is the number of protein types in the pathway model). The eigenvector components along the stiffest direction, ξ_s, may be used for the assessment of the relative importance of certain genes/proteins in a pathway, as follows, $w_j^{(2)} = |\xi_{sj}|$.

Taking into account the considerations above, we arrive at the following formula, $PAS_p^{(1,2)} = \sum ARR_{np} \cdot BTIF_n \cdot w_n^{(1,2)} \cdot \log(CNR_n)$. Here, the Boolean flag of *BTIF* (*beyond tolerance interval flag*) indicates that the expression level for the gene n for the given sample is different enough from the respective expression level for the reference sample or set of the reference samples. In our studies we stipulated that to be significantly pathologic any gene for a cancer patient must be at least by 50% higher or 50% lower expressed compared to the average value for the reference set of samples, and, at the same time, its expression level should differ by more than two standard deviations from the average of the reference set.

To check if the introduction of the weight (importance) coefficients, either sensitivity-based, $w^{(1)}$, or stiffness-based, $w^{(2)}$, makes any significant difference, we have performed the verification on the example of the EGFR pathway, we have performed the verification on the example of the EGFR pathway. For these two sets of weighting factors, as well as without them, we have performed a computational experiment with nine datasets on the results for microchip investigation of nine samples taken from patients with a high-grade glioblastoma [57]. Our findings suggest that the cloud of values for the ratio of $\frac{PAS^{(1)}_{\text{EGFR}}}{PAS_{\text{EGFR}}}$(where $PAS_{\text{EGFR}}$is the *PAS* value for the EGFR pathway with all importance factors equal to 1) lies within the interval of (0.7 ± 0.3), whereas the ratios of $\frac{PAS^{(2)}_{\text{EGFR}}}{PAS_{\text{EGFR}}}$ are slightly higher and may be assessed as (1.0 ± 0.6).

Unfortunately, the overall number of signaling pathways, which were characterized in terms of only activation/inhibition relationships between the different proteins, is significantly higher than the number of the pathways that have been quantitatively described using the kinetic models. That is why for many pathways in our database the evaluation of weighting factors $w^{(1,2)}$ was impossible. However, we have performed some tests for the stochastic robustness analysis of the proposed formula for *PAS* [30]. During this testing, we have introduced the extra randomly perturbation factors, w_n, which were used as multiplication coefficients for each logarithm of relative gene expression. In our computational experiment, the distribution for w_n was logarithmically normal; namely, they were calculated as follows, $w_n = 2^{x_n}$, where x_n are normally distributed random numbers with the expected value of

$M = 0$ standard deviation $\sigma = 0.5$. The random perturbation factors w_n were applied to one of samples of the glioblastoma gene expression dataset [57]. Importantly, although the perturbation was done independently 98 times with independent weighting factors w_n, for each gene, the values of standard deviation for the set of alternate *PAS* (*APAS*) were not big enough to bias the proportional trend between the average perturbed and unperturbed *PAS* for each signaling pathway [30].

4.3 The Transition to Signalome-Based Description Restores Correlation Between the Same Samples Investigated by Different Methods (Cumulative Effect)

Previous studies, e.g., [58–60] revealed one important discouraging feature of full-transcriptome investigations. If one applies *different experimental methods* (e.g., microarray hybridization and next-generation sequencing) for *the same samples*, little or no correlation may be observed at the level of distinct genes.

We checked if transition from the gene-based to pathway-based approach, e.g., our OncoFinder system, can restore the correlations between the same biosamples. During this checking procedure, we assigned the untreated cell culture samples for the datasets GSE36244 [58] and GSE41588 [59], and healthy lung samples for the dataset GSE37765 [60] as the "normal" or control states. To decrease the batch effects, all the microarray results were quantile normalized according to [61]. The NGS data were normalized using the method DESeq [62]. To avoid the divergence when calculating the log-fold-change values, we skipped all the genes and gene products in RNA-seq and microarray datasets that contained zero intensities.

For further normalization of the transcriptional data to the control samples, we calculated the case-to-normal ratio (*CNR*). When comparing the normalized expression logarithms between the NGS and microarray expression data, we detected small or moderate correlation for all the datasets under investigation (Fig. 7, Table 3). These results suggest that there is a considerable gap between the different experimental platform data.

In contrast, for the OncoFinder-processed data and pathway activation strength (PAS), we detected clear-cut correlations between the NGS and microarray gene expression datasets (Fig. 7, Table 3). The correlation coefficients for PAS were greater than for the CNR with only three outliers out of 16 samples. This finding evidences that the PAS calculation algorithm produces far more congruent results compared to the initial gene expression signatures between the microarray and NGS datasets.

Importantly, both NGS and microarray hybridization strategies may produce a large number of errors through the stages of RNA purification, library preparation and amplification, hybridization and sequencing, and finally mapping and annotation of the reads and reading the array [63–65]. It is hard to identify the errors and to find out what type of experimental assay provides more accurate data for each individual gene. It is important to minimize the errors

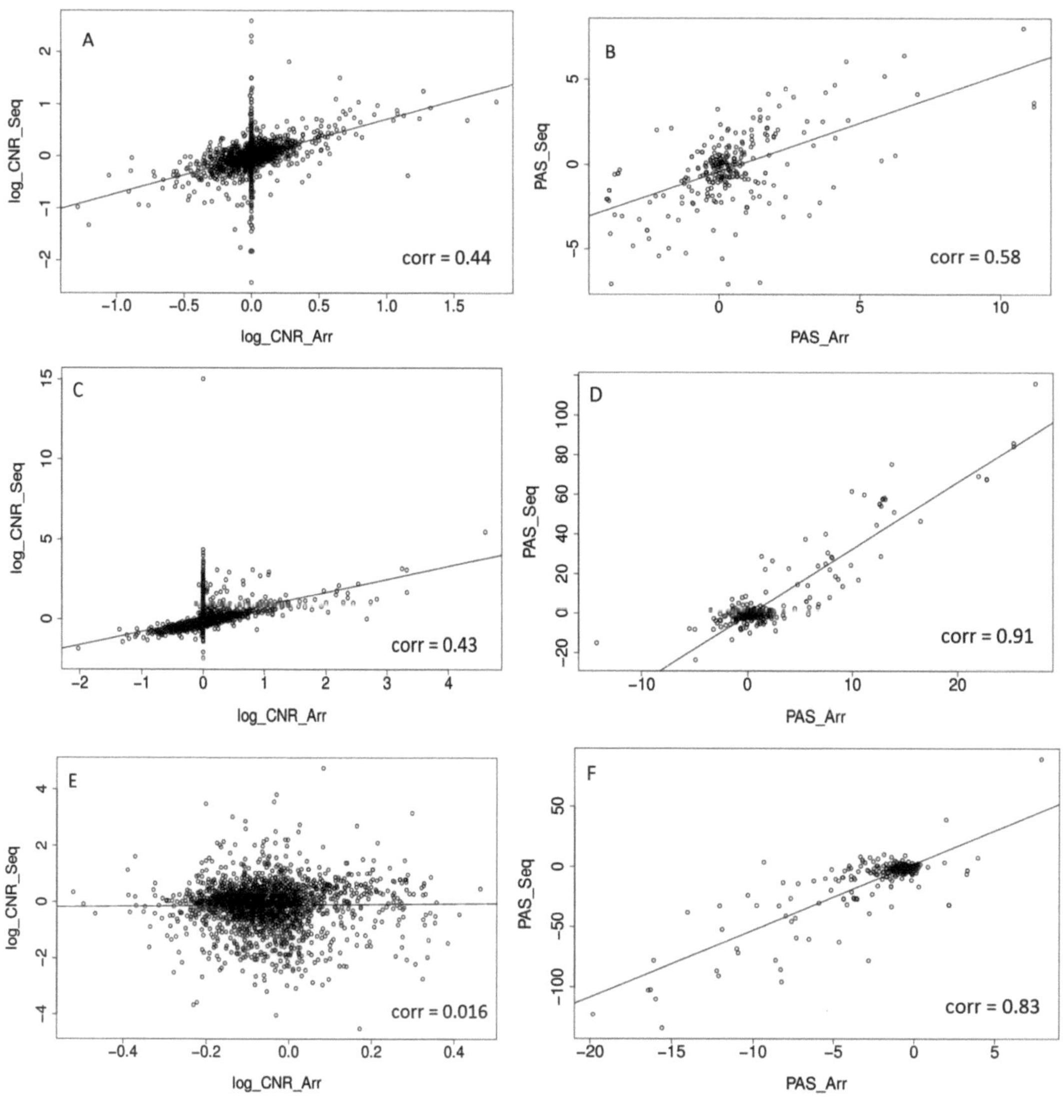

Fig. 7 Clouds of values obtained using the RNA next-generation sequencing vs. RNA microarray analysis methods. *Upper row* (**a**, **b**): cell replica 1, 24 h after BaP treatment from the HepG2 cells, dataset GSE36244 [58]. *Middle row* (**c**, **d**): treatment with 5 μM of 5-Aza and cell replica 1 from the HT-29 cells, dataset GSE41588 [59]. *Lower row* (**e**, **f**): sample P8 from the lung adenocarcinoma dataset GSE37765 [60]. *Left column* (**a**, **c**, **e**): values of logarithmic *CNR* for each gene. *Right column* (**b**, **d**, **f**): values of *PAS*

in the transcriptomic data and, theoretically, quantitative real-time PCR might provide a solution as a reference gene expression measure. However, the existing PCR platforms do not allow making high-throughput, transcriptome-scale experiments. Our approach makes it possible to surmount this obstacle as, unlike the original data, the outgoing *PAS* values are highly congruent among the NGS and microarray data. This effect of the

Table 3
Correlation coefficients between values obtained using the RNA microarray analysis and RNA sequencing methods for the HepG2 cells dataset GSE36244 [58], HT-29 cells dataset GSE41588 [59], and lung adenocarcinoma dataset GSE37765 [60].

Sample		Transcriptome level (logarithmic *CNR* for different genes), C_g	Signalome level (*PAS* value for different pathways), C_p
GSE36244, 24 h after BaP treatment	Replica 1, 12 h	0.60	0.68
	Replica 2, 12 h	0.57	0.66
	Replica 1, 24 h	0.44	0.58
	Replica 2, 24 h	0.41	0.57
GSE41588, 5 μM or 10 μM of 5-Aza	Replica 1, 5 μM	0.43	0.91
	Replica 1, 10 μM	0.44	0.48
	Replica 2, 5 μM	0.50	0.90
	Replica 2, 10 μM	0.48	−0.032
	Replica 3, 5 μM	0.54	0.77
	Replica 3, 5 μM	0.43	0.75
GSE37765	P1	0.16	0.30
	P3	0.018	0.17
	P4	0.069	0.37
	P5	−0.068	−0.032
	P6	0.075	0.0042
	P8	0.016	0.83

OncoFinder algorithm is most likely due to its cumulative nature. The PAS value is formed by the addition of multiple individual members, each representing a gene product involved in the pathway. The concentration of each individual gene product can be measured with a certain error, which is clearly seen when untreated NGS vs. array data are compared, but a combination of a large number of these concentration members into a signalome-oriented network apparently diminishes an overall error, as reflected by the better correlation records.

We conclude that this feature of *PAS* makes it possible to more accurately measure the changes in the functional states of the cellular/tissue transcriptome and interactome across the many microarray and NGS platforms, which makes OncoFinder a method of choice for many applications including genetics, physiology, biomedicine, and molecular diagnostics.

References

1. Marshall CJ (1995) Specificity of receptor tyrosine kinase signaling: transient versus sustained extracellular signal-regulated kinase activation. Cell 80:179–185
2. Hanahan D, Weinberg RA (2000) The hallmarks of cancer. Cell 100:57–70
3. Hanahan D, Weinberg RA (2011) Hallmarks of cancer: the next generatio. Cell 144: 646–674
4. Kreeger PK, Lauffenburger LA (2010) Cancer systems biology: a network modeling perspective. Carcinogenesis 31:2–8

5. Kholodenko BN, Demin OV, Moehren G, Hoek JB (1999) Quantification of short term signaling by the epidermal growth factor receptor. J Biol Chem 274:30169–30181
6. Moehren G, Markevich NI, Demin O, Kiyatkin A et al (2002) Temperature dependence of the epidermal growth factor receptor signaling network can be accounted for by a kinetic model. Biochemistry 41:306–320
7. Markevich NI, Hoek JB, Kholodenko BN (2004a) Signaling switches and bistability arising from multisite phosphorylation in protein kinase cascades. J Cell Biol 164:353–359
8. Markevich NI, Moehren G, Demin O, Kiyatkin A et al (2004b) Signal processing at the Ras circuit: what shapes Ras activation patterns? Syst Biol (Stevenage) 1:104–113
9. Kiyatkin A, Aksamitiene E, Markevich NI, Borisov NM et al (2006) Scaffolding protein Grb2-associated binder 1 sustains epidermal growth factor-induced mitogenic and survival signaling by multiple positive feedback loops. J Biol Chem 281:19925–19938
10. Borisov N, Aksamitiene E, Kiyatkin A et al (2009) Systems-level interactions between insulin-EGF networks amplify mitogenic Signaling. Mol Syst Biol 5:256
11. Kuzmina NB, Borisov NM (2011) Handling complex rule-based models of mitogenic cell signaling (On the example of ERK activation upon EGF stimulation). Int Proc Chem Biol Environ Eng 5:76–82
12. Faeder JR, Blinov LM, Goldstein B, Hlavacek WS (2005) Combinatorial complexity and dynamical restriction of network flows in signal transduction. Syst Biol (Stevenage) 2:5–15
13. Morton-Firth MJ, Shimizu TS, Bray D (1999) A free-energy-based stochastic simulation of the Tar receptor complex. J Mol Biol 286:1059–1074
14. Le Novere N, Shimizu TS (2001) STOCHSIM: modelling of stochastic biomolecular processes. Bioinformatics 17:575–576
15. Danos V, Laneve C (2004) Formal molecular biology. Theor Comput Sci 325:69–110
16. Feret J, Danos V, Krivine J, Harmer R, Fontana W (2009) Internal coarse-graining of molecular systems. Proc Natl Acad Sci U S A 106 (2009):6453–6458
17. Lok L, Brent R (2005) Automatic generation of cellular reaction networks with Moleculizer 1.0. Nat Biotechnol 23:131–136
18. Mallavarapu A, Thomson M, Ullian B, Gunawardena J (2008) Programming with models: modularity and abstraction provide powerful capabilities for systems biology. J R Soc Interface 6:257–270
19. Borisov NM, Markevich NI, Hoek JB, Kholodenko BN (2005) Signaling through receptors and scaffolds: independent interactions reduce combinatorial complexity. Biophys J 89:951–966
20. Borisov NM, Markevich NI, Hoek JB, Kholodenko BN (2006) Trading the micro-world of combinatorial complexity for the macro-world of protein interaction domains. Biosystems 83:152–166
21. Conzelmann H, Saez-Rodriguez J, Sauter T, Kholodenko BN, Gilles ED (2006) A domain-oriented approach to the reduction of combinatorial complexity in signal transduction networks. BMC Bioinformatics 7:4
22. Borisov NM, Chistopolsky AS, Faeder JR, Kholodenko BN (2008) Domain-oriented reduction of rule-based network models. IET Syst Biol 2:342–351
23. Kholodenko B, Kiyatkin A, Bruggeman F, Sontag E et al (2003) Untangling the wires: a strategy to trace functional interactions in signaling and gene networks. Proc Natl Acad Sci U S A 20:12841–12846
24. Daniels BC, Chen YJ, Sethna JP, Gutenkunst RN, Myers CR (2008) Sloppiness, robustness and evolvability in systems biology. Curr Opin Biotechnol 19:389–395
25. Gao S, Wang X (2007) TAPPA: topological analysis of pathway phenotype association. Bioinformatics 23:3100–3102
26. Ibrahim MA, Jassim S, Cawthorne MA, Langlands K (2012) A topology-based score for pathway enrichment. J Comput Biol 19:563–573
27. Draghici S, Khatri P, Tarca AL et al (2007) A systems biology approach for pathway level analysis. Genome Res 17:1537–1545
28. Tarca AL, Draghici S, Khatri P, Hassan SS et al (2009) A novel signaling pathway impact analysis. Bioinformatics 25:75–82
29. Zhavoronkov AA, Buzdin AA, Garazha AV, Borisov NM, Moskalev AA (2014) Signaling pathway cloud regulation for in silico screening and ranking of the potential geroprotective drugs. Front Genet 5:49
30. Buzdin AA, Zhavoronkov AA, Korzinkin MB et al (2014) OncoFinder, a new method for the analysis of intracellular signaling pathway activation using transcriptomic data. Front Genet 5:55
31. Spirin PV, Lebedev TD, Orlova NN, Gornostaeva AS (2014) Silencing AML1-ETO gene expression leads to simultaneous activation of both pro-apoptotic and proliferation signaling. Leukemia 28:2222–2228
32. Buzdin AA, Zhavoronkov AA, Korzinkin MB et al (2014) The OncoFinder algorithm for

minimizing the errors introduced by the high-throughput methods of transcriptome analysis. Front Mol Biosci 1:8

33. Borisov NM, Terekhanova NV, Aliper SM, Venkova LS et al (2014) Signaling pathways activation profiles make better markers of cancer than expression of individual genes. Oncotarget 5:10198–10205
34. Lezhnina K, Kovalchuk O, Zhavoronkov AA, Korzinkin MB et al (2014) Novel robust biomarkers for human bladder cancer based on activation of intracellular signaling pathways. Oncotarget 5:9022–9032
35. Zhu Q, Izumchenko E, Aliper A, Makarev E et al (2015) Pathway activation strength (PAS) is a novel independent prognostic biomarker for cetuximab sensitivity in colorectal cancer patients. Hum Genome Var 2:15009
36. Venkova LS, Aliper AM, Suntsova M, Kholodenko R et al (2015) Combinatorial high-throughput experimental and bioinformatics approach identifies molecular pathways linked with the sensitivity to anticancer target drugs. Oncotarget 6:27227–27238
37. Shepelin D, Korzinkin M, Vanyushina A, Aliper A (2015) Molecular pathway activation features linked with transition from normal skin to primary and metastatic melanomas in human. Oncotarget 7:656–670
38. Gillespie DT (1976) A general method for numerically simulating the stochastic time evolution of coupled chemical reactions. J Comput Phys 22:403–434
39. GEO Profiles, a National Center of Biotechnology Information database. http://www.ncbi.nlm.nih.gov/geo/. Accessed 08 March 2016.
40. Cunnick JM, Dorsey JF, Munoz-Antonia T, Mei L, Wu J (2000) Requirement of SHP2 binding to Grb2-associated binder-1 for mitogen-Activated Protein Kinase Activation in response to lysophosphatidic acid and epidermal growth factor. J Biol Chem 275:13842–13848
41. Montagner A, Yart A, Dance M, Perret B et al (2005) A novel role for Gab1 and SHP2 in epidermal growth factor-induced Ras activation. J Biol Chem 280:5350–5360
42. Chan PC, Chen YL, Cheng CH et al (2003) Src phosphorylates Grb2-associated binder 1 upon hepatocyte growth factor stimulation. J Biol Chem 278:44075–44082
43. Ingham RJ, Holgado-Madruga M, Siu C, Wong AJ, Gold MR (1998) The GAB protein is a docking site for multiple proteins involved in signaling by the B cell antigen receptor. J Biol Chem 273:30630–30637
44. Songyang Z, Shoelson SE, McGlade J, Olivier P et al (1994) Specific motifs recognized by the SH2 Domains of Csk, 3BP2, fps/fes, GRB-2, HCP, SHC, Syk, and Vav. Mol Cell Biol 14:2777–2785
45. Saxton TM, Cheng AM, Ong SH, Lu Y et al (2001) Gene dosage-dependent functions for phosphotyrosine-Grb2 signaling during mammalian tissue morphogenesis. Curr Biol 11:662–670
46. Mattoon DR, Lamothe B, Lax I, Schlessinger J (2004) The docking protein Gab1 is the primary mediator of EGF-stimulated activation of the PI-3K/Akt cell survival pathway. BMC Biol 2:24
47. Rodrigues GA, Falasca M, Zhang Z, Ong SH, Schlessinger J (2000) A novel positive feedback loop mediated by the docking protein Gab and phosphatidylinositol 3-kinase in epidermal growth factor receptor signaling. Mol Cell Biol 20:1448–1459
48. Shepherd PR, Withers DJ, Siddle K (1998) Phosphoinositide 3-kinase: the key switch mechanism in insulin signaling. Biochem J 333:471–940
49. Vanhaesebroeck B, Alessi DR (2000) The PI3K-PDK1 connection: more than just a road to PKB. Biochem J 346:561–576
50. Ravichandran KS, Lorenz U, Shoelson SE, Burakoff SJ (1995) Interaction of Shc with Grb2 regulates association of Grb2 with mSOS. Mol Cell Biol 15:593–600
51. Dhillon AS, Meikle S, Yazici Z, Eulitz M, Kolch W (2002) Regulation of Raf-1 activation and signalling by dephosphorylation. EMBO J 21:64–71
52. Wellbrock C, Karasarides M, Marais R (2004) The RAF proteins take center stage. Nat Rev Mol Cell Biol 5:875–885
53. Kolch W (2005) Coordinating ERK/MAPK signaling through scaffolds and inhibitors. Nat Rev Mol Cell Biol 6:827–837
54. Paz K, Hemi R, LeRoith D, Karasik A et al (1997) A molecular basis for insulin resistance. J Biol Chem 272:29911–29918
55. Gu H, Neel BG (2003) The 'Gab' in signal transduction. Trends Cell Biol 13:122–130
56. Johnston AM, Pirola L, van Obberghen E (2003) Molecular mechanisms of insulin receptor substrate protein-mediated modulation of insulin signaling. FEBS Lett 546:32–36
57. Griesinger AM, Birks DK, Donson AM, Amani V et al (2013) Characterization of distinct immunophenotypes across pediatric brain tumor types. J Immunol 191:4880–4888
58. Van Delft J, Gaj S, Lienhard J, Albrecht MW et al (2012) RNA-seq provides new insights in

the transcriptome responses induced by the carcinogen benzo[a]pyrene. Toxicol Sci 130:427–439

59. Xu X, Zhang Y, Williams J, Antoniou E et al (2013) Parallel comparison of Illumina RNA-Seq and Affymetrix microarray platforms on transcriptomic profiles generated from 5-aza-deoxy-cytidine treated HT-29 colon cancer cells and simulated datasets. BMC Bioinformatics 14:S1
60. Kim SC, Jung Y, Park J, Cho S et al (2013) A high-dimensional, deep-sequencing study of lung adenocarcinoma in female never-smokers. PLoS One 8:e55596
61. Bolstad BM, Irizarry RA, Astrand M, Speed TP (2003) A comparison of normalization methods for high density oligonucleotide array data based on variance and bias. Bioinformatics 19:185–193
62. Anders S, Huber W (2010) Differential expression analysis for sequence count data. Genome Biol 11:R106
63. Chalaya T, Gogvadze E, Buzdin A, Kovalskaya E, Sverdlov ED (2004) Improving specificity of DNA hybridization-based methods. Nucleic Acids Res 32:e130
64. Shugay M, Britanova OV, Merzlyak EM, Turchaninova MA et al (2014) Towards error-free profiling of immune repertoires. Nat Methods 11:653–655
65. Buzdin AA, Lukyanov SA (2007). Nucleic acids hybridization modern applications. ISBN: 978-1-4020-6039-7

Chapter 4

Bioinformatics Meets Biomedicine: OncoFinder, a Quantitative Approach for Interrogating Molecular Pathways Using Gene Expression Data

Anton A. Buzdin, Vladimir Prassolov, Alex A. Zhavoronkov, and Nikolay M. Borisov

Abstract

We propose a biomathematical approach termed OncoFinder (OF) that enables performing both quantitative and qualitative analyses of the intracellular molecular pathway activation. OF utilizes an algorithm that distinguishes the activator/repressor role of every gene product in a pathway. This method is applicable for the analysis of any physiological, stress, malignancy, and other conditions at the molecular level. OF showed a strong potential to neutralize background-caused differences between experimental gene expression data obtained using NGS, microarray and modern proteomics techniques. Importantly, in most cases, pathway activation signatures were better markers of cancer progression compared to the individual gene products. OF also enables correlating pathway activation with the success of anticancer therapy for individual patients. We further expanded this approach to analyze impact of micro RNAs (miRs) on the regulation of cellular interactome. Many alternative sources provide information about miRs and their targets. However, instruments elucidating higher level impact of the established total miR profiles are still largely missing. A variant of OncoFinder termed MiRImpact enables linking miR expression data with its estimated outcome on the regulation of molecular processes, such as signaling, metabolic, cytoskeleton, and DNA repair pathways. MiRImpact was used to establish cancer-specific and cytomegaloviral infection-linked interactomic signatures for hundreds of molecular pathways. Interestingly, the impact of miRs appeared orthogonal to pathway regulation at the mRNA level, which stresses the importance of combining all available levels of gene regulation to build a more objective molecular model of cell.

Key words Systems biology, Bioinformatics, Intracellular molecular pathways, Gene expression, Transcriptomics, Proteomics, Epigenetics, micro RNA, Cancer biomarkers, Sensitivity to drug treatment

1 Introduction

Intracellular molecular pathways (IMPs), including signaling, DNA repair, metabolic and cytoskeleton reorganization pathways, regulate all major cellular events in health and disease [1–3]. Changes in their activity may reflect various differential conditions such as

Tatiana V. Tatarinova and Yuri Nikolsky (eds.), *Biological Networks and Pathway Analysis*, Methods in Molecular Biology, vol. 1613, DOI 10.1007/978-1-4939-7027-8_4, © Springer Science+Business Media LLC 2017

differences in physiological state, aging, disease, treatment with drugs, infections, media composition, additives and nutrients, hormones, etc. Many bioinformatic tools have been developed recently that analyze IMPs. Today, hundreds of IMPs and related gene product interaction maps are cataloged that show sophisticated relationships between the individual molecules in various databases such as UniProt [4], HPRD [5], QIAGEN SABiosciences [6], WikiPathways [7], Ariadne Pathway Studio [8], SPIKE [9], Reactome [10], KEGG [11], HumanCyc [12], etc. The information about activation of IMPs can be obtained from the massive proteomic or transcriptomic data. Although the proteomic level may be somewhat closer to the biological function of IMPs, the transcriptomic level of studies today is far more feasible in terms of performing experimental tests and analyzing the data. The transcriptomic methods such as next-generation sequencing (NGS) or microarray analysis of RNA can routinely determine expression levels for all or virtually all human genes [13]. Transcriptome profiling may be performed for the minute amount of the tissue sample, not necessarily fresh, but also for the clinical formalin-fixed, paraffin-embedded (FFPE) tissue blocks [14].

However, until recently, it remained challenging to efficiently do the high-throughput quantification of pathway activation for the individual biological samples. Several biomathematical approaches were published to measure pathway activation based on large-scale gene expression data, either transcriptomic or proteomic. For example, Khatri et al. [15] classified those methods into three major groups: Over-Representation Analysis (ORA), Functional Class Scoring (FCS), and Pathway Topology (PT)-based approaches. ORA-based methods calculate if the pathway is significantly enriched with differentially expressed genes [16–18]. These methods have many limitations, as they ignore all non-differentially expressed genes and do not take into account many gene-specific characteristics. FCS-based approaches partially tackle aforementioned limitations by calculating fold change-based scores for each gene and then combining them into a single pathway enrichment score [19–21]. PT-based analysis also takes into account topological characteristics of each given pathway, assigning additional weights to the genes (for a review, *see* [22]). Recently, to account for gene expression variability within a pathway, another set of differential variability methods has been developed [23]. Differential variability analysis determines a group of genes with a significant change in variance of gene expression between case and control groups [24]. This approach was further extended and applied on the pathway level [23, 25, 26].

In 2014, we published a new biomathematical method for pathway analysis, termed OncoFinder [27]. Based on kinetic models that use the “low-level” approach of mass action law, OncoFinder performs quantitative and qualitative enrichment analyses of the signaling pathways. For each sample investigated, it performs a case-control pairwise comparison and calculates the Pathway Activation Strength (PAS), a value that serves as a qualitative measure of pathway activation. Unlike most other methods, this approach determines if the signaling pathway is significantly up- or down-regulated compared to the reference. Negative and positive overall PAS values correspond to an inhibited or activated state of signaling pathway [27]. OncoFinder is also, to our knowledge, a unique PAS-calculating method, which was reported to provide output data with significantly reduced noise introduced by the experimental transcriptome profiling systems [28]. This approach was shown to be efficient in finding new biomarkers for various human diseases [29–33] and in modeling melanoma development [34], regression of neuroblastoma [35], immunity and apoptosis [36], stimulation by the nutrients [37], and acquiring resistance against drug treatment in leukemia cells [38]. Furthermore, the same rationale was employed for pathway activity calculations based on micro RNA (miR) expression data. The related technique, termed MiRImpact, enables linking miR with its estimated outcome on the regulation of molecular pathways [39]. Finally, OncoFinder was used to link molecular pathway activation features with the sensitivity of cancer cells to drugs, at both cell culture and patient levels [40–42]. Here, we review selected applications of this technology to human molecular biomedicine.

2 The OncoFinder Algorithm

The OncoFinder algorithm operates with the calculation of the Pathway Activation Strength (PAS), a value that serves as a qualitative measure of a molecular pathway activation. The formula for the PAS calculation accounts for gene expression data and for information on the protein interactions in a pathway, namely, individual protein activator or repressor roles in a pathway [27]. This is also important to identify control sample or a group of control samples, which will be used as the norms for PAS calculation. The gene expression data under investigation are compared against control/normal gene expression profiles. For microarray gene expression data, previous data normalization may be required, such as quantile normalization [43]. The positive value of PAS indicates abnormal activation of a molecular pathway compared to norms, and the negative value—its downregulation, whereas zero PAS scores represent unaffected pathways acting similarly in case and in normal samples. Briefly, the enclosing algorithm utilizes the following formula to evaluate pathway activation, where summation is made for all the genes, whose products participate in a pathway:

$$PAS_p = \sum_n ARR_{np} \cdot BTIF_n \cdot \log(CNR_n).$$

Here, the *case-to-normal ratio, CNRn*, is the ratio of expression levels for a gene *n* in the sample under investigation to the same average value for the control group of samples. For each gene *i*, case-to-normal ratio (CNR_i) is calculated for the respective concentrations of mRNA or for protein concentrations, depending on the origin of input data (shown for mRNA):

$$CNR_i = \frac{\text{Case}_{\text{mRNA}}\text{Signal}_i}{\text{Norm}_{\text{mRNA}}\text{Signal}_i}.$$

In most applications, for each CNR value, a Boolean flag of BTIF (beyond tolerance interval flag) was applied, which equals 1 when the CNR value passed, and 0 when the CNR value did not pass both or either one of the two criteria of significantly differential expression: first, the expression level for the sample must fit outside the tolerance interval for norms, with $p < 0.05$, and second, the value of CNR must differ from 1 by at least 1.5-fold. The second criterion is the discrete value of *ARR* (*activator/repressor role*) that reflects the functional role of a gene product n in a pathway [27]. For each gene of a pathway *p*, its activator-repressor role ($ARR_{i,p}$) is defined, which depends on the functional role of this gene product in a pathway:

$$ARR_{i,p} = \begin{cases} -1, & \text{repressor} \\ -0.5, & \text{repressor} > \textit{activator} \\ 0, & \text{neither} \\ 0.5, & \text{activator} > \textit{repressor} \\ 1, & \text{activator} \end{cases}.$$

For the calculations, databases including up to ~270 signaling, ~360 metabolic, and ~300 other intracellular molecular pathways were used in the published reports. As the knowledge bases for building OncoFinder-compatible pathway datasets, the resources such as QIAGEN SABiosciences, WikiPathways [7], Ariadne Pathway Studio [8], Reactome [10], KEGG [11], HumanCyc [12], and others may be used, depending on the user's preferences.

3 Modification of OncoFinder Algorithm for micro RNA Expression Analysis (MiRImpact Algorithm)

MiRImpact biomathematical algorithm was built to enable quantization of the effects, caused by the changes in overall miR concentrations, on the activity of intracellular molecular pathways [39]. The algorithm was created on the basis of a rationale previously published for the OncoFinder method [27]. For each miR, a case-to-normal ratio is calculated for the respective miR concentrations ($miCNR_j$):

$$miCNR_j = \frac{\text{Case\_microRNA\_Signal}_j}{\text{Norm\_microRNA\_Signal}_j}.$$

The miR beyond tolerance interval flag ($miBTIF_j$) marker determines if the difference between case and norm is significant:

$$miBTIF_j = \begin{cases} 0, miCNR_j \text{ belongs to microRNA\_tolerance\_interval} \\ 1, miCNR_j \text{ doesn't belong to microRNA\_tolerance\_interval.} \end{cases}$$

The unique coefficient termed miR involvement index ($miII$) determines, if a given mRNA transcript of a gene i is a molecular target of a miR j:

$$miII_{j,i} = \begin{cases} 0, \text{ target} \\ 1, \text{not target} \end{cases}.$$

The value of miR-defined activation strength of a pathway p ($miPAS_p$) is calculated according to the following:

$$miPAS1_p = -\sum_i ARR_{ip} \cdot DIF_i \cdot \log(miCNR_i) \sum_j miII_{ij}.$$

Similarly to OncoFider, a positive value of $miPAS_p$ indicates activation, whereas a negative one indicates repression of a pathway p, calculated based on the available miR expression data.

In the initial application of this method, we took the previously published Oncofinder signaling database featuring 2725 unique genes and 271 signaling pathways [29, 38]. These data are needed to identify genes involved in each pathway and their functional roles expressed by ARR values. To find out mII indexes, a database covering target gene product specificities of miRs is needed. We used the most recent available updates of the two alternative knowledge bases on miRs and their experimentally validated targets: miRTarBase [44] and Diana TarBase [45]. Both databases include information on more than 50 thousands of molecular interactions of miRs with target mRNA molecules, in case of miRTarBase—for 18 species, in case of Diana-TarBase—for 24 species, including human. The most commonly used experimental approaches for validating molecular targets of miRs are luciferase reporter assay, Western blots, and next-generation sequencing approaches. This information is manually curated by the database developers based on published literature on functional experimental studies of miRs [44, 45]. The target specificities of miRs catalogued there cover, respectively, 72 and 18% of the genes listed in the OncoFinder database, which was taken in MiRImpact for the analysis of the molecular pathways (Table 1).

Table 1
Characteristics of validated miR target databases, based on the data collected from miRTarBase, Diana TarBase, and OncoFinder pathway databases

Data base	miRTarBase	Diana TarBase
Number of miRs targeting gene products from OncoFinder database	596	183
Number of individual records	12103	3006
Number of target genes in OncoFinder database	1968	497

For the set of experimental human bladder cancer samples, we observed a weak, but still statistically significant, correlation between the miPAS data calculated for both databases (Fig. 1). However, the high level of noise reflects a big difference between their content and completeness. The results obtained suggest that the method MiRImpact may be compatible with various databases collecting data on miR specificities and on their particular activities [39]. This means that the future developments based on the MiR-Impact method may utilize any kind of new miR target databases, either based on computational prediction, or on experimental validation of miR interactions. Similarly, the enclosed OncoFinder database of signaling pathways may be updated, extended, or replaced by another database of molecular pathways, in a user-definitive way. Furthermore, knowledge of the qualitative aspects of molecular interactions between miRs and their targets, and between the molecules participating in molecular pathways, may be used to tune the databases to assign specific weighting coefficients to each miR and/or gene product. The mathematical algorithm used here is rather universal and can be employed to trace also metabolic, cytoskeleton rearrangement, DNA repair, and other types of intracellular molecular pathways, in any organism or species of the interest. The apparently seen correlation between the data calculated using miRTarBase and Diana-TarBase suggests that the algorithm works in the same manner for both miR target databases. We compared the obtained results with the literature data on the impact of particular miRs on the respective signaling pathways. For the data calculated using the miRTarBase, we observed a greater congruence between the experimental and the literature data (in 47% of the cases), whereas for Diana-TarBase, the data were compatible in only 23% of the cases. We suggest, therefore, that the miRTarBase is currently a database of choice for the estimation of molecular pathways regulation by miRs in humans [39]. Finally, we propose that other types of noncoding

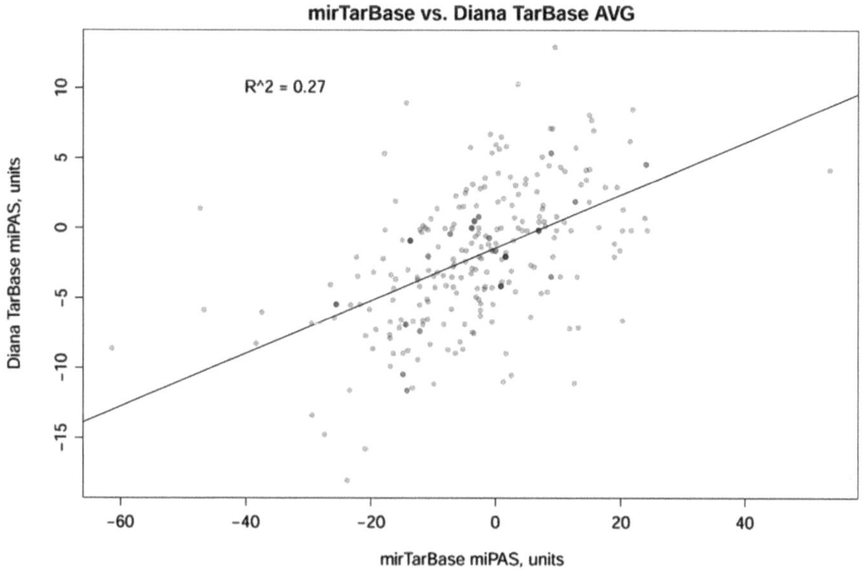

Fig. 1 Comparison of microRNA Pathway Activation Strength (miPAS) values calculated using miRTarBase and Diana TarBase databases of miR targets, for an averaged miR expression between all the bladder cancer samples under investigation. The resulting virtual sample is the result of averaging of miR expression measured by deep sequencing for eight bladder cancer samples. The results for each individual sample showed correlation coefficients varying between 0.06 and 0.53 with the mean value of 0.26

RNAs than miRs can be also analyzed using the MiRImpact method, when their regulatory roles and target\effector gene products are known.

4 Intracellular Pathways Activation Profiles Make Better Markers than Expression of Individual Genes

Identification of reliable and accurate molecular markers of cancer remains one of the major challenges of contemporary biomedicine. Thousands of reports have been published communicating new RNA, protein, and non-protein biochemical biomarkers sensitive to cancer development [46]. Most of these markers represent products of individual gene expression at the RNA or protein levels. Some of them are widely used in clinical practice, but there remains an overall unsolved problem of finding new cancer biomarkers with enhanced specificity and sensitivity scores compared to the existing ones. Another aspect of the same problem deals with the shortage of the cancer type-specific molecular markers, e.g., melanoma-specific, bladder or pancreatic cancer-specific, etc. Association of the marker expression with the success of the medical treatment may provide clues to a more efficient, patient-oriented cancer treatment therapy [47].

We applied OncoFinder to profile gene expression datasets for the nine human cancer types including bladder cancer, basal cell carcinoma, glioblastoma, hepatocellular carcinoma, lung adenocarcinoma, oral tongue squamous cell carcinoma, primary melanoma, prostate cancer, and renal cancer, totally 292 cancer and 128 matching normal tissue samples taken from the Gene expression omnibus (GEO) repository [48]. We profiled activation of 82 signaling pathways that involve ~2700 gene products. Based on the comparison of the cancer vs normal tissue transcriptomic data, we obtained the PAS profiles characteristic of the above cancer types. We next calculated the area-under-curve (AUC) values [49] for the PAS scores of each of the pathways under investigation. The AUC value is the universal characteristics of biomarker robustness and it is dependent on the sensitivity and specificity of a biomarker. It correlates positively with the biomarker quality and may vary in an interval from 0.5 till 1. The AUC threshold for discriminating good and bad biomarkers is typically 0.7 or 0.75. The entries having greater AUC score are considered good-quality biomarkers and vice versa [50]. The AUC values were calculated when comparing each cancer type against the remaining eight cancer types. Enhanced AUC values here meant that the corresponding signaling pathway is a good biomarker distinguishing an individual cancer type from the others (Table 2). This kind of AUC score will be referred here as AUC1. In parallel, we also calculated the analogous AUC scores for the individual gene products involved in the pathways (namely, for the values of lg CNR for them). For each of these 2726 human gene products, we next calculated the average AUC scores characteristic of each signaling pathway/cancer type, referred here as AUC2. AUC1 reflects the quality of PAS as the biomarker for a given signaling pathway, and AUC2 is the integral characteristics of the biomarker quality for the expression of the genes that are involved in the same pathway. The outline of the data analysis is shown in Fig. 2. The results showed that among the good-quality biomarkers (AUC cut-off value 0.75) the values for AUC1 were higher than for the AUC2 for all cancer types (Table 2). For example, in all cancer types there were only 14 AUC2 (gene expression) markers, in contrast to 160 AUC1 (pathway activation) markers (Table 2). Moreover, for 10 of these 14 AUC2 markers, the corresponding AUC1 values were greater (Table 2), thus suggesting the stronger biomarker potential of the AUC1 (pathway activation) markers. This was true for 9/9 of the cancer types tested [48]. These results evidence that the PAS values can be used as a new type of cancer biomarkers, superior to the traditional gene expression biomarkers [48].

Importantly, these data also evidence that the pathway activation strength (PAS)- based biomarkers may serve efficiently to distinguish the different cancer types. Among the 82 signaling pathways profiled in this assay, 75 showed a potential to serve as

Table 2
Comparison of the AUC1 and AUC2 scores calculated for 81 intracellular signaling pathways for nine human cancer types based on the transcriptomic data

Cancer type	AUC1 > 0.75[a]	AUC2 > 0.75[b]	AUC1 > AUC2[c]	AUC2 > AUC1[d]
Basal cell carcinoma	23	0	23	0
Bladder cancer	10	9	8	4
Glioblastoma	59	5	59	0
Hepatocellular	7	0	7	0
Lung adenocarcinoma	21	0	21	0
Oral tongue squamous cell carcinoma	2	0	2	0
Primary melanoma	13	0	13	0
Prostate cancer	16	0	16	0
Renal cancer	10	0	10	0
Total	161	14	159	4

[a]Number of signaling pathways where AUC1 > 0.75
[b]Number of signaling pathways where AUC2 > 0.75
[c]Number of signaling pathways where AUC1/2 > 0.75, and AUC1 > AUC2
[d]Number of signaling pathways where AUC1/2 > 0.75, and AUC2 > AUC1

the strong cancer type-specific biomarkers with the AUC greater than 0.75 [48]. For each cancer type, the number of these PAS biomarkers (AUC > 0.75) varied from 2 till 59 (Table 2). This suggests that during cancer progression the signaling pathway regulation is a more uniform process rather than the activation of certain individual genes. Indeed, an intracellular signaling pathway is a complex regulatory network that may include hundreds of different gene products [51]. Theoretically, expression of every gene in this network may have an influence on the overall functioning of the signaling pathway. Alterations in the expression profiles of many different genes can, therefore, lead to a similar result of a pathway activation or suppression during cancer development [52].

Strong increase in Notch signaling (avg PAS ~ 9) denotes glioma, mild upregulation of RNA polymerase II complex activity (avg PAS ~ 1.4)—basal cell carcinoma, moderate decrease in IP3 signaling (avg PAS ~ −1.9)—lung adenocarcinoma, etc. [48]. It may be seen that any investigated tissue type has its unique profile of statistically significant pathway activation features, which provides a potent instrument for further analysis and specific targeting of various cancer types in the future.

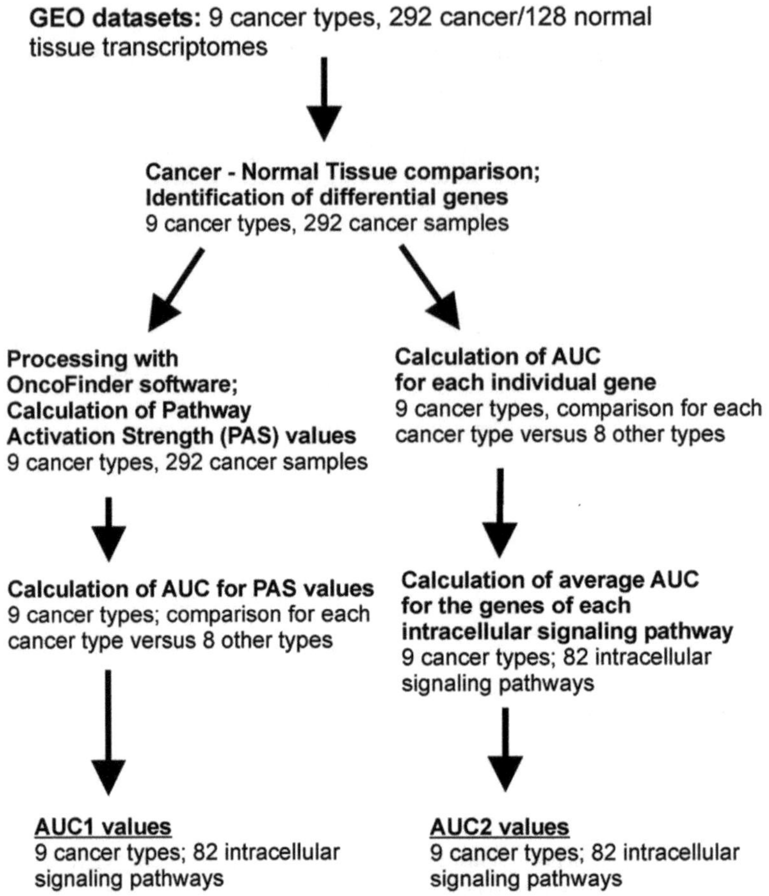

Fig. 2 Outline of the bioinformatics procedures used to calculate AUC1 and AUC2 values in various cancer-type transcriptomes

5 Messenger RNA and micro RNA Pathway Activation Markers of Human Bladder Cancer

Bladder cancer (BC) is the second most frequent urological cancer and the ninth most common of all cancers. Approximately 356,000 new BC cases are reported annually worldwide [53], with the highest incidences in countries where the dominant population is Caucasoid [54]. BC accounts for 3.1 and 1.8% of the overall cancer mortality in males and females, respectively.

Early diagnosis is a prerequisite for successful BC treatment. Existing methods are, in general, not efficient for detecting BC in its early stages; as a result, there is an urgent need and opportunity to develop novel diagnostic tools that would efficiently detect early-stage BC [29]. Moreover, associating marker expression with successful medical treatment may provide clues to a more efficient, patient-oriented cancer treatment therapy [55].

Using Illumina HT12v4 microarrays and NGS methods, we profiled mRNA and micro RNA (miR) expression, respectively, in 17 experimental cancer and seven non-cancerous bladder tissue samples [29, 39]. We analyzed activations of 271 intracellular signaling pathways and found 44 signaling pathways that serve as excellent new biomarkers of BC at the mRNA level, supported by the high AUC values >0.75 [29]. Among these 44 PAS biomarkers, 10 (23%) were upregulated and 34 (77%) were downregulated in BC. Eight differential PAS biomarkers (18%) represented independent regulatory networks, whereas the rest, 36 (82%), were terminal branches of larger molecular signaling pathways. The up/downregulation of the 44 differential pathways seen in the BC samples could lead to contradictory effects on the survival and proliferation of cancer cells [29]. Information in the literature indicates that seven (16%) of the changes in the affected pathways promote cancer cell survival, while 12 (27%) of the changes exert negative effects on cancer cells. The rest of the pathways play contradictory roles in cancer cells, which prevents us from unambiguous labeling them as "positive" or "negative" regulators of BC progression [29].

Interestingly, overall pathway activation profiles obtained using OncoFinder for mRNA regulation level, and using MiRImpact for miR regulation level, differed dramatically. This was reflected by the apparent differences between the PAS and miPAS scores [39]. At the level of miPAS scoring, the results depended greatly on the database used to establish molecular targets of miRs (miRTarBase or Diana-TarBase). Previously, we identified 44 molecular signaling pathways that may serve as potent biomarkers of BC. For 21 of them, we found literature data connecting miR expression and pathway activation abnormalities in cancer [39]. Based on our own experimental analysis, for miRTarBase we observed congruence with finding of pathway up/downregulated state in 10/21 molecular pathways, and for Diana-Tarbase—in only 5/21 pathways [39]. The remaining pathways that did not coincide with both miRTarBase- and Diana-Tarbase-based versions of MiRImpact were either apparently inconclusively (bidirectionally) regulated in BC, or were unchanged according to miPAS data [39]

We next compared pathway activation signatures for the 44 above characteristic BC-associated pathways at the mRNA and miR levels. In the case of miRTarBase version, 20 pathways had contrary trends, and only 10 had common trends at the miPAS and PAS levels. For Diana-Tarbase version, nine pathways had contrary trends, and ten pathways—common trends on mRNA and miR regulation levels [39]. This suggests that the regulation of many characteristic BC-linked pathways differs dramatically at the mRNA and miR levels. For 11 and 6 characteristic pathways we observed, respectively, common and contradictory trends in pathway regulation using miRTarBase and Diana-Tarbase databases. Pathways commonly upregulated according to both databases were ILK pathway_wound healing and mTOR_Pathway_VEGF_pathway activation. Downregulated pathways were two branches of AHR pathway: AHR_Pathway_C_Myc_Expression and AHR_Pathway_-Cath_D_Repression, a terminal branch of CREB pathway (CREB_Pathway_Gene_Expression), a branch of Glucocorticoid receptor pathway (Glucocorticoid Receptor Pathway Cell cycle arrest), two branches of ILK pathway: ILK_Pathway_Cell_motility and ILK_Pathway_G2_phase_arrest regulation, a branch of JAK-STAT pathway (JAK mStat Pathway JAK degradation), and an RNA Polymerase II Complex Pathway. Five pathways were unchanged at the level of miR regulation, according to both databases [39].

A fraction of consensus data obtained using both databases demonstrates that three molecular pathways, previously shown to be aberrantly regulated at the mRNA level, are congruently regulated at the miR level as well. These are the branches of the integrin-linked kinase (ILK) signaling pathway, responsible for the cell motility and wound healing, and a branch of the mTOR pathway, responsible for the activation of VEGF signaling [39].

A similar figure was seen when comparing miPAS values for both miRTarBase and Diana-Tarbase versions of MiRImpact, for all available pathways. Comparison of pathway activation features at the mRNA and miR levels also showed quite distinct peculiarities in terms of variation between the individual samples. We observed relatively uniform regulation of pathways at the mRNA level, with relatively small number of pathways showing significant variations between the individual samples [39]. In contrast, at the level of miR regulation, the apparently observed differences between the samples were significantly stronger, as established for both miRTarBase and Diana-Tarbase databases [39]. In the latter cases, the majority of the pathways were also strongly differential between the normal and cancer samples. These peculiarities of miPAS scores suggest that they may be more sensitive compared to the PAS values to discriminate between the individual cancer samples. This may be highly beneficial for finding new diagnostic markers, e.g., linked with the individual sensitivity of patient to treatment.

Finally, using both above-mentioned miR target databases, we demonstrated that at least for the human BC tissues, the intracellular pathway regulation at the miR level differs greatly from that at the mRNA level, thus showing orthogonal dependencies for the extents of pathway activation (Fig. 3). This characteristic trend was seen for all individual samples, and for the averaged samples shown in Fig. 3, as well. Of note, many molecular pathways showing zero PAS scores at the same time had quite distinct miPAS scores (Fig. 3). This lack of correlation shown for both alternative

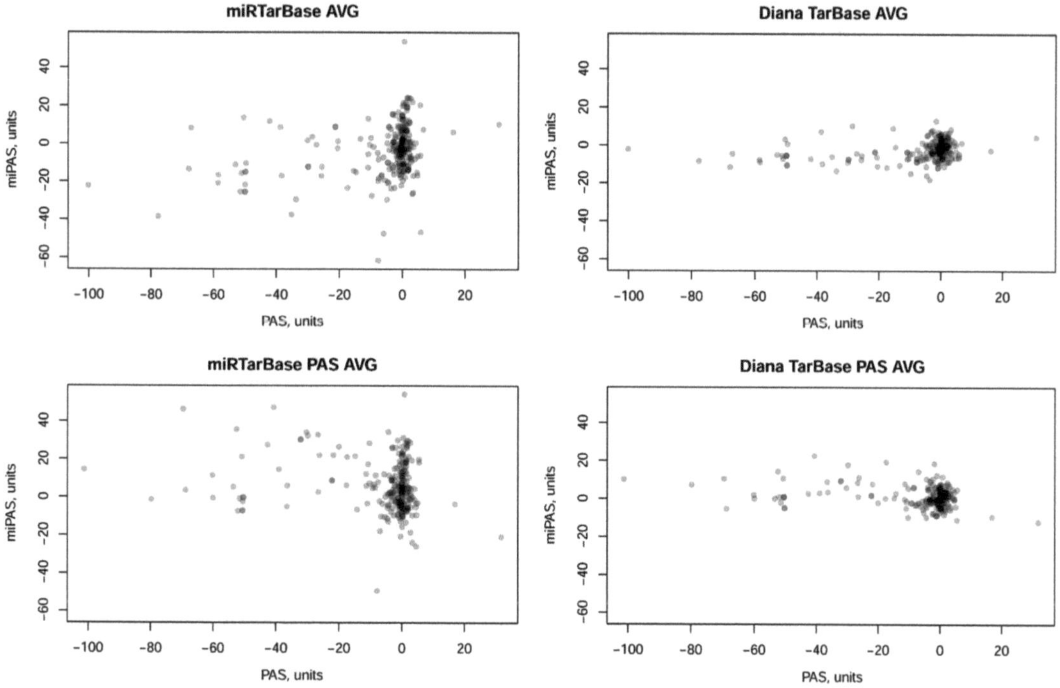

Fig. 3 Pathway Activation Strength (PAS) versus microRNA Pathway Activation Strength (miPAS) for an averaged miR and mRNA expression between all the bladder cancer samples (BC) under investigation. The resulting virtual sample is the result of averaging of miR expression measured by deep sequencing and mRNA expression measured using microarrays. "AVG" samples were averaged at the level of individual mRNA/miR expression across all tested BC samples, whereas "PAS AVG" was averaged at the level of PAS/miPAS values across all BC samples

databases clearly suggests that transcriptional profiling at the mRNA level alone may be not sufficient to estimate the activation of molecular pathways [39]. So far, we cannot quantitatively compare the effects of PAS and miPAS scores on the pathway activation. We presume that this will be done in the future by comparing high-throughput miR, mRNA, and proteomic expression data, at the level of molecular pathways. To this end, a combination of MiR-Impact approach communicated here and of OncoFinder technique published previously may provide a feasible methodological solution. The MiRImpact method would provide information on the activation of molecular pathways at the miR level, whereas OncoFinder—at the whole-transcriptome mRNA and proteomic levels. In addition, ribosome profiling data may be processed with these bioinformatic tools to uncover crosstalk between mRNA concentration, quantitative measure of protein translation efficiency, and final protein concentrations [39].

6 Molecular Pathway Activation Features Linked with Transition from Normal Skin to Primary and Metastatic Melanomas in Human

Melanoma is a type of skin cancer formed from melanocytes, skin cells that produce the pigment melanin. Melanomas are very active in forming metastases, and if not diagnosed at the early stage, the survival prognosis is poor. Melanoma accounts for 75% of deaths related to skin cancer [56]. Development of melanomas is commonly caused by mutations from UV-linked DNA damage [57] and by inherited genetic factors like highly penetrant loss-of-function mutations in tumor suppressor genes [58]. About 40% of human melanomas contain activating mutations of the B-Raf protein, resulting in constitutive signaling through the Raf to MAP kinases growth signaling pathways [59]. The presence of multiple melanocytic nevi, a genetic trait compounded by sun exposure, also increases the risk of developing melanoma, although the transition from benign nevi to melanoma does not usually occur and what triggers this change is largely unknown.

To learn more about the mechanisms that induce melanoma and cause it to progress, we performed high-throughput analysis of melanoma-related intracellular molecular networks including 592 signaling and metabolic pathways. We profiled a total of 478 transcriptomes consisting of 132 human primary melanoma, 222 metastatic melanoma, 103 normal skin, and 21 nevi samples [34]. The normalized gene expression data were next processed using the OncoFinder algorithm to establish pathway activation strength (PAS) profiles. To assess the functional relations between the investigated groups of samples, we built hierarchical clustering heatmaps with Ward method using Euclidean distance for all samples and all investigated molecular pathways and observed rather uncertain clustering features hardly distinguishing between the four sample classes [34]. To increase the resolution of clustering methods and to identify features that distinguish the above functional groups, we applied a selection of machine learning classifier algorithms, including Random Forest (RF) Support Vector Machines (SVM) with Linear and Radial kernels, Partial Least Squares (PLS) and Generalized linear regression with Glmnet regularization. Prior to classification, we filtered for small deviation and collinearity to prevent using two highly correlated variables when one would suffice. Overall, the SVM family classifiers showed the best results compared to other models. Such approaches allowed us to achieve ~0.94 average balanced accuracy of a 4-class problem (classification into four groups: Skin, Nevi, Primary, and Metastatic melanoma) using only metabolic pathways and ~0.94 average balanced accuracy using only signaling pathways [34]. In accordance with their vague transitional state, the most difficult group for all the classifiers used were nevi, for which the classifiers showed lowest combinations of sensitivity (0.4–0.8) and balanced accuracy (0.7–0.9) [34]. Other groups formed significantly more clear-cut clusters, which corresponded to their physiologically distinct states.

For each statistical model, we identified the top 30 metabolic and top 30 signaling pathways, distinguishing the two classes, which unifies different techniques of measuring importance between different models. Next, the top pathways were intersected and a list of consensus pathways was established (Tables 3 and 4). The consensus records included 25 metabolic and 19 signaling pathways for two different models of melanoma development, the first occurring via transitional state of the nevus (Skin → Nevus → Melanoma) and the second not involving nevus (Skin → Primary Melanoma → Metastatic Melanoma). To test the classification power of these top pathways, we built a new hierarchical clustering heatmap with the Ward method, using Euclidean distance for all samples and top investigated molecular pathways with supporting Principal Components Analysis (PCA) projections plots (Fig. 4). These top pathways enabled significantly better discrimination between the groups, as evidenced by PCA projections plots for all pathways (Fig. 4a) compared to plots for the selected top pathways (Fig. 4b). Next, we used these top pathways in the same 4-type prediction model as before. Results for the best model (SVM Linear model) confirmed adequacy of the classifier pathway selection and showed an averaged balanced accuracy of ~0.93, very close to the model with full pathways [34].

On the heatmap and PCA projection plots, the samples corresponding to nevi formed a cloudy group and clustered either with each other or diffusely between primary melanoma and normal skin samples. In agreement with previous reports, this suggests that nevi form a complicated group of highly variable samples, which frequently correspond to the intermediate state between normal skin and primary melanoma [60]. The top classifier elements included 25 metabolic and 19 signaling pathways. For all of these signaling pathways, association with melanoma was reported previously in the literature. However, for the metabolic pathways, this was not the case, and previous reports on the association with melanoma were not found for the following: Allopregnanolone biosynthesis, L-carnitine biosynthesis, Zymosterol biosynthesis (inhibited in melanoma), D-myo-inositol hexakisphosphate biosynthesis (activated in primary, inhibited in metastatic melanoma), Fructose 2,6 bisphosphate synthesis and dephosphorylation, Resolvin D biosynthesis (activated in melanoma). Thus, we identified six novel associations between activation of metabolic molecular pathways and progression of melanoma [34].

We found 25 metabolic and 19 signaling pathways that were good-quality characteristic discriminators between the classes of normal skin, nevus, primary melanoma, and serotonin metastatic melanoma (Tables 3 and 4). We considered two general models of

Table 3
Top metabolic pathways implicated in progression of melanoma

Pathway	Nevus vs skin	Pr. Mel vs skin	Met. Mel vs skin	Met.Mel vs Pr.Mel	Primary vs nevus
Allopregnanolone biosynthesis	UP	DOWN	DOWN	DOWN	DOWN
Citrulline-nitric oxide cycle	UP	DOWN	DOWN	DOWN	DOWN
dTMP ide novoi biosynthesis mitochondrial	DOWN	UP	UP	UP	UP
L-carnitine biosynthesis	UP	DOWN	DOWN	DOWN	DOWN
5-Aminoimidazole ribonucleotide biosynthesis	DOWN	UP	UP	UP	UP
Eumelanin biosynthesis	UP	UP	UP	DOWN	DOWN
Putrescine biosynthesis II	DOWN	DOWN	UP	UP	UP
Pyrimidine deoxyribonucleosides salvage	DOWN	UP	UP	UP	UP
Spermine and spermidine degradation I	UP	DOWN	DOWN	DOWN	DOWN
Superpathway of tryptophan utilization	UP	DOWN	DOWN	UP	DOWN
Tryptophan degradation X mammalian via tryptamine	UP	DOWN	DOWN	DOWN	DOWN
1D-imyoi-inositol hexakisphosphate biosynthesis V from Ins134P3	UP	UP	DOWN	DOWN	UP
D-mannose degradation	UP	UP	UP	UP	DOWN
Fructose 26-bisphosphate synthesis, dephosphorylation	UP	UP	UP	DOWN	DOWN
Histamine biosynthesis	UP	DOWN	DOWN	DOWN	DOWN
Inosine-5-phosphate biosynthesis	UP	UP	UP	UP	DOWN
Melatonin degradation II	UP	DOWN	DOWN	DOWN	DOWN
Pyrimidine deoxyribonucleosides degradation	UP	UP	UP	DOWN	UP
Resolvin D biosynthesis	UP	UP	UP	DOWN	UP
Retinoate biosynthesis I	DOWN	DOWN	DOWN	UP	UP
Superpathway of steroid hormone biosynthesis	UP	DOWN	DOWN	DOWN	DOWN
tRNA charging	UP	UP	UP	UP	UP
UDP-N-acetyl-D-galactosamine biosynthesis II	UP	UP	UP	UP	DOWN
Valine degradation	DOWN	DOWN	DOWN	UP	DOWN
Zymosterol biosynthesis	UP	DOWN	DOWN	DOWN	DOWN

UP or DOWN indicates positive and negative difference between the state of interest (nevus, primary, and metastatic melanoma) and skin in median PAS value, respectively

Table 4
Top signaling pathways implicated in progression of melanoma

Pathway	Nevus vs skin	Pr. Mel vs skin	Met. Mel vs skin	Met.Mel vs Pr.Mel	Pr. Mel. vs nevus
Fas signaling pathway (negative)	DOWN	UP	UP	UP	UP
cAMP pathway (glycolysis)	UP	DOWN	DOWN	UP	DOWN
CD40 pathway (cell survival)	UP	UP	UP	UP	UP
AKT pathway (protein synthesis)	UP	DOWN	DOWN	DOWN	DOWN
ATM pathway (apoptosis, senescense)	DOWN	UP	UP	UP	UP
BRCA1 main pathway	UP	UP	UP	UP	UP
cAMP pathway (endothelial cell regulation)	UP	DOWN	DOWN	DOWN	DOWN
cAMP pathway (myocardial contraction)	DOWN	DOWN	DOWN	DOWN	DOWN
cAMP pathway (protein retention)	DOWN	UP	UP	UP	UP
Caspase cascade (apoptosis)	UP	DOWN	DOWN	DOWN	DOWN
CD40 pathway (IKBs degradation)	UP	UP	UP	UP	UP
DDR pathway apoptosis	DOWN	UP	UP	UP	UP
Glucocorticoid receptor pathway (cell cycle arrest)	UP	DOWN	DOWN	DOWN	DOWN
HGF pathway (PKC pathway)	UP	UP	UP	UP	DOWN
HIF1-alpha main pathway	UP	UP	UP	UP	UP
JNK pathway (insulin signaling)	UP	DOWN	DOWN	DOWN	DOWN
mTOR pathway (VEGF pathway)	DOWN	DOWN	UP	UP	DOWN
PAK pathway (myosin activation)	DOWN	DOWN	DOWN	DOWN	DOWN
Ubiquitin proteasome pathway (degraded Protein)	DOWN	UP	UP	UP	UP

UP or DOWN indicates positive and negative difference between the states of interest (nevus, primary, and metastatic melanoma) and skin in median PAS value, respectively

melanoma formation and transformation including transitions (1) Skin → Nevus → Primary melanoma → Metastatic melanoma) and nevus-independent model, (2) Skin → Primary Melanoma → Metastatic Melanoma (Fig. 5). In both transition axes, HIF1-alpha and BRCA1 pathways were gradually increasing when moving from normal state to metastatic melanoma [34].

Transition from normal skin to nevi compared to primary melanoma was very peculiar because it included activation of histamine, allopregnanolone, and citrulline—NO cycle biosynthesis

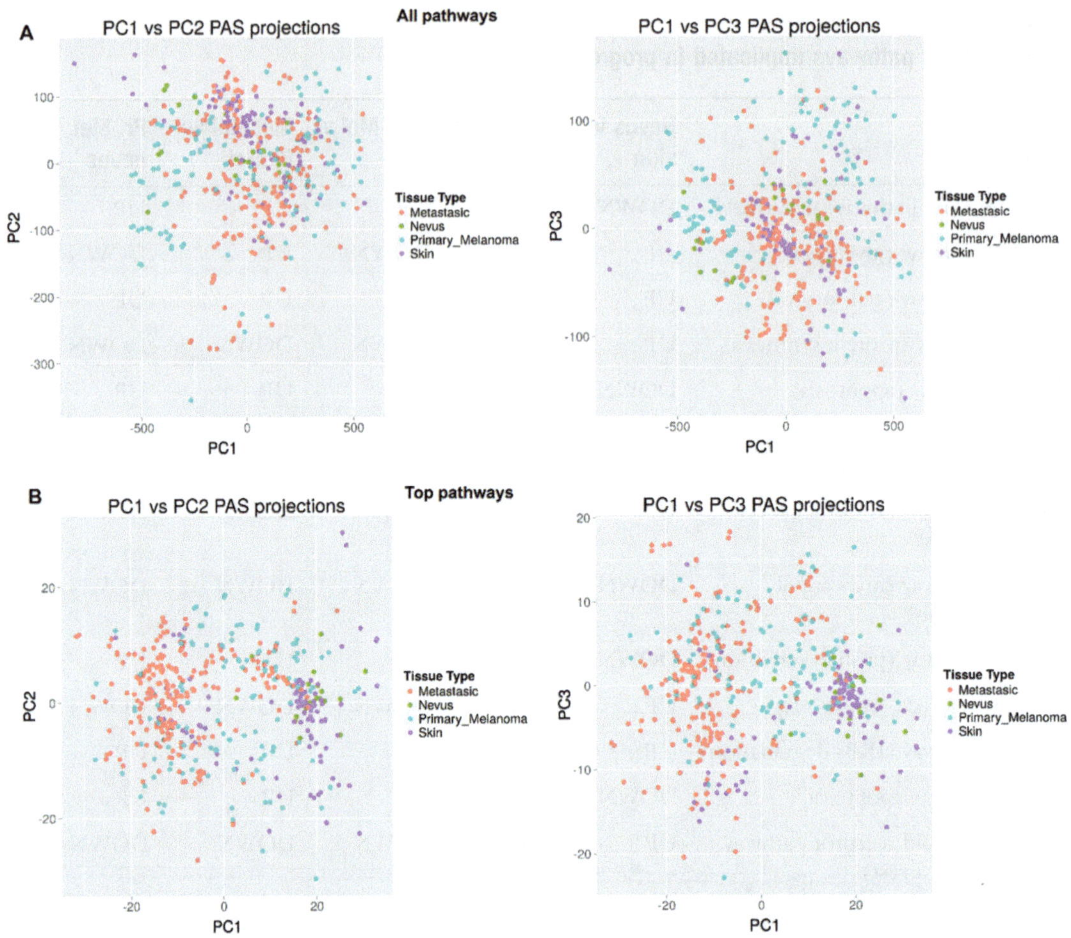

Fig. 4 Scatterplots for principal component analysis of melanoma-related transcriptomes. (**a**) Results built for all metabolic and signaling pathways. (**b**) Results built for top characteristic metabolic (*right*) and signaling (*left*) pathways

pathways. Eumelanin biosynthesis and BRCA1, HIF1-alpha signaling pathways were also activated. Several pathways were also suppressed in nevi, in contrast to primary and metastatic melanomas; these included putrescin biosynthesis, valine degradation, and the senescence/apoptotic branch of the ATM pathway.

Transition from normal skin to primary melanoma was characterized by upregulation of the eumelanin biosynthesis pathway, BRCA1, HIF1-alpha pathways, senescence/apoptotic branch of the ATM pathway, cell death-promoting Fas signaling pathways, and the cell survival-promoting branch of the CD40 pathway. In turn, pathways of putrescine, histamine, allopregnanolone, steroid hormone, and citrulline–NO cycle biosynthesis and of valine degradation were inhibited in primary melanoma compared to skin.

Transition from nevus to primary melanoma showed upregulation of the BRCA1, HIF1-alpha pathways, senescence/apoptotic

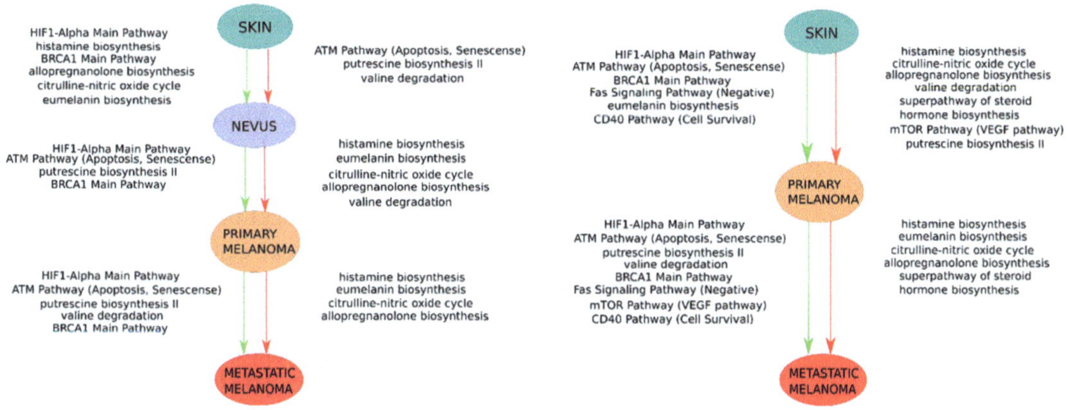

Fig. 5 Schematic representation of two alternative models of melanoma progression built in this study. One model comprises transition from skin to primary melanoma versus "nevus" stage (*left panel*), the second—direct transition from skin to primary melanoma (*right panel*). *Green arrows* indicate activated molecular pathways, *red arrows*—suppressed pathways

branch of the ATM pathway, and putrescine biosynthesis pathway. Inhibited pathways were histamine, allopregnanolone, eumelanin biosynthesis, and citrulline–NO cycle biosynthesis and of valine degradation.

In turn, transition from primary to metastatic melanoma comprised upregulation of BRCA1, HIF1-alpha pathways, the senescence/apoptotic branch of the ATM pathway, putrescine biosynthesis, and valine degradation pathways. Inhibited pathways were histamine, allopregnanolone, eumelanin biosynthesis, and citrulline–NO cycle biosynthesis (Tables 3 and 4) [34].

Finally, we applied the Weighted Correlation Network Analysis (WGCNA) method to identify similar regulation patterns between the molecular pathways. We found that molecular pathways form 14 distinguishable clusters, each characterized by concordant activation signatures of the enclosing pathways. In some instances, congruent activation for the pathways forming the same clusters could be explained by the structural similarities between the cluster-forming pathways [34]. However, for the majority (10 out of 14) of clusters, pathways were combined not due to similar gene content, but rather because of the true functional coordination between the cluster members. This common regulation of various molecular pathways was a novel finding and merits to be analyzed in detail in further studies [34].

7 Molecular Pathways as Predictors of Response to Anticancer Therapeutics

For over six decades, chemotherapy has been a key treatment for many types of cancer, often with high rates of success. For example, the use of cisplatin-containing regiments in the treatment of testicular cancer turned ~100% mortality to ~90–95% disease-specific survival observed nowadays [61]. However, many individual cases and types of cancer remain incurable or even unresponsive using standard chemotherapy approaches. Moreover, chemotherapy generally causes severe side effects, which significantly decrease the quality of life of a patient [62]. The chemical compounds included in standard chemotherapy cocktails have numerous molecular targets in cancerous and normal cells, which makes it difficult to simulate and predict the activity of drug to an individual patient based on the molecular data, and in standard practice clinicians routinely use clinical or morphological predictive factors such as stage, grade, proliferative activity, etc. [63]. These predictive factors are typically very inaccurate and not applicable for tracing the individual patient response to chemotherapy drugs and regimens.

To address specific activities of certain functionally relevant proteins and their aggregates frequently observed in cancer, a new generation of anticancer drugs was generated that target one or a few specific molecules in a cell [64]. This class of drugs consists mostly of specific monoclonal antibodies (Mabs) and low molecular weight kinase-inhibitor molecules (Nibs). The emergence of target drugs was beneficial for the treatment of several cancer types. For example, trastuzumab (anti-HER2 monoclonal antibody) and several other new anti-HER2 medications at least doubled median survival time in patients with metastatic HER2-positive breast cancer and improved 5-year survival in early stage disease to ~90–95% [65]. Interestingly, before the introduction of trastuzumab, HER2-positive cancers had the worst prognoses across all breast cancer subtypes, whereas now the situation is reverted [66]. Patients with melanoma (deadly skin cancer type) for decades had no treatment opportunities except dacarbazine chemotherapy, which resulted in <10% chance of very short-lasting (~5–6 months) response and median survival less than a year. Now, in the case of BRAF-mutated tumor, they can receive vemurafenib (anti-BRAF target drug) and have ~50% chance of response [67], or, irrespectively of BRAF mutation, ipilimumab (immune checkpoint inhibitor) with ~20% chance of long-term (>5 years) disease control [68].

Importantly, the results of clinical trials clearly suggest that for many drugs considered inefficient for the treatment of a given cancer type, a tiny fraction of the patients exists to whom these drugs can be of a significant benefit. For example, no benefit was seen in large randomized studies in a cohort of unselected patients with non-small cell lung cancer after the introduction of anti-EGFR drugs (gefitinib and erlotinib). But it was observed that ~10–15% of the patients who participated in these studies survived unpredictably long. Further investigation revealed that all these patients had activating mutation of EGFR and that this mutation may predict response to the EGFR-targeting drugs. Indeed, contemporary studies showed that patients with EGFR-mutated tumors have the strongest advantage with these types of target therapy [69]. In the case of colorectal cancer, discovery of the role of KRAS mutation in the resistance to the EGFR-targeting antibody (cetuximab or panitumumab) helped to identify a group of patients who can benefit from this kind of treatment (patients with wild-type KRAS). Moreover, further studies demonstrated that for KRAS-mutated tumors (~40% of colorectal cancer), anti-EGFR antibodies cause harm and decrease survival [70].

It is of great priority, therefore, to identify accurate predictive markers of target drug efficacy. Several clinical tests have been used to identify optimal personalized cancer treatments. However, most of these predictor features profile only several biomarkers, cover only a minor fraction of target drugs, and are limited to a particular type of cancer. Somewhat more universal methods are required to rank the maximum number of existing drugs [41].

7.1 Cell Culture-Based Model

We compared molecular pathway activation features linked with the sensitivity of human cell cultures to four target anticancer drugs routinely used for treatment of renal carcinoma and other cancers: Pazopanib, Sunitinib, Sorafenib, and Temsirolimus [40]. To this end, we obtained pathway activation strength (PAS) signatures for experimental group of samples including 11 human cell lines grown and profiled in our laboratory, and for a database linked with "Genomics of Drug Sensitivity in Cancer" [71] project and including transcriptomes of 227 different human cell lines. In both projects, the half maximal inhibitory concentration (IC50) was measured for the above four anticancer drugs, which is a measure of the effectiveness of these drugs in inhibiting cell growth, proliferation, and viability. The IC50 features were further compared with the PAS signatures of both experimental and GDS cell lines, and lists of molecular pathways showing significant ($p < 0.05$) correlation between PAS profiles and IC50 were generated. We next overlapped these lists of characteristic experimental and GDS datasets, and identified a set of molecular pathways linked with sensitivity to drugs and common to both datasets. These pathways included both intracellular signaling and metabolic pathways, and in general had multiple direct and indirect connections with the molecular targets of the respective drugs, thus explaining their association with the drug efficiency. Outline of the experimental and bioinformatic procedures utilized in this study is shown in Fig. 6.

Overall results of OncoFinder analysis depend significantly on what sample or group of samples is taken as the control. To ensure the suboptimal control will not bias the results, we applied multiple simultaneous controls for calculating PAS scores in our experiments, and took separately 11 control gene expression datasets

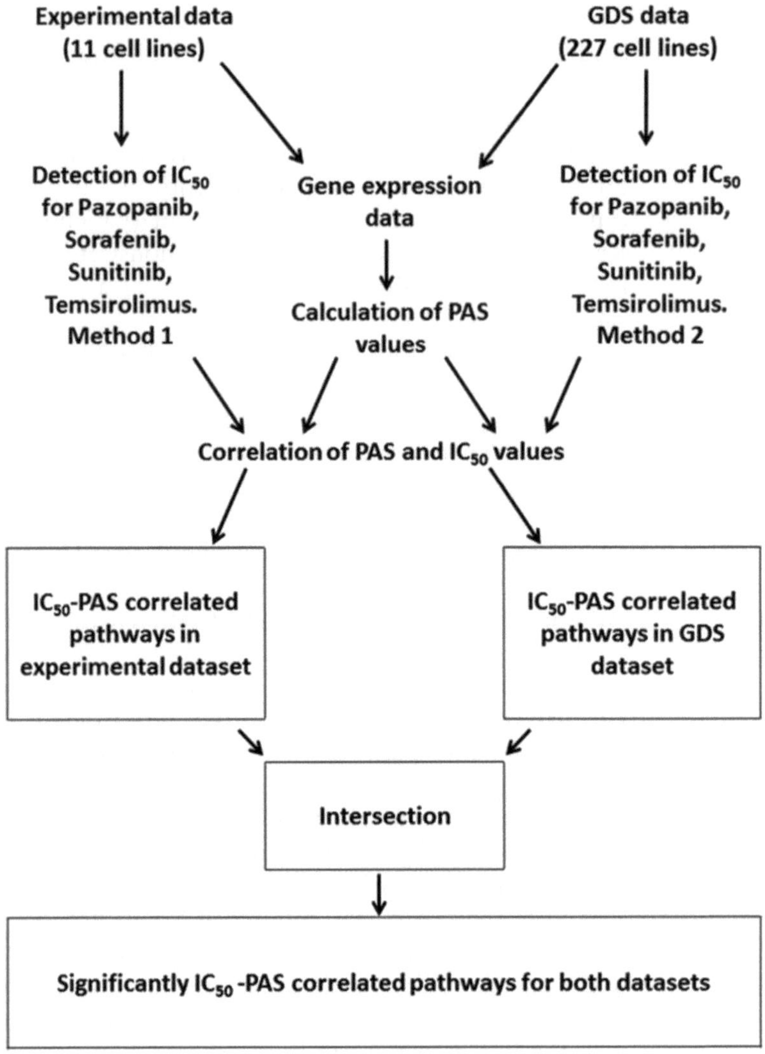

Fig. 6 Outline of the procedures used to identify drug sensitivity-linked pathways in cell cultures

corresponding to different normal human tissues profiled on the same platform as the experimental sampling [40]. The results for 272 signaling and 321 metabolic pathways were obtained for each sample, being normalized separately on each of the 11 control datasets.

We next analyzed GDS project gene expression data deposited at ArrayExpress database. This database accumulates data on gene expression in 707 human cell lines along with the corresponding IC50 values measured for 140 chemical components, including the four drugs under investigation. We calculated PAS values for these transcriptomes, for the same set of signaling and metabolic pathways. For the normalization of transcriptomes prior to processing through the OncoFinder algorithm, we used three independent gene expression datasets taken from GEO database that were obtained using the same experimental platform, corresponding to three normal human tissues [40].

To find out dependences between PAS and IC50 signatures, we calculated correlation coefficient values according to Pearson's product moment correlation coefficient, separately for the experimental and the GDS datasets, for all the normalization methods used. The statistical threshold $p < 0.05$ was used to filter significant vs nonsignificant correlations. We identified a number of pathways showing significant positive or negative correlation between PAS and IC50 values for the above four drugs. A positive correlation between PAS and IC50 values means that the greater is the pathway activation score, the bigger is the half-inhibitory drug concentration, and the lower is the drug efficiency. Negative correlation, in contrast, means increase of the drug efficiency with the increase of PAS value. We next compared significantly correlated pathways from both datasets and found 13, 1, 5, and 7 overlapping molecular pathways for Pazopanib, Sunitinib, Sorafenib, and Temsirolimus, respectively (Fig. 7) [40].

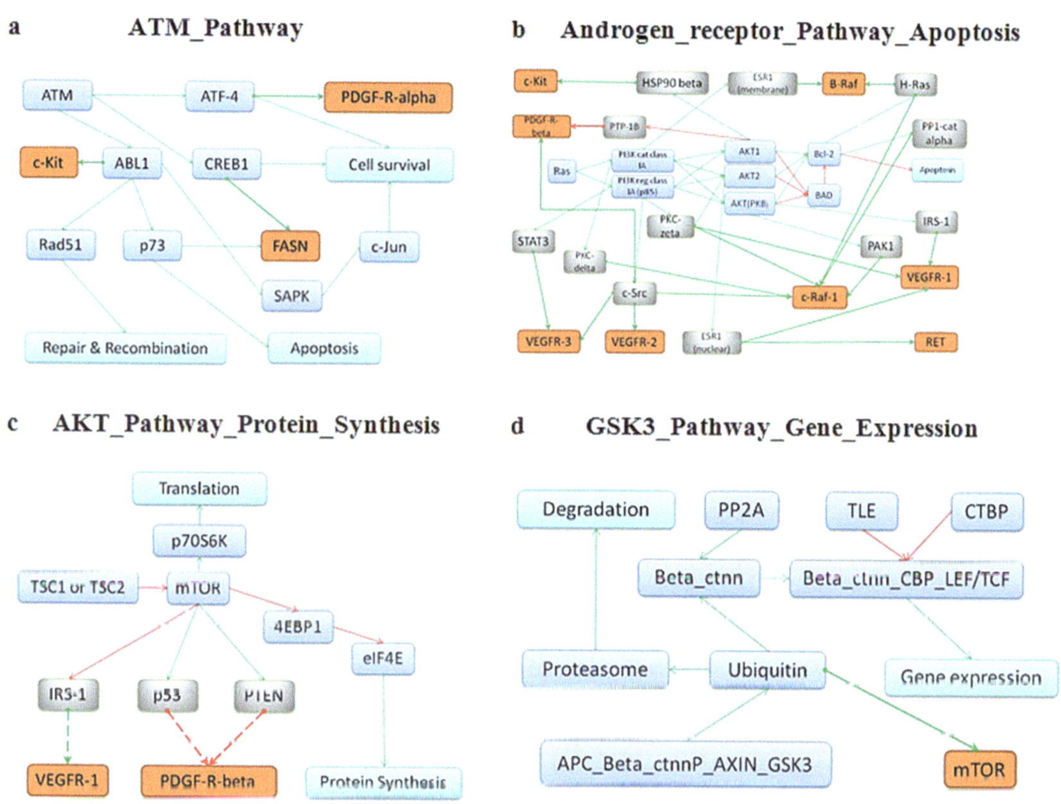

Fig. 7 Schematic representation of the respective drug targets in the overall architecture of molecular interactions for the top pathways correlating with response to Pazopanib (**a**), Sorafenib (**b**), Sunitinib (**c**), and Temsirolimus (**d**). Protein targets of the respective drugs are shown in *orange*, intermediate molecules between pathway members and drug targets—in *gray*, and pathway members—in *blue*

7.2 Target Drug Mechanism-Based Model

We developed a novel approach for choosing an optimal personalized treatment for cancer patients based on high-throughput gene expression profiling of tumor samples [41]. We introduced a Drug Score (DS) index as a measure of effectiveness of a drug in a patient based on the rationale that a drug needs to compensate for the changes in pathway activation/deactivation associated with cancer progression. Clinical trials data were used to validate this scoring system. We compared the distribution of the predicted drug efficacy scores for five drugs (Sorafenib, Bevacizumab, Cetuximab, Sorafenib, Imatinib, Sunitinib) and seven cancer types (Clear Cell Renal Cell Carcinoma, Colon cancer, Lung adenocarcinoma, non-Hodgkin Lymphoma, Lung Adenocarcinoma, Thyroid cancer, and Sarcoma) with the available clinical trials data for the respective drugs and cancer types. The proportion of tumors for which high drug scores were calculated with the proposed algorithm correlated significantly with the percent of responders to a drug treatment (Pearson's correlation 0.77, $p = 0.023$).

7.2.1 Drug Scoring Algorithm

OncoFinder algorithm is based on the processing of Pathway Activation Strength (PAS) signatures of the cancer tissues under investigation. According to OncoFinder method, PAS is calculated using expression values of individual genes to investigate activation/deactivation of intracellular signaling pathways [27]. To construct a scoring function for a drug in a patient, or DS, we defined the following indicators:

-*AMCF* flag (*activation-to-mitosis conversion factor*) shows if the pathway activation promotes or inhibits mitosis and cell survival:

$$AMCF_p = \begin{cases} 1, \text{pathway p promotes mitosis} \\ -1, \text{pathway p inhibits mitosis} \end{cases}$$

DTI (drug-target index):

$$\begin{aligned} DTI_{dt} &= I(\text{drug } d \text{ affects target protein } t) \\ &= \begin{cases} 0, \text{drug } d \text{ does NOT affect target } t \\ 1, \text{drug } d \text{ affect target } t \end{cases} \end{aligned}$$

NII (node involvement index):

$$\begin{aligned} NII_{tp} &= I(\text{protein } t \text{ is involved in pathway } p) \\ &= \begin{cases} 0, & \text{protein } t \text{ is NOT involved in pathway } p \\ 1, & \text{protein } t \text{ is involved in pathway } p \end{cases} \end{aligned}$$

-*DS*, which estimates the ability of a drug d to turn cancer-related pathological changes in the transcriptome of a tumor back to normal state, is defined as follows:

$$DS_d = \sum_t DTI_{dt} \sum_p NII_{tp} AMCF_p\, PAS_p.$$

In other words,

$$DS_d = \sum_t I(\text{drug } d \text{ affects protein } t)$$

$$\sum_p I(\text{protein } t \text{ is involved in the pathway } p) AMCF_p\, PAS_p.$$

Briefly, DS can be understood as a sum of Pathway Activation Scores (PAS) for the pathways in which the targets of a drug are involved. The same PAS can be summed up several times if a drug targets multiple proteins involved in the pathway.

The given formula for DS is, in principle, applicable for all target drugs, including small molecule inhibitors (Nibs) and monoclonal antibodies (Mabs) [41].

7.2.2 Validation of the Drug Scoring Algorithm Based on Tumor Expression Profiling and Clinical Trials Data

We calculated DS for 113 anticancer target drugs for different cohorts of patients with different cancer types [41]. We investigated gene expression in a total of 371 samples of tumors and control sets of corresponding normal tissues for seven cancer types: Clear Cell Renal Cell Carcinoma, Colon cancer, Lung adenocarcinoma, non-Hodgkin Lymphoma, Thyroid cancer, and Sarcoma [41]. To investigate whether the DS successfully predicts treatment efficacy, we analyzed publically available clinical trials data from the ClinicalTrials database (clinicaltrials.gov) and different human cancer transcriptomes extracted from the Gene Expression Omnibus (GEO) database [72]. We checked if the number of patients responding and not responding to a treatment with a particular drug in a particular cancer type could be explained by the distribution of DS for that drug in patients with the particular cancer type. We assumed that the higher number of drug responders among the clinically investigated group of particular cancer patients should correspond to higher Drug Scores for the patients with same cancer type. Using cut-off value DS = 250, we next calculated the percent of patients from a transcriptional profiling study showing greater DS values. We observed that the fraction of patients with high DS correlated significantly with response rates in the respective clinical trials (Pearson's correlation 0.77, $p = 0.023$) (Fig. 8).

Unlike other approaches to ranking drugs for personalized cancer treatment, the algorithm suggested here does not require preliminary data on somatic mutations in tumors, and thus substantially reduces the costs of analysis. While identifying the presence of mutations causing loss and gain of function of regulatory proteins is frequently an important step in predicting clinical outcome and treatment efficiency (e.g., *BRAF* V600E mutation), we show here that a transcriptome-only approach also has the power to

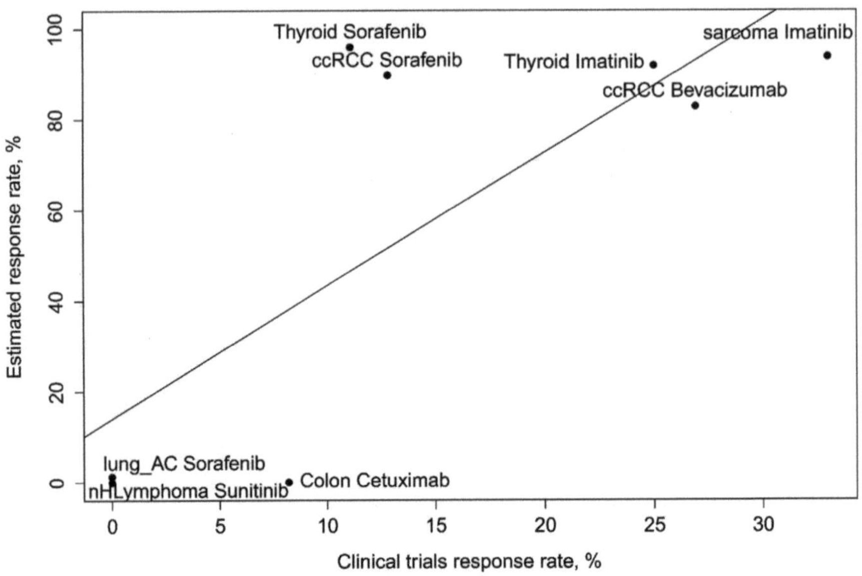

Fig. 8 Scatter plot showing the percent of patients with a particular cancer type responding to a particular treatment (*x*-axis) in a clinical trial versus the percent of patients with a particular cancer type having the Drug Score for the particular drug above an arbitrary chosen cut-off value (250) (*y*-axis). ccRCC stands for Clear Cell Renal Cell Carcinoma, nHLymphoma for non-Hodgkin Lymphoma, lung AC for lung adenocarcinoma

detect these changes at the gene expression level for downstream targets of the mutated regulator [41]. Theoretically, the expression data may provide even more biologically meaningful results, as reliable methods for prediction of particular somatic mutations (e.g., gain-of-function) do not exist to date, and many mutations have limited or no phenotypic manifestations, depending heavily on the enclosing genomic context [73].

To investigate the ability of our transcriptome-based drug-scoring approach to distinguish between tumors harboring different driver mutations, we explored gene expression in melanoma patients. Vemurafenib is a target drug that is effective for melanoma tumors with V600E gain-of-function mutation in *BRAF* gene. We compared DS for patients with wild-type and V600E *BRAF* melanomas [41]. We demonstrated that the percent of patients for whom Vemurafenib was expected to be beneficial (those having a positive DS for this drug) was significantly higher for the cohort of BRAF V600E-mutated tumors (p(Fisher) = 0.042, Fig. 9).

The reason why an expression-based approach works well in this case is likely due to the ability to detect expression changes introduced by transcriptional reprogramming driven by the molecular consequences of V600E *BRAF* mutation.

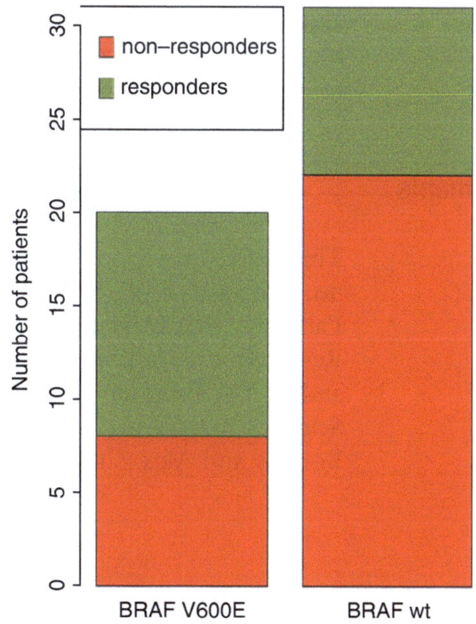

Fig. 9 Cohort of tumors with BRAF V600E mutation (*left bar*) had significantly higher proportion of patients for whom Vemurafenib was predicted to be beneficial compared to a cohort with wild-type BRAF (*right bar*). *Red bars* show predicted nonresponders and *green bars* show predicted responders (having nonzero DS for Vemurafenib)

8 Validation

Here, we present a biomathematical method OncoFinder that has a potential to be universal tool for the analysis of intracellular molecular pathways and for predicting drug efficacy via characterization of specific patterns in intracellular signaling. It may have wide applicability, not only across the range of cancer types, but also to individual samples toward the goal of personalized cancer treatment. Unlike most part of other approaches to drug scoring in cancer, the current method does not require data on somatic mutations in tumors, thus substantially reducing the costs of an assay. Rather, it relies on advanced gene expression analysis. Although the presence of mutations causing loss and gain of function of certain regulator proteins is an important factor in the prediction of clinical outcome and treatment efficacy, a transcriptome-only approach will still potentially detect these changes as expression changes in downstream targets of the mutated regulator. Moreover, because reliable methods for predicting the effects of many specific somatic mutations (e.g., gain of function) do not yet exist, results based on expression data may be more biologically meaningful. The approach we report here is platform-independent, i.e., any kind of high-throughput proteomic and transcriptomic data may be used to estimate expression of gene products.

Acknowledgments

This work was supported by the Russian Science Foundation grant no. 14-14-01089 (for V.Prassolov and Anton Buzdin), by the Pathway Pharmaceuticals (Hong-Kong) and First Oncology Research and Advisory Center (Russia) Joint Research Initiative and by the Program of the Presidium of the Russian Academy of Sciences "Dynamics and Conservation of Genomes" (for Nikolay Borisov and Alex Zhavoronkov).

References

1. Blagosklonny MV (2013) MTOR-driven quasi-programmed aging as a disposable soma theory: blind watchmaker vs. intelligent designer. Cell Cycle 12:1842–1847
2. Demidenko ZN, Blagosklonny MV (2011) The purpose of the HIF-1/PHD feedback loop: to limit mTOR-induced HIF-1α. Cell Cycle 10:1557–1562
3. Blagosklonny MV (2011) The power of chemotherapeutic engineering: arresting cell cycle and suppressing senescence to protect from mitotic inhibitors. Cell Cycle 10:2295–2298
4. UniProt Consortium (2011) Ongoing and future developments at the Universal Protein Resource. Nucleic Acids Res 39:D214–B219
5. Mathivanan S, Periaswamy B, Gandhi T, Kandasamy K et al (2006) An evaluation of human protein-protein interaction data in the public domain. BMC Bioinformatics 7:S19
6. Pathway central, a Qiagen portal. https://www.qiagen.com/ro/shop/genes-and-pathways/pathway-central/. Accessed 15 Mar 2016
7. Bauer-Mehren A, Furlong LI, Sanz F (2009) Pathway databases and tools for their exploitation: benefits, current limitations and challenges. Mol Syst Biol 5:290
8. Nikitin A, Egorov S, Daraselia N, Mazo I (2003) Pathway studio—the analysis and navigation of molecular networks. Bioinformatics 19:2155–2157
9. Elkon R, Vesterman R, Amit N (2008) SPIKE—a database, visualization and analysis tool of cellular signaling pathways. BMC Bioinformatics 9:110
10. Haw R, Stein L (2014) The Reactome pathway knowledgebase. Nucleic Acids Res 42: D472–D477
11. Nakaya A, Katayama T, Itoh M, Hiranuka K et al (2013) KEGG OC: a large-scale automatic construction of taxonomy-based ortholog clusters. Nucleic Acids Res 41:D353–D357
12. HumanCyc: encyclopedia of human genes and metabolism. http://www.humancyc.org/. Accessed 15 Mar 2016
13. Vivar JC, Pemu P, McPherson R, Ghosh S (2013) Redundancy control in pathway databases (ReCiPa): an application for improving gene-set enrichment analysis in Omics studies and "Big data" biology. OMICS 17:414–422
14. Eikrem O, Beisland C, Hjelle K, Flatberg A et al (2016) Transcriptome sequencing (RNA-seq) enables utilization of formalin-fixed, paraffin-embedded biopsies with clear cell renal cell carcinoma for exploration of disease biology and biomarker development. PLoS One 11:e0149743
15. Khatri P, Sirota M, Butte AJ (2012) Ten years of pathway analysis: current approaches and outstanding challenges. PLoS Comput Biol 8: e1002375
16. Khatri P, Drăghici S (2005) Ontological analysis of gene expression data: current tools, limitations, and open problems. Bioinformatics 21:3587–3595
17. Khatri P, Draghici S, Ostermeier GC, Krawetz SA (2002) Profiling gene expression using onto-express. Genomics 79:266–270
18. Zeeberg BR, Feng W, Wang G, Wang MD et al (2003) GoMiner: a resource for biological

interpretation of genomic and proteomic data. Genome Biol 4:R28

19. Barry WT, Nobel AB, Wright FA (2005) Significance analysis of functional categories in gene expression studies: a structured permutation approach. Bioinformatics 21:1943–1949
20. Subramanian A, Tamayo P, Mootha VK, Mukherjee S et al (2005) Gene set enrichment analysis: a knowledge-based approach for interpreting genome-wide expression profiles. Proc Natl Acad Sci USA 102:15545–15550
21. Tian L, Greenberg SA, Kong SW, Altschuler J et al (2005) Discovering statistically significant pathways in expression profiling studies. Proc Natl Acad Sci USA 102:13544–13549
22. Mitrea C, Taghavi Z, Bokanizad B, Hanoudi S et al (2013) Methods and approaches in the topology-based analysis of biological pathways. Front Physiol 4:278
23. Afsari B, Geman D, Fertig EJ (2014) Learning dysregulated pathways in cancers from differential variability analysis. Cancer Inform 13 (Suppl 5):61–67
24. Ho JW, Stefani M, dos Remedios CG, Charleston MA (2008) Differential variability analysis of gene expression and its application to human diseases. Bioinformatics 24:i390–i398
25. Eddy JA, Hood L, Price ND, Geman D (2010) Identifying tightly regulated and variably expressed networks by Differential Rank Conservation (DIRAC). PLoS Comput Biol 6: e1000792
26. Zhang J, Li J, Deng HW (2009) Identifying gene interaction enrichment for gene expression data. PLoS One 4:e8064
27. Buzdin AA, Zhavoronkov AA, Korzinkin MB et al (2014) OncoFinder, a new method for the analysis of intracellular signaling pathway activation using transcriptomic data. Front Genet 5:55
28. Buzdin AA, Zhavoronkov AA, Korzinkin MB et al (2014) The OncoFinder algorithm for minimizing the errors introduced by the high-throughput methods of transcriptome analysis. Front Mol Biosci 1:8
29. Lezhnina K, Kovalchuk O, Zhavoronkov AA, Korzinkin MB et al (2014) Novel robust biomarkers for human bladder cancer based on activation of intracellular signaling pathways. Oncotarget 5:9022–9032
30. Aliper AM, Frieden-Korovkina VP, Buzdin A et al (2014) Interactome analysis of myeloid derived suppressor cells in murine models of colon and breast cancer. Oncotarget 5:11345–11353
31. Aliper AM, Csoka AB, Buzdin A, Jetka T et al (2015) Signaling pathway activation drift during aging: Hutchinson-Gilford Progeria Syndrome fibroblasts are comparable to normal middle-age and old-age cells. Aging (Albany NY) 7:26–37
32. Makarev E, Cantor C, Zhavoronkov A, Buzdin A et al (2014) Pathway activation profiling reveals new insights into age-related macular degeneration and provides avenues for therapeutic interventions. Aging (Albany NY) 6:1064–1075
33. Alexandrova E, Nassa G, Corleone G, Buzdin A et al (2016) Large-scale profiling of signalling pathways reveals an asthma specific signature in bronchial smooth muscle cells. Oncotarget 7(18):25150–25161. [Epub ahead of print]
34. Shepelin D, Korzinkin M, Vanyushina A, Aliper A et al (2016) Molecular pathway activation features linked with transition from normal skin to primary and metastatic melanomas in human. Oncotarget 7:656–670
35. Lebedev TD, Spirin PV, Suntsova MV, Ivanova AV et al (2015) Receptor tyrosine kinase KIT may regulate expression of genes involved in spontaneous regression of neuroblastoma. Mol Biol (Mosk) 49:1052–1055
36. Ram DR, Ilyukha V, Volkova T, Buzdin A et al (2016) Balance between short and long isoforms of cFLIP regulates Fas-mediated apoptosis in vivo. Proc Natl Acad Sci USA 113:1606–1611
37. Vishniakova KS, Babizhaev MA, Aliper AM, Buzdin AA et al (2014) Stimulation of proliferation by carnosine: cellular and transcriptome approaches. Mol Biol 48:824–833
38. Spirin PV, Lebedev TD, Orlova NN, Gornostaeva AS (2014) Silencing AML1-ETO gene expression leads to simultaneous activation of both pro-apoptotic and proliferation signaling. Leukemia 28(11):2222–2228
39. Artcibasova AV, Korzinkin MB, Sorokin MI, Shegay PV et al (2016) MiRImpact, a new bioinformatic method using complete microRNA expression profiles to assess their overall influence on the activity of intracellular molecular pathways. Cell Cycle 15(5):689–698
40. Venkova LS, Aliper AM, Suntsova M, Kholodenko R et al (2015) Combinatorial high-throughput experimental and bioinformatics approach identifies molecular pathways linked with the sensitivity to anticancer target drugs. Oncotarget 6:27227–27238
41. Artemov A, Aliper A, Korzinkin M, Lezhnina K et al (2015) A method for predicting target

drug efficiency in cancer based on the analysis of signaling pathway activation. Oncotarget 6:29347–29356

42. Zhu Q, Izumchenko E, Aliper AM, Makarev E (2015) Pathway activation strength is a novel independent prognostic biomarker for cetuximab sensitivity in colorectal cancer patients. Hum Genome Var 2:15009
43. Bolstad BM, Irizarry RA, Astrand M, Speed TP (2003) A comparison of normalization methods for high density oligonucleotide array data based on variance and bias. Bioinformatics 19:185–193
44. Hsu SD, Tseng YT, Shrestha S, Lin YL et al (2014) miRTarBase update 2014: an information resource for experimentally validated miRNA-target interactions. Nucleic Acids Res 42:D78–D85
45. Vergoulis T, Vlachos IS, Alexiou P, Georgakilas G et al (2012) TarBase 6.0: capturing the exponential growth of miRNA targets with experimental support. Nucleic Acids Res 40: D222–D229
46. Keshaviah A, Dellapasqua S, Rotmensz N, Lindtner J et al (2007) CA15-3 and alkaline phosphatase as predictors for breast cancer recurrence: a combined analysis of seven International Breast Cancer Study Group trials. Ann Oncol 18:701–708
47. Blagosklonny MV (2012) Common drugs and treatments for cancer and age-related diseases: revitalizing answers to NCI's provocative questions. Oncotarget 3:1711–1724
48. Borisov NM, Terekhanova NV, Aliper SM, Venkova LS et al (2014) Signaling pathways activation profiles make better markers of cancer than expression of individual genes. Oncotarget 5:10198–10205
49. Swets JA, Green DM, Getty DJ, Swets JB (1978) Signal detection and identification at successive stages of observation. Percept Psychophys 23:275–289
50. Boyd JC (1997) Mathematical tools for demonstrating the clinical usefulness of biochemical markers. Scand J Clin Lab Invest Suppl 227:46–63
51. Munshi A, Ramesh R (2013) Mitogen-activated protein kinases and their role in radiation response. Genes Cancer 4:401–408
52. Morgenroth A, Vogg AT, Ermert K et al (2014) Hedgehog signaling sensitizes glioma stem cells to endogenous nano-irradiation. Oncotarget 5:5483–5493
53. Parkin DM, Bray F, Ferlay J, Pisani P (2005) Global cancer statistics, 2002. CA Cancer J Clin 55:74–108
54. Ploeg M, Aben KK, Kiemeney LA (2009) The present and future burden of urinary bladder cancer in the world. World J Urol 27:289–293
55. Zabolotneva AA, Zhavoronkov AA, Shegay PV, Gaifullin NM (2013) A systematic experimental evaluation of microRNA markers of human bladder cancer. Front Genet 4:247
56. Jerant AF, Johnson JT, Sheridan CD, Caffrey TJ (2000) Early detection and treatment of skin cancer. Am Fam Physician 62:357–368, 375–376, 381–382
57. El Ghissassi F, Baan R, Straif K, Grosse Y (2009) A review of human carcinogens—part D: radiation. Lancet Oncol 10:751–752
58. Halachmi S, Gilchrest BA (2001) Update on genetic events in the pathogenesis of melanoma. Curr Opin Oncol 13:129–136
59. Davies MA, Samuels Y (2010) Analysis of the genome to personalize therapy for melanoma. Oncogene 29:5545–5555
60. Elder D (1999) Tumor progression, early diagnosis and prognosis of melanoma. Acta Oncol 38:535–547
61. Hanna N, Einhorn LH (2014) Testicular cancer: a reflection on 50 years of discovery. J Clin Oncol 32:3085–3092
62. Svensson L, Finlay BB, Bass D et al (1991) Symmetric infection of rotavirus on polarized human intestinal epithelial (Caco-2) cells. J Virol 65:4190–4197
63. Zhukov NV, Tjulandin SA (2008) Targeted therapy in the treatment of solid tumors: practice contradicts theory. Biochemistry 73:605–618
64. Sawyers C (2004) Targeted cancer therapy. Nature 432:294–297
65. Nahta R, Esteva FJ (2007) Trastuzumab: triumphs and tribulations. Oncogene 26:3637–3643
66. Onitilo AA, Engel JM, Greenlee RT, Mukesh BN (2009) Breast cancer subtypes based on ER/PR and Her2 expression: comparison of clinicopathologic features and survival. Clin Med Res 7:4–13
67. Chapman PB, Hauschild A, Robert C, Haanen JB et al (2011) Improved survival with vemurafenib in melanoma with BRAF V600E mutation. N Engl J Med 364:2507–2516
68. Prieto PA, Yang JC, Sherry RM, Hughes MS (2012) CTLA-4 blockade with ipilimumab: long-term follow-up of 177 patients with metastatic melanoma. Clin Cancer Res 18:2039–2047
69. Gridelli C, De Marinis F, Di Maio M, Cortinovis D et al (2011) Gefitinib as first-line treatment for patients with advanced non-small-cell

lung cancer with activating epidermal growth factor receptor mutation: review of the evidence. Lung Cancer 71:249–257
70. Grothey A, Lenz HJ (2012) Explaining the unexplainable: EGFR antibodies in colorectal cancer. J Clin Oncol 30:1735–1737
71. Yang W, Soares J, Greninger P, Edelman EJ et al (2013) Genomics of Drug Sensitivity in Cancer (GDSC): a resource for therapeutic biomarker discovery in cancer cells. Nucleic Acids Res 41:D955–D956
72. GEO Profiles, a National Center of Biotechnology Information database. http://www.ncbi.nlm.nih.gov/geo/. Accessed 16 Mar 2016
73. Reva B, Antipin Y, Sander C (2011) Predicting the functional impact of protein mutations: application to cancer genomics. Nucleic Acids Res 39:e118

Chapter 5

Strategic Integration of Multiple Bioinformatics Resources for System Level Analysis of Biological Networks

Mark D'Souza, Dinanath Sulakhe, Sheng Wang, Bing Xie, Somaye Hashemifar, Andrew Taylor, Inna Dubchak, T. Conrad Gilliam, and Natalia Maltsev

Abstract

Recent technological advances in genomics allow the production of biological data at unprecedented tera- and petabyte scales. Efficient mining of these vast and complex datasets for the needs of biomedical research critically depends on a seamless integration of the clinical, genomic, and experimental information with prior knowledge about genotype-phenotype relationships. Such experimental data accumulated in publicly available databases should be accessible to a variety of algorithms and analytical pipelines that drive computational analysis and data mining.

We present an integrated computational platform Lynx (Sulakhe et al., Nucleic Acids Res 44: D882–D887, 2016) (http://lynx.cri.uchicago.edu), a web-based database and knowledge extraction engine. It provides advanced search capabilities and a variety of algorithms for enrichment analysis and network-based gene prioritization. It gives public access to the Lynx integrated knowledge base (LynxKB) and its analytical tools via user-friendly web services and interfaces. The Lynx service-oriented architecture supports annotation and analysis of high-throughput experimental data. Lynx tools assist the user in extracting meaningful knowledge from LynxKB and experimental data, and in the generation of weighted hypotheses regarding the genes and molecular mechanisms contributing to human phenotypes or conditions of interest. The goal of this integrated platform is to support the end-to-end analytical needs of various translational projects.

Key words High-throughput genomics, Systems biology, Bioinformatics, Data mining, Network analysis

1 Introduction

Understanding the genetic architecture underlying complex biological phenomena and heritable multigene disorders is one of the major goals of human genetics in the next decade. Advances in whole genome sequencing and the success of high-throughput functional genomics help to supplement conventional reductionist biology with systems-level approaches, thus allowing researchers to

Tatiana V. Tatarinova and Yuri Nikolsky (eds.), *Biological Networks and Pathway Analysis*, Methods in Molecular Biology, vol. 1613, DOI 10.1007/978-1-4939-7027-8_5, © Springer Science+Business Media LLC 2017

study biology and medicine as complex networks of interacting genetic and epigenetic factors in relevant biological contexts. This integrative approach holds the promise of unveiling hitherto unexplored levels of molecular organization and biological complexity. It also holds the key to deciphering the multigene patterns of inheritance that predispose individuals to a wide array of genetic diseases. Numerous studies have identified genes associated with many rare single gene (Mendelian) developmental disorders, but only limited progress has been made in finding the underlying causes for autism, schizophrenia, diabetes, predisposition to cancer, and cardiovascular diseases. The reason is that these diseases display complex patterns of inheritance and may result from many genetic variations, each contributing only weak effects to the disease phenotype. Identification of causative disease genes or genetic variations within the myriad of susceptibility loci identified in linkage and association studies is difficult because these loci may contain hundreds of genes. Fortunately, recent advances in biological science have provided new perspectives into the study of complex heritable disorders. These advances include: (1) high-throughput integrative genomics and informatics; (2) networks-based view of human disorders; and (3) emergence of "phenomics," and a notion of interrelatedness of diseases and disease traits. These approaches offer a strategy for system-level exploration of complex clinical phenotypes in relevant biological contexts. They utilize expertise from the fields of genomics, molecular biology, bioinformatics, and clinical studies to develop integrative models of molecular events driving the emergence of cellular and organismal phenotypes. At the basic science level this research seeks to understand the nascent properties of interacting molecular networks and how they relate to biological complexity. At the application level, identifying combinations of interacting genes that underlie complex genetic disorders is the practical first step in moving from today's genetic understanding to the era of individualized medicine. Interpretation of the genetic architecture of common diseases will afford presymptomatic testing of individuals at risk for common disorders, gradually shifting the practice of medicine from a "reactive" science to a "predictive" science. It will also allow state-of-the-art technologies such as high-throughput genetic screening to advance drug discovery and development.

However, the extraction of meaningful information from an avalanche of available biomedical information requires seamless integration of data and services across the analytical workflows. These workflows start from the raw experimental data and include multiple analytical steps leading to the generation of high-confidence hypotheses regarding molecular mechanisms contributing to the phenotypes of interest. Each step of such a pipeline generates additional annotations utilized by the subsequent steps of analysis or displayed to the user to aid in manual investigation of the data. The nature of contemporary biology dictates the need for the use of multiple data sources and distributed analytical services developed by a number of scientific groups. This distributed research paradigm calls for an integrated analytical platform to address the end-to-end requirements of translational projects. Such a platform requires advanced computational technologies that will ensure fast and reliable movement of terabytes of data, provide on-demand scalable computational resources and guarantee security and provenance of every analytical step. The need for a profound integration of data and services was expressed in many publications [1–3]. A number of large-scale initiatives were launched to bring together information resources and make them available to the scientific community [4–6].

2 Methods

2.1 Analysis of Biological Networks

Recent advances in functional genomics have allowed for the deciphering of millions of inter-relationships between gene products. This data is used for the development of integrative network-based models. The systems-level analysis offered by these models holds the promise of uncovering biological mechanisms underlying development and differentiation, and driving the emergence of complex phenotypes, including human disease [7, 8]. There is an urgent scientific need for support of contextual, comparative, and evolutionary analyses for the studies of biological systems. The comparative analysis of contextual networks will support a number of scientific directions, such as the following:

2.2 Comparative Analysis of Contextual Models of Biological Processes

The systems biology approach is contextual by definition. It studies the emergent behavior of self-organizing biological systems in the relevant spatial and temporal biological contexts. Indeed, the insights offered by contextual analyses are essential to further progress in biology: (a) the comparative analysis of developmental networks will establish a foundation for the fields of embryology and developmental biology and provide insights into the pathogenesis of developmental disorders; (b) tissue-specific gene expression plays a fundamental role in metazoan biology and is an important aspect of many complex diseases. The comparative analysis of tissue-specific networks is essential for the selection of potential drug targets where it will allow a reduction in the number of side effects for the developed drugs by selecting cellular components specific to the targeted tissue [9–11].

2.2.1 Types of Comparative Networks Analysis

Conceptually, network comparison is the process of contrasting two or more interaction networks, representing different species, conditions, interaction types, or time points. It aims at providing answers to a number of fundamental biological questions regarding

Table 1
Modes and biological goals of network comparison

Mode and applications	Biological goals	Current networks: comparisons and limitations
Network integration: Comparisons of the networks based on the *different data types*	Identification of functional modules supported by several interaction data types and data provenance estimates; Studies of interrelations between data types; Prediction of molecular interactions	No agreed-upon way to combine scores over different networks; Not associated with the knowledge bases and networks-reconstruction tools
Network Alignment: Comparisons of the networks based on the *same data types*	Context-specific comparisons (e.g., different developmental stages, health and disease, tissue-specific networks); Identification of evolutionarily conserved pathways and sub-networks across multiple species; Prediction of the molecular interactions and gene functions based on the inter-species comparisons; Validation of the applicability of animal models	Limited to few species; Evolutionary and provenance information is not factored in the analysis; Not associated with the knowledge bases and networks-reconstruction tools
Network querying: Identification of the sub-network modules in a network	Identification of redundant and conserved functional modules within and between species; Meta-data-based queries and knowledge transfer	No support for hierarchical networks queries; No evolutionary-based scoring

the evolution and modular organization of biological systems, as well as their development and functionality under a variety of normal and pathophysiological conditions (*see* Table 1). It may also be used for measuring and increasing data provenance.

Sharan and Ideker [12] have postulated three types, or modes, of comparative methods:

(a) ***Network integration*** is the process of combining several networks, based on interactions of *different types* (e.g., protein-protein interaction networks, biological pathways, and text-mining information) over the same set of elements, to study their interrelations. Network integration can be used for predicting protein interactions and discovering protein modules supported by interactions of different types. The main conceptual difference from network alignment is that the integrated networks are defined on the same set of elements;

(b) ***Network alignment*** is the process of global comparison between two networks *of the same type*, identifying sub-networks and regions of similarity and dissimilarity.

Network alignment is commonly applied to detect sub-networks that are conserved across species or vary from one developmental stage to another; and

(c) **Network querying**, in which a given network is searched for sub-networks that are similar to a sub-network query of interest. This basic database search operation is aimed at transferring biological knowledge within and across species.

2.3 Evolutionary Analysis of Contextual Models

Unlike the evolutionary analysis of individual genes, the exploration of the evolutionary modifications of molecular networks allows us to understand the emergence of new phenotypes as concerted changes in network topology, as well as the functionality of multiple components of a biological system. *Systems-level evolutionary analysis is especially important for biomedical studies*, where the majority of experiments are performed on model organisms. Comparative and evolutionary network analysis can help us estimate the applicability of animal models to human studies and validate the knowledge transfer across species. Moreover, the comparative analysis of molecular networks reconstructed from tissue- and cell-specific gene expression experiments will provide the basis for the identification of functional modules characteristic for various biological contexts. This information will shed light onto mechanisms of genetic epistasis, robustness, and adaptation of biological systems in health and disease [13].

2.4 Comparative Phenomics

In recent years, it has become increasingly evident that human diseases are related to each other and share common phenotypic features, molecular mechanisms, and common genetic determinants. As demonstrated by multiple studies, the disease phenome should be regarded as a network of interrelated diseases and disease traits rather than a list of distinct disease entities [8, 14–16]. It is now widely accepted by the scientific community that comparative analysis of disease network models for different disorders will provide new insights into the etiology, pathogenesis, and classification of the diseases, and will assist in the development of new therapeutic strategies.

Undeniably, comparative and contextual network analysis offers exciting new opportunities for biomedical research on a new integrative systems level. The need for the comparative analysis of contextual networks has been expressed in a growing number of publications [12, 17]. As it was stated by Beltrao [18] "In the same way that comparative genomics has resulted in an impressive leap forward in our understanding of genome evolution, we argue that combining and comparing different cellular interaction data are crucial for our understanding of the evolutionary process."

2.5 Context-Specific Networks

Understanding of cellular responses specific to cell or tissue type, gender, developmental stage, or environmental conditions is of paramount significance for the development of efficient diagnostic and therapeutic strategies [19–22]. Context-specific studies performed for cancer [23–25], cardiac disorders [26], and various developmental processes [27–29] have underscored the importance of predictive models of focus disorders. For example, in their study of the lung-specific pathways used by the influenza virus Shaefer et al. [30] demonstrated that context consistency correlates with the experimental reliability of PPIs, which allows generating high-confidence tissue- and function-specific subnetworks. Shao et al. [31] have emphasized the importance of contextual analysis for drug design, stating that "Substantial effort in recent years has been devoted to analyzing data-based large-scale biological networks, which provide valuable insight into the topologies of complex biological networks but are rarely context specific and cannot be used to predict the responses of cell signaling proteins to specific ligands or compounds."

3 Existing Tools

A number of excellent bioinformatics platforms and tools have been developed to support various steps of analysis of high-throughput data and prioritization of genomic variants [32–34]. These include, but not limited to, GeneMANIA [35], STRING [36, 37], ToppGene [38], ENDEAVOUR [39] widely used by the scientific community. The eXtasy platform developed by Sifrim et al. [40] prioritizes mutations for follow-up validation studies by integrating variant-impact and haploinsufficiency predictions with phenotype-specific information. Another scientific environment, SPRING [41], has been designed to facilitate the prioritization of pathogenic non-synonymous SNVs associated with disorders whose genetic bases are either partly known or completely unknown. It is achieved by integrating the results of analyses by multiple publicly available and developed-in-house bioinformatics tools. There are other analytical platforms, such as Jannovar [42], KGGSeq [43], MToolBox [44], and FamAnn [45]. Moreover, multiple resources support the analysis of noncoding regions and their regulatory roles [46]. Most of these existing resources, understandably, address either the analysis of coding sequences or the characterization of noncoding regions.

3.1 Overview of the Resources for Evolutionary and Context-Specific Networks Analysis

A number of groups have implemented cross-species and contextual analysis of biological data. The first efforts to perform a large-scale comparison of PPI networks of *Saccharomyces cerevisiae* against other microbial species, such as *Helicobacter pylori,* to predict previously uncharacterized PPIs were performed by Matthews et al. 2001 [47] and Yu et al. 2004 [48]. Quantitative analysis of genetic interactions was initially accomplished in budding yeast [49]. Recently, Bandyopadhyay et al. [9] have developed differential epistasis mapping (dE-MAP) a strategy for the quantitative and differential mapping of genetic networks. A similar approach has been used to demonstrate changes in genetic interactions in a lower-throughput format [50], and in *Drosophila melanogaster* to map genetic interactions using RNAi in different genetic backgrounds [51]. Troyanskaya et al. [10, 11] stressed the need for contextual analysis of biological networks and developed a Bayesian approach for context-sensitive integration and query-based recovery of biological process-specific networks. This approach was applied to *Saccharomyces cerevisiae* to demonstrate that leveraging contextual information can significantly improve the precision of network predictions, including assignment for uncharacterized genes. Lage et al. [52] have studied the link between tissue-specific gene expression and pathological manifestations in human diseases and cancers. They created a disease-tissue covariation matrix of high-confidence associations of >1000 diseases to 73 tissues.

3.2 Overview of the currently Available Integrated Global Networks.

A number of resources taking a meta-analysis approach include STRING v9.1 [36, 37], GeneMANIA [53], ConsensusPathDB [54], I2D [55], VisANT [56], hPRINT [57], HitPredict [58], IMID [59] and IMP [60]. A number of text-mining resources and databases provide context to biological data, such as text-mining engines EnvMine [61], BioContext [62], splice variants databases (e.g., SpliceMiner[63] and ASD [64]), and the Primate Embryo Gene Expression Resource in embryology PREGER [65]. These resources may provide a significant aid in the development of context-specific network models.

4 Lynx—an Integrated Platform for Network-Based Analysis of Translational Data

Here, we present an example of a project-driven integrated computational platform Lynx. The goal of this scalable platform is to support the end-to-end analytical requirements of individual translational projects. Working with multiple translational projects allowed us to identify crosscutting shared computational and analytical requirements. These projects have converged toward well-defined standard steps of analysis of translational data as represented in Fig. 1. Sections below will describe the steps involved in translational data analysis in greater detail.

4.1 Lynx Annotation and Knowledge Extraction Engine

Lynx [66] (http://lynx.ci.uchicago.edu) is an integrated bioinformatics platform for annotation and analysis of high-throughput biomedical data. The platform supports both hypothesis-based and discovery-based approaches to predict the genetic factors and networks associated with phenotypes of interest. It provides a

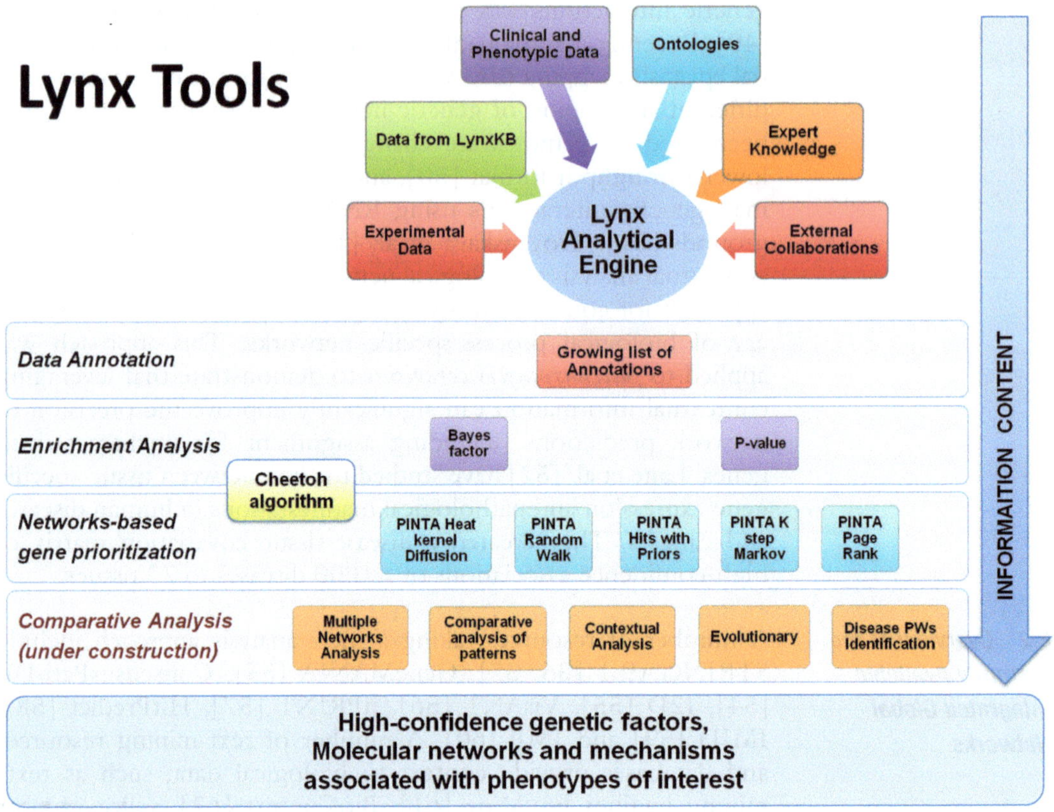

Fig. 1 Annotation and analysis steps in Lynx

knowledge extraction engine and a supporting knowledge base (LynxKB) that combines various classes of information from over 35 public databases and private collections.

Lynx receives user data as genomic variants whose coding and noncoding signals have been previously characterized by some external tool, for example RViewer [67]. Lynx knowledge retrieval engine offers advanced search capabilities and a variety of algorithms for gene enrichment analysis and network-based gene prioritization. Lynx's XML schema-driven annotation service supports extraction of annotations for an individual object (e.g., a gene) or batch queries (e.g., list of genes) from LynxKB. Annotations include among other things associated pathways, diseases, phenotypes, molecular interactions, Gene Ontology categories, and toxicogenomic information displayed according to the user's preferences. All information related to the objects is easily accessible via user interface and available for download in tab-delimited, XML, or JSON formats (Web Services).

Lynx gene enrichment analysis supports Bayes factor and p-value estimates for the identification of functional categories over-represented in the query data sets (*see* B. Xie et al. [68] for more details). Lynx enrichment analysis is based on a large variety of features, such as Gene Ontology terms, toxicogenomic information, and tissues. It also allows the exclusive analysis against the symptoms-level phenotypes and associated non-coding signals from VISTA [69] (e.g., enhancers and clusters of transcription factors binding sites). Lynx also supports context-sensitive enrichment analysis (e.g., against genes expressed on a particular developmental stage or in a particular tissue) that may substantially increase the accuracy of the results.

Additionally, Lynx integrates five network propagation algorithms (simple random walk, heat kernel diffusion, PageRank with priors, HITS with priors, and K-step Markov) as initially developed in the gene prioritization tool PINTA [70]. These algorithms were modified for Lynx to replace continuous gene expression data with binary data from seed genes. This modification accommodates the use of a variety of weighted data types for gene prioritization including ranked gene to phenotype associations, weighted canonical pathways, gene expression, results of sequencing analyses, and others. STRING v9.1 [36] is used as the underlying protein interaction network. Networks-based gene prioritization facilitates prioritization of promising candidate genes from large gene sets or even from the entire genome to provide a preliminary step for network reconstruction. Lynx Service Oriented Architecture provides public access to LynxKB and its analytical tools via user-friendly web services and interfaces.

Since the last release the Lynx workbench has been supplemented with a number of new tools. These include Cheetoh [71], a unique feature-and-network-based gene- prioritization tool and NetLynx (in press), a tool for the reconstruction of co-expression networks.

The current release of LynxKB includes additional information as described in Sulakhe et al. [66]. We have integrated these new datasets within the results of existing analytical tools (e.g., enrichment analysis tool) and the new tools (e.g., Cheetoh algorithm) [71, 72]. Integration of this information also enhances data annotation in Lynx.

Lynx's usage has been increasing steadily with thousands of users each month accessing the platform for annotation and analysis of high-throughput biomedical data.

4.2 Lynx Design and Components

Lynx provides a one-stop solution for generating weighted hypotheses regarding the genes or molecular mechanisms contributing to the phenotypes of interest (Fig. 1). It supports annotations and analyses of the following data: (1) various types of experimental results, such as gene expression, NGS, GWAS, CNV data, etc.; (2) data extracted from LynxKB via search and annotation engines; and (3) lists of genes provided by the user.

Lynx contains the following major components: (1) Lynx annotation engine consisting of Integrated Lynx Knowledge Base (LynxKB) and Knowledge extraction services; (2) Lynx analytical workbench that includes tools for features-based gene enrichment analysis, feature-and-network-based gene prioritization, and reconstruction of co-expression networks; and (3) user-friendly web interface for accessing the annotations and analytical tools.

4.2.1 Updates to Lynx Analytical Workbench

Updates to statistical enrichment analysis. Lynx enrichment analysis allows identification of functional categories over-represented in the query datasets, thus assisting users in formulating hypotheses regarding the molecular mechanisms involved in the phenomena under study. Two singular enrichment analysis algorithms, Bayes factor and *P*-value estimates, are used in our pipeline for this purpose (*see* Xie et al. [68]). Enrichment analysis in Lynx is based on a large variety of features obtained from multiple sources, as well symptoms-level phenotypes and associated noncoding signals as mentioned in our previous publication (1). Several new feature categories, including inter alia Pubmed (UniProt and NCBI GeneRifs), UniProt Keywords, and InterPro Domains, are introduced in the current release to enable literature and protein function-oriented discovery. The results of the Lynx enrichment analysis can now be filtered and utilized by our new prioritization tool, Cheetoh, to perform the feature and network-based gene prioritization.

Updates to Lynx gene prioritization and prediction of molecular mechanisms. Gene prioritization identifies promising candidate genes and sets of genes relevant to molecular mechanisms contributing to a phenotype or a condition of interest extracted from a large set of genes or even from the entire genome. It can also serve as a preliminary step for network reconstruction. In addition to the previously described PINTA network-based gene prioritization [70, 73, 74], Lynx now contains Cheetoh, a network-and-feature-based gene prioritization tool. These prioritization tools perform distinct but complementary analyses suitable for the scientific goals of an investigation, as outlined below.

Cheetoh. A list of genes submitted to the Cheetoh algorithm first undergoes enrichment analysis to identify and score over-represented functional categories. The results of the enrichment analysis are passed to the Cheetoh algorithm as node features. Cheetoh integrates these enrichment analysis results with the underlying network structure as edge features through the Conditional Random Field (CRF) model. It further ranks the genes in the whole genome by global inference scores on the CRF model. Please refer to Xie et al. [71, 72] for a detailed description of the Cheetoh algorithm and its performance evaluation and validation procedures. The output of the tool consists of 1000 top ranked genes ordered by ascending Bonferroni (multiple testing correction) corrected *P*-values based on all user-selected categories as well as rankings and corrected *P*-values from individual category. The results are available both for viewing via the Lynx interactive interface as well as for downloading. The resulting top-ranked genes can be used in both hypothesis and discovery-based approaches to identify a small set of high-confidence candidate genes relevant to user's interests or to explore larger sets of high-ranking genes to identify molecular mechanisms associated with the conditions under investigation. Moreover, the user can increase the resolution of the analyses by choosing particular categories of interest from among a collection of the enrichment analysis categories to enable customized prioritization. For general-purpose gene prioritization, the combination of Gene ontology (Molecular Function/ Biological Process/Cellular Component), phenotype, and pathway categories are recommended. Users are advised to use Cheetoh in cases when (1) pre-existing knowledge is available, such as a list of validated genes or highly differentially expressed (DE) genes, associated with phenotype or condition of interest and (2) the network associated with the input list of genes is sparse or input genes are poorly annotated.

PINTA. In contrast to Cheetoh, Pinta is an unsupervised gene prioritization tool, which propagates the input information in the form of genes and associated scores or gene expression values through the gene–gene interaction networks. It accepts gene lists annotated with experimental values (e.g., gene expression results, differential expression values, scored sets of candidate genes, etc.) that are factored into the analytical procedure.

Users are encouraged to use PINTA when the scoring for the input genes is available, such as reliability scores, differential expression values, and the strength of association to the phenotypes. Since this information propagated through the network can determine whether a gene's neighborhood is functionally related to the input gene set, it could further identify promising candidate genes and sub-networks, even if no knowledge is available about the disease or phenotype under consideration. Please refer to [70, 73, 74] for a detailed description of PINTA, its comparison with the other similar tools, and rigorous validation procedures.

NetLynx. Reconstruction of co-expression networks has proved to be a promising approach to the investigation of system-level properties. Lynx now contains NetLynx, a co-expression-based network prediction tool to rank the interactions between each pair of genes with respect to their gene expression profiles. NetLynx uses a well-established method for modeling gene expression correlations as a multivariate Gaussian distribution with an L1 norm penalty. A comparison of NetLynx with the Pearson-correlation-based and mutual-information-based methods demonstrated its good performance (manuscript in press). NetLynx may be used for the reconstruction of co-expression networks utilizing a user-input threshold to infer the final gene co-expression network. The resulting co-expression networks can be annotated through Lynx annotation resources and then further analyzed by Lynx workbench tools for enrichment analysis and gene prioritization.

Lynx customized workflows. Lynx aims to support various scientific scenarios by offering flexible analytical workflows containing complementary tools. Lynx workflows allow users to explore biological data, accessible via search engine as well as specialized gene pages. Lynx user interface allows easy navigation between Lynx tools as well as external tools, such as RaptorX [75] and VISTA RViewer [67]. This flexibility enables the user to create workflows suitable for his/her research goals. An iterative application of Lynx analytical tools can also help users validate hypotheses or discover new mechanisms hidden in the data.

Data and analytical web services. The integrated data and annotations, as well as the various analytical tools, are presented to users via the web interface. The service-oriented architecture enables other users/groups to leverage our work and integrate it within their own research tools and platforms. Other public systems such as UCSC Genome Browser [76] and RViewer provide external links to Lynx annotation pages. Databases such as DBDB [77] use Lynx RESTful web service interface for annotation of genomic data. End users can download the datasets of interest and results of analysis from the web interface.

5 Conclusions

We present an updated Lynx integrated knowledge base and analytical workbench designed to support discovery and hypothesis-based approaches for the analysis of high-throughput genomic data. Lynx integrates the main downstream analyses, such as gene annotation; gene set enrichment analysis, various algorithms for gene prioritization and network reconstruction within one engine, based on a large knowledge base. Two newly added tools, Cheetoh and NetLynx, further expand our platform's analytical repertoire.

Future developments planned for Lynx include (a) the support of the isoforms-based reconstruction of contextual biological networks; (b) the expansion of the Lynx workbench to allow the identification and characterization of networks modules and integration of additional data types (e.g., epigenetic data) in network-based models of phenotypes of interest.

References

1. Chen J et al (2013) Translational biomedical informatics in the cloud: present and future. Biomed Res Int 2013:658925
2. Payne PR, Embi PJ, Sen CK (2009) Translational informatics: enabling high-throughput research paradigms. Physiol Genomics 39 (3):131–140
3. Ranganathan S et al (2011) Towards big data science in the decade ahead from ten years of InCoB and the 1st ISCB-Asia Joint Conference. BMC Bioinformatics 12(Suppl 13):S1
4. Boyd LB et al (2011) The caBIG® Life Science Business Architecture model. Bioinformatics 27(10):1429–1435
5. Hillman-Jackson, J., et al. (2012) Using Galaxy to perform large-scale interactive data analyses. Curr Protoc Bioinformatics. Chapter 10: p. Unit10.5.
6. Schuler R et al (2012) A flexible, open, decentralized system for digital pathology networks. Stud Health Technol Inform 175:29–38
7. Ideker T, Krogan NJ (2012) Differential network biology. Mol Syst Biol 8:565
8. Koyutürk M (2010) Algorithmic and analytical methods in network biology. Wiley Interdiscip Rev Syst Biol Med 2(3):277–292
9. Bandyopadhyay S et al (2010) Rewiring of genetic networks in response to DNA damage. Science 330(6009):1385–1389
10. Chikina MD et al (2009) Global prediction of tissue-specific gene expression and context-dependent gene networks in *Caenorhabditis elegans*. PLoS Comput Biol 5(6):e1000417
11. Myers CL, Troyanskaya OG (2007) Context-sensitive data integration and prediction of biological networks. Bioinformatics 23 (17):2322–2330
12. Sharan R, Ideker T (2006) Modeling cellular machinery through biological network comparison. Nat Biotechnol 24(4):427–433
13. Takemoto K, Kihara K (2013) Modular organization of cancer signaling networks is associated with patient survivability. Biosystems 113(3):149–154
14. Ideker T, Sharan R (2008) Protein networks in disease. Genome Res 18(4):644–652
15. Kiemer L, Cesareni G (2007) Comparative interactomics: comparing apples and pears? Trends Biotechnol 25(10):448–454
16. Nibbe RK et al (2011) Protein protein interaction networks and subnetworks in the biology of disease. Wiley Interdiscip Rev Syst Biol Med 3(3):357–367
17. Blank MC et al (2011) Multiple developmental programs are altered by loss of Zic1 and Zic4 to cause Dandy-Walker malformation cerebellar pathogenesis. Development 138(6):1207–1216
18. Beltrao P, Ryan C, Krogan NJ (2012) Comparative interaction networks: bridging genotype to phenotype. Adv Exp Med Biol 751:139–156
19. Black DL, Grabowski PJ (2003) Alternative pre-mRNA splicing and neuronal function. Prog Mol Subcell Biol 31:187–216
20. Ellis JD et al (2012) Tissue-specific alternative splicing remodels protein-protein interaction networks. Mol Cell 46(6):884–892
21. Greene CS et al (2015) Understanding multicellular function and disease with human tissue-specific networks. Nat Genet 47 (6):569–576
22. Yap K, Makeyev EV (2013) Regulation of gene expression in mammalian nervous system through alternative pre-mRNA splicing coupled with RNA quality control mechanisms. Mol Cell Neurosci 56:420–428
23. Biamonti G et al (2014) The alternative splicing side of cancer. Semin Cell Dev Biol 32:30–36
24. Kaida D, Schneider-Poetsch T, Yoshida M (2012) Splicing in oncogenesis and tumor suppression. Cancer Sci 103(9):1611–1616
25. Zhang J, Manley JL (2013) Misregulation of pre-mRNA alternative splicing in cancer. Cancer Discov 3(11):1228–1237
26. Wells QS et al (2013) Whole exome sequencing identifies a causal RBM20 mutation in a large pedigree with familial dilated cardiomyopathy. Circ Cardiovasc Genet 6(4):317–326
27. Stallings-Mann M, Radisky D (2007) Matrix metalloproteinase-induced malignancy in mammary epithelial cells. Cells Tissues Organs 185(1–3):104–110
28. Sumithra B, Saxena U, Das AB (2016) Alternative splicing within the Wnt signaling pathway: role in cancer development. Cell Oncol (Dordr) 39(1):1–13
29. Yabas M, Elliott H, Hoyne GF (2016) The role of alternative splicing in the control of immune homeostasis and cellular differentiation. Int J Mol Sci 17(1):3
30. Schaefer MH et al (2013) Adding protein context to the human protein-protein interaction network to reveal meaningful interactions. PLoS Comput Biol 9(1):e1002860
31. Shao H et al (2013) Systematically studying kinase inhibitor induced signaling network

signatures by integrating both therapeutic and side effects. PLoS One 8(12):e80832
32. Cordero F et al (2012) Large disclosing the nature of computational tools for the analysis of next generation sequencing data. Curr Top Med Chem 12(12):1320–1330
33. Hong H et al (2013) Critical role of bioinformatics in translating huge amounts of next-generation sequencing data into personalized medicine. Sci China Life Sci 56(2):110–118
34. Wang S, Xing J (2013) A primer for disease gene prioritization using next-generation sequencing data. Genomics Inform 11 (4):191–199
35. Warde-Farley D et al (2010) The GeneMANIA prediction server: biological network integration for gene prioritization and predicting gene function. Nucleic Acids Res 38(Web Server issue):W214–W220
36. Franceschini A et al (2013) STRING v9.1: protein-protein interaction networks, with increased coverage and integration. Nucleic Acids Res 41(Database issue):D808–D815
37. Szklarczyk D et al (2011) The STRING database in 2011: functional interaction networks of proteins, globally integrated and scored. Nucleic Acids Res 39(Database issue): D561–D568
38. Chen J et al (2009) ToppGene Suite for gene list enrichment analysis and candidate gene prioritization. Nucleic Acids Res 37(Web Server issue):W305–W311
39. Tranchevent LC et al (2008) ENDEAVOUR update: a web resource for gene prioritization in multiple species. Nucleic Acids Res 36(Web Server issue):W377–W384
40. Sifrim A et al (2013) eXtasy: variant prioritization by genomic data fusion. Nat Methods 10 (11):1083–1084
41. Wu J, Li Y, Jiang R (2014) Integrating multiple genomic data to predict disease-causing nonsynonymous single nucleotide variants in exome sequencing studies. PLoS Genet 10(3): e1004237
42. Jäger M et al (2014) Jannovar: a java library for exome annotation. Hum Mutat 35 (5):548–555
43. Li MX et al (2012) A comprehensive framework for prioritizing variants in exome sequencing studies of Mendelian diseases. Nucleic Acids Res 40(7):e53
44. Calabrese C et al (2014) MToolBox: a highly automated pipeline for heteroplasmy annotation and prioritization analysis of human mitochondrial variants in high-throughput sequencing. Bioinformatics 30(21):3115–3117
45. Yao J et al (2014) FamAnn: an automated variant annotation pipeline to facilitate target discovery for family-based sequencing studies. Bioinformatics 30(8):1175–1176
46. Li X, Montgomery SB (2013) Detection and impact of rare regulatory variants in human disease. Front Genet 4:67
47. Matthews LR et al (2001) Identification of potential interaction networks using sequence-based searches for conserved protein-protein interactions or "interologs". Genome Res 11(12):2120–2126
48. Yu H et al (2004) Annotation transfer between genomes: protein-protein interologs and protein-DNA regulogs. Genome Res 14 (6):1107–1118
49. Mewes HW et al (2011) MIPS: curated databases and comprehensive secondary data resources in 2010. Nucleic Acids Res 39(Database issue):D220–D224
50. St Onge RP et al (2007) Systematic pathway analysis using high-resolution fitness profiling of combinatorial gene deletions. Nat Genet 39 (2):199–206
51. Bakal C et al (2008) Phosphorylation networks regulating JNK activity in diverse genetic backgrounds. Science 322(5900):453–456
52. Lage K et al (2008) A large-scale analysis of tissue-specific pathology and gene expression of human disease genes and complexes. Proc Natl Acad Sci U S A 105(52):20870–20875
53. Zuberi K et al (2013) GeneMANIA prediction server 2013 update. Nucleic Acids Res 41(Web Server issue):W115–W122
54. Kamburov A et al (2013) The ConsensusPathDB interaction database: 2013 update. Nucleic Acids Res 41(Database issue): D793–D800
55. Niu Y, Otasek D, Jurisica I (2010) Evaluation of linguistic features useful in extraction of interactions from PubMed; application to annotating known, high-throughput and predicted interactions in I2D. Bioinformatics 26 (1):111–119
56. Hu Z et al (2013) VisANT 4.0: Integrative network platform to connect genes, drugs, diseases and therapies. Nucleic Acids Res 41(Web Server issue):W225–W231
57. Elefsinioti A et al (2011) Large-scale de novo prediction of physical protein-protein association. Mol Cell Proteomics 10(11): M111–010629
58. Patil A, Nakai K, Nakamura H (2011) HitPredict: a database of quality assessed protein-protein interactions in nine species. Nucleic Acids Res 39(Database issue):D744–D749

59. Balaji S et al (2012) IMID: integrated molecular interaction database. Bioinformatics 28 (5):747–749
60. Wong AK et al (2012) IMP: a multi-species functional genomics portal for integration, visualization and prediction of protein functions and networks. Nucleic Acids Res 40 (Web Server issue):W484–W490
61. Tamames J, de Lorenzo V (2010) EnvMine: a text-mining system for the automatic extraction of contextual information. BMC Bioinformatics 11:294
62. Gerner M et al (2012) BioContext: an integrated text mining system for large-scale extraction and contextualization of biomolecular events. Bioinformatics 28(16):2154–2161
63. Kahn AB et al (2007) SpliceMiner: a high-throughput database implementation of the NCBI Evidence Viewer for microarray splice variant analysis. BMC Bioinformatics 8:75
64. Thanaraj TA et al (2004) ASD: the Alternative Splicing Database. Nucleic Acids Res 32(Database issue):D64–D69
65. Latham KE (2006) The Primate Embryo Gene Expression Resource in embryology and stem cell biology. Reprod Fertil Dev 18(8):807–810
66. Sulakhe D et al (2016) Lynx: a knowledge base and an analytical workbench for integrative medicine. Nucleic Acids Res 44(D1): D882–D887
67. Lukashin I et al (2011) VISTA Region Viewer (RViewer)–a computational system for prioritizing genomic intervals for biomedical studies. Bioinformatics 27(18):2595–2597
68. Xie B, et al (2012) Prediction of candidate genes for neuropsychiatric disorders using feature-based enrichment. Proceedings of the ACM conference on bioinformatics, computational biology and biomedicine, Association for Computing Machinery, pp 564–566
69. Frazer KA et al (2004) VISTA: computational tools for comparative genomics. Nucleic Acids Res 32(Web Server issue):W273–W279
70. Nitsch D et al (2011) PINTA: a web server for network-based gene prioritization from expression data. Nucleic Acids Res 39(Web Server issue):W334–W338
71. Xie B et al (2015) Disease gene prioritization using network and feature. J Comput Biol 22 (4):313–323
72. Xie B, et al (2013) Conditional random field for candidate gene prioritization. Proceedings of the international conference on bioinformatics, computational biology and biomedical informatics, Association for Computing Machinery, p 700
73. Dubchak I et al (2014) An integrative computational approach for prioritization of genomic variants. PLoS One 9(12):e114903
74. Nitsch D et al (2010) Candidate gene prioritization by network analysis of differential expression using machine learning approaches. BMC Bioinformatics 11:460
75. Källberg M et al (2012) Template-based protein structure modeling using the RaptorX web server. Nat Protoc 7(8):1511–1522
76. Rosenbloom KR et al (2015) The UCSC genome browser database: 2015 update. Nucleic Acids Res 43(Database issue): D670–D681
77. Mirzaa GM et al (2014) The developmental brain disorders database (DBDB): a curated neurogenetics knowledge base with clinical and research applications. Am J Med Genet A 164A(6):1503–1511

Chapter 6

Functional Analysis of OMICs Data and Small Molecule Compounds in an Integrated "Knowledge-Based" Platform

Alexey Dubovenko, Yuri Nikolsky, Eugene Rakhmatulin, and Tatiana Nikolskaya

Abstract

Analysis of NGS and other sequencing data, gene variants, gene expression, proteomics, and other high-throughput (OMICs) data is challenging because of its biological complexity and high level of technical and biological noise. One way to deal with both problems is to perform analysis with a high fidelity annotated knowledgebase of protein interactions, pathways, and functional ontologies. This knowledgebase has to be structured in a computer-readable format and must include software tools for managing experimental data, analysis, and reporting. Here, we present MetaCore™ and Key Pathway Advisor (KPA), an integrated platform for functional data analysis. On the content side, MetaCore and KPA encompass a comprehensive database of molecular interactions of different types, pathways, network models, and ten functional ontologies covering human, mouse, and rat genes. The analytical toolkit includes tools for gene/protein list enrichment analysis, statistical "interactome" tool for the identification of over- and under-connected proteins in the dataset, and a biological network analysis module made up of network generation algorithms and filters. The suite also features Advanced Search, an application for combinatorial search of the database content, as well as a Java-based tool called Pathway Map Creator for drawing and editing custom pathway maps. Applications of MetaCore and KPA include molecular mode of action of disease research, identification of potential biomarkers and drug targets, pathway hypothesis generation, analysis of biological effects for novel small molecule compounds and clinical applications (analysis of large cohorts of patients, and translational and personalized medicine).

Key words Pathway analysis, Functional analysis, Systems biology, Signaling and metabolic networks, Biological networks, "Knowledge-based" platform, Interactome, Causal reasoning

1 Introduction

Steady introduction of high-throughput methods in experimental biology since the late 1990s created a need for novel techniques for data analysis. First, the sheer volume of data points in a single OMICs assay was non-comprehendible for the biologists. Indeed, how could one connect thousands to tens of thousands of differentially expressed genes at once in a biologically meaningful way? A reductionist discipline, modern experimental biology, typically

Tatiana V. Tatarinova and Yuri Nikolsky (eds.), *Biological Networks and Pathway Analysis*, Methods in Molecular Biology, vol. 1613, DOI 10.1007/978-1-4939-7027-8_6, deals with much smaller systems of several proteins, probably linked into a complex or a one—two pathways system. Second, OMICs data is notoriously "noisy." By different estimates, 50–70% of all microarray expression or yeast-two-hybrid interactions are false-positives and false-negatives [1]. One needs tools that adhere to some sort of "gold standard" to access this noise. Third, different types of OMICs data cannot be compared directly, as they have very little overlap [2]. In order to address these issues, systems biology must be further developed, evolving in parallel with OMICs technologies. A comprehensive analytical system should combine a large knowledge base of experimental literature as well as a toolkit for OMICs data management and analysis.

Recent advances in next-generation sequencing (NGS) brought several new powerful and more precise methods to capture full-scale genomic, transcriptomic, and epigenomic molecular changes that are no more tied to microarray physical size and capacity. DNA sequencing might give all possible gene variants and identify germ line as well as somatic mutations on both gene-coding regions (whole exome sequencing) as well as on gene regulatory regions in intergenic areas (whole genome sequencing). RNA sequencing allows identifying specific alternative transcript isoforms and gene fusions expression. Chromatin immuneprecipitation (ChIP) sequencing allows identifying genome regions where proteins like transcription factors bind with DNA. Bisulfite sequencing shows DNA methylation events occurred on whole genome.

Described technologies allow scientific groups to produce big datasets in a rapid and relatively cheap way that might form a multifactor disease profile for each patient. Oncology is the first focus area for the new data generation paradigm. The Cancer Genome Atlas (TCGA) was developed by NIH and designed as a publically available collection of multi-OMICs datasets for 33 cancers with thousands patients in each cohort (https://tcga-data.nci.nih.gov). For almost all patients, DNA-seq, RNA-seq, CNV, microarray expression, DNA methylation, and clinical meta-data are stored. TCGA research network published multiple studies of this dataset, the most recent are [3–6], however number of studies published by other scientific groups is much bigger (PubMed search shows several hundreds).

The key assumption beyond knowledge-based data analysis is that high-throughput data can only be de-convoluted within the framework of an underlying biology, i.e., accumulated knowledge of biologically relevant interactions between molecular entities (genes, proteins, and compounds). In all living organisms, such interactions are grouped into higher level structures such as processes, pathways, mechanistic signaling, and metabolic networks, as well as genetic "causative" networks interconnecting disease biomarkers, relations within protein complexes, and groups of complexes. All this information has to be collected from the original sources in computer-readable form and assembled into a semantically consistent knowledge database. Due to the overwhelming nature of complex biological systems and the fact that peer-review articles are not standardized, it is generally accepted that manual expert curation is necessary for the task. It requires a large and well-trained annotation team working for several years and an advanced annotation technology for collecting such a comprehensive knowledge base.

Over the last several years, many knowledge-based methods of analysis of OMICs data were developed, which can be divided into three main categories: pathway analysis [7, 8], biological interactions networks [9–11], interactome analysis [12–17]. Classic Enrichment analysis is a "low resolution" tool that consists of dividing the gene/protein list of interest into "entities" of a functional ontology such as cellular processes, disease, or toxicity categories and ranking these entities based on relative saturation with the genes/proteins from the list. Enrichment analysis of pathways might take into account network properties like protein and genetic interactions between molecules and identify if experimentally derived gene properties (like abundance of gene product) are concordant with signal transduction effect, e.g., activation or inhibition interactions and properties of pathway itself [18]. Existing knowledge about pathways and cellular processes is not universal, so the number of proteins contained in pathway databases differ and do not cover the whole genome, just overlaying DEGs on pathways. Therefore, content from one database alone may not be sufficient to ensure complete disease mechanism understanding (*see* Table 1 for summary).

Table 1
Summary of pathway database gene and molecular interactions coverage

Database	Genes	Interactions
MetaCore™	7317	30,186
KEGG	7086	N/A
Reactome	9622	9865
NCI PID	2626	14,000
WikiPathways	9584	9758
NetPath	1053	11,446

Both numbers show data available only through pathway maps analysis. MetaCore also contains 1,700,000 interactions not visualized on pathway but accessible through network analysis algorithms which extend our pathway knowledge and identify putative signal transduction ways and cross-talks

Large amount of published molecular interactions (including protein, RNA, gene regulation by transcriptional factors, and more) which extends our knowledge about possible molecular processes in the cell and though allow going beyond pathway diagram analysis. Such whole set of molecular interactions made network biology analysis very popular field. Biological networks link genes and proteins into interconnected structures of nodes and edges (networks) of different topology, using the uploaded genes/proteins as "seed" nodes and the interaction content from the knowledge base as edges [19–21]. Networks can then be visualized in different environments, such as the popular tool Cytoscape [22]. Integrated platforms such as MetaCore and Key Pathway Advisor have internal network generation and visualization tools. Networks provide the highest resolution of analysis at the level of individual proteins (isoforms in some cases) and individual protein interactions. Modern network tools calculate and visualize the most relevant sub-networks using different algorithms and filters. Finally, network properties might be used for node prioritization procedures that calculate the general interconnectivity within the uploaded gene/protein list (Interactometopology), such as node distribution, number of interactions per node (degree), the average length of the path between the nodes, etc.

Advances in NGS and overwhelming data flow for molecular alterations of different nature caused development of new approaches that utilize or combine several discussed above and applied to data integration paradigm. For such tasks when gene variant, gene expression, and other alteration data need to be analyzed simultaneously to identify the most promising driver genes algorithmic rules should be modified to account biologically diverse nature of data spots [23–25].

The methods of functional analysis are realized in several dozen public domain and commercial programs. The vast majority of these tools are specialized, designed for bioinformaticians and require programming skills to be used effectively. Only a few systems managed to adapt knowledge-based functional analysis for a broader audience of end users, biologists, and chemists. These include the later versions of commercial integrated platforms such as Pathway Studio (Elsevier B.V., https://www.elsevier.com/solutions/pathway-studio-biological-research), Ingenuity Pathway Analysis (Qiagen, www.ingenuity.com), and MetaCore, including Key Pathway Advisor (Clarivate Analytics, http://clarivate.com/life-sciences/discovery-and-preclinical-research/metacore/). These commercial integrated suites feature intuitive GUI's for non-programmers, parsers for uploading OMICs data and gene lists, large proprietary databases of interactions and other annotated knowledge, advanced search, pathway editing, and reporting capabilities. All three suites are well integrated with third-party tools in bio- and chemoinformatics and evolve rapidly in a highly competitive environment. Here, we present our Systems Biology suite of apps, focusing on non-interactions content and tools. The protein interactions database MetaCore and Key Pathway Advisor works on top of manually curated set of molecular interaction that captures interactions effect, the biochemical mechanism (binding, catalysis, transcriptional regulation and 20 more) from corresponding research articles.

2 Materials

2.1 Overview

MetaCore consists of the knowledge base, dubbed MetaBase, and six analytical modules:

- MetaCore—a main platform for the analysis of OMICs and other biological experimental data. MetaCore includes tools for gene list enrichment analysis, multi-experiment comparison, interactome analysis, and biological networks (algorithms and filters).
- Genomic Analysis Toolkit (GAT)—a MetaCore module for the analysis of Next Generation Sequencing data. GAT includes tools for patient cohorts' comparison, gene variant data annotation, and filtering.
- Key Pathway Advisor (KPA)—a standalone comprehensive one-clickworkflow that combines ease-of-use interface, modern causal reasoning interactome analysis, and enrichment analysis approaches and biomarker and drug target identification capabilities. The application is designed specifically for molecular biologists with enhanced visualization of network biology for more intuitive analysis.
- MetaDrug—a "systems pharmacology" MetaCore module designed for the analysis of medicinal chemistry data (structures and assays). MetaDrug predicts biological effects of novel drug-like compounds, including indications, side effects, and human toxicity.
- Pathway Map Creator—a standalone Java-editing application coupled with MetaBase and MetaCore. Pathway Map Creator enables generation of custom pathway maps from scratch, editing of standard maps from the MetaCore collection, and conversion of networks into pathway view.
- Advanced Search—a Java application for combinatorial Boolean search of the MetaBase content. It is a companion app for the MetaCore.

2.2 Content (Knowledge Base)

Annotated content of MetaCore and KPA consists of two domains: (1) binary molecular interactions and gene-disease associations and (2) higher level, multi-protein structures such as pathways,

pathways maps, and process networks. Both domains are inter-linked into an Oracle database with 107 tables, with entities linked by semantically consistent ontologies.

2.2.1 Protein Interactions and Gene-Disease Associations

The MetaCore knowledge base includes over 1,720,000 molecular interactions and 200,000 gene-disease associations. The method is described in detail in its chapter in the Interactions section of this book.

2.2.2 Pathways and Functional Ontologies

MetaCore features 12 different functional ontologies used for gene list enrichment analysis, by network algorithms and prioritization of experimental data.

- Signaling pathways. These are linear multi-step chains of consecutive interactions, typically consisting of a ligand-receptor interaction, an intra-cellular signal transduction cascade between receptor (R) and transcription factor (TF), and, finally, TF–target gene interaction. Signaling pathways are mainly used by network generation algorithms and only visualized on networks.
- Metabolic pathways. These are multi-step chains of metabolic reactions, linked into functionally self-sufficient linear chains and cycles. Fragments of metabolic pathways are shown as static images reachable from the protein pages. Metabolic pathways are also used for network generation and visualized on the networks.
- Canonical pathways maps. Maps are the main level of pathway visualization in MetaCore and KPA. Maps represent interactive images drawn in Java-based Pathway Map Creator and typically contain three to six pathways. There are over 1.600 maps in MetaCore, comprehensively covering human signaling and metabolism, certain diseases, and some drug targets mechanisms. Pathway maps are primarily used as an ontology for enrichment analysis.
- Canonical pathway maps folders. All canonical maps are assembled into a hierarchical tree folder structure. The folders structure can be visualized in a Browser mode and from enrichment analysis distributions.
- Process network models. This ontology represents reconstruction of main signaling and metabolic processes in the cell, such as a "cell cycle checkpoints" or "innate immune response." The manually built process networks typically have over 100 nodes (proteins) belonging to certain normal cellular processes. The edges are selected from MetaBase content.
- GO processes. These are a GUI-supported representation of the Gene Ontology (GO) collection of cellular processes, which comes with GO tree structure and access to proteins and interactions within a process. This ontology is updated

with GO standard updates. GO processes are mostly used in enrichment analysis and for prioritization of genes on the built networks.

- GO molecular functions. A GUI-supportedontology of standard protein functions from GO. It is mostly used in enrichment analysis.
- Disease biomarkers. These are a collection of genes genetically linked to over 500 diseases and conditions, supported by the hierarchical disease tree and GUI for gene retrieval. Disease biomarkers are mostly used in enrichment analysis.
- Disease network models. GeneGo reconstruction of disease mechanisms in a form of manually built networks. These are mechanistic networks linking the disease-associated genes via physical and functional protein interactions.
- Toxicity networks. GeneGo reconstruction of toxicity mechanisms in a form of manually built networks. These are mechanistic networks linking genes associated with a particular toxicity endpoint via physical and functional protein interactions.

3 Methods

MetaCore is used in three main modes: Browser, Combinatorial search, and for Analysis and Editing. KPA is designed for gene expression data analysis.

3.1 Browser

The content of functional ontologies, gene, protein, and compounds annotations can be accessed from multiple pages in MetaCore's different applications. The main content browsing tab menu includes Canonical Pathway Maps as a separate entry, as well as Process Networks and Disease Networks. Process Networks and Disease Networks can be opened up from enrichment analysis distribution or called from the main menu. Annotations for genes and proteins are available by either clicking on an object on maps or networks, or found by search genes/proteins. The gene/protein pages contain links to outside databases such as Swissprot, EntezGene, etc., has information on protein isoforms, gene variants, as well as information on SNPs and mutations, etc. The Compound page is common for all exobiotics and endogenous metabolites in the database, and includes pharmacological information such as prime and secondary indication, toxicity, drug-drug interactions, drug-target interactions, etc.

3.2 Advanced Search

This is a Java application for combinatorial (Boolean) search within MetaCore platform. This advanced search allows for retrieval of specific information from the knowledge database by generation of complex queries via a simple, user-friendly interface. This is all done without using the tools, rather it uses the embodied controlled vocabulary of terms and ontology trees. The user can combine attributes of the objects in the database such as "protein class," "molecular function," "subcellular localization," as well as functional ontologies such as "GO process," Canonical Pathway Map," "Disease Networks," etc. into a query. A typical query results in a list of genes, proteins, or compounds. The lists can be saved internally to the Data Manager for further research such as networks analysis, or it can be exported as Excel file. Some examples of queries in Advanced Search are listed below:

Find all **drugs** for **kinases** involved in **apoptosis** in **breast cancer.**

Find all **kinases** implicated in **breast cancer.**

Find all **biomarkers** which are **kinases.**

Find **genes** for **colorectal cancer** which are **not** involved in **breast cancer.**

Find all **ligandsorreceptors** in **inflammatory response.**

Find **genes** for **fatty acid metabolism** which are **nothuman.**

3.3 Data Upload and Analysis

MetaCore is designed for the analysis of a large variety of gene/protein/compound lists, small molecules structures, and "high-throughput" experimental data often collectively referred to as "OMICs data." The data types include microarray and SAGE "genome-wide" gene expression, SNP arrays genotyping data, DNA sequencing data (methylation, gene copy number, somatic mutations, and SNPs), proteomics, and metabolomics (both NMR and MS data). On the chemistry side, MetaCore handles SMILES strings, Brutto formulas, molecular weights, and structures. Experimental datasets, as well as gene, protein, and compound lists, are analyzed in a similar way, by matching gene/protein/compound IDs from the datasets with the internal MetaBase IDs. These IDs are then used as seed nodes for network and interactome analysis and as a gene/protein list for enrichment analysis. Experimental numerical data such as level of gene expression on a microarray, protein abundance measured by NMR, or a metabolite concentration in body liquids are visualized as histograms or as a solid circle of gradient intensity on pathway maps and networks accordingly (Fig. 1a, b). The list of objects with matching numerical data can be exported from maps, networks, and ontology entries (Fig. 1c).

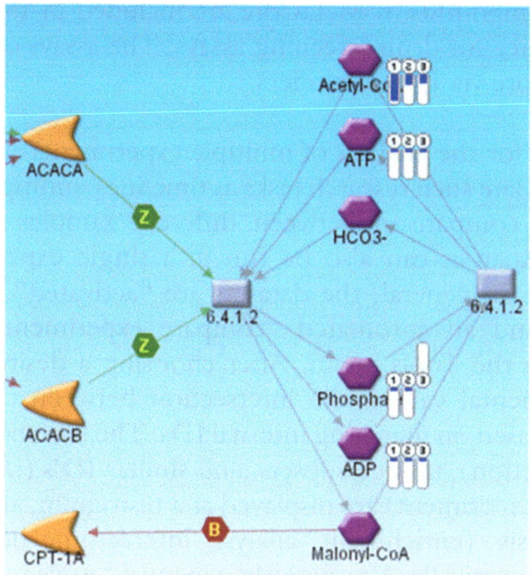

A

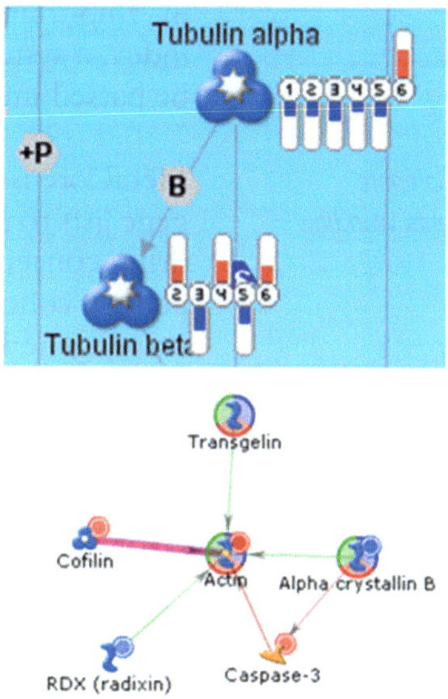

B

#	Gene Symbol	Protein	Class	Protein name	Signal	P_value	Signal	P_value
1	ACTC1	ACTC_HUMAN		Actin, alpha cardiac muscle 1	0.484	0		
2	CA6	CAH6_HUMAN		Carbonic anhydrase 6 precursor				
3	CASP3	CASP3_HUMAN		Caspase-3 precursor	0.333	0		
4	CFL1	COF1_HUMAN		Cofilin-1	0.325	0		
5	CLU	CLUS_HUMAN		Clusterin precursor	0.414	0	0.372	0
6	CRYAB	CRYAB_HUMAN		Alpha-crystallin B chain	-0.525	0		
7	CTSB	CATB_HUMAN		Cathepsin B precursor				
8	FABP4	FABPA_HUMAN		Fatty acid-binding protein, adipocyte				
9	FABP5	FABPE_HUMAN		Fatty acid-binding protein, epidermal	-0.405	0		
10	FABP7	FABPB_HUMAN		Fatty acid-binding protein, brain	-0.49	0	-0.81	0
11	KRT15	K1C15_HUMAN		Keratin, type I cytoskeletal 15	-0.646	0	0.5	0

C

Fig. 1 Mapping numerical data on pathway maps and networks. (**a**) Metabolic concentrations in blood of atherosclerosis mice on a map. (**b**) Microarray gene expression data superimposed on the network nodes (invasive breast cancer human data). (**c**) Data export file with expression data from the network

3.3.1 Data Parsers

Most of experimental data, such as gene expression, are uploaded by a universal parser that recognizes most common systems of gene/protein/compound IDs. The majority of commercial microarrays including Illumina, Affymetrix, ABI, and GE Healthcare for human, mouse, rat, dog, bovine, and chimpanzee are recognized directly. Gene variant parser allows uploading VCF files or gene variant lists that have to contain chromosome position and reference/alternative alleles change for each gene variant (SNP, MVP, deletion, insertion). The metabolic parser is designed for uploading endogenous small molecule compounds and recognizes AC numbers, SMILES strings, molecular weights, and KEGG IDs. Xenobiotic compounds are uploaded with the help of the integrated Accord module (Accelrys) in a form of SDF and MOL files. The chemical structures can also be drawn using the ChemDraw plug-in. Importantly, all compounds in MetaCore are included in ISIS index, a system of choice for drug screening assays. The assays can be parsed into MetaCore via ISIS identifiers.

3.3.2 Compare Experiments Workflow

MetaCore is designed for the analysis of multiple experiments and gene lists to accommodate such research tasks as timelines, multiple drug concentrations, comparison between different samples in patient cohort, etc. Analysis can also be run in a single experiment/gene list mode. In general, the datasets are "activated" in the Data Manager, and an automated "compare experiments" workflow is chosen in the Tools menu. After choosing a desired threshold for experimental values, the intersection between the datasets is calculated based on matching internal IDs. The common subset of IDs (intersection), unique subsets, and similar ID's (i.e., present in all but one experiment) are displayed as a histogram, and the three-step analysis (enrichment analysis–interactome–networks) is then run automatically. Alternatively, a manual "Advanced biomarkers" feature could be applied for a large scope of logical operations between the datasets (nonredundant union, "either or" operation, subtraction of different types).

3.3.3 Standard Data Analysis Overview

The uploaded experiments or gene/protein/compound lists are subjected to several stages of systems biology analysis:

- *The experimental set(s) are custom filtered* according to the user's needs. Filters include gene expression in human tissues and cellular organelles, matching with orthologs in ten organisms, specific cellular processes, etc. In addition, the uploaded gene lists can be normalized against microarray content or a custom dataset.
- *Enrichment analysis (EA)* in multiple functional ontologies. EA is a "classical" tool that shows relative prevalence of genes from certain cellular processes, pathways, diseases, etc. in the uploaded dataset(s).
- The *interactomeanalysis* feature calculates relative connectivity (number of interactions) of individual proteins/genes within the set compared to the whole database. Proteins are divided by protein classes such as transcription factors, receptors, ligands (secreted proteins), kinases, phosphatases, proteases, and endogenous metabolic enzymes. Connectivity can be calculated for individual datasets and between the datasets.
- *Network analysis.* Genes/proteins in the dataset(s) can be connected to each other via protein interactions, forming signaling and metabolic networks. The network topology and composition vary depending on chosen algorithms, filters, and purpose of analysis. Networks provide the highest resolution among functional analysis tools.

- *Key Pathway Advisor (KPA)* works with several levels at once providing data filtering; interactome analysis by reverse causal reasoning algorithm that reconstructs gene expression regulatory pathways on network and though identifies Key Hubs (regulator molecules with predicted activation or inhibition status); synergy enrichment analysis for experimental data and found Key Hubs. For the list of experimental genes and Key Hubs search for known biomarkers and drugs designed to treat a disease that user selects.
- *Genomic Analysis Toolkit (GAT)* provides capabilities to identify potential driver variants from the initial VCF files prior to systems biology analysis. It provides wide capabilities to compare data from different cohorts of patients, family trio analysis and perform comprehensive annotation and filtering. Each gene variant is mapped on the genome to identify its type, class, gene region, previously reported biomarkers, external gene variant data bases (dbSNP, 1000 genomes, ESP, dbNSFP), disease, pathways, etc. More than 40 fields with different information are associated with each gene variant providing rich capabilities for filtering.
- The capabilities of the *MetaDrug* module also include QSAR models that can be used for prediction of toxicity, activity and physio-chemical properties of novel compounds, and prediction of human metabolites for heterocyclic compounds of differing structure. Now, we will consider in more detail the basic analysis steps.

3.3.4 Dataset Filters

The content of any uploaded dataset can be focused depending on the purpose of the analysis and the experimental conditions. Data filters in MetaCore include tissue specificity, presence in body liquids, specific cellular processes, diseases, and pathways. The data can be normalized against standard gene lists as well as the content of the main types of microarrays used in gene expression experiments. Gene variant filters provide following additional filters functional class (missense, nonsense, UTRs, splice-sites, etc.), Functional prediction scores for different algorithms (SIFT, Mutation Tester, MutationAssessor, etc.), evolutionary conservation scores (GERP, PhyloP, etc.), by presence in public gene variant data bases and population frequencies (dbSNP, 1000 genomes, ESP), presence in Thomson Reuters Gene Variant Data Base (a rich collection of gene variant biomarkersassociated with diseases and treatment responses).

3.3.5 Dataset/Gene List Enrichment Analysis (EA)

In MetaCore, the EA module calculates the probability of a random intersection between the uploaded dataset and an ontology's subfolder (say Cell Cycle) based on a hypergeometric distribution. The p-value essentially represents the probability of a particular mapping arising by chance, given the numbers of genes in the set of all genes on maps/networks/processes, genes on a particular map/network/process, and genes in your experiment. The negative natural logarithm of the *p*-value is displayed so that a larger bar represents a higher significance. The histogram is automatically sorted by selecting an option in the "Sorting method" drop-down menu. The "Statistically significant" option sorts the histogram by the maximum −ln(*p*-value), the "Differentially affected" option sorts the histogram by the standard deviation of the −ln(*p*-value) among the experiments, and the "Similarity" option sorts the histogram by the standard deviation of the −ln(*p*-value) divided by the mean of the −log(*p*-value). The False Discovery Rate (FDR) correction procedure is standard. FDR threshold can be custom changed or switched off. There are two important issues in the EA calculation:

- Functional ontologies. EA analysis is only as informative as the ontology behind it. Using only one ontology (for instance, GO molecular functions) provides a rather insufficient overview of large datasets. For instance, GO processes help little in the evaluation of a toxicogenomics expression dataset, for which a specialized ontology of toxic categories and pathological processes is needed. In MetaCore, 12 different functional ontologies (*see* Subheading 2) are used for the comprehensive EA overview.
- Standard datasets and normalization. EA calculates relative enrichment of a dataset on a background of a larger database of IDs the set of interest is part of. For instance, a subset of genes differentially expressed in breast cancer has to be "normalized" to the gene ID content of the microarray it was generated on. The subset of gene IDs is also part of a larger database of IDs it has to be normalized against. In MetaCore, normalization is calculated against three levels of the database standard arrays and custom defined standard sets.

3.3.6 Interactome Analysis Module

Interactome calculates relative connectivity between the proteins in the uploaded dataset/list of interest (local interactome) compared to general connectivity within the interactions database (global interactome). The module's procedures evaluate general topological parameters of the local interactome and identify interactome neighborhoods around individual proteins divided into protein classes: transcription factors, receptors, ligands, kinases, phosphatases, proteases, and enzymes of endogenous metabolism.

Evaluation of Interactome Topology

This procedure calculates the main properties of the local interactome defined as the compilation of all interactions between the genes/proteins within the uploaded list/experiment. The topological parameters include

- *Degree of nodes.* The number of links (interactions) connected to a node (protein) gives the node's degree. Since our network is directed, the nodes are characterized by in and out-degree, giving the number of outgoing and incoming interactions.
- *Average shortest path.* The shortest distance between two nodes is the number of links (interactions) along the shortest path(s). The average shortest path is the average over the shortest paths for all node pairs in the network. When we calculate the shortest paths for a subset of nodes (the set of proteins for colon and breast cancer) in the global network, we also consider paths crossing through nodes that are not part of the subset.
- *Average clusteringcoefficient.* The clustering coefficient captures to what degree node's neighbors are connected. It is defined as $C_i = \frac{2n_i}{k_i(k_i-1)}$, where n_i is the number of links among the k_i neighbors of node i. As $k_i(k_i-1)/2$ is the maximum number of such links, the clustering coefficient is a number between 0 and 1. The average clustering coefficient is obtained by averaging over the clustering coefficient of individual nodes. A network with a high clustering coefficient is characterized by highly connected sub-graphs. Statistical significance of network parameters can be evaluated by p-values (*see* **Note 1**).

Evaluation of Significantly Over (Under)-Connected Proteins in the Gene/Protein List of Interest

It is widely accepted and shown in multiple studies that proteins that are more critical in a given dataset (for instance, drug targets, disease-related proteins, etc.) have more connections within the dataset than expected on random. In MetaCore, we realized this observation in a statistical tool that evaluates relative connectivity of proteins of different types. The interactions between proteins within a set of data are retrieved from the database and compared with the number of connections in the global interactome. The goal of this analysis is to identify proteins with statistically significant large and small numbers of interactions within the dataset of interest, between any two datasets and between the dataset and all the proteins in the database (Fig. 2). Statistical significance is assigned by using the cumulative hypergeometric distribution as follows: $p(k) - \sum_{i=k}^{D} P(i, D, n, N)$, where

$$P(k, D, n, N) = \frac{\binom{D}{k}\binom{N-D}{n-k}}{\binom{N}{n}}$$

N—the number of proteins (protein-based network objects) in our

global interactome extracted from MetaCore.

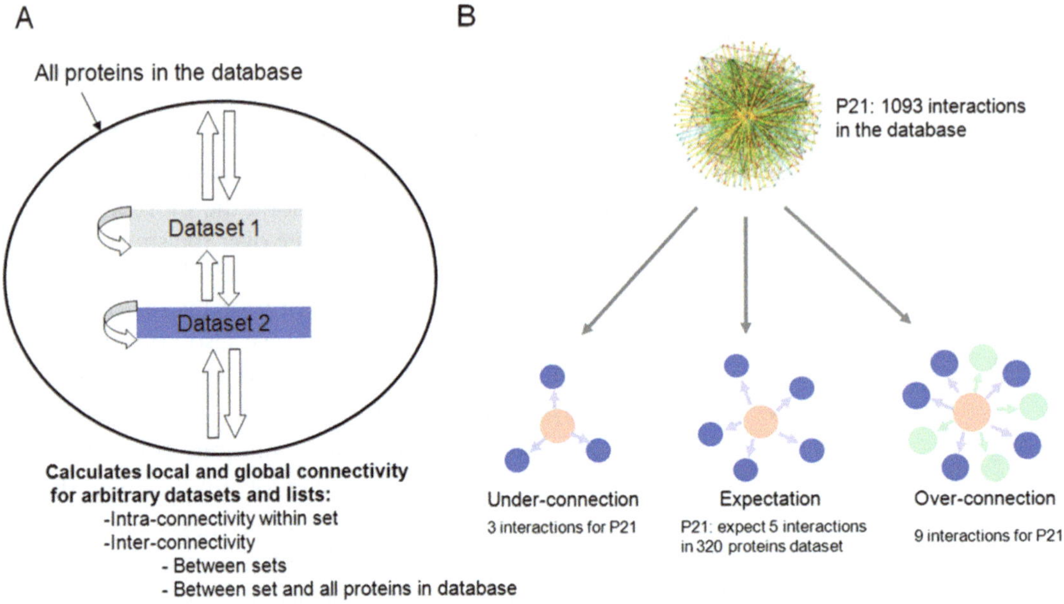

Fig. 2 Interactome analysis of OMICs datasets and gene lists. (**a**) The general schema of interactions inside the set, between the sets, and between the set and "global interactome." (**b**) The "over" and "uner"-connectivity phenomenon. The hub (P21 protein from MetaCore database, marked pink) is expected to be linked with five other proteins in the hypothetical dataset of 320 genes (*purple circles*), but in reality it can be linked with nine genes (*purple* and *green circles*), or three genes (*purple circles*). In these cases, it will be considered "over" connected or "under" connected

n—number of proteins derived from the sets of genes of interest.

D—the degree of a given protein in the global interactome database.

k—the degree of a given protein within the set of interest.

The *p*-value calculated above gives the probability of observing *k* or more interactions of a given protein (with degree *D* in global network) by random chance within the set of interest (of size (*n*)).

The probability of observing under-connected proteins can be calculated by $1-p(k)$.

The input lists of genes were converted to protein-based network objects which have been used in our analysis. The resulting network objects sets were divided into subsets based on their molecular function (receptors, ligands, etc.).

3.3.7 Network Analysis Tools

In MetaCore, networks are built using proteins, genes, and compounds (network objects) from a user's list as seed nodes and MetaBase as the source of interactions as links between them. As the seed lists are different, the networks are unique for the uploaded datasets and chosen conditions. The same dataset can be networked in different ways, depending on chosen network parameters. The network toolbox features network algorithms and filters enabling generation of networks specific for cellular processes, species, orthologs, cellular processes, expression in human tissues, mechanisms of interactions, and effects.

Network Generation Algorithms

Network algorithms use the uploaded network objects (converted from gene and protein IDs) as seed nodes, link them together by pulling interactions from the database, and display the built networks on the screen. There are seven basic algorithms and two additional variants in MetaCore.

- Direct Interactions (DI) identifies the islands of nodes corresponding to genes/proteins from the user's list directly connected to each other. Each connection represents a direct, experimentally confirmed, physical interaction between the objects.
- Shortest Paths (SP) algorithm connects the chosen objects by the shortest network paths (smallest possible number of directed one-step interactions) using standard Dijkstra's shortest paths algorithm. A user can constrain the paths' length by using the pull-down menu under Options.
- Analyze Network (AN) algorithm starts with building a super network by applying a simplified version of the "Auto Expand" algorithm to the initial list of objects. The network, which is never visualized, connects all objects from the input list with all other objects. In the next step, this large network is "divided" into smaller fragments of chosen size, from 2 to 100 nodes. This is done in a cyclical manner, i.e., fragments are created sequentially one by one. Edges used in a fragment are never reused in subsequent fragments. Nodes may be reused, but with different edges leading to them in different fragments. The end result of the AN algorithm is a list of overlapping multiple networks (usually ~30), which can be prioritized based on five parameters: the number of nodes from the input list among all nodes on the network, the number of canonical pathways on the network, and three statistical parameters: *p*-value, *z*-score, and *g*-score (*see* **Note 2**).
- Analyze network (Transcription Factors—TFs) and Analyze network (Receptors). Both algorithms start with creating two lists of objects expanded from the initial list: the list of transcription factors and the list of receptors. Next, the algorithm calculates the shortest paths from the receptors to TFs. Then, the shortest paths are prioritized in a similar way. The first algorithm, AN(TFs), connects every TF with the closest receptor by all shortest paths and delivers one specific network per TF in the list. Similarly, the second algorithm AN(R) delivers a network consisting of all the shortest paths from a receptor in the list to the closest TF; one network per receptor. Since all the edges, and therefore, paths are directional, the resulted networks are not reciprocal.

  Every network built by the AN algorithm may be optionally enriched with the receptor's ligands and the TF's targets. The networks may be grouped, and merged within every group. Namely, if we are building one network for every transcription factor, then all such networks with the same receptors are grouped and merged within each group.

- Transcription regulation (TR). This algorithm starts with a small sub-network that consists of the initial list of objects plus all the "immediate transcription factors" for those initial objects, i.e., the objects that are linked to at least one of the initial objects by an edge of the "transcription regulation" type. Then, a separate network is built around every such transcription factor, using the Auto Expand algorithm with "upstream" option and limiting to the objects from the initial list. Then the transcription factor's targets from the initial list are added to network. The algorithm delivers a list of networks, one per transcription factor.
- Auto-expand (AE). AE algorithm creates sub-networks around *every* object from the uploaded list. The expansion halts when the sub-networks intersect. The objects that do not contribute to connecting sub-networks are automatically truncated.
- Expand by One Interaction. This algorithm builds one-stepsub-networks around *any* object from the list and finds "islands" of nodes from the user's list connected by no more than two bridging objects.

Network Filters and Options

These tools allow researchers to choose the input list of seed objects and the interactions space for the edges in accordance with the customer's preference. The seed objects filters include tissue expression, subcellular localizations (organelles), species (human, mouse, and rat), orthologs (yeast, fly, worm, chicken, dog, bovine, chimpanzee), protein types (29 types, such as receptors, kinases, ion-gated channels, etc.). The "edges" options include interaction mechanisms (19 types), confidence level (physical vs. indirect), interaction weights (can be adjusted). Also, a user has a control over every object in the seed list: he/she can remove or add objects and specify the type of interactions coming in and out of the object.

Network Statistics

The network statistics function calculates specific network features (conversion and diversion hubs, general hubs, longest pathways on the network, etc.) and exports the network content in Excel format (*see* **Note 3**).

3.4 MetaDrug Tools

The MetaDrug module of the platform is designed for prediction of biological effects of small molecules (heterocyclic) compounds of arbitrary structure. Essentially, it uses cheminformatics tools for the conversion of a compound structure to a list of proteins—possible targets and metabolizing enzymes that are then processed via functional analysis as with any other gene/protein list. MetaDrug has several chemistry tools not found elsewhere in the platform.

3.4.1 Metabolites Prediction Tool

The first step in the conversion of a chemical structure into a protein list is to split the molecule into a series of predicted human metabolites. It is well established that in many cases the active (and often toxic) ingredients in many drug molecules are their metabolites that human (mostly liver) enzymes break down the original molecules onto. MetaDrug uses a set of 90 empirical rules based on manual curation of xenobiotic metabolism literature and metabolism prioritization algorithms to deduce Phase I and Phase II metabolites.

3.4.2 QSAR Models

In the next step, the compound structure and the predicted metabolites are tested for bioactivity by calculation of quantitative structure-activity relationship (QSAR) models. MetaDrug uses the ChemTree modeling module developed by Golden Helix Inc. (http://www.goldenhelix.com) for model generation. There are over 100 models in MetaDrug for the evaluation of a compound's physio-chemical properties, reactivity, metabolic hepatotoxicity (phase I and II drug metabolism), general toxicity (Herg, transporters, etc.), as well as activity on potential drug-able targets (Fig. 3).

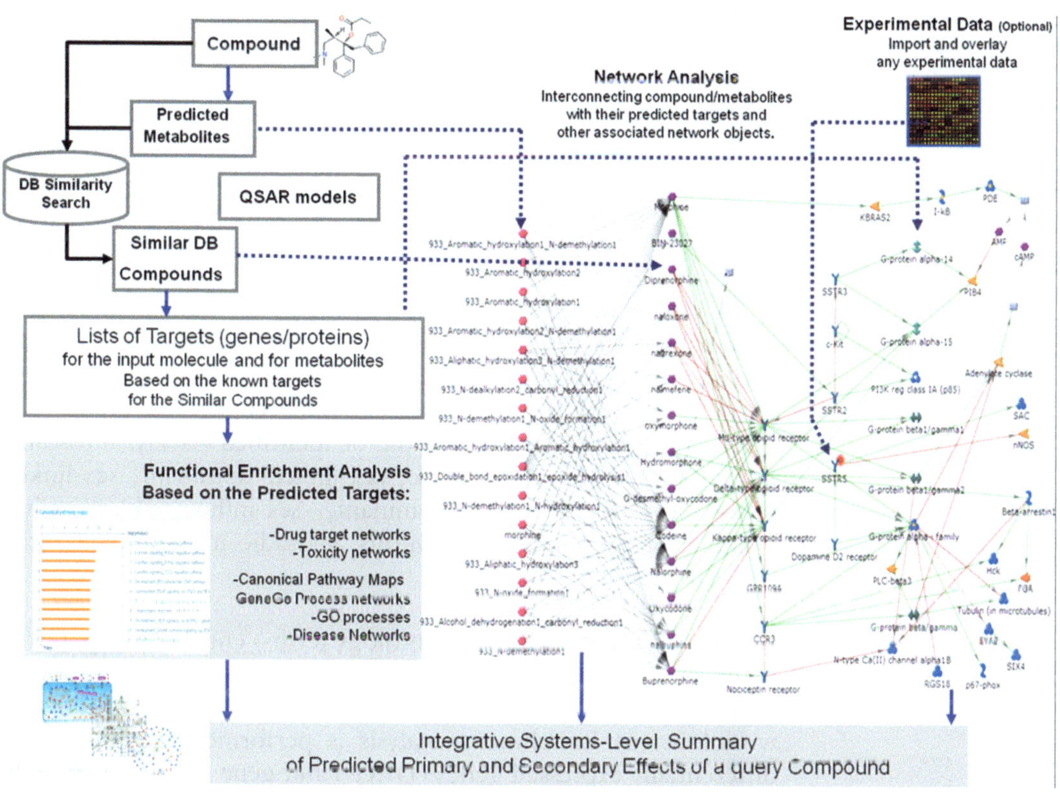

Fig. 3 General schema of functional analysis of small molecules compounds in MetaDrug module. A compound of arbitrary structure is divided onto human metabolites using empirical rules, and all structures are evaluated for activity, toxicity, and physico-chemical parameters by >100 QSAR models. The original compound and metabolites are then used as a query against 780,000 compounds with known activity. Similar compounds retrieved from the database and their protein targets are then subjected to enrichment and network analysis

Some models are built around specific proteins, Phase II drug metabolism enzymes, transporters, membrane and nuclear receptors, kinases, etc. These proteins can be selected by a user for the follow-upfunctional analysis.

3.4.3 Chemical Similarity Search and Assembly of Protein Target List

The uploaded compounds and their metabolites are screened against the chemistry content of MetaBase by chemical structure resemblance and sub-structure search. A Tanimoto coefficient is used as the similarity parameter for the sub-structure search. We also use GVK MediChem database (*see* a chapter in this book for details) as an annotated source of chemistry bioactivity data. The Accord module from Accelrys is used for similarity calculations.

3.5 Genomic Analysis Tools (GAT)

Two main prefiltering analyses in GAT are Cohort Analysis (CA) and Trio Analysis (TA). CA main purpose is to compare two sample groups (e.g., disease-affected patients versus healthy patients) and calculate statistical significance of the gene variants for case cohort against the control cohort. At least ten samples (at least five samples for each case and control cohorts) are required for CA input. It is also possible to have only 5 samples in case group and select 1 of 1000 Genomes and Exome Sequencing Projects population datasets as control. CA calculates *p*-values of statistical significance using Cochran-Armitage Trend Test [26]. Resulting *p*-values indicate the usual probabilistic interpretation of each GV association with case cohort. Variants with appropriate *p*-values (e.g., lower than 0.01 threshold) could be filtered out using Genomic Variant Filter.

TA main purpose is to compare the pattern of the gene variants in the child sample compared to the parental samples (e.g., inherited rare diseases) and calculate the inheritance pattern of these variants. At least three samples are required for TA input. The following gene variant types could be identified via TA: autosomal dominant, autosomal recessive, sex-linked dominant, sex-linked recessive, sex-influenced dominant, sex-influenced recessive, Y-linked, compound heterozygous, sporadic mutation inherited not via classical mechanisms.

3.6 Key Pathway Advisor

Key Processes are defined as ontology terms/entities (i.e., pathway maps) enriched with both input genes and corresponding topologically significant Key Hubs. They are identified by the following workflow. (a) Enrichment analysis is performed for the list of differentially expressed genes (DEGs) and gene variants if submitted. Statistically significant ontology entities (enrichment p-value < 0.001) for differentially expressed genes are identified. Enrichment is calculated for several Thomson Reuters' proprietary functional ontologies. Key Hubs are calculated using a Causal Reasoning approach (if DEGs associated with expression values *see* **Note 4**) or Over-connectivity analysis (if DEGs uploaded without expression values). For further analysis, statistically significant hubs with p-value < 0.001 are identified. Statistically significant Ontology Entities are identified for both the list of differentially expressed genes and the list of corresponding Key Hubs. The enrichment synergy method offered for comparing datasets that are functionally relevant but poorly overlapping at the gene level. If genes derived from different datasets may populate the very same pathway or process, this suggests that they are functionally complimentary. To determine whether two distinct gene lists cooperatively alter a certain cellular pathway or process, we calculate the synergy between them by **ontology enrichment**. An ontology term (pathway or process) is considered **synergistic** if the enrichment p-value for the nonredundant union of compared gene lists is lower than p-values for individual lists. More significant enrichment for the union is reflected in the functional connectivity of two gene lists and their complementary effect on the pathway. The final list of synergistic ontology entities includes all ontology terms with synergistic expression pattern for the union of DEGs and Key Hubs and p-value < 0.001. KPA workflow is schematically shown in Fig. 4.

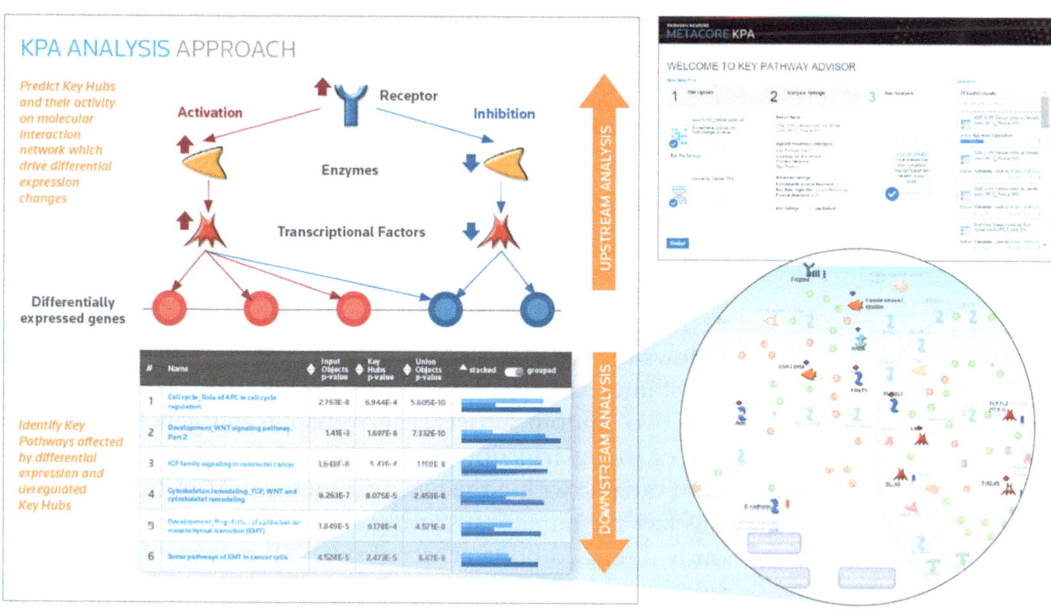

KPA enables researchers to easily perform upstream and downstream analysis of gene expression – even if they have little to no bioinformatics experience.

KPA helps researchers generate key pathway maps with input genes and Key Hubs identified.

Fig. 4 Key Pathway Advisor provides a system that analyzes molecular activity of high-throughput gene expression profiles. Working from published molecular biology studies that are manually curated by the Clarivate Analytics editorial team, KPA creates a hypothesis about abnormal transcription factor activity and upstream signaling cascades that are potentially causing the differential gene expression

3.7 Pathway Map Creator

Pathway Map Creator is a standalone Java application available within MetaCore. It enables creation of custom pathway maps using the MetaBase interactions content and 300 dpi imaging capability. The tool is described in detail in the MetaMiner (Cystic Fibrosis) chapter in the Applications section of this book.

4 Notes

1. **Evaluation of significance (*p*-value) for topological properties for "local interactome"—over-connectivity analysis.** The protein-protein interactions for the gene/protein list of interest are uploaded from MetaBase, followed by calculation of topological properties (average degree, clustering coefficient, and shortest paths) for the list. The topological parameters are then compared with those for the entire collection of interactions in the database (global interactome). Statistical significance of the differences between local and global interactomes can be evaluated by generation of lists of randomly picked genes, the size of the list of interest and calculation of the topological properties for random lists 1000–10,000 times. For example, if one had a subset of ten genes we would calculate the average degree of these ten genes and generate 10,000 sets of genes (of size ten) by randomly picking genes from the experimentally analyzed set and count how many times our set of interest gives larger degree than the randomly generated sets. If our set of interest has a larger average degree than 9500 of the random sets (and respectively smaller average degree than 500 of the random sets), one can assign a *p*-value of 0.05 (i.e., 500/10,000), that is, our set has significantly large average degree at $p = 0.05$ significance level.
2. **Prioritization of AN networks**. Prioritization within the list of AN networks can be based on different parameters, but follows the same procedure that we will describe next. A data set of interest (e.g., the list of all pre-filtered nodes) is divided into two random subsets that overlap in this general case. The size of the intersection between the two sets represents a random variable within the hypergeometric distribution. We apply this fact for numerical scoring and prioritization of the previously discussed node-centered small SP networks. Let us consider a general set size of N with R marked objects/events (e.g., the nodes with expression data). The probability of a random subset of size of n which includes r marked events/objects is described by the distribution.

$$P(r,n,R,N) = \frac{C_R^r \cdot C_{N-R}^{n-r}}{C_N^n} = \frac{C_n^r \cdot C_{N-n}^{R-r}}{C_N^R}$$
$$= \frac{R! \cdot (N-R)!}{N!} \cdot \frac{n! \cdot (N-n)!}{r! \cdot (R-r)!} \cdot \frac{1}{(n-r)! \cdot (N-R-n+r)!}$$

The mean of this distribution is equal to the following:

$$\mu = \sum_{r=0}^{n} r \cdot P(r,n,R,N) = \frac{n \cdot R}{N} = n \cdot q,$$

where $q = R/N$ defines the ratio of marked objects.

The dispersion of this distribution is described as follows:

$$\sigma^2 = \sum_{r=0}^{n} r^2 \cdot P(r,n,R,N) - \mu^2 = \frac{n \cdot R \cdot (N-n) \cdot (N-R)}{N^2 \cdot (N-1)}$$
$$= n \cdot q \cdot (1-q) \cdot \left(1 - \frac{n-1}{N-1}\right)$$

It is essential that these equations are invariant in terms of exchange of n for R. This means that the subset and the marked sets are equivalent and symmetrical. Importantly, in the cases of $r > n$, $r > R$ or $r < R + n - N$, $P(r, n, R, N) = 0$.

We will use the following z-scoring for comparison and prioritization of node-specific SP sub-networks.

$$z - \text{score} = r - n \frac{\frac{R}{N}}{\sqrt{n\left(\frac{R}{N}\right)\left(1 - \frac{R}{N}\right)\left(1 - \frac{n-1}{N-1}\right)}} = \frac{r - \mu}{\sigma}$$

where,

N is the total number of nodes after filtration;

R is the number of nodes in the input list or the nodes associated with experimental data;

n is the number of the nodes in the network;

r is the number of the network's nodes associated with experimental data or included in the input list;

μ and σ are, respectively, the mean and dispersion of the hypergeometric distribution as described above.

3. ***P*-value and evaluation of statistical significance of networks.** For a network of a certain size, we can evaluate its statistical significance based on the probability of its assembly from a random set of nodes of identical or similar size to the input list. We can also evaluate the relevance of the network based on biological processes (defined as a subset of the network nodes associated with the particular process) or any other

subset of nodes. For example, let us consider a complete set of nodes on the network, divided into two overlapping subsets. These subsets represent the nodes linked to a certain predefined node list, e.g., the list of nodes belonging to Gene Ontology (GO) cellular processes, or a list of genes expressed in a certain tissue. Generally, these subsets are different but overlapping. Assuming that the intersection between the two subsets is large enough and nonrandom (we do not consider a situation when the intersection is small but nonrandom), the null-hypothesis states that the subsets are independent and, therefore, the size of the intersection satisfies a hypergeometric distribution. The alternative hypothesis states that there is positive correlation between the subsets. Based on these assumptions, we can calculate a p-value as the probability of intersection of the given or a larger size network from two random subsets from the same set.

$$\begin{aligned} p\text{-val}(r, n, R, N) &= \sum_{i=\max(r, R+n-N)}^{\min(n,R)} P(i, n, R, N) \\ &= \frac{R! \cdot n! \cdot (N-R)! \cdot (N-n)!}{N!} \\ &\times \sum_{i=\max(r, R+n-N)}^{\min(n,R)} \frac{1}{i! \cdot (R-i)! \cdot (n-i)! \cdot (N-R-n+i)!} \end{aligned}$$

4. **Causal reasoninganalysis. Causal Reasoning** is a shortest-path-based method aimed at the identification of upstream regulators that cause gene expression changes observed in transcriptomics data [17, 27]. Causal Reasoning relies on a directed network that is annotated with activation and inhibition edges as well as biological mechanisms (transcription regulation). Causal Reasoning identifies candidates (hypotheses) in the network that can be reached via a predefined maximum shortest path length from the differentially expressed genes. Candidates are scored based on the number of differentially expressed genes that can be reached via the shortest paths and the correctness of the regulation. The correctness is assessed based on the activation and inhibition edges along the paths and the expected and observed direction of fold changes of the differentially expressed genes.

   The significance of the predictions made by a hypothesis is assessed using a binomial test based on the following information:

   (a) k—the sum of correct predictions.

   (b) n—the sum of correct and incorrect predictions.

(c) The p-value is calculated as probability to get k successes in n predictions using binomial trials with $p = 0.5$.

$$p\text{-value} = \binom{n}{k} p^k (1 - p)^{n-k}$$

(d) p-values are assigned in the score matrix and hypotheses above the p-value threshold are filtered out of the score matrix.

References

1. Salwinski L, Eisenberg D (2003) Computational methods of analysis of protein–protein interactions. Curr Opin Struct Biol 13:377–382
2. Kemmeren P et al (2002) Protein interaction verification and functional annotation by integrated analysis of genome-scale data. Mol Cell 9:1133–1143
3. Ceccarelli M, Barthel FP et al (2016) Molecular profiling reveals biologically discrete subsets and pathways of progression in diffuse glioma. Cell 164(3):550–563
4. The Cancer Genome Atlas Network (2015) The molecular taxonomy of primary prostate cancer. Cell 163(4):1011–1025
5. The Cancer Genome Atlas Network (2015) Comprehensive molecular characterization of papillary renal cell carcinoma. N Engl J Med 374(2):135–145
6. Ciriello G, Gatza ML et al (2015) Comprehensive molecular portraits of invasive lobular breast cancer. Cell 163(2):506–519
7. Khatri P, Sirota M, Butte AJ (2012) Ten years of pathway analysis: current approaches and outstanding challenges. PLoS Comput Biol 8 (2):e1002375
8. Jin L, Zuo X-Y, Su W-Y et al (2014) Pathway-based analysis tools for complex diseases: a review. Genomics Proteomics Bioinformatics 12(5):210–220
9. Yook SH, Oltvai ZN, Barabási AL (2004) Functional and topological characterization of protein interaction networks. Proteomics 4 (4):928–922
10. Barabasi AL, Oltvai Z (2004) Network biology: understanding the cell's functional organization. Nat Rev Genet 5(2):101–113
11. Bader S, Kühner S, Gavin AC (2008) Interaction networks for systems biology. FEBS Lett 582(8):1220–1224
12. Nitsch D, Gonçalves JP, Ojeda F, de Moor B, Moreau Y (2010) Candidate gene prioritization by network analysis of differential expression using machine learning approaches. BMC Bioinformatics 11:460
13. Hsu C-L, Huang Y-H, Hsu C-T, Yang U-C (2011) Prioritizing disease candidate genes by a gene interconnectedness-based approach. BMC Genomics 12(Suppl 3):S25
14. Köhler S, Bauer S, Horn D, Robinson PN (2008) Walking the interactome for prioritization of candidate disease genes. Am J Hum Genet 82(4):949–958
15. Vanunu O, Magger O, Ruppin E, Shlomi T, Sharan R (2010) Associating genes and protein complexes with disease via network propagation. PLoS Comput Biol 6(1):e1000641
16. Chen J, Aronow BJ, Jegga AG (2009) Disease candidate gene identification and prioritization using protein interaction networks. BMC Bioinformatics 10:73
17. Chindelevitch L, Ziemek D, Enayetallah A et al (2012) Causal reasoning on biological networks: interpreting transcriptional changes. Bioinformatics 28:1114–1121
18. Li X, Shen L, Shang X, Liu W (2015) Subpathway analysis based on signaling-pathway impact analysis of signaling pathway. PLoS One 10(7): e0132813
19. Ulitsky I, Krishnamurthy A, Karp RM, Shamir R (2010) DEGAS: de novo discovery of dysregulated pathways in human diseases. PLoS One 5(10):e13367
20. Leiserson MDM, Vandin F, Wu H-T et al (2015) Pan-cancer network analysis identifies combinations of rare somatic mutations across pathways and protein complexes. Nat Genet 47 (2):106–114
21. Hendrix W, Rocha AM, Padmanabhan K et al (2011) DENSE: efficient and prior knowledge-driven discovery of phenotype-associated protein functional modules. BMC Syst Biol 5:172
22. Shannon P, Markiel A et al (2003) Cytoscape: a software environment for integrated models of biomolecular interaction networks. Genome Res 13(11):2498–2504

23. Paull EO, Carlin DE, Niepel M et al (2013) Discovering causal pathways linking genomic events to transcriptional states using tied diffusion through interacting events (TieDIE). Bioinformatics 29(21):2757–2764
24. Suthram S, Beyer A, Karp RM, Eldar Y, Ideker T (2008) eQED: an efficient method for interpreting eQTL associations using protein networks. Mol Syst Biol 4:162
25. Vaske CJ, Benz SC, Sanborn JZ et al (2010) Inference of patient-specific pathway activities from multi-dimensional cancer genomics data using PARADIGM. Bioinformatics 26(12): i237–i245
26. Purcell S et al (2007) PLINK: a tool set for whole-genome association and population-based linkage analyses. Am J Hum Genet 81 (3):559–575
27. Pollard J Jr, Butte AJ, Hoberman S, Joshi M, Levy J, Pappo J (2005) A computational model to define the molecular causes of type 2 diabetes mellitus. Diabetes Technol Ther 7(2):323–336

Chapter 7

Extracting the Strongest Signals from Omics Data: Differentially Expressed Pathways and Beyond

Galina Glazko, Yasir Rahmatallah, Boris Zybailov, and Frank Emmert-Streib

Abstract

The analysis of gene sets (in a form of functionally related genes or pathways) has become the method of choice for extracting the strongest signals from omics data. The motivation behind using gene sets instead of individual genes is two-fold. First, this approach incorporates pre-existing biological knowledge into the analysis and facilitates the interpretation of experimental results. Second, it employs a statistical hypotheses testing framework. Here, we briefly review main Gene Set Analysis (GSA) approaches for testing differential expression of gene sets and several GSA approaches for testing statistical hypotheses beyond differential expression that allow extracting additional biological information from the data. We distinguish three major types of GSA approaches testing: (1) differential expression (DE), (2) differential variability (DV), and (3) differential co-expression (DC) of gene sets between two phenotypes. We also present comparative power analysis and Type I error rates for different approaches in each major type of GSA on simulated data. Our evaluation presents a concise guideline for selecting GSA approaches best performing under particular experimental settings. The value of the three major types of GSA approaches is illustrated with real data example. While being applied to the same data set, major types of GSA approaches result in complementary biological information.

Key words Omics data, Gene set analysis approaches, Hypotheses testing, Self-contained, Competitive, Differential expression, Differential co-expression, Differential variability

1 Introduction

Biological systems are living proofs of Aristotle's idea that the whole is greater than the sum of its parts. For example, cell is a product of synergistic actions of its constituents (genes, proteins, metabolites, just to name a few). Together with cellular environment this synergy defines what we call the cell type (e.g., a stem cell or dendritic cell). At the level of cell's key molecules (nucleic acids and proteins) the idea of synergy also holds true the following: genes work together in biological pathways, proteins form protein complexes, that is genes and proteins are organized in functional

Tatiana V. Tatarinova and Yuri Nikolsky (eds.), *Biological Networks and Pathway Analysis*, Methods in Molecular Biology, vol. 1613, DOI 10.1007/978-1-4939-7027-8_7, units acting overall differently than a single gene or a single protein would. Thus, when an investigator studies omics data, the idea to consider functional units instead of individual components comes naturally to mind. In fact, this idea was first employed for the analysis of gene expression data more than a decade ago [1]. Analyzing microarray data from diabetics vs. healthy controls Mootha and colleagues [1] did not find a single gene to be differentially expressed. However, when genes were analyzed at the pathway level using Gene Set Enrichment Analysis (GSEA) approach, it was found that genes involved in oxidative phosphorylation showed reduced expression in diabetics although the average decrease per gene was only 20% [1]. There were two reasons behind the success of the pathway analysis approach in this case. First, the number of hypotheses to test by arranging genes into pathways is dramatically reduced, which leads to the increase in power. Second, in metabolic diseases such as diabetes changes in gene expression are moderate and therefore can be overlooked by using methods focusing on each gene individually. These two reasons explain why pathway analysis has become the method of choice in analyzing omics data in general and expression data in particular. Nowadays, we also recognize yet another important reason to employ pathway (gene set) analysis for omics data. Gene Set Analysis (GSA) approaches provide flexibility to test different statistical hypotheses, thus increasing the biological interpretability of experimental results. Here, we briefly review main Gene Set Analysis (GSA) approaches for testing differential expression of gene sets and several GSA approaches for testing statistical hypotheses beyond differential expression, which allow extracting additional biological information from the data.

We distinguish the three major types of GSA approaches that test statistically and biologically different hypotheses: (1) differential expression (DE), (2) differential variability (DV), and (3) differential co-expression (DC) of gene sets between two phenotypes. All major types of GSA approaches can be univariate (gene-level) or multivariate (accounting for intergene correlations). The chapter is organized as follows: In the first part of Subheading 2, we discuss GSA approaches developed for identification of differentially expressed pathways applicable for the analysis of microarrays and RNA-seq data (GSA-DE). The traditional GSA-DE framework aims to identify pathways with significant changes in mean gene expressions and it is well understood. In the second part of Subheading 2, DV analysis in application to gene sets (GSA-DV) is considered. The analysis of differential variability (DV) is somewhat appreciated with regards to individual genes, when the aim is to find genes with significant changes in expression variance between two phenotypes [2–6]. It was shown that many statistically significant DV genes are relevant to disease development and that DV is an indication of changes in gene regulation [2, 3]. Moreover, it was found that there are genes showing consistently higher across-sample variability in tumors of different origin as compared to normal samples [7]. These DV genes can serve as a robust molecular signature for multiple cancer types [7, 8]. Given the evidence that DV genes may play an important role in observed phenotypes, and given the popularity of GSA approaches one would expect there are many approaches implementing GSA-DV test. Our group was the first to suggest extending the DV analysis to a multivariate GSA-DV case using multivariate statistical test [9, 10]. In the same publication we further demonstrated that for three different cancer types GSA-DV approach was able to identify cancer-specific pathways, while pathways identified using conventional GSA-DE approaches were shared between the three cancer types. Thus, GSA-DV approach provides additional biological information beyond GSA-DE. It should be noted that there are other approaches claiming to perform GSA-DV test, e.g., DIRAC and EVA [11], but because they compare variability in gene ranks within a pathway between two phenotypes rather than variance estimates, these approaches are out of the scope of this chapter. We discuss two principally different GSA-DV approaches: (1) nonparametric multivariate GSA-DV approach, "radial" Kolmogorov-Smirnov (RKS) [9] and (2) new gene-level GSA-DV test we suggest here for the first time. This gene-level GSA-DV approach applies Fisher Method (FM) [12] for combining *P*-values from gene-level *F*-test for differential variability [3]. It should be noted that currently GSA-DV approaches are applicable only to microarray data, because RNA-seq read counts are most frequently modeled with Negative Binomial distribution that has complex dependence between mean and variance. In the third part of Subheading 2, GSA approaches estimating differential co-expression of gene sets between two phenotypes (GSA-DC) are considered. In a pathway, genes are working together, i.e., they form a co-expression network. For finding DC pathways GSA-DC approaches with or without network inference step can be employed. The most general GSA-DC approach with a network inference step is based on a Gaussian Graphical Model (GGM) [13]. In this approach, the network structure of a pathway for each phenotype is estimated and the null hypothesis to test is that the network structure across phenotypes is the same [13]. The network inference step per se is challenging because there are too many ways of estimating network structure. For example, the implementation of network inference in Bioconductor package `nethet` (that provides two-sample testing in GGMs) includes several options, such as the Graphical Lasso (GL) [14], the Meinshausen-Buhlmann approach [15], and the approach proposed by Schafer and Strimmer based on shrinkage estimation of the covariance matrix [16]. Needless to say, the `nethet` results for networks comparison will vary significantly depending on the algorithm selected for the network inference step. In addition, many approaches for network inference (e.g., GGM) require the assumption of normality that may or may not be met in the real data. This is why we present in this review only GSA-DC approaches that do not require a network inference step. The simplest GSA-DC approach, the gene sets co-expression analysis (GSCA) [17] is purely univariate. GSCA calculates the Euclidian distance between two correlation vectors (constructed from diagonal matrices of pairwise correlations for different conditions) and the significance of the difference is estimated using permutation test. The gene sets net correlations analysis (GSNCA) [18] assesses multivariate changes in the gene co-expression network between two conditions but does not require network inference step. Net correlation changes are estimated by introducing for each gene a weight factor that characterizes its cross-correlations in the co-expression networks. Weight vectors in both conditions are found as eigenvectors of correlation matrices with zero diagonal elements. Gene sets net correlations analysis (GSNCA) tests the hypothesis that for a gene set there is no difference in the gene weight vectors between two conditions [18]. The Co-expression Graph Analysis (CoGA) identifies co-expressed gene sets by statistically testing the equality in the spectral distributions [19]. For each phenotype CoGA constructs a full network from pairwise correlations between gene expressions. Then the structural properties of the two networks are compared by applying Jensen-Shannon divergence as a distance measure between the graph spectrum distributions [19, 20]. All methods are supplied with the implementation reference if available.

In Subheading 3, we first present a comparative power analysis and Type I error rates for different approaches in each major type of GSA on simulated data. Second, the value of applying the three major types of GSA approaches is illustrated with real data example, where these approaches provide different biological information obtained on the same data set.

2 Methods

2.1 Gene Set Analysis Approaches for Testing Differential Expression (GSA-DE)

There are many GSA-DE approaches readily distinguished based on the null hypothesis they test. According to Goeman and Buhlmann [21] the formulation can be either *self-contained* or *competitive*. *Self-contained* approaches compare whether a gene set is differentially expressed between different conditions, while *competitive* (e.g., GSEA) approaches compare a gene set against its complement that contains all genes except genes in the set [21, 22]. *Self-contained* approaches can be (1) univariate, in a sense that they use gene-level tests for GSA and combine univariate statistics for individual genes into a single test score [10, 23, 24]; and (2) multivariate, when a multivariate statistic is used to address the null

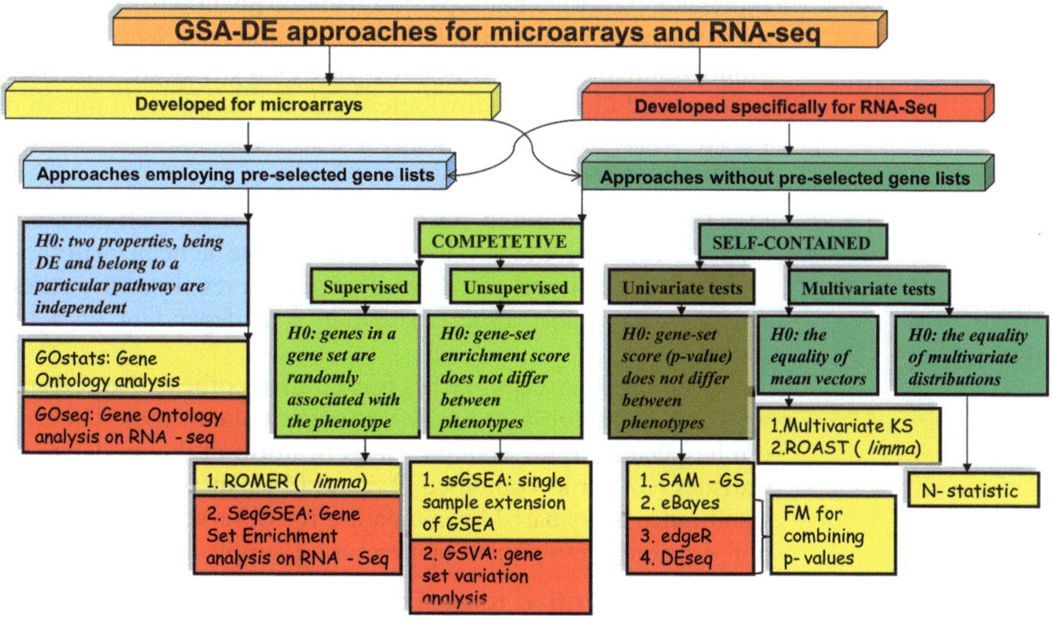

Fig. 1 Schematic overview illustrating the breakup of the GSA-DE methods into different categories based on the null hypotheses they test

hypothesis. In a real biological setting, moderate [25] and extensive [26] correlations between genes in gene sets are well documented [27] and that may result in a decrease of the power for gene-level tests compared to multivariate tests [24, 27–29]. In turn, *competitive* GSA approaches can be (1) "supervised," when the class labels are known; or (2) "unsupervised," when the enrichment score is computed for each gene set and individual sample [30]. For GSA-DE the "supervised" term indicates that the samples classification is known, while the "unsupervised" term indicates that the samples classification is unknown [30]. A number of review articles concerning the different aspects of GSA-DE approaches developed for microarrays data analysis have been published [21, 23, 31–36].

To summarize, GSA-DE approaches that test intrinsically statistically different null hypotheses developed thus far are: *self-contained* (univariate, multivariate) and *competitive* (supervised, unsupervised). Figure 1 illustrates different null hypotheses tested by various GSA-DE approaches together with R packages implementing each test. For the sake of generality, all power and Type I error rate estimates for GSA-DE approaches are presented for simulated RNA-seq counts.

Null Hypotheses

Consider two different biological phenotypes, with n_1 samples of measurements for the first and n_2 samples of the same measurements for the second. Let the two random vectors of $X = (X_1, \ldots, X_{n1})$ and $Y = (Y_1, \ldots, Y_{n2})$ represent the measurements of p gene expressions (constituting a pathway) in two phenotypes where X_i is the ith p-dimensional sample in one phenotype and Y_i is the ith p-dimensional sample in the other phenotype. Let X, Y be independent and identically distributed with the distribution functions F_x, F_y, mean vectors $\overline{\mu}_x$ and $\overline{\mu}_y$, and $p \times p$ positive-definite and symmetric covariance matrices Σ_x and Σ_y.

H_0 for self-contained tests. For multivariate self-contained tests we consider the problem of testing the general hypothesis H_0: $F_x = F_y$ against an alternative $F_x \neq F_y$, or a restricted hypothesis H_0: $\overline{\mu}_x = \overline{\mu}_y$ against an alternative $\overline{\mu}_x \neq \overline{\mu}_y$ depending on a test statistic.

Gene-level GSA approaches test a null hypothesis that the gene set-associated score does not differ between phenotypes. The score can be calculated, for example, as an L_2-norm of the moderated t-statistics [37] or as combined P-values [24]. In all cases statistical significance is evaluated by comparing the observed score with the null distribution, obtained by permuting sample labels.

H_0 for competitive tests. The Gene Set Enrichment Analysis (GSEA) method [1, 38] is one of the most widely used competitive approaches. As a local test statistic, it uses a signal-to-noise ratio and a weighted Kolmogorov-Smirnov as a global test statistic (enrichment score, normalized to factor out the gene set size dependence) [34, 38]. Assuming a null distribution F_0^{perm} induced by permuting sample labels, GSEA evaluates significance of the global test statistic ζ_k^{GSEA} by estimating nominal P-value from F_0^{perm} [34, 38]. Thus, GSEA tests the null hypothesis that the genes in a gene set are randomly associated with the phenotype.

Most competitive GSA approaches are supervised, in a sense that sample labels are known (that is, there are at least two different phenotypes). Recently, the concept of unsupervised GSEA where an enrichment score is computed for each gene set and individual sample was introduced [30]. Essentially, unsupervised GSEA transforms a matrix of gene expressions across samples into a matrix of gene sets enrichment scores across the same samples. It makes the choice of null hypothesis flexible and context dependent. For example, Barbie et al. [39] use unsupervised competitive GSEA to test the null hypothesis that the Spearman correlation between gene sets enrichment scores is zero, while Hazelmann et al. [30] test the hypothesis that gene set enrichment score does not differ between two phenotypes.

SELF-CONTAINED GENE-LEVEL TESTS FOR GSA

Gene-level tests for GSA can be easily designed in three steps: (1) select a gene-level score based on a univariate test statistic (e.g., a value of t-test), (2) transform a score (e.g., take an absolute value of t-statistic, or consider its P-value), and (3) summarize gene-level scores into a gene set statistic (e.g., take an average of transformed scores or use combining *P*-values approach) [10, 23, 24].

Gene-level GSA-DE tests that combine genes *P*-values. Gene-level tests for GSA that combine *P*-values from individual tests for microarray data were studied in [40]. As a gene-level test, the authors used an *F*-statistic for the correlation between the gene expression and phenotype ($F = (N - 2)[r^2/(1 - r^2)]$ (not to confuse with *F*-test) and compared several approaches for combining *P*-values: Fisher's method (FM) [12], Stouffer's method (SM) [41], tail strength (TS) [42], and a modified tail strength statistic (MTS) [40]. It was found that FM outperformed all the other methods for combining *P*-values [40].

Gene-level tests for GSA that combine *P*-values from individual tests for RNA-seq data were studied in [24]. In what follows, we briefly reiterate the conclusions from comparative power and Type I error rate analyses of different gene-level GSA tests [24]. There are two popular univariate tests specifically developed for RNA-seq data that rely on Negative Binomial model for read counts: edgeR [43] and DESeq [44]. Empirical Bayes method eBayes [45] correctly identifies hypervariable genes and can be adapted for RNA-seq data through VOOM normalization [46]. When applied correctly the gene-level test does not per se influence the performance of a gene-level GSA approach as much as the procedure used to combine univariate statistics into a single test score does [24]. Among many approaches available for combining *P*-values from gene-level tests, we have shown that, similar to the results for microarray data, the safest option is to use FM [24, 47]. Here, for comparative power and Type I error rate estimates eBayes in combination with FM is selected.

Gene-level GSA-DE test that combines statistics. In the analysis of microarrays, shrinking the standard error of a test statistic (e.g., a *t*-test) in testing DE of individual genes improves the power of the test. Several shrinkage approaches at the level of individual genes were suggested, including the Significance Analysis of Microarrays (SAM) test [48], the regularized *t*-test [49], and the moderated *t*-test [50]. In particular, an extension of SAM test to gene set analysis (SAM-GS) has been demonstrated to outperform several conventional self-contained tests and even the original competitive GSEA approach for microarray data [10, 37, 51, 52].

SAM GS can be applied to RNA-seq count data by using the VOOM normalization [46] prior to the test to find the log-scale counts per million (CPM) of the raw counts normalized for library sizes. The test statistic is the L_2-norm of the moderated *t*-statistics for the gene expressions:

$$T_{\text{SAM-GS}} = \sum_{i=1}^{p} \left(\frac{X_i - Y_i}{s_i + s_0} \right)^2$$

where $\bar{X}_i$ and $\bar{Y}_i$ are respectively the mean expression levels for gene i under phenotypes X and Y, s_i is a pooled standard deviation over the samples in the two phenotype, s_0 is a small positive constant to adjust for small variability, and p is the number of genes in the gene set.

SELF-CONTAINED MULTIVARIATE TESTS FOR GSA

Based on their high power and popularity we consider two multivariate test statistics.

N-statistic. N-statistic [53, 54] tests the most general hypothesis H: $F_x = F_y$ against a two-sided alternative $F_x \neq F_y$:

$$N_{n_1 n_2} = \frac{n_1 n_2}{n_1 + n_2}\left[\frac{1}{n_1 n_2}\sum_{i=1}^{n_1}\sum_{j=1}^{n_2} L(X_i, Y_j) - \frac{1}{2n_1^2}\sum_{i=1}^{n_1}\sum_{j=1}^{n_2} L(X_i, X_j) - \frac{1}{2n_2^2}\sum_{i=1}^{n_1}\sum_{j=1}^{n_2} L(Y_i, Y_j)\right]^{1/2}$$

Here, we consider only $L(X, Y) = X - Y$, the Euclidian distance in R^p. N-statistic was applied to microarray data and was shown to outperform other univariate and multivariate GSA-DE tests under different parameter settings [10, 28]. After VOOM normalization [46] N-statistic can also be applied to RNA-seq data and also was shown to outperform other GSA-DE tests [24, 47].

ROAST. In the context of microarray data, a parametric multivariate rotation gene set test (ROAST) has become popular for the self-contained GSA approaches [55]. ROAST uses the framework of linear models and tests whether for all genes in a set, a particular contrast of the coefficients is nonzero [55]. It can account for correlations between genes and has the flexibility of using different alternative hypotheses, testing whether the direction of changes in mean is *up*, *down*, or *mixed* (up or down) [55]. For microarrays it was shown that when correlations are low ROAST performance is similar to N-statistic [10]. Using ROAST with RNA-seq count data requires proper normalization. The VOOM normalization [46] was proposed specifically for this purpose where log counts per million, normalized for library size are used. In addition to counts normalization, VOOM calculates associated precision weights that can be incorporated into the linear modeling process within ROAST to eliminate the mean-variance trend in the normalized counts [46].

SUPERVISED COMPETETIVE TESTS FOR GSA

ROMER. The first competitive GSA test for microarray data analysis GSEA [1] was developed a decade ago. The original GSEA was sensitive to the gene set size and the influence of other gene sets [56], so it was subsequently upgraded into GSEA-P that used a correlation-weighted KS statistic, an improved enrichment normalization, and an FDR-based estimate of significance [34, 38]. For the sake of simplicity, we will only consider the GSEA version implemented in Bioconductor package `limma` function ROMER (the rotation testing using mean ranks) [57]. ROMER is a parametric method developed originally for microarray data and uses the framework of linear models [46] and rotations instead of permutations (*see* ref. 55 for more detail). In contrast to ROAST, the `limma` implementation of ROMER does not incorporate the weights, estimated by VOOM into the linear modeling process to account for the mean-variance trend in the data.

UNSUPERVISED COMPETETIVE TESTS FOR GSA

The goal of unsupervised competitive approaches is to characterize the degree of expression enrichment of a gene set in each sample within a given data set [39]. The term "competitive" is reminiscent of the way the enrichment score is calculated: as a function of gene expression inside and outside the gene set.

Gene set variation analysis (GSVA). GSVA can be applied to microarray expression values or RNA-seq counts. Depending on the data type, expression values (counts) are first transformed using a Gaussian (or discrete Poisson) kernel into expression-level statistics [30]. The sample-wise enrichment score for a gene set is calculated using KS-like random walk statistic. An enrichment statistic (GSVA score) can be calculated as its maximum deviation from zero over all genes (similar to the original GSEA) or as the difference between the largest positive and negative deviations from zero (*see* ref. 30 for more detail).

Single sample extension of GSEA (ssGSEA). The difference between GSVA and ssGSEA stems from the way an enrichment score is calculated. In ssGSEA the enrichment score for a gene set under one sample is calculated as a sum of the differences between two weighted empirical cumulative distribution functions of gene expressions inside and outside the set [39]. The approach, together with GSVA, is implemented in the Bioconductor `GSVA` package [30].

2.2 Gene Set Analysis Approaches for Testing Differential Variability (GSA-DV)

It is well recognized that multivariate statistics have more power than univariate in the case of GSA-DE when intergene correlations are high [24, 27–29]; however, in the case of GSA-DV, this question was not studied at all. Here, we address this shortcoming by providing comparative power analysis for RKS, N-statistic, and gene-level approach for GSA-DV (see below).

Null Hypotheses

H_0 for GSA-DV. While H_0 for RKS is the same general hypothesis tested, e.g., by N-statistic, namely H_0: $F_x = F_y$, an alternative in this case is not $F_x \neq F_y$ or $\overline{\mu}_x \neq \overline{\mu}_y$ but $\overline{\sigma}_x \neq \overline{\sigma}_y$, i.e., differences in scale. N-statistic tests an alternative $F_x \neq F_y$. Because this general alternative implicitly includes inequality of variances for distribution functions F_x and F_y, N-statistic can also capture differences in scale, so if

H_0 is rejected by N-statistic the true alternative is unknown. N-statistic is included in comparative power analysis for GSA-DV.

Gene-level GSA-DV approach we suggest here tests a null hypothesis that the gene-set-associated score does not differ between phenotypes. The score here is calculated by applying FM to combine *P*-values from gene-level *F*-test of the equality of two variances.

Gene-level GSA-DV test that combines genes *P*-values. To find genes with significant variability we suggest using *F*-test, similar to what was described for individual genes by Ho and colleagues [3]. Gene-level GSA-DV test is designed by combining *P*-values of individual *F*-tests for genes in a pathway. Because for gene level GSA-DE FM was found to be the best performing approach for combining *P*-values among many others [24, 40] FM is also applied here to combine *P*-values of *F*-tests. This method tests the alternative hypothesis that there are genes DV between two phenotypes.

Radial Kolmogorov Smirnov (RKS). The basic operational procedure employed in the univariate Kolmogorov-Smirnov test is to sort pooled observations in ascending order. The difficulty in extending this procedure to multivariate observations is that the notion of a sorted list cannot be immediately generalized [9]. Friedman and Rafsky suggested overcoming this difficulty using the Minimum Spanning Trees (MSTs) [9]. The multivariate generalization of KS ranks multivariate observations based on their MST. The purpose of MST ranking is to obtain the strong relation between observations differences in ranks and their distances in R^p. The ranking algorithm can be designed specifically to confine a particular alternative hypothesis more power. The general scheme is to root MST tree at a node with the largest geodesic distance and then rank the nodes in the "height directed preorder" traversal of the tree. If one is interested in a test with high power toward changes in the variance structure of the distribution, the ranking is implemented differently, aiming to give higher ranks to more distant points in R^p. That is, MST tree is rooted at the node with the smallest geodesic distance (centroid) and nodes with the largest depths are assigned higher ranks [9]. This "radial" Kolmogorov-Smirnov (RKS) test is sensitive to alternatives having similar mean vectors but differences in scale. The test statistic considering N samples under two phenotypes X and Υ is the maximum absolute difference

$$D = \left| \frac{s_X^{(i)}}{N_X} - \frac{s_\Upsilon^{(i)}}{N_\Upsilon} \right|$$

where $s_X^{(i)}$ and $s_\Upsilon^{(i)}$ are respectively the number of observations in X and Υ ranked lower than i, $1 \leq i \leq N$, N_X and N_Υ are respectively the number of samples under phenotypes X and Υ. The null distribution of the test statistic is estimated by a permutation procedure and P-value is defined as

$$P_{\text{value}} = \frac{\sum_{k=1}^{N_{\text{perm}}} I\left[D_{\text{perm}}(k) \geq D_{\text{obs}}\right] + 1}{N_{\text{perm}} + 1}$$

where $D_{\text{perm}}(k)$ is the test statistic of permutation k, D_{obs} is the observed test statistic from the original data, N_{perm} is the number of permutations, and I is the indicator function. RKS is implemented in Bioconductor package Gene Set Analysis in R (GSAR) [10, 18].

2.3 Gene Set Analysis Approaches for Testing Differential Co-Expression (GSA-DC)

Null Hypotheses

Each individual GSA-DC approach we consider has its own null hypothesis (see below).

Gene Sets Co-Expression Analysis (GSCA). Briefly, GSCA works as follows [17]. For all $p(p\text{-}1)/2$ gene pairs, GSCA calculates inter-gene correlations under the two biological conditions. The test statistic is the Euclidean distance, adjusted for the size of a gene set,

$$D_{\text{GSCA}} = \sqrt{\frac{1}{p(p-1)/2} \sum_{k=1}^{p(p-1)/2} \left(\rho_k^{(1)} - \rho_k^{(2)}\right)^2}$$

where k is the index of the gene pair within the gene set and $\rho_k^{(i)}$ denotes the correlation of gene pair k in condition i. GSCA tests the hypothesis H_0: $D_{\text{GSCA}} = 0$ against the alternative H_1: $D_{\text{GSCA}} \neq 0$.

Gene Sets Net Correlations Analysis (GSNCA). In order to quantitatively characterize the importance of gene i in a correlation network, we introduce a weight (w_i) and set w_i to be proportional to a gene's cross-correlation with all the other genes in the gene set [24]. Then, the objective is to find a weight vector w, which achieves equality between a gene weight and the sum of its weighted cross-correlations for all genes simultaneously. Thus, genes with high cross-correlations will have high weights that may indicate their regulatory importance. This problem can be formulated as a system of linear equations

$$\mathrm{w}_i = \sum_{j \neq i} w_j r_{ij}, \quad 1 \leq i \leq p$$

where r_{ij} is the absolute correlation coefficient between genes i and j, and p is the gene set size. Equivalently, this system of linear equations can be represented in the matrix form

$$(R - I)w = w$$

where R is the correlation matrix. This is an eigenvector problem that has a unique solution when the eigenvalue $\lambda_{(R-I)} = 1, w > 0$. Because the matrix $(R - I)$ is not guaranteed to have eigenvalue

$\lambda_{(R-I)} = 1$, we introduce a multiplicative factor, γ, which ensures a proper scaling for eigenvalues and solves the following problem:

$$\gamma(R - I)w = w$$

The unique solution w is an eigenvector of matrix $\gamma(R - I)$ corresponding to $\lambda_{\gamma(R-I)} = 1$ [24]. As a test statistic, w_{GSNCA}, we use the L_1 norm between the scaled weight vectors $w^{(1)}$ and $w^{(2)}$ (each vector is multiplied by its norm to scale the weight factor values around one) between two conditions,

$$w_{\text{GSNCA}} = \sum_{i=1}^{p} \left| w_i^{(1)} - w_i^{(2)} \right|$$

This statistic tests the hypothesis H_0: $w_{\text{GSNCA}} = 0$ against the alternative H_1: $w_{\text{GSNCA}} \neq 0$. P-values for the test statistic are obtained by comparing the observed value of the test statistic to its null distribution, which is estimated using a permutation approach. GSNCA is implemented in Bioconductor package GSAR [10, 18].

Co-expression Graph Analyzer (CoGA). Let $G = (V, E)$ be an undirected graph with the adjacency matrix A. The *spectrum* of G is a set of eigenvalues of its adjacency matrix A [20]. The spectrum of a graph describes several of its structural properties, such as diameter, number of walks, and cliques. Takahashi and colleagues [20] suggested that the graph spectrum distribution is a better characterization of graph's properties than conventionally used measures such as number of edges, average path length, and clustering coefficient. Co-expression Graph Analyzer (CoGA) constructs co-expression graphs and identifies differentially co-expressed gene sets by testing the equality of the spectral distributions for two graphs by calculating Jensen-Shannon divergence between spectral densities of two adjacency matrices [19]. Let Θ measure the distance between structural properties of two graphs. CoGa tests H_0: $\Theta = 0$ against H_1: $\Theta > 0$ [19]. CoGA is implemented in Biocnductor package CoGA [20].

3 Data Analysis

3.1 Comparative Power Analysis and Type I Error Rate: Simulation Setup

Simulation Setup for GSA-DE

Due to the increasing popularity of RNA-seq data as compared to microarrays the simulation setup here is presented in the context of RNA-seq data. It is conventionally assumed that RNA-seq count data follow Poisson or Negative Binomial (NB) distribution. Here, the count for gene i in sample j is modeled by a random variable C_{ij} with NB distribution

$$C_{ij} \sim \text{NB}(\text{mean} = \mu_{ij}, \text{var} = \mu_{ij}(1 + \mu_{ij}\varphi_{ij})) = \text{NB}(\mu_{ij}, \varphi_{ij})$$

where μ_{ij} and φ_{ij} are respectively the mean and dispersion parameters of gene i in sample j. For each gene, a vector of realistic values of mean count, dispersion, and gene length information (μ_i, φ_i, L_i) is randomly picked from a pool of vectors derived from a real RNA-seq dataset. As a real dataset, we selected a subset of the Pickrell et al. [58] dataset of sequenced cDNA libraries generated from 69 lymphoblastoid cell lines that were derived from Yoruban Nigerian individuals. Samples from 58 unrelated individuals were considered (29 males and 29 females). Dispersion parameters for individual genes were estimated using the Bioconductor `edgeR` package [43].

Type I error rate. To simulate the null hypothesis H_0: $F_x = F_y$, we generated a dataset consisting of N samples (equally separated between two different phenotypes) and $S = 1000$ nonoverlapping gene sets, each of size p. The randomly selected parameter vector (μ_i, φ_i, L_i) is used to generate counts from the Negative Binomial distribution for gene i under all the samples in the dataset. Gene length information is used for expression normalization if necessary. To examine the effects of different sample and gene set sizes, we estimated Type I error rate under different parameter settings. We chose $p \in \{16, 60, 100\}$ and $N \in \{10, 20, 40, 60\}$. Type I error rate for a statistical test is calculated as the proportion of gene sets detected by the test. The results were averaged over ten independent datasets to obtain more stable estimates.

Detection Power. A differentially expressed (DE) gene set in real data may include up-regulated, down-regulated, and equally regulated genes between two phenotypes. To mimic real data we introduce three simulation parameters: β, the proportion of gene sets in the dataset that have truly DE genes; γ, the proportion of genes, truly DE in each gene set; and FC, the fold change in gene counts between two phenotypes. We consider $\beta \in \{0.05, 0.25\}$, $\gamma \in \{0.125, 0.25, 0.5\}$, and $FC \in [1.2, 3]$. Two different biological conditions are represented by two groups of samples with equal size $N/2$ where $N = 40$. Under each condition, $S = 1000$ nonoverlapping gene sets were formed, each consists of $p = 16$ random realizations from the Negative Binomial distribution. The power for all statistical methods was estimated by testing the hypothesis H_0: $\mu_x = \mu_y$ (or H_0: $FC = 1$) against the alternative H_1: $\mu_x \neq \mu_y$ (or H_1: $FC \neq 1$) for all gene sets. For each of the $(1 - \beta)S$ non-DE gene sets p random realizations of $NB(\mu_i, \varphi_i)$ were sampled, $1 \leq i \leq p$, under both phenotypes. For each of the βS gene sets that have truly DE genes, half of the γp DE genes in each gene set were up-regulated and half were down-regulated between the two phenotypes. Specifically, $\gamma p/2$ random realizations from $NB(\mu_i, \varphi_i)$ and $NB(FC\mu_i, \varphi_i)$ were sampled respectively under phenotype 1 and phenotype 2 for $1 \leq i \leq \gamma p/2$ and another $\gamma p/2$ random realizations from $NB(FC\,\mu_i, \varphi_i)$ and

$NB(\mu_i, \varphi_i)$ were sampled respectively under phenotype 1 and phenotype 2 for $(\gamma p/2) + 1 \leq i \leq \gamma p$.

Simulation Setup for GSA-DV

Typically, RNA-seq counts are modeled using Poisson or Negative Binomial (NB) distribution. Since in the case of Poisson distribution variance is equal to the mean and in the case of NB distribution variance depends on the mean, there is no GSA-DV test for RNA-seq data. Therefore, we present simulation setup assuming multivariate normal distribution of gene expressions that is a standard assumption for microarray data.

Type I error rate. We generated two samples of equal size, $N/2$ from the p-dimensional normal distribution $N(\mathbf{0}, \mathrm{I}_{p\times p})$ where $\mathrm{I}_{p\times p}$ is a $p \times p$ identity matrix and p represent the gene set size. 1000 non-overlapping gene sets were generated and Type I error rate for a statistical test is calculated as the proportion of gene sets detected by the test. We consider $p \in \{20, 60, 100\}$ and $N \in \{20, 40, 60\}$.

Detection Power. In a real gene set, the proportion of DV genes, the amount of difference in variance, and the intergene correlation vary. Therefore, three parameters: γ, the proportion of genes truly DV in a gene set, σ, the fold change in variance, and r, the strength of intergene correlation were introduced. We examine how these parameters influence the power of different tests. Two groups of samples of equal size, $N/2$ from p-dimensional normal distributions $N(\mathbf{0}, \Sigma_x)$ and $N(\mathbf{0}, \Sigma_y)$ to represent two biological phenotypes were generated. We consider the relationship between the covariance and correlation matrices where the correlation matrix $R = D^{-1}\Sigma D^{-1}$, $D = \sqrt{\mathrm{diag}(\Sigma)}$ and Σ is the covariance matrix.

Let Σ_x and Σ_y be $p \times p$ positive definite and symmetric covariance matrices. The diagonal elements of Σ_x are equal to 1 and off-diagonal elements are equal to r. Matrix Σ_y is defined as

$$\Sigma_y = \begin{bmatrix} A & B \\ C & D \end{bmatrix}$$

where A is a $\gamma p \times \gamma p$ matrix with $A_{ij} = \sigma$ for $i = j$ and $A_{ij} = r\sigma$ for $i \neq j$, B and C are respectively $\gamma p \times (1\text{-}\gamma)p$ and $(1\text{-}\gamma)p \times \gamma p$ matrices where $B_{ij} = C_{ij} = \sqrt{\sigma}\, r$ for all i and j, and D is a $(1-\gamma)p \times (1-\gamma)p$ matrix with $D_{ij} = 1$ for $i = j$ and $D_{ij} = r$ for $i \neq j$. We consider the parameters $\gamma \in \{0.25, 0.5, 0.75, 1\}$, $r \in \{0.1, 0.5, 0.9\}$, $\sigma \in [1, 5]$, $p = 20$, and $N = 40$.

Simulation Setup for GSA-DC

Since GSA-DC approaches are not yet frequently applied to RNA-seq data here again the simulation setup is presented for microarray data, assuming multivariate normal distribution of gene expressions. Let X and Υ be independent p-dimensional vectors with distribution functions $F_x = N(\mathbf{0}, \Sigma_x)$ and $F_y = N(\mathbf{0}, \Sigma_y)$.

Type I error rate. To simulate the null hypothesis H_0: $\Sigma_x = \Sigma_y$, we generated two samples of equal size, $N/2$ from the p-dimensional normal distribution $N(\mathbf{0}, I_{p\times p})$ where $I_{p\times p}$ is a $p \times p$ identity matrix. We generated 1000 gene sets and Type I error rate for a statistical test is the proportion of gene sets detected by the test. We consider $p \in \{20, 100, 200\}$ and $N \in \{20, 40, 60\}$.

Detection Power. In a real biological setting, the proportion of co-expressed genes in a gene set varies and intergene correlations vary in strength. Therefore, two parameters: γ, the proportion of genes truly co-expressed in a gene set, and r, the strength of intergene correlation were introduced. We examine how these parameters influence the power of different tests. We simulated two groups of samples of equal size, $N/2$ ($N = 40$) from p-dimensional normal distributions $N(\mathbf{0}, \Sigma_x)$ and $N(\mathbf{0}, \Sigma_y)$ to represent two biological phenotypes where $p \in \{20, 100, 200\}$. We test the null hypothesis H_0: $\Sigma_x = \Sigma_y$, against the alternative $\Sigma_x \neq \Sigma_y$. To ensure that Σ_x and Σ_y are positive-definite and symmetric, two different scenarios for the alternative hypothesis were studied.

First, Σ_x was set to $I_{p\times p}$ and Σ_y was set such that its elements are

$$\sigma_{ij} = \begin{cases} r & i \neq j, \forall i, j \leq \gamma p \\ 0 & i \neq j, \forall i, j > \gamma p \\ 1 & i = j. \end{cases}$$

We consider $\gamma \in \{0.25, 0.5, 0.75, 1\}$ and $r \in \{0.1, 0.2, \ldots, 0.9\}$. Figure 2 (parts A and B) depicts the covariance matrices Σ_x and Σ_y under this scenario for $p = 20$ and $\gamma = 0.25$. Dark and light colors represent high and low correlations, respectively. This design presents a gene set with low intergene correlations under one

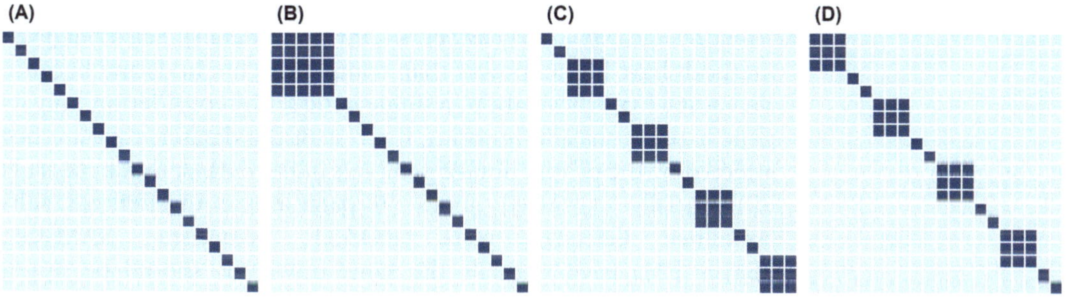

Fig. 2 The correlation matrices of the two simulation setups with sample size $N = 40$ and gene set size $p = 20$. Parts (**A**) and (**B**) respectively represent the correlation matrices of two conditions when the alternative hypothesis of setup 1 is true and $\gamma = 0.25$. Parts (**C**) and (**D**) respectively represent the correlation matrices of two conditions when the alternative hypothesis of setup 2 is true, $\beta = 0.25$ and $\gamma = 0.6$. *Dark* and *light colors* respectively represent high and low correlation coefficients

phenotype (Fig. 2A) and one group of highly co-expressed genes under the second phenotype (Fig. 2B).

Second, both Σ_x and Σ_y are set such that they have diagonal blocks of equal size βp, where β is the ratio of block size to gene set size. For each of the diagonal blocks, the first scenario is reproduced. Therefore, each diagonal block has $\gamma\beta p$ genes with intergene correlation specified by r while all the other genes in the block have zero correlations. The locations of the $\gamma\beta p$ co-expressed genes inside each block are assigned differently for Σ_x and Σ_y under the alternative hypothesis. While for Σ_x these genes occupy the upper-left corner of the block, for Σ_y they occupy the lower-right corner. Figure 2 (C, D) depicts this scenario for $p = 20$, $\beta = 0.25$, and $\gamma = 0.6$. Dark and light colors represent high and low correlations, respectively. Depending on γ, the diagonal blocks in Σ_x and Σ_y may have a few common genes (when $\beta > 0.5$) or may be exclusive (when $\beta \leq 0.5$). We consider the case $\beta = 0.25$ and let $\gamma = 0.6, 0.4$, and 0.5 respectively when $p = 20, 100$, and 200 to allow $\gamma\beta p$ to be an integer number. These settings yield diagonal blocks of 3, 10, and 25 genes respectively when $p = 20, 100$, and 200. All intergene correlations outside the diagonal blocks are set to zero. This setup presents a gene set with low intergene correlations except for a selected group of highly co-expressed genes and the membership of the genes in this group is changing between the two phenotypes.

3.2 Comparative Power Analysis and Type I Error Rate: Results

Results for GSA-DE

Type I error rate. Table 1 presents the estimates of the attained significant levels for all GSA tests considered ($\alpha = 0.05$). Overall, self-contained and competitive tests control Type I error rate near nominal $\alpha = 0.05$. For more detailed discussion of Type I error rates for self-contained and competitive GSA-DE tests, *see* [47].

Power. Figure 3 presents the power estimates when H_1: $\mu_x \neq \mu_y$ is true ($N = 20$, $p = 16$) (*see* ref. 47 for more detail). Self-contained methods have higher power than competitive methods and because they test a hypothesis about a single gene set by considering only its gene expressions and ignoring the rest of the dataset, they are not affected by the proportion of gene sets in the dataset that have truly differentially expressed genes (β parameter). Overall, all self-contained GSA-DE tests (ROAST, N-statistic, SAM-GS, eBayes_FM) have virtually the same power. It should be noted that the simulation setup here does not include intergene correlations. This is why there is no difference in power of multivariate and univariate self-contained approaches. For simulation setup that includes intergene correlations, we refer the reader to [10, 28]. The power of ROMER demonstrates dependence on the proportion of truly DE genes in a gene set (parameter γ). While the power is relatively low at $\gamma = 0.125$, it increases drastically at higher γ

Table 1
Estimated Type I error rates for GSA-DE methods, $\alpha = 0.05$

		Self-contained			N-statistic		SAM-GS		ROAST	
		Competitive			GSVA		ssGSEA		ROMER	
Method placement		Combined *P*-value			eBayes_FM		–		–	
	$P = 16$				$P = 60$		$P = 100$			
$N = 10$	Self.	0.049	0.044	0.084	0.048	0.045	0.042	0.048	0.045	0.041
	Comp.	0.025	0.042	0.047	0.017	0.047	0.050	0.013	0.045	0.047
	Comb.	0.047	–	–	0.042	–	–	0.044	–	–
$N = 20$	Self.	0.052	0.046	0.044	0.055	0.050	0.047	0.051	0.055	0.050
	Comp.	0.040	0.047	0.051	0.038	0.041	0.054	0.037	0.050	0.053
	Comb.	0.048	–	–	0.051	–	–	0.054	–	–
$N = 40$	Self.	0.054	0.054	0.051	0.047	0.047	0.044	0.050	0.053	0.055
	Comp.	0.051	0.044	0.050	0.057	0.048	0.045	0.060	0.049	0.052
	Comb.	0.051	–	–	0.047	–	–	0.055	–	–
$N = 60$	Self.	0.051	0.051	0.052	0.046	0.047	0.048	0.049	0.054	0.054
	Comp.	0.060	0.046	0.051	0.061	0.051	0.049	0.066	0.047	0.050
	Comb.	0.052	–	–	0.046	–	–	0.055	–	–

values. Competitive methods have slightly lower power for higher values of β especially ROMER. This observation can be explained by the fact that competitive methods are influenced by adding more genes to the dataset where adding non-DE genes enhances their power [36], while adding DE genes may decrease it.

The lack of power under all settings demonstrated by unsupervised competitive methods (especially GSVA) can be explained by the sample-wise ranking they perform to calculate the enrichment scores for gene sets [47]. While half of the genes in the gene set are up-regulated under one phenotype, the other half are up-regulated under the other phenotype. This setup maintains a stable enrichment score for the gene set under all samples and hence the gene set is found non-DE between the two phenotypes. When all DE genes in the gene set are up-regulated under one phenotype only, samples under that phenotype would have had higher gene set enrichment scores compared to the samples under the second phenotype. To substantiate this explanation with simulation results, we consider two hypothetical cases of expression patterns in a gene set consisting of 16 genes. In the first case, all DE genes in a gene set are up-regulated in phenotype 1 compared to phenotype 2. These genes normally have higher ranks in samples under phenotype 1 compared to samples under phenotype 2, and hence the gene set has

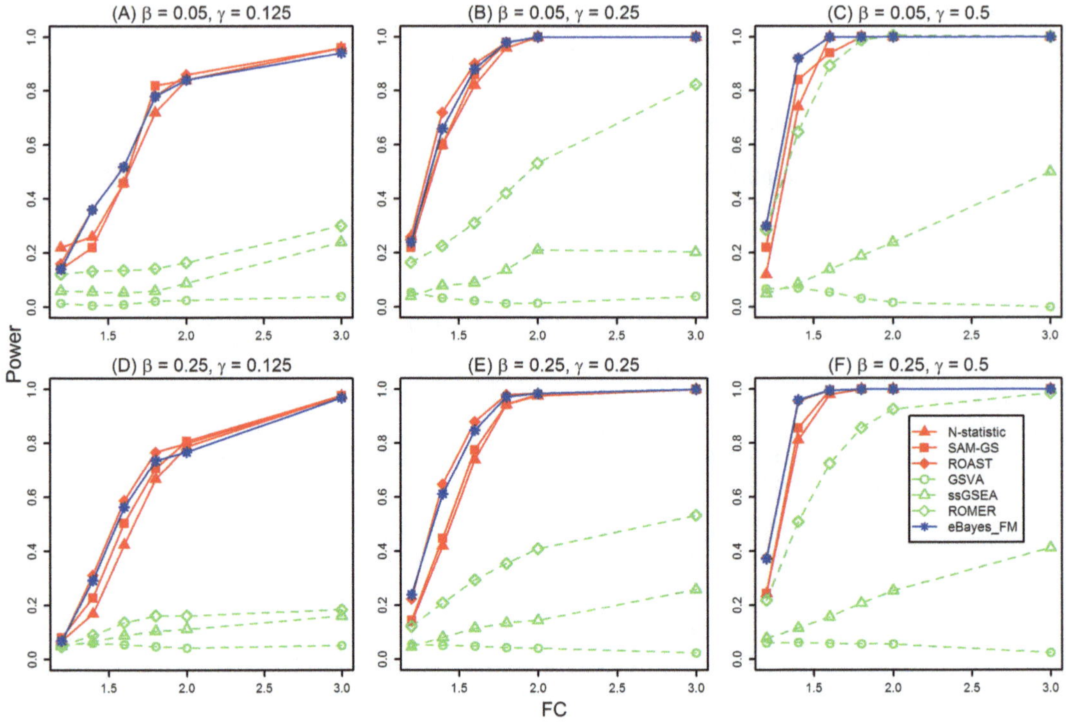

Fig. 3 The power of different DE tests to detect differential expression between two phenotypes of samples when the alternative hypothesis $\overline{\mu}_x \neq \overline{\mu}_y$ is true with different settings (values of β, γ and FC). The gene set size is $p = 16$ and the sample size in each group is $N/2$ ($N = 20$). Half of the $\gamma \times p$ DE genes in a gene set are up-regulated under one phenotype and the other half are up-regulated under the other phenotype

higher enrichment score under phenotype 1 as compared to phenotype 2. This case is expected to demonstrate high power as shown in Fig. 4. Consider the second case where DE genes in a gene set are equally divided into up-regulated genes between phenotype 1 and phenotype 2, similar to the simulation setup that produced Fig. 3. While the up-regulated genes under phenotype 1 have higher ranks under phenotype 1 as compared to phenotype 2, the up-regulated genes under phenotype 2 are exactly the opposite. This case yields high (however lower than the first case) enrichment score for the gene set under all samples. Due to the expected small difference (if any) in average enrichment score between the two phenotypes, low power is expected (*see* Fig. 4). Since it is more likely to have both up-regulated and down-regulated genes between two phenotypes in a real gene set than having all up-regulated or down-regulated genes, the power of supervised competitive methods is likely to be consistently lower than other methods for real expression data. It should be noted that the authors of the ssGSEA method expected their enrichment score to be slightly more robust and more sensitive to

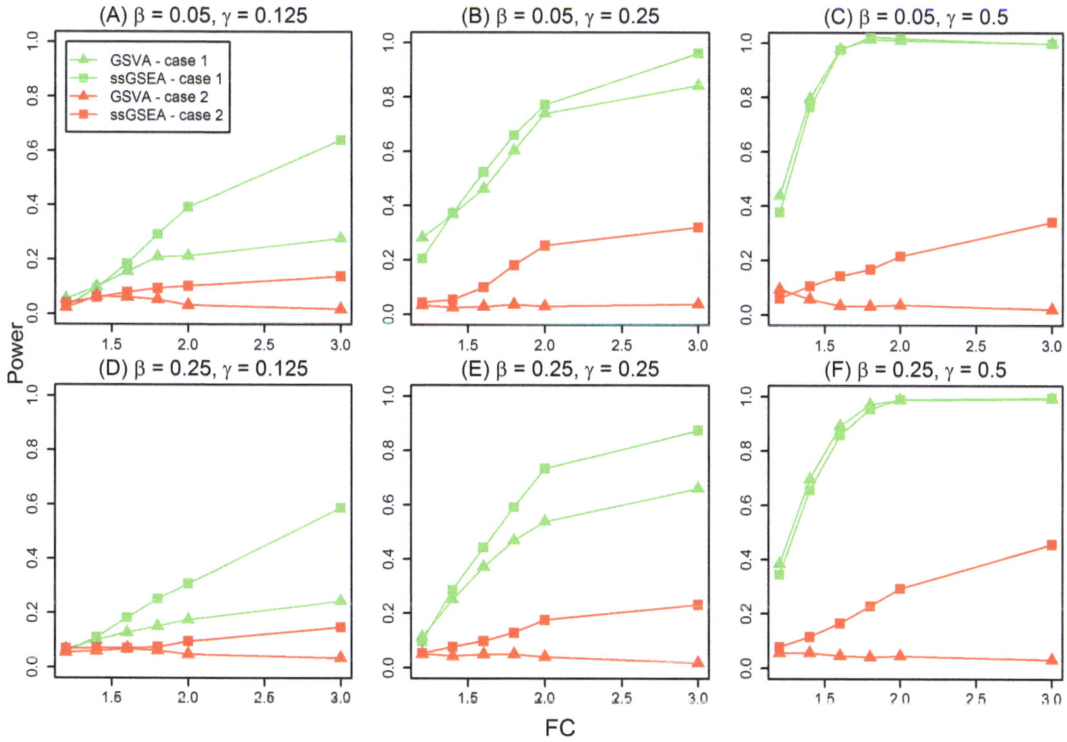

Fig. 4 The power of unsupervised competitive tests (GSVA and ssGSEA) to detect differences between two phenotypes when the alternative hypothesis $\overline{\mu}_x \neq \overline{\mu}_y$ is true with different settings (values of β, γ, and FC). The gene set size $p = 16$ and the sample size in each group is $N/2$ ($N = 20$). In case 1, all the γp DE genes in a gene set are up-regulated in phenotype 1 as compared to phenotype 2. In case 2, half of the γp DE genes in a gene set are up-regulated in phenotype 1 and the other half are up-regulated in phenotype 2. Both GSVA and ssGSEA have much higher power under case 1

Table 2
Estimated Type I error rates for GSA-DV methods, $\alpha = 0.05$

Method	N-statistic			*F*-test			RKS		
p	20	60	100	20	60	100	20	60	100
$N = 20$	0.050	0.047	0.050	0.044	0.040	0.043	0.036	0.025	0.049
$N = 40$	0.052	0.060	0.045	0.057	0.040	0.052	0.047	0.038	0.039
$N = 60$	0.053	0.035	0.049	0.053	0.054	0.056	0.041	0.038	0.034

differences in the tails of the distributions compared to the Kolmogorov–Smirnov-like statistic [39]. The simulation results in Fig. 4 confirm this expectation.

Results for GSA-DV

Type I error rate. Table 2 presents the estimates of the attained significant levels for all GSA-DV approaches considered ($\alpha = 0.05$).

Overall, RKS test is slightly more conservative than N-statistic and gene-level GSA-DV test that combines *F*-tests *P*-values with FM.

Power. Figure 5 presents the power estimates for the three GSA-DV approaches considered against the alternative hypothesis $\overline{\sigma}_x \neq \overline{\sigma}_y$. It appears that in contrast with GSA-DE approaches, where multivariate tests always outperform univariate tests when correlation increases, multivariate N-statistic and RKS have lower power than gene-level GSA-DV test that combines *F*-test *P*-values with FM in all settings. Gene-level GSA-DV test has the highest power, RKS test has an intermediate power, and N-statistic has the lowest power in all settings (Fig. 5). This pattern can be explained by the fact that it is much easier to satisfy the alternative hypothesis tested by the gene-level GSA-DV under our simulation setup than the alternatives tested by both N-statistic and RKS. N-statistic and RKS both test H_0: $F_x = F_y$, with different alternatives $F_x \neq F_y$ and $\overline{\sigma}_x \neq \overline{\sigma}_y$, respectively. Thus, the rejection of H_0 in the case of N-statistic can happen when $\overline{\mu}_x \neq \overline{\mu}_y$, $\overline{\sigma}_x \neq \overline{\sigma}_y$ or other higher order moments of F_x, F_y are not equal. The rejection of H_0 in the case of RKS test is supposedly happened when $\overline{\sigma}_x \neq \overline{\sigma}_y$, but not necessary so because the RKS test is just "more sensitive" to "differences in scale" as compared to "shift differences" [9]. It means that both tests are sensitive to not strictly one alternative, while gene-level GSA-DV test that combines *F*-test *P*-values with FM is sensitive to only the case when genes in a gene set are DV genes between two conditions. Figure 6 illustrates this point by showing the estimated power when the alternative hypothesis $\overline{\mu}_x \neq \overline{\mu}_y$ is true. The power trend is just the opposite of the trend presented in Fig. 5. Here, N-statistic has the highest power, RKS test has an intermediate power, and gene-level GSA-DV test has the lowest power in all settings (Fig. 6).

Results for GSA-DC

Type I error rate. Table 3 presents the estimates of the attained significant levels for the three GSA-DC tests considered ($\alpha = 0.05$). Overall, all tests control Type I error rate near nominal $\alpha = 0.05$.

Power. Figure 7 presents power estimates under the first simulation setup (see the simulation setup for GSA-DC) for different parameter settings. For each parameter setting, the results are obtained from 1000 independent gene sets. First, consider the case when only 25% of genes in a gene set are co-expressed ($\gamma = 0.25$). This case is highly plausible in real expression data since only a few genes in a gene set are expected to be highly co-expressed [25, 27]. GSNCA has the highest power followed respectively by GSCA and CoGA for all settings ($p = \{20, 100, 200\}$). Second, consider the case when 50% of genes in a gene set are co-expressed ($\gamma = 0.5$). While all tests show similar power when the size of gene set is relatively small ($p = 20$), GSNCA outperforms both GSCA and CoGA when the size of gene set is relatively large

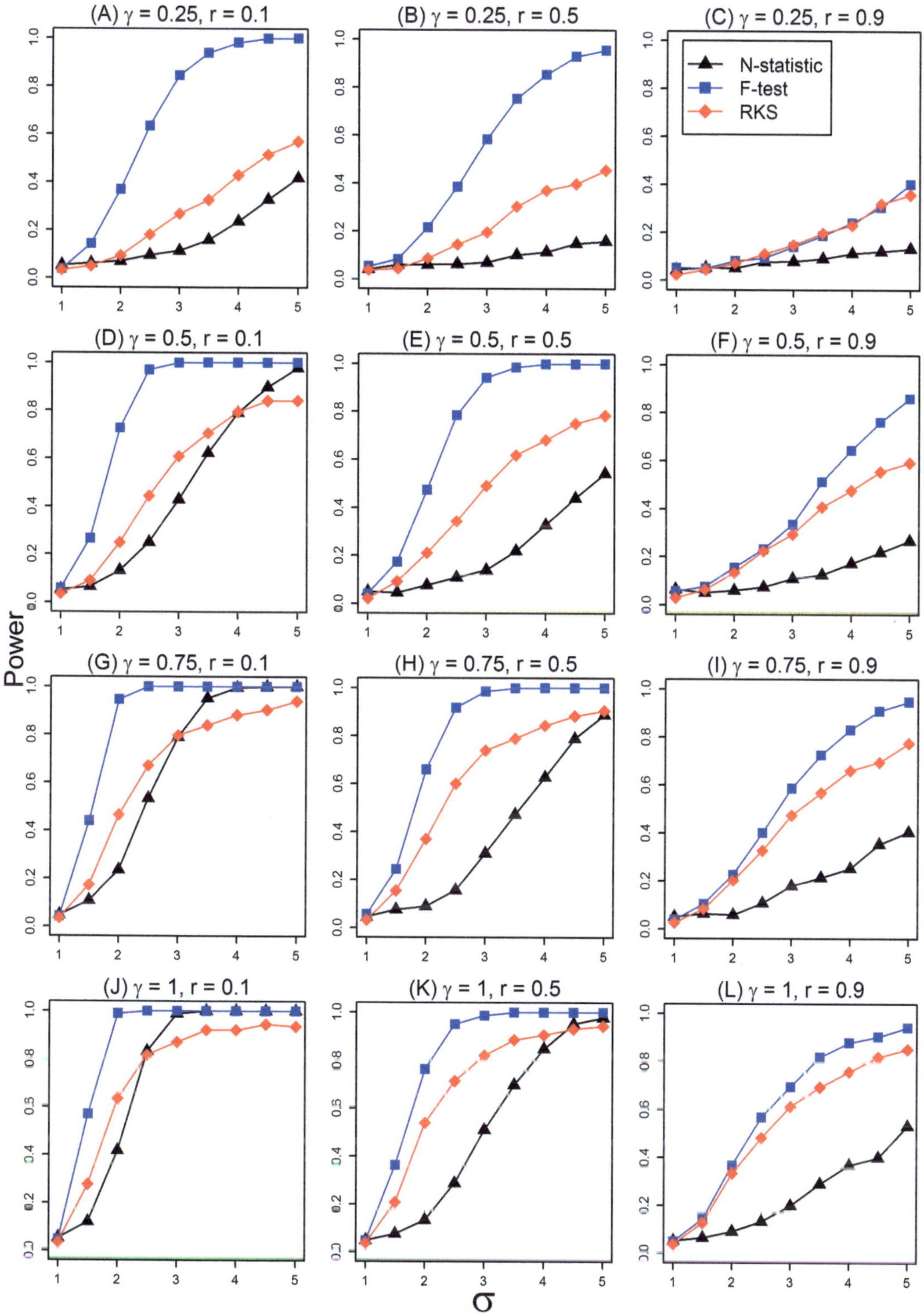

Fig. 5 The power of three GSA-DV tests to detect differential expression between two phenotypes of samples when the alternative hypothesis $\overline{\sigma}_x \neq \overline{\sigma}_y$ is true with different settings (values of β, γ, and σ). The gene set size $p = 20$ and the sample size in each group is $N/2$ ($N = 20$)

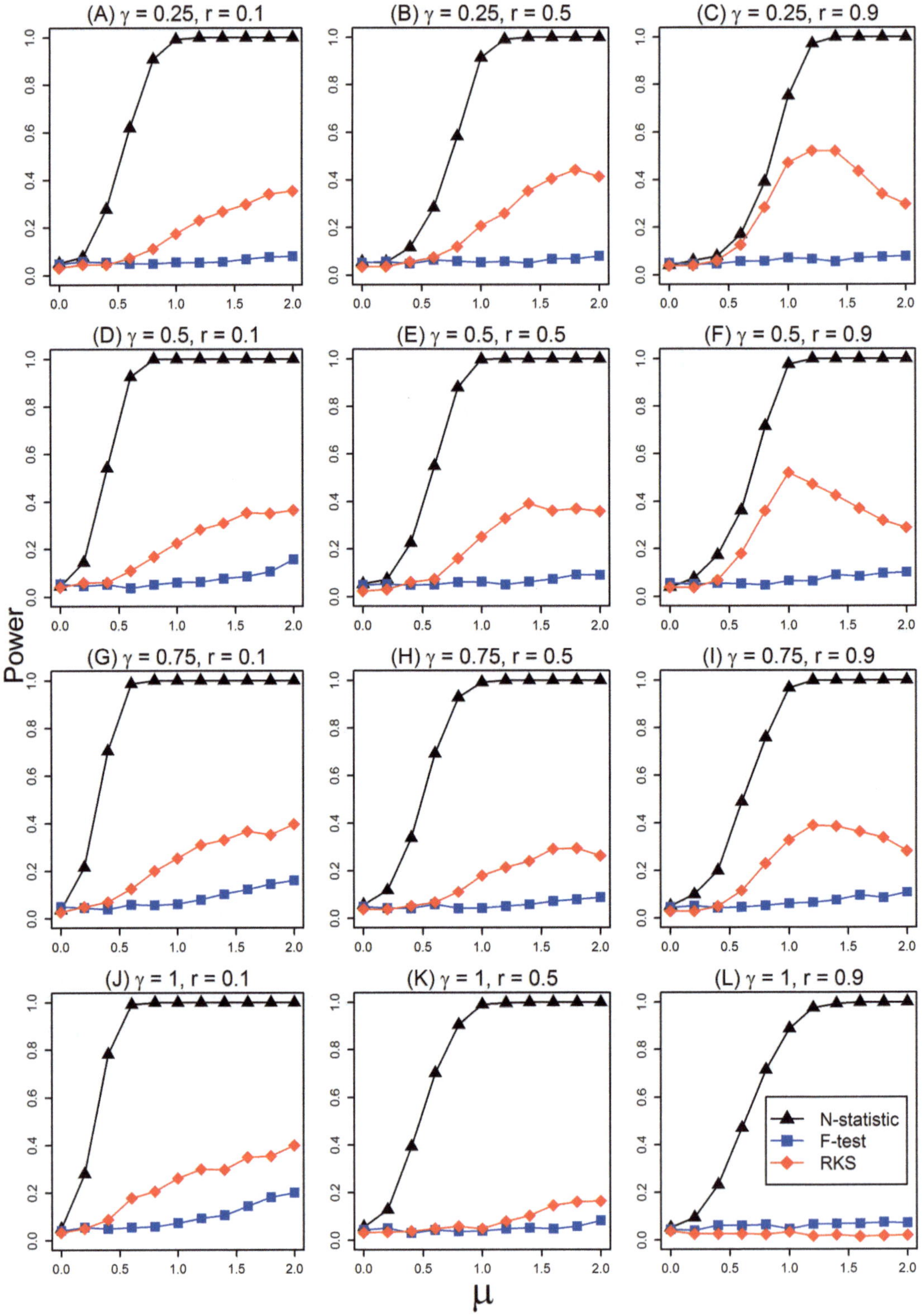

Fig. 6 The power of three GSA-DV tests to detect differential expression between two phenotypes of samples when the alternative hypothesis $\overline{\mu}_x \neq \overline{\mu}_y$ is true with different settings (values of β, γ, and σ). The gene set size $p = 20$ and the sample size in each group is $N/2$ ($N = 20$)

Table 3
Estimated Type I error rates for GSA-DC methods, $\alpha = 0.05$

Method	GSCA			GSNCA			CoGA		
p	20	100	200	20	100	200	20	100	200
$N = 20$	0.057	0.050	0.044	0.052	0.042	0.045	0.053	0.048	0.056
$N = 40$	0.046	0.045	0.059	0.036	0.051	0.051	0.043	0.052	0.050
$N = 60$	0.052	0.049	0.047	0.054	0.047	0.054	0.043	0.048	0.050

($p = 100$ and $p = 200$). Third, consider the case when 75% of genes in a gene set are co-expressed ($\gamma = 0.75$). While GSCA and CoGA outperform GSNCA when the size of gene set is relatively small ($p = 20$), all tests have virtually the same power when the number of genes is relatively large ($p = 100$ and $p = 200$). Fourth, consider the case when 100% of genes in a gene set are co-expressed ($\gamma = 1$). GSCA and CoGA have similar power and GSNCA has virtually no power.

The statistic used in GSCA depends on the average pairwise correlation difference between the two phenotypes. Hence, power increases when γ becomes higher as shown in Fig. 7. Similar argument can be applied to CoGA where larger γ causes larger changes in the spectral distribution of the correlation matrix in one phenotype as compared to the other. When intergene correlation (r) is uniformly low in one phenotype and uniformly high in another phenotype ($\gamma = 1$ and r is high), eigenvectors corresponding to the largest eigenvalues for both correlation matrices remain unchanged while the eigenvalues (spectral distribution) change. Therefore, GSNCA does not detect changes regardless of the value of r when $\gamma = 1$, while CoGA shows high power. This case illustrates the fundamental difference between GSNCA and both GSCA and CoGA. Both GSCA and CoGA detect any differences in pairwise correlations, while GSNCA detects differences in the co-expression structure, i.e., when some pairwise correlations change relative to others in the same phenotypes. The greatest change in the co-expression structure between two phenotypes in the first simulation setup occurs when $\gamma = 0.5$ and hence GSNCA is expected to show highest power as shown in Fig. 7.

Figure 8 presents power estimates under the second simulation setup (see simulation setup for GSA-DC) for different parameter settings. When $p = 20$ and $\gamma = 0.6$ (diagonal block size $\gamma\beta p = 3$), GSCA outperforms GSNCA. When $p = 100$ and $\gamma = 0.4$ (diagonal block size $= 10$), both GSCA and GSNCA show similar power. When $p = 200$ and $\gamma = 0.5$ (diagonal block size $= 25$), GSNCA outperforms GSCA. The increment in the size of the diagonal block of differential correlations results in increased detection

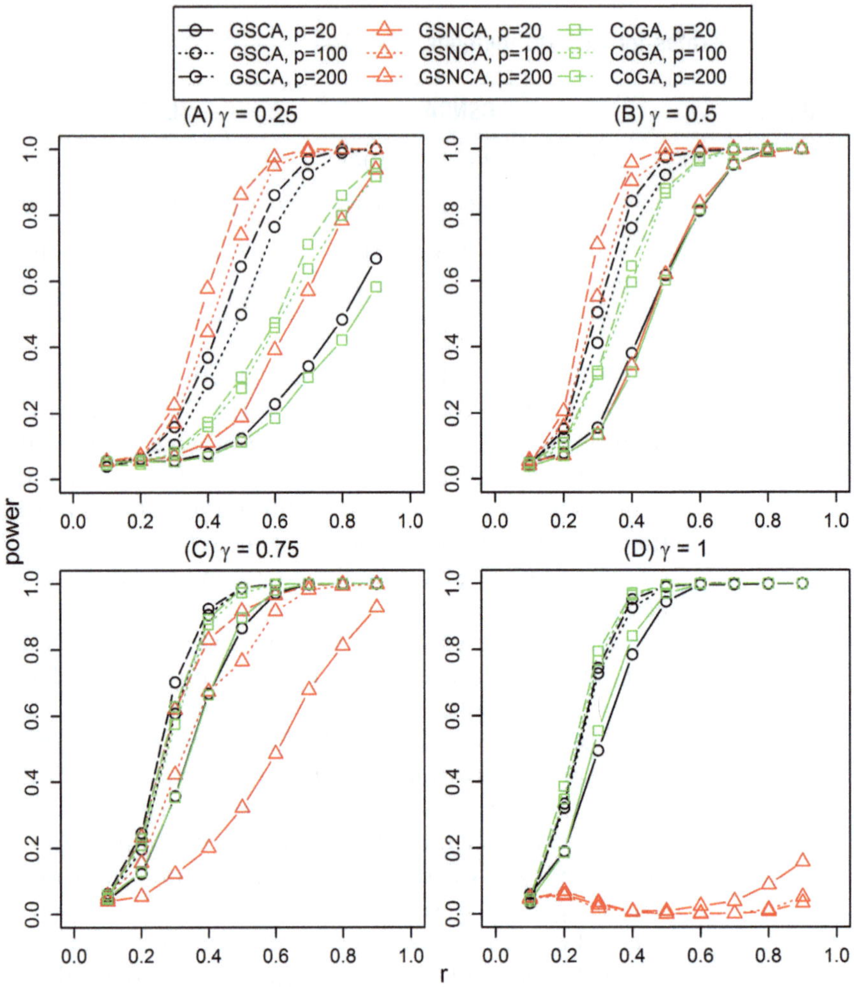

Fig. 7 The power of three GSA-DC tests to detect differential expression between two phenotypes of samples when the alternative hypothesis of the first simulation setup is true with different settings (values of γ and r). The gene set size $p = \{20, 100, 200\}$ and the sample size in each group is $N/2$ ($N = 40$)

power when the gene set size increases. When $p = 200$ and $\gamma = 0.5$, power follows similar pattern to what has been shown in Fig. 7 when $\gamma = 0.5$, i.e., GSNCA outperforms GSCA. The difference in power between GSNCA and GSCA when $p = 20$ and $\gamma = 0.6$ follows a similar pattern to what has been observed in Fig. 7 when $\gamma = 0.75$ and could be attributed to the correlation matrix in one phenotype moving closer to a uniformly high correlation pattern. CoGa has almost no power for all settings. This is explained by the fact that unlike eigenvectors the eigenvalues remain unchanged when the number of pairwise intergene correlations with value r remains unchanged but the set of pairwise correlations having value r differs between phenotypes.

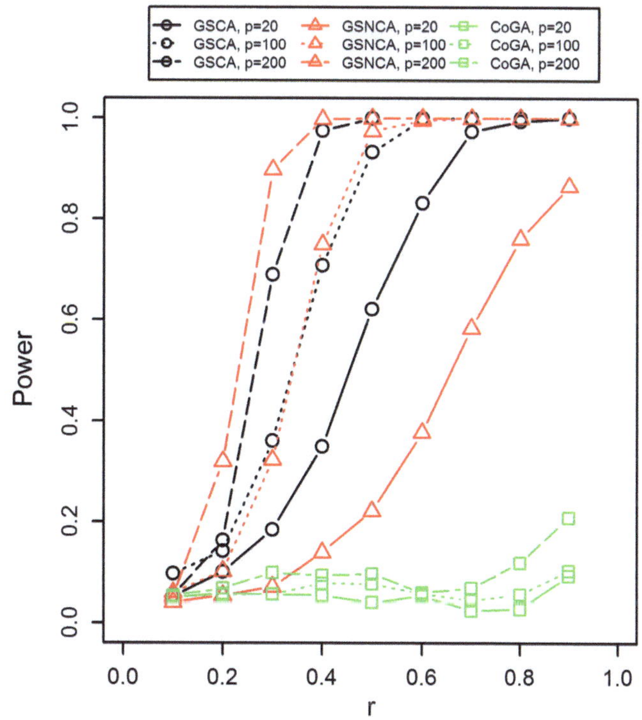

Fig. 8 The power of three GSA-DC tests to detect differential expression between two phenotypes of samples when the alternative hypothesis of the second simulation setup is true with different settings (values of β, γ, and r). The gene set size $p = \{20, 100, 200\}$ and the sample size in each group is $N/2$ ($N = 40$)

3.3 Application to Expression Data

We illustrate the use of GSA-DE, GSA-DV, and GSA-DC tests applied to the NCI-60 cell lines (p53) dataset. The p53 dataset comprises 50 samples of NCI-60 cell lines differentiated based on the status of the TP53 gene: 17 cell lines carrying normal (wild type, WT) TP53 gene and 33 cell lines carrying mutated TP53 (MUT) [38, 59]. For this data set, probe-level intensities were quantile normalized and transformed to the log scale. Gene sets were taken from the C2 pathways set of the molecular signature-database (MSigDB version 5.1) [38, 60, 61]. Pathways with less than 10 or more than 500 genes were discarded and the resulted dataset comprised 4256 gene sets.

Results for GSA-DE

To find pathways, differentially expressed between cancer cell lines with and without p53 mutation we applied SAM-GS. We choose SAM-GS because it tests a fairly simple null hypothesis, namely whether the difference in moderated t-statistics averaged over all pathway genes, is zero between two phenotypes. SAM-GS detected 44 gene sets at the given significance level ($P < 0.001$) (Table 4). All but one detected pathways were significantly enriched with p53

Table 4
Pathways differentially expressed between $p53^{WT}$ and $p53^{MUT}$ cell lines[a)]

	Gene set name	Size	TP53 targets	$P_{hypergeo}$	Target genes
1	KEGG_P53_SIGNALING_PATHWAY	53	34	1.1E-26	SFN_TSC2_CDK1_RCHY1_IGFBP3_SERPINB5_PPM1D_BID_MDM4_BAX_MDM2_TP53I3_PMAIP1_ATM_ATR_CHEK2_APAF1_CDKN2A_RRM2_CDKN1A_BBC3_GADD45A_TNFRSF10B_DDB2_SIAH1_PTEN_CDK2_CHEK1_SERPINE1_TP53_CCNE2_CCNB1_CCNG1_CCNG2
2	KEGG_AMYOTROPHIC_LATERAL_SCLEROSIS_ALS	48	12	9.3E-05	MAPK11_BID_MAPK13_BAX_DAXX_MAPK14_MAPK12_APAF1_GPX1_BCL2_BCL2L1_TP53
3	BIOCARTA_CHEMICAL_PATHWAY	22	12	4.7E-09	PTK2_BCL2_BID_BCL2L1_STAT1_PRKCA_APAF1_BAX_CASP6_TP53_ATM_PARP1
4	BIOCARTA_ATM_PATHWAY	19	15	1.4E-14	JUN_CHEK2_MAPK8_MRE11A_BRCA1_RELA_CHEK1_MDM2_ABL1_CDKN1A_TP53_GADD45A_ATM_RAD51_NFKBIA
5	BIOCARTA_CERAMIDE_PATHWAY	22	6	3.4E-03	BCL2_MAPK8_RELA_BAX_MAPK3_MAPK1
6	BIOCARTA_HIVNEF_PATHWAY	56	17	1.6E-07	CHUK_CFLAR_PRKCD_PTK2_BID_MDM2_CASP6_RB1_DAXX_PRKDC_NFKBIA_MAPK8_RELA_APAF1_TRAF1_BCL2_PARP1
7	BIOCARTA_P53HYPOXIA_PATHWAY	21	17	1.0E-16	MAPK8_CSNK1D_CSNK1A1_EP300_BAX_HIF1A_HIC1_MDM2_TAF1_CDKN1A_TP53_NQO1_GADD45A_HSP90AA1_IGFBP3_ATM_RPA1
8	BIOCARTA_IL22BP_PATHWAY	13	2	2.3E-01	STAT1_STAT6
9	BIOCARTA_P53_PATHWAY	16	12	2.0E-11	BCL2_E2F1_APAF1_BAX_CDK2_MDM2_RB1_CDKN1A_TP53_GADD45A_ATM_PCNA
10	BIOCARTA_BAD_PATHWAY	26	6	8.3E-03	IGF1R_BCL2_BCL2L1_BAX_MAPK3_MAPK1
11	SA_G1_AND_S_PHASES	13	6	1.3E-04	E2F1_CDK2_MDM2_TP53_CDKN2A_CDKN1A
12	SA_PROGRAMMED_CELL_DEATH	10	6	2.0E-05	BCL2_BAX_BCL2L1_BAK1_BID_APAF1

13	PID_P73PATHWAY	69	41	5.0E-30	TP53I3_PML_GRAMD4_WT1_EP300_GDF15_RCHY1_BAX_MDM2_PLK3_CCNB1_S100A2_AFP_MAPK11_KAT5_SERPINE1_WWOX_BRCA2_CCNA2_PIN1_MYC_CDK1_MAPK14_BBC3_ABL1_CDK2_PLK1_SFN_RELA_ITCH_SP1_RB1_RAD51_TP63_CHEK1_BAK1_UBE4B_CCNE2_BUB1_FOXO3_CDKN1A
14	PID_HDAC_CLASSIII_PATHWAY	18	11	4.1E-09	FOXO1_CREBBP_HDAC4_BAX_TP53_XRCC6_FOXO3_EP300_KAT2B_CDKN1A_TUBB2A
15	PID_REG_GR_PATHWAY	78	36	1.9E-21	BAX_NR4A1_MAPK8_HDAC1_STAT1_HDAC2_EGR1_NCOA2_GSK3B_SMARCD1_RELA_SUMO2_TP53_SMARCA4_TBP_SFN_MDM2_MAPK3_CREBBP_EP300_FOS_MAPK9_MAPK10_NR3C1_HSP90AA1_TSG101_MAPK1_MAPK14_MAPK11_NCOA1_SMARCC1_JUN_AFP_CREB1_CDKN1A_CDK5
16	PID_P53_DOWNSTREAM_PATHWAY	110	97	1.6E-99	DDIT4_FDXR_SERPINB5_IGFBP3_GPX1_ATF3_MAP4K4_BNIP3L_TSC2_BCL2_TP63_TP53_MMP2_TNFRSF10D_SP1_BBC3_PRKAB1_TYRP1_CEBPZ_NFYB_MLH1_PCNA_SMARCA4_BAK1_JUN_CARM1_VCAN_TAF9_AFP_CSE1L_IRF5_PRDM1_NFYA_CCNG1_MDM2_MET_PTEN_TFDP1_CAV1_CCNB1_CX3CL1_ARID3A_PML_DDX5_BDKRB2_TP53I3_HIC1_GADD45A_TGFA_APC_NFYC_SERPINE1_PRMT1_BTG2_SH2D1A_BAX_TRIAP1_RB1_VDR_KAT2A_TRRAP_TNFRSF10B_EP300_HTT_NDRG1_MSH2_PPP1R13B_DDB2_CASP10_GDF15_LIF_CASP6_EGFR_PLK3_SNAI2_DKK1_CTSD_EPHA2_COL18A1_RCHY1_PCBP4_SFN_BCL2L2_E2F1_BCL6_BID_S100A2_BCL2L1_DUSP5_CDKN1A_APAF1_CREBBP_TP53BP2_HDAC2_MCL1_EDN2_PMAIP1
17	PID_RXR_VDR_PATHWAY	25	9	3.0E-05	NR4A1_VDR_NCOA1_THRB_RARA_TGFB1_SREBF1_BCL2_MED1
18	PID_TAP63_PATHWAY	48	29	7.6E-22	EP300_NQO1_SERPINB5_YWHAQ_S100A2_CDKN1A_PML_BBC3_GADD45A_CHUK_TP63_SP1_MDM2_NOC2L_PMAIP1_ITCH_PRKCD_IGFBP3_GPX2_TFAP2C_TP53I3_CDKN2A_VDR_PLK1_BAX_IKBKB_FDXR_GDF15_ABL1

(continued)

Table 4
(continued)

	Gene set name	Size	TP53 targets	$P_{hypergeo}$	Target genes
19	PID_P53_REGULATION_PATHWAY	46	43	1.0E-46	NEDD8_PPM1D_HIPK2_CHEK2_TP53_TRIM28_CCNG1_HUWE1_CSNK1E_SKP2_ATM_CSE1L_CSNK1D_CSNK1A1_CDK2_PIN1_CHEK1_MDM4_KAT5_CCNA2_SMYD2_EP300_KAT2B_KAT8_DAXX_DYRK2_RPL11_PRKCD_MDM2_CDKN2A_ATR_ABL1_CSNK1G2_GSK3B_CREBBP_MAPK14_MAPK9_RPL23_USP7_RCHY1_UBE2D1_YY1_MAPK8
20	PID_RB_1PATHWAY	58	28	1.3E-17	SMARCB1_CDKN1A_JUN_CREBBP_TBP_SMARCA4_MAPK14_TFDP1_DNMT1_CTBP1_PAX3_MET_MAPK11_SKP2_RBBP4_ABL1_EP300_HDAC3_CDKN2A_CCNA2_E2F1_CDK2_HDAC1_TAF1_MAPK9_MDM2_RB1_CEBPB
21	REACTOME_SIGNALING_BY_ERBB2	85	19	5.4E-06	CDKN1A_CHUK_CREB1_EGFR_FOXO1_FOXO3_MTOR_GRB2_NR4A1_HSP90AA1_MDM2_PRKCA_PRKCD_MAPK1_MAPK3_PTEN_RPS27A_TSC2_CDK1
22	REACTOME_PI3K_EVENTS_IN_ERBB2_SIGNALING	36	12	3.6E-06	CDKN1A_CHUK_CREB1_EGFR_FOXO1_FOXO3_MTOR_GRB2_NR4A1_MDM2_PTEN_TSC2
23	REACTOME_PI3K_AKT_ACTIVATION	31	10	3.3E-05	CDKN1A_CHUK_CREB1_FOXO1_FOXO3_MTOR_NR4A1_MDM2_PTEN_TSC2
24	REACTOME_AKT_PHOSPHORYLATES_TARGETS_IN_THE_CYTOSOL	11	4	5.5E-03	CDKN1A_CHUK_MDM2_TSC2
25	REACTOME_GAB1_SIGNALOSOME	30	12	3.7E-07	CDKN1A_CHUK_CREB1_EGFR_FOXO1_FOXO3_MTOR_GRB2_NR4A1_MDM2_PTEN_TSC2
26	REACTOME_DOWNSTREAM_SIGNAL_TRANSDUCTION	81	18	1.0E-05	CDKN1A_CHUK_CREB1_FOXO1_FOXO3_MTOR_GRB2_NR4A1_MDM2_PRKCA_PRKCD_MAPK1_MAPK3_PTEN_STAT1_STAT6_TSC2_CDK1

27	REACTOME_PIP3_ACTIVATES_AKT_SIGNALING	22	10	8.7E-07	CDKN1A_CHUK_CREB1_FOXO1_FOXO3_MTOR_NR4A1_MDM2_PTEN_TSC2
28	REACTOME_INTRINSIC_PATHWAY_FOR_APOPTOSIS	26	13	4.3E-09	E2F1_BBC3_APAF1_NMT1_PMAIP1_MAPK8_BAK1_BAX_BCL2_BCL2L1_BID_TFDP1_TP53
29	GAZDA_DIAMOND_BLACKFAN_ANEMIA_MYELOID_UP	26	4	1.0E-01	BAX_TNFRSF10B_DDB2_SRSF1
30	GARGALOVIC_RESPONSE_TO_OXIDIZED_PHOSPHOLIPIDS_BLACK_UP	20	3	1.6E-01	PLAGL1_CDKN1A_UBB
31	NUNODA_RESPONSE_TO_DASATINIB_IMATINIB_DN	13	6	1.3E-04	CDKN1A_IKBKB_CASP10_STAT6_STAT1_BAX
32	GALLUZZI_PERMEABILIZE_MITOCHONDRIA	38	19	1.0E-12	NR4A1_BBC3_PRKCD_SERPINB5_FDXR_BAK1_SIVA1_BCL2L1_PPID_BAX_BID_STK11_MAPK8_ABL1_TP53_MCL1_BCL2_PMAIP1_GSK3B
33	DUTTA_APOPTOSIS_VIA_NFKB	30	9	1.5E-04	BAX_TP53_MDM2_CFLAR_BCL2L1_AFP_TNFRSF10B_TRAF1_BCL2
34	GALLUZZI_PREVENT_MITOCHONDIAL_PERMEABILIZATION	20	9	3.4E-06	BCL2_MAPK14_MCL1_BCL2L1_BAK1_MUC1_TXN_BAX_BCL2L2
35	SCHAVOLT_TARGETS_OF_TP53_AND_TP63	14	11	6.1E-11	CDKN1A_SERPINB5_BAX_TP53I3_SFN_PMAIP1_EPHA2_VCAN_FDXR_MDM2_PCNA
36	AMUNDSON_DNA_DAMAGE_RESPONSE_TP53	15	8	2.4E-06	LIF_DDB2_MDM2_CDKN1A_CCNG1_CTSD_BTG2_PPM1D
37	FLECHNER_BIOPSY_KIDNEY_TRANSPLANT_OK_VS_DONOR_DN	25	5	2.8E-02	BAX_FOS_XRCC6_MYC_EIF2AK2
38	MA_MYELOID_DIFFERENTIATION_UP	35	9	5.6E-04	BNIP3L_PPM1G_CDKN1A_UBB_MT1A_RB1_CCT5_MDM2_NCL
39	GENTILE_UV_LOW_DOSE_UP	24	8	1.6E-04	CDKN1A_SOX4_BTG2_FDXR_SAT1_GDF15_PMAIP1_CSNK1G2
40	INGA_TP53_TARGETS	15	12	5.3E-12	MDM2_SFN_BAX_TNFRSF10B_FOS_PCNA_CCNG1_PMAIP1_BBC3_IGFBP3_CDKN1A_GADD45A

(continued)

Table 4 (continued)

	Gene set name	Size	TP53 targets	$P_{hypergeo}$	Target genes
41	DELLA_RESPONSE_TO_TSA_AND_BUTYRATE	19	5	8.8E-03	HSPB1_CDKN1A_PRKCD_NR4A1_GADD45A
42	ONO_FOXP3_TARGETS_DN	34	5	8.8E-02	PARP1_CCNA2_CDKN1A_E2F1_BCL2
43	WARTERS_RESPONSE_TO_IR_SKIN	37	13	7.2E-07	MDM2_DDB2_FDXR_TRIAP1_PPM1D_TP53TG1_BTG2_GDF15_BBC3_CDKN1A_CCNG1_STRA13_SERPINB5
44	WARTERS_IR_RESPONSE_5GY	21	12	2.3E-09	MDM2_CDKN1A_GDF15_FDXR_BBC3_DDB2_PPM1D_PLK3_ATF3_BTG2_PCNA_GADD45A

[a]) *P*-values were calculated using hypergeometric test. The names of p53 target genes for a pathway are listed

target genes (Table 4). This is not a surprise because if the expression level of a regulator changes, so do the levels of the regulated genes, leading to significant differences in the average expression of pathways, enriched with p53 targets.

Results for GSA-DV

To find pathways with differential variability between cancer cell lines with and without p53 mutation, the gene-level test combining *F*-test *P*-values was applied. It detected only three pathways, between WT and MUT phenotypes at a significance level $P < 0.001$. These pathways are "BANDRES RESPONSE TO CARMUSTIN WITHOUT MGMT 24HR UP," "BANDRES RESPONSE TO CARMUSTIN MGMT 48HR UP," and "ONGUSAHA BRCA1 TARGETS DN." These pathways are not significantly enriched in p53 targets. The first two pathways represent cellular response to carmustine treatment that involves regulation of complex pathways responsible for cell death [62]. All of them employ directly or indirectly expression of p53 gene and expectedly mutation in this gene results in different variability in these pathways. The "ONGUSAHA BRCA1 TARGETS DN" pathway consists of BRCA1 target genes [63] and because the p53 protein regulates BRCA1 transcription, mutation in p53 interferes with gene' functions, in particular regulation of BRCA1. This may cause indirect mixed effects on regulation of BRCA1 targets.

Results for GSA-DC

To find pathways, differentially co-expressed between cancer cell lines with and without p53 mutation GSNCA test was applied. GSNCA detected only four pathways differentially co-expressed between two phenotypes at a significance level $P < 0.001$. Two of them ("KEGG PEROXISOME," "REACTOME NOREPINEPHRINE NEUROTRANSMITTER RELEASE CYCLE") are related to crucial metabolic processes such as fatty acid oxidation, biosynthesis of ether lipids, and free radical detoxification and release of noradrenalin synaptic vesicle. One is related to changes in DNA methylation and histone acetylation ("ZHONG RESPONSE TO AZACITIDINE AND TSA UP") and one with changes in gene expressions related to intercellular matrix ("PEDERSEN METASTASIS BY ERBB2 ISOFORM 4"). These pathways are also not significantly enriched in p53 target genes.

In addition to detecting DC pathways GSNCA identifies hub genes—genes with the largest weights in each pathway. Hub genes provide useful biological information beyond the test result that a pathway is differentially co-expressed between two conditions. For example, pathway KEGG PEROXISOME (Fig. 9) presents genes that play key roles in redox signaling and lipid homeostasis. For p53 wild-type data, hub gene is MVK (mevalonate kinase Fig. 9A), which encodes the peroxisomal enzyme mevalonate kinase, a key

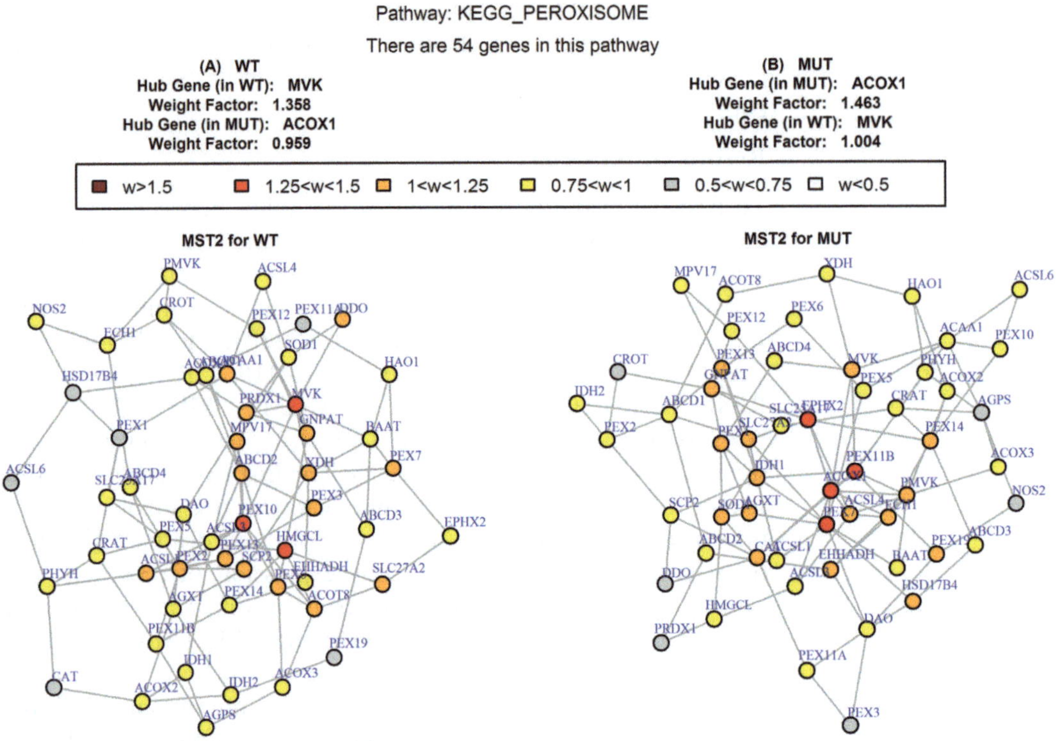

Fig. 9 MST2s of pathway "KEGG PEROXISOME" co-expression network under both (**A**) wild-type (WT) and (**B**) mutated (MUT) p53 phenotypes

early enzyme in isoprenoid and sterol synthesis. When p53 is mutated (Fig. 9B), hub gene becomes ACOX1 (acyl-coenzyme A oxidase 1, palmitoyl) that is the first enzyme of the fatty acid beta-oxidation pathway which catalyzes the desaturation of acyl-CoAs to 2-transenoyl-CoAs. That is in p53 MUT phenotype a shift from isoprenoid and sterol synthesis to fatty acid beta-oxidation pathway may happen. For more discussion of hub genes the reader is referred to [18].

4 Conclusion

In this chapter, we provided an in-depth review of univariate and multivariate Gene Set Analysis approaches (GSA-DE, GSA-DV, GSA-DC) for testing different statistical hypotheses. A comparative power analysis and Type I error rate estimates for different approaches in each major type of GSA provide concise guidelines for selecting GSA approaches that are best performing under particular experimental settings. An example was presented applying the methods GSA-DE, GSA-DV, GSA-DC on a p53 data set. This analysis demonstrated that different GSA types are allowing to obtain new and complementary biological information for the same underlying data set.

Acknowledgments

We would like to thank **Bárbara Macías Solís** for proof reading of the manuscript. Support has been provided in part by the Arkansas INBRE program, with grants from the National Center for Research Resources (P20RR016460) and the National Institute of General Medical Sciences (P20 GM103429) from the National Institutes of Health. Large-scale computer simulations were implemented using the High Performance Computing (HPC) resources at the UALR Computational Research Center supported by the following grants: National Science Foundation grants CRI CNS-0855248, EPS-0701890, MRI CNS-0619069 and OISE-0729792.

References

1. Mootha VK et al (2003) PGC-1alpha-responsive genes involved in oxidative phosphorylation are coordinately downregulated in human diabetes. Nat Genet 34(3):267–273
2. Bar HY, Booth JG, Wells MT ((2012)) A mixture-model approach for parallel testing for unequal variances. Stat Appl Genet Mol Biol 11(1.) p. Article 8
3. Ho JW et al (2008) Differential variability analysis of gene expression and its application to human diseases. Bioinformatics 24(13): i390–i398
4. Hulse AM, Cai JJ (2013) Genetic variants contribute to gene expression variability in humans. Genetics 193(1):95–108
5. Mar JC et al (2011) Variance of gene expression identifies altered network constraints in neurological disease. PLoS Genet 7(8): e1002207
6. Xu Z et al (2011) Antisense expression increases gene expression variability and locus interdependency. Mol Syst Biol 7:468
7. Bravo HC et al (2012) Gene expression antiprofiles as a basis for accurate universal cancer signatures. BMC Bioinform 13:272
8. Dinalankara W, Bravo HC (2015) Gene expression signatures based on variability can robustly predict tumor progression and prognosis. Cancer Informat 14:71–81
9. Friedman JH, Rafsky LC (1979) Multivariate generalizations of the Wald-Wolfowitz and Smirnov two-sample tests. Ann Stat 7 (4):697–717
10. Rahmatallah Y, Emmert-Streib F, Glazko G (2012) Gene set analysis for self-contained tests: complex null and specific alternative hypotheses. Bioinformatics 28 (23):3073–3080
11. Afsari B, Geman D, Fertig EJ (2014) Learning dysregulated pathways in cancers from differential variability analysis. Cancer Informat 13 (Suppl 5):61–67
12. Fisher R (1932) Statistical methods for research workers. Oliver and Boyd, Edinburg
13. Stadler N, Mukherjee S (2015) Multivariate gene-set testing based on graphical models. Biostatistics 16(1):47–59
14. Friedman J, Hastie T, Tibshirani R (2008) Sparse inverse covariance estimation with the graphical lasso. Biostatistics 9(3):432–441
15. Meinshausen N, Bühlmann P (2006) High-dimensional graphs and variable selection with the lasso. Ann Stat 34(3):1436–1462
16. Schafer J, Strimmer K (2005) A shrinkage approach to large-scale covariance matrix estimation and implications for functional genomics. Stat Appl Genet Mol Biol 4(1): Article 32
17. Choi Y, Kendziorski C (2009) Statistical methods for gene set co-expression analysis. Bioinformatics 25(21):2780–2786
18. Rahmatallah Y, Emmert-Streib F, Glazko G (2014) Gene Sets Net Correlations Analysis (GSNCA): a multivariate differential coexpression test for gene sets. Bioinformatics 30 (3):360–368
19. Santos Sde S et al (2015) CoGA: an R package to identify differentially co-expressed gene sets by analyzing the graph spectra. PLoS One 10 (8):e0135831
20. Takahashi DY et al (2012) Discriminating different classes of biological networks by analyzing the graphs spectra distribution. PLoS One 7(12):e49949
21. Goeman JJ, Buhlmann P (2007) Analyzing gene expression data in terms of gene sets:

methodological issues. Bioinformatics 23 (8):980–987
22. Tian L et al (2005) Discovering statistically significant pathways in expression profiling studies. Proc Natl Acad Sci U S A 102 (38):13544–13549
23. Ackermann M, Strimmer K (2009) A general modular framework for gene set enrichment analysis. BMC Bioinform 10(1):47
24. Rahmatallah Y, Emmert-Streib F, Glazko G (2014) Comparative evaluation of gene set analysis approaches for RNA-Seq data. BMC Bioinform 15(1):397
25. Montaner D et al (2009) Gene set internal coherence in the context of functional profiling. BMC Genomics 10:197
26. Gatti DM et al (2010) Heading down the wrong pathway: on the influence of correlation within gene sets. BMC Genomics 11:574
27. Tripathi S, Emmert-Streib F (2012) Assessment method for a power analysis to identify differentially expressed pathways. PLoS One 7 (5):e37510
28. Glazko GV, Emmert-Streib F (2009) Unite and conquer: univariate and multivariate approaches for finding differentially expressed gene sets. Bioinformatics 25(18):2348–2354
29. Wang X et al (2011) Linear combination test for hierarchical gene set analysis. Stat Appl Genet Mol Biol 10(1.) Article 13
30. Hanzelmann S, Castelo R, Guinney J (2013) GSVA: gene set variation analysis for microarray and RNA-seq data. BMC Bioinform 14:7
31. Khatri P, Sirota M, Butte AJ (2012) Ten years of pathway analysis: current approaches and outstanding challenges. PLoS Comput Biol 8 (2):e1002375
32. Maciejewski H (2014) Gene set analysis methods: statistical models and methodological differences. Brief Bioinform 15(4):504–518
33. Nam D, Kim SY (2008) Gene-set approach for expression pattern analysis. Brief Bioinform 9 (3):189–197
34. Tamayo P et al (2012) The limitations of simple gene set enrichment analysis assuming gene independence. Stat Methods Med Res 25 (1):472–487
35. Tarca AL, Bhatti G, Romero R (2013) A comparison of gene set analysis methods in terms of sensitivity, prioritization and specificity. PLoS One 8(11):e79217
36. Tripathi S, Glazko GV, Emmert-Streib F (2013) Ensuring the statistical soundness of competitive gene set approaches: gene filtering and genome-scale coverage are essential. Nucleic Acids Res 41(7):e82
37. Dinu I et al (2007) Improving gene set analysis of microarray data by SAM-GS. BMC Bioinform 8:242
38. Subramanian A et al (2005) Gene set enrichment analysis: a knowledge-based approach for interpreting genome-wide expression profiles. Proc Natl Acad Sci U S A 102 (43):15545–15550
39. Barbie DA et al (2009) Systematic RNA interference reveals that oncogenic KRAS-driven cancers require TBK1. Nature 462 (7269):108–112
40. Fridley BL, Jenkins GD, Biernacka JM (2010) Self-contained gene-set analysis of expression data: an evaluation of existing and novel methods. PLoS One 5(9)
41. Stouffer S, DeVinney L, Suchmen E (1949) The American soldier: adjustment during army life, vol 1. Princeton University Press, Princeton, NJ
42. Taylor J, Tibshirani R (2006) A tail strength measure for assessing the overall univariate significance in a dataset. Biostatistics 7 (2):167–181
43. Robinson MD, McCarthy DJ, Smyth GK (2010) edgeR: a Bioconductor package for differential expression analysis of digital gene expression data. Bioinformatics 26 (1):139–140
44. Anders S, Huber W (2010) Differential expression analysis for sequence count data. Genome Biol 11(10):R106
45. Smyth G (2005) Limma: linear models for microarray data. In: Smyth G, Gentleman R, Carey V, Dudoit S, Irizarry R, Huber W (eds) Bioinformatics and computational biology solutions using r and bioconductor. Springer, New York, pp 397–420
46. Law CW et al (2014) Voom: precision weights unlock linear model analysis tools for RNA-seq read counts. Genome Biol 15(2):R29
47. Rahmatallah Y, Emmert-Streib F, Glazko G (2016) Gene set analysis approaches for RNA-seq data: performance evaluation and application guideline. Brief Bioinform 17 (3):393–407
48. Tusher VG, Tibshirani R, Chu G (2001) Significance analysis of microarrays applied to the ionizing radiation response. Proc Natl Acad Sci U S A 98(9):5116–5121
49. Baldi P, Long AD (2001) A Bayesian framework for the analysis of microarray expression data: regularized t -test and statistical inferences of gene changes. Bioinformatics 17 (6):509–519
50. Smyth GK (2004) Linear models and empirical Bayes methods for assessing differential

expression in microarray experiments. Stat Appl Genet Mol Biol 3:3
51. Dinu I et al (2009) Gene-set analysis and reduction. Brief Bioinform 10(1):24–34
52. Liu Q et al (2007) Comparative evaluation of gene-set analysis methods. BMC Bioinform 8:431
53. Baringhaus L, Franz C (2004) On a new multivariate two-sample test. J Multivar Anal 88:190–206
54. Klebanov L et al (2007) A multivariate extension of the gene set enrichment analysis. J Bioinforma Comput Biol 5(5):1139–1153
55. Wu D et al (2010) ROAST: rotation gene set tests for complex microarray experiments. Bioinformatics 26(17):2176–2182
56. Damian D, Gorfine M (2004) Statistical concerns about the GSEA procedure. Nat Genet 36(7):663. author reply 663
57. Ritchie ME et al (2015) limma powers differential expression analyses for RNA-sequencing and microarray studies. Nucleic Acids Res 43 (7):e47
58. Pickrell JK et al (2010) Understanding mechanisms underlying human gene expression variation with RNA sequencing. Nature 464(7289):768–772
59. Olivier M et al (2002) The IARC TP53 database: new online mutation analysis and recommendations to users. Hum Mutat 19 (6):607–614
60. Liberzon A et al (2011) Molecular signatures database (MSigDB) 3.0. Bioinformatics 27 (12):1739–1740
61. Wu D, Smyth GK (2012) Camera: a competitive gene set test accounting for inter-gene correlation. Nucleic Acids Res 40(17):e133
62. Bandres E et al (2005) Gene expression profile induced by BCNU in human glioma cell lines with differential MGMT expression. J Neuro-Oncol 73(3):189–198
63. Ongusaha PP et al (2003) BRCA1 shifts p53-mediated cellular outcomes towards irreversible growth arrest. Oncogene 22 (24):3749–3758

Chapter 8

Search for Master Regulators in Walking Cancer Pathways

Alexander E. Kel

Abstract

In this chapter, we present an approach that allows a causal analysis of multiple "-omics" data with the help of an "upstream analysis" strategy. The goal of this approach is to identify master regulators in gene regulatory networks as potential drug targets for a pathological process. The data analysis strategy includes a state-of-the-art promoter analysis for potential transcription factor (TF)-binding sites using the TRANSFAC® database combined with an analysis of the upstream signal transduction pathways that control the activity of these TFs. When applied to genes that are associated with a switch to a pathological process, the approach identifies potential key molecules (master regulators) that may exert major control over and maintenance of transient stability of the pathological state. We demonstrate this approach on examples of analysis of multi-omics data sets that contain transcriptomics and epigenomics data in cancer. The results of this analysis helped us to better understand the molecular mechanisms of cancer development and cancer drug resistance. Such an approach promises to be very effective for rapid and accurate identification of cancer drug targets with true potential. The upstream analysis approach is implemented as an automatic workflow in the geneXplain platform (www.genexplain.com) using the open-source BioUML framework (www.biouml.org).

Key words Upstream analysis, Promoter analysis, Pathway analysis, Microarray data, ChIP-seq, RNA-seq, Pathway rewiring

1 Introduction

Gene regulatory networks of cancer cells are currently subject of very intense studies. Successful diagnosis and treatment of cancer still remains difficult mainly due to our poor understanding of underlying molecular mechanisms and respective gene regulatory networks involved in the pathogenesis of cancer [1]. The elucidation of these mechanisms for identification of novel drug targets and promising biomarkers has therefore been a major focus in cancer research in recent years. However, this is quite a challenging task, since the multiplicity of pathways involved [2, 3]. And another, even more challenging task is to understand the high "plasticity" of regulatory networks governing pathological transformations, growth and survival of cancer cells. Plasticity of cancer

Tatiana V. Tatarinova and Yuri Nikolsky (eds.), *Biological Networks and Pathway Analysis*, Methods in Molecular Biology, vol. 1613, DOI 10.1007/978-1-4939-7027-8_8, © Springer Science+Business Media LLC 2017

gene regulatory networks is characterized, first, by big amount of individual genetic and epigenetic variations carried by cancer cells leading to a huge diversity of respective variants of the corresponding networks. Most difficult for understanding are those changes in the gene regulatory networks that happened in the course of development of the cancer in time, and in response to anti-cancer therapy. These so-called pathway rewiring events are characterized by changes in the set of active components of the cellular network and inter-connections between them.

Nevertheless, intensive studies of the molecular phenotypes of a big amount of individual cancer cases [4] led researchers to the formulation of typical cancer subtypes, concluding that according to such subtypes the particular cancers become "rewired" in consistent and rather predictable ways. It appeared that gene regulatory networks of such cancer subtypes often contain nodes called "master regulators" that represent important regulatory molecules in the signal transduction hierarchy which are necessary and sufficient to achieve and maintain a certain tumor state [5]. Therefore, revealing such master regulators is an important task since they may serve as potential drug targets as well as good biomarkers characterizing particular cancer subtypes.

Numerous "-omics" studies on various cancer samples offer the possibility to mine these high-throughput data by applying existing computational tools. Big collections of different "-omics" data are deposited in databases such as ArrayExpress [6] or Gene Expression Omnibus (GEO) [7], as well as in derived sets of differentially expressed genes (DEG) (expression signatures) that can be found in such databases as the Expression Atlas [8], the Mouse Expression Database (GXD) [9] and others. Direct identification of the most significant expression changes in such experiments can be used for selection of potential drug targets or identification of cancer biomarkers, as it is done with the help of various statistical methods and classification methods in many studies today [10]. However, the high inter-sample variation of the gene expression and extreme misbalance between relatively small number of samples analyzed and, very often, astronomic number of features (genes and their combinations) used for the training of the classification algorithms leads to a very low reproducibility of such mere statistical findings in independent experiments, especially those developed in specific patient populations [11]. More refined analysis of the molecular mechanisms of disease is usually done by mapping the DEG sets to Gene Ontology (GO) categories or to KEGG pathways, for instance by GSEA (gene set enrichment analysis) [12, 13]. Since such an approach provides only a very limited clue to the causes of the observed phenomena, we introduced earlier a novel strategy, the "upstream analysis" approach for the causal interpretation of expression changes [14–17].

The strategy of "upstream analysis" consists of two consecutive data analysis steps: (1) identification of transcription factors (TFs) potentially regulating DEGs through sequence analysis of promoters and enhancers of the genes under study; (2) reconstruction of signaling pathways that may activate identified TFs and search of master-regulators at the top of such pathways. The promoter and enhancer analysis step is done with the help of the TRANSFAC® database [18] and the site identification algorithms, Match [19] and CMA [20]. The pathway analysis step is done with the help of the TRANSPATH® database [21], one of the first signaling pathway databases available, and special graph search algorithms [14]. Recently, we introduced an important enhancement to the upstream analysis approach [22], which helps to search for master-regulators in the condition of so-calledwalking pathways.

With the term "walking pathways" we emphasize the fluctuating nature of the cancer pathway rewiring mechanism. A vast amount of evidence confirms that during initiation and development of tumors the structure of gene regulation and signal transduction pathways in the tumor cells is often drastically changing due to the variety of genomic and epigenomic alterations as well as due to the gross changes in gene expression [23]. Such alterations lead to changes in the pathways when a certain part of a pathway becomes inactive (disappears), but other parts or pathway cross-talks become hyper-active (appear), significantly changing the connectivity of the signaling paths and gene regulation circuits in the cells. In order to be able to "catch" such walking pathways and to reconstruct the most probable wiring in a particular cellular situation, we introduced an enhancement into the previously published network analysis algorithm [17]. We added a new graph-weighting schema to the algorithm of master-regulator search that enables it to incorporate proteomics or gene expression data by adding "context nodes" lists that push the graph search toward those nodes that are expressed in the cell in the particular cellular situation.

Currently, multiple "-omics" data are generated worldwide that measure in various cancer samples several gene expression profiles as well as various epigenomic signatures of DNA methylation and modifications of chromatin. Analysis of such data is very important for deciphering cancer mechanisms. In this work, first, we analyzed one set of multi-omics data consisting of gene expression data and data from ChIP-seq experiments on nucleosome methylation. The data were generated on two cell lines, one being malignant and the other not. With the help of our analysis strategy we were able to identify master regulators in signal transduction pathways potentially responsible for the transformation of the cells into the malignant state. In this example, we demonstrated the principle of "walking pathways" as a useful model for understanding the molecular mechanisms of carcinogenic transition between two relatively stable cellular states.

In the second example, we applied the walking pathway principle to a real case study. We analyzed a multi-omics data consisting of transcriptomics and ChIP-seq data that were generated on a colon cancer cell line that developed resistance against one of the most widely used chemotherapies—methotrexate (MTX). Our case study was devoted to deciphering the mechanisms of development of cancer cell resistance against MTX to identify possible ways to suppress such resistance by interacting with specific molecular targets. Emergence of resistance to MTX of various cancer cells is one of the most important problems in the long-term application of this drug. Methotrexate (MTX) is a folate antagonist, which kills the proliferating cell by binding tightly to the enzyme dihydrofolate reductase (DHFR). Due to this binding the pathway of de novo DNA synthesis is blocked [24]. However, continued administration of MTX to patients often results in the emergence of drug-resistance [25]. Our analysis helped not only to decipher the potential mechanisms of resistance but also suggested potent drug targets for possibly combating resistance mechanisms in these cells.

In summary, in this work, we applied the upstream analysis strategy to multi-omics data of different complexity, which leads us to a better understanding of the molecular mechanisms triggering malignant transformation as well as the mechanisms of emergent resistance of cancer cells to chemotherapy and propose promising drug targets and biomarkers.

The upstream analysis strategy is implemented in the geneXplain platform (www.genexplain.com) [17] using the open-source BioUML framework (www.biouml.org) in the form of an automatic workflow that wire together several algorithms performing the two analysis steps outlined above. Below in this chapter we will present the details of this workflow and will briefly describe its user interface and input parameters.

2 Data

2.1 Microarray Data, Differential Expression Analysis

For the analysis of gene expression profiles of a malignant cancer cell line and comparing it with a non-malignant cell line, we used publicly available microarray data from Gene Expression Omnibus (NCBI, Bethesda, MD, USA), data entry GSE75168. The study was done on breast tissue-derived cell lines: normal-like cell line, MCF10A, and transformed breast cancer cell line MCF7 [26]. In total, six RNA-seq data sets (three for MCF10A and three for MCF7) were generated by the authors with the help of the Illumina HiSeq 1500 instrument. The raw RNA-seq fastq files were preprocessed by the authors of the original paper [26] using the standard procedure of adapter removal, low base quality trimming (with FASTQ Quality Trimmer 1.0.0), alignment to genome build hg38 (using Tophap2), and read quantification (using HTSeq-count with genecode annotation v21). We downloaded the table with the read count data from GEO entry and further analyzed it with the geneXplain platform. The Limma (Linear Models for Microarray Data) method was applied to define fold changes of genes and to identify the significantly expressed genes using a Benjamini-Hochberg adjusted p-value cutoff (≤ 0.05) [27].

For the analysis of gene expression changes in MTX-resistant cells, we took publicly available microarray data from Gene Expression Omnibus, data entry GSE11440 [28]. The authors analyzed the transcriptome of the colon cancer HT29 cells that were MTX-sensitive and compared them to MTX-resistant cells generated from the same cell line. In total, six Affymetrix microarray experiments were done, three biological replicas for the sensitive cells and three replicas for the resistant cells.

Raw microarray data of MTX-resistant and sensitive cells, the latter being used as control in our study, were normalized and background corrected using RMA (Robust Multi-array Average). The Limma method was applied then as well to define fold changes of genes and to identify the significantly expressed genes using a Benjamini-Hochberg adjusted p-value cutoff (≤ 0.05) [27].

2.2 ChIP-Seq Data

ChIP-seq data for the breast cancer cell lines MCF7 and MCF10A were downloaded from the links provided in Gene Expression Omnibus (GEO), data entry GSE69377. This study examines the distribution of H3K4me3 and H3K4ac histone modification associated with active chromatin in these cell types. In our further analysis, we took the H3K4me3 signal only as the most commonly accepted mark of active chromatin.

In the second study of MTX resistance in colon cancer cell line HT29, we analyzed ChIP-seq data of the CDK8 kinase mediator complex, which is known to be over-expressed in colorectal cancer [29]. We analyzed data from a study investigating genome-wide localization of CDK8 in human colorectal cancer cell line HT29. The data were extracted from Gene Expression Omnibus, data entry GSE53602. In that study, Genomic DNA was enriched by chromatin immunoprecipitation (ChIP) and analyzed by Solexa sequencing. ChIP was performed using an antibody against CDK8.

In both studies, we have downloaded the raw NGS sequences of ChIP-seq experiments from the SRA repository (http://www.ncbi.nlm.nih.gov/sra), and analyzed them with the help of the geneXplain platform. The ChIP seq sequence reads were mapped to the human genome build hg19 with the use of the genome mapper Bowtie [30] with default parameters. The peak-calling program MACS [31] (using almost all default parameters, except parameter "Enrichment ratio," which was set to value 5.0 to achieve a higher number of peaks) was applied then to the obtained alignments of ChIP-seq data from the MCF7 cell line with data from the MCF10A cell line as control. As a result, we obtained 13,738 peaks of H3K4me3 histone modifications with the length varying from 349 bp up to 2000 bp. In the colon cancer cell line HT29 MACS returned 29,400 peaks of CDK8 complex binding in the whole human genome. In the case of the peaks for the CDK8 kinase mediator complex, we further trimmed the peaks to the length of 400 bp around the summit of the identified peaks, since the peaks were generally longer in comparison to the peaks obtained from the H3K4me3 ChIP-seq experiments.

3 Methods

3.1 Analysis of Enriched Transcription Factor Binding Sites

Transcription factor binding sites in promoters of differentially expressed genes were analyzed using known DNA-binding motifs described in the TRANSFAC® library, release 2014.4 (BIOBASE, Wolfenbüttel, Germany) (http://genexplain.com/transfac). The motifs are specified using position weight matrices (PWMs) that assign weights to each nucleotide in each position of the DNA-binding motif for a transcription factor or a group of them.

The geneXplain platform provides tools to identify transcriptionfactor-binding sites (TFBS) that are enriched in the promoter regions under study as compared to a background sequence set such as promoters of genes that were not differentially regulated under the condition of the experiment. We denote study and background sets briefly as Yes and No sets. The algorithm for TFBS enrichment analysis, called F-Match, has been described in [14]. Briefly, as it has been described in detail previously [17], the procedure finds a critical value (a threshold) for the score of each PWM in the library that maximizes the Yes/No ratio R_{YN} as defined in Eq. (1) under the constraint of statistical significance.

$$R_{\mathrm{YN}} = \frac{\#Sites_{\mathrm{Yes}}/\#Sites_{\mathrm{No}}}{\#Seq_{\mathrm{Yes}}/\#Seq_{\mathrm{No}}} \tag{1}$$

In Eq. (1), *#Sites* and *#Seq* are the sites and sequences counted in Yes and No sets. A high Yes/No ratio indicates strong enrichment of binding sites for a given PWM in the Yes sequences. The statistical significance is computed as follows:

$$\begin{aligned} P(X \geq x) &= \sum_{n=x}^{N} \binom{N}{n} \cdot p^n \cdot (1-p)^{N-n} \\ p &= \#Seq_{\mathrm{Yes}}/\left(\#Seq_{\mathrm{Yes}} + \#Seq_{\mathrm{No}}\right) \\ N &= \#Sites_{\mathrm{Yes}} + \#Sites_{\mathrm{No}} \\ n &= \#Sites_{\mathrm{Yes}} \end{aligned} \tag{2}$$

In the geneXplain platform, such a binding site enrichment analysis is carried out as part of a dedicated workflow. We consider for further analysis only those TFBSs that achieved a Yes/No ratio > 1 and a p-value <0.01. The workflow further maps the matrices to respective transcription factors, and generates visualizations of all results. In the current work, we have modified the workflow by considering not only promoter sequences of a standard length of 1100 bp (-1000 to $+100$), but also sequences of potential enhancers and silencers derived from combined transcriptomics and epigenomics data as described below.

3.2 Finding Master Regulators in Networks

We searched for master regulator molecules in signal transduction pathways upstream of the identified transcription factors using geneXplain platform tools. The master-regulator search uses the TRANSPATH® database (http://genexplain.com/transpath) [21]. A comprehensive signal transduction network of human cells is built by the network analysis module of the geneXplain platform on the basis of reactions annotated in TRANSPATH®. The main algorithm of the master regulator search has been described earlier [14]. The goal of the algorithm is to find nodes in the global signal transduction network that may potentially regulate the activity of the set of transcription factors found at the previous step of analysis. Such nodes are considered most potent drug targets, since any influence on such a node may switch the transcriptional programs of hundreds of genes that are regulated by the respective TFs. In our analysis, we have run the algorithm with a maximum radius of ten steps upstream of the TFs.

3.2.1 Basic Algorithm of the Master Regulator Search

The basic algorithm of the master regulator search was described previously [16, 17]. Here, we present a short summary of the algorithm. The signal transduction network is represented as a weighted graph, which is defined as $\mathbf{\Theta}=(M, E, S)$, where M is the set of nodes (all molecules in the database), E is the set of edges (all reactions between molecules in the database), and $S : E \rightarrow \boldsymbol{R}^{+} \cap \{0\}$ is the cost function that defines a non-negative value for every edge. In the simplest variant of the algorithm, the initial values of the cost functions for each direct reaction are defined as 1.0. So, the cost of any path through the consequent reactions will be equal to the number of reactions. In our application of the algorithm in the geneXplain platform, we used a weighted graph, which encodes the types of reactions (direct or indirect) in TRANSPATH® into different edge weights (costs).

The core of the algorithm is Dijkstra's shortest-path algorithm [32]. The upstream search starts from each molecule of the input set (subset M_x of M; e.g., transcription factors found in the previous step) and constructs the shortest-path to all nodes i of M, limited by a specified radius. After evaluating all nodes of M_x the algorithm calculates the number of visits N_i for each node i of M. This corresponds to the number of molecules of the input set M_x that can receive the signal from the node *i*. The values N_i are required to calculate the "master-regulator" score. The higher the value N_i the better chances molecule *i* has to transduce its signal to the molecules of the initial list M_x. However, to use the value of N_i as a straightforward "master-regulator" score would be too simplistic since it would rank those molecules high that are in general highly connected within the network and thus are likely to be false positives (so-called hub molecules). So, in our score, we incorporate the full number of TF nodes in the whole network M that can be reached from the molecule *i*. The score is computed for each potential master regulator and reflects a certain balance between sensitivity and specificity of signal transduction from this master node to the downstream effector TFs:

$$S(k) = \sum_{k=1}^{k_{\max}} \frac{M_k}{\left(1 + \kappa \cdot \frac{N_k}{N_{\max,k}}\right) \cdot M_{\max,k}} \qquad (3)$$

where k is the radius of pathway steps from the master node to the effector nodes, M_k is the number of input TFs reached by a signal from the master node within k steps, and N_k is the total number of all potential TFs in the database reached by a signal from the master node within k steps. $M_{\max,k}$ and $N_{\max,k}$ are the highest values among all possible master regulator nodes which help to normalize the score to a (0,1)-interval. The higher this score, the more sensitive and more specific this master regulator is for the set of input TFs. The parameter κ is a user-defined penalty, which is set by default to 0.1.

Incorporation of proteomics or gene expression data into the analysis of master regulators is done through application of the so-called Context Algorithm, which is implemented in the geneXplain platform.

3.2.2 Context Algorithm

When we have additional experimental information about the activity of certain components of the signal transduction network (additional context information) of the cells, we can use this information to adapt the search for master regulators. To do that, the algorithm encodes this additional context information as modified edge costs in the signaling network. For instance, the proteomics data give information about proteins that are expressed in the cell. Gene expression data can also provide a proxy for such data. By putting a certain threshold on the expression signal (obtained by microarrays or through RNA-seq), we can obtain a set of genes, which are expressed in the particular cellular condition, and we can assume the proteins encoded in those genes are also present in the cells. We call such a list of proteins "context nodes." The idea of the approach is to attract the key node search (e.g., the underlying Dijkstra algorithm for shortest paths) toward context nodes by decreasing the costs of those edges that are close to the context nodes in the network. There are two major parameters in this algorithm: (1) Attraction power (**gravity**) of the shortest-paths toward context nodes. The gravity is achieved by decreasing the costs of the incoming and outgoing edges in the graph for the nodes corresponding to the context proteins. A zero cost of the edges gives the maximal "gravity" for the nodes. During the search the path that follows through this node will be always "cheaper" than other alternative paths. This way we push the search algorithm to construct its paths maximally through the context proteins when possible. (2) **"Decay"** factor. It is necessary to extend the attraction power of the "gravity" to a wider area of the network around the context proteins to attract the search algorithm to go as close as possible to the context proteins, in case there is no path that goes through the context gene directly. To enable such "long distance" attraction power we distribute the gravity to edges of the next neighboring nodes. This pushes the shortest path algorithm to find alternative paths still at least close to the context proteins. The gravity strength reduces with increasing distance from a decay factor. The decay factor is set by default to 0.1 and can be changed by the user by setting it to a value between 0 and 1.

The context algorithm is based on creating a graph G consisting of gravity factors (0.1), which reflect both aspects described above of modifying edge costs. Graph G is used to modify the costs of S of original graph $\boldsymbol{\Theta}$ resulting in S' which can be then used for any subsequent key node analysis or possibly other shortest path-based analyses.

Usually, we use the list of proteins encoded by up-regulated genes as the "context nodes." With this we direct the algorithm of master-regulator search toward those paths through the signal transduction network that go maximally through those proteins that are expressed by the genes known to be upregulated in this type of cells. The algorithm does not exclude completely the other paths through proteins that were not found to be up-regulated, just because their concentration might be below the detection limit. Therefore, they may well be active in the cells and may participate in the transduction of the relevant signals. Nevertheless, the proteins that were detected as upregulated are considered with higher weights in the algorithm and contribute more to directing the search toward master regulators. Even better source of information for the "context nodes" could be proteomics or phosphoproteomics data albeit not always being available.

3.3 GUI of geneXplain Platform

GeneXplain platform is an online tool, which is available upon free registration at the URL: http://platform.genexplain.com. When users login into the geneXplain platform for the first time, a window opens that contains the following five areas 1–5: (*see* Fig. 1)

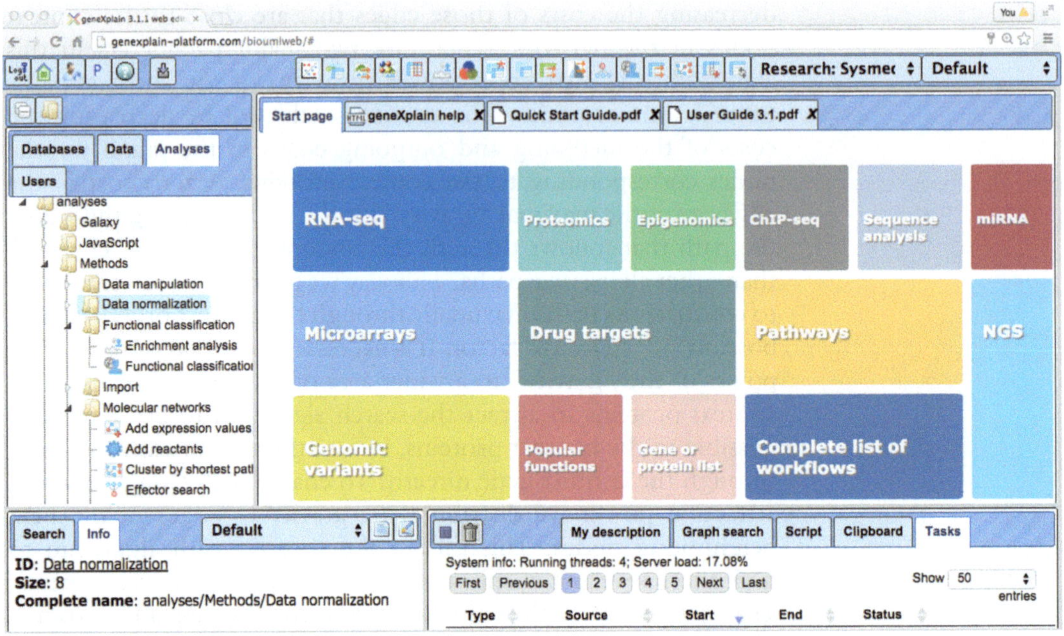

Fig. 1 User interface of the BioUML/geneXplain platform after first user login. Practically any "omics" data can be analyzed in the platform

1. The Work Space is the main part of the window. The Start page presents a couple of predefined workflows and methods.
2. In the Tree Area the user finds the collection of databases, the uploaded data files, and the available analyses methods under the corresponding tabs.
3. The Info Box gives the user information about the data file or analysis method that she may select with a single click in the Tree Area. The user can also select the data resource to search in.
4. The Operations Field provides additional analysis options under the different tabs in a context-dependent manner.
5. The General Control Panel (tool bar), on top of the different areas, shows a context-dependent set of icons for the available operations.

The geneXplain platform provides a number of predefined workflows for analysis of various omics data. The user can find the list of workflows available for particular data type after clicking on the respective area at the Work Space. For instance, by clicking on the "Microarrays" area the user will see a list of workflows (*see* Fig. 2) organized according to their consequent use. The list starts with a data-loading utility and ends with a most complex workflow "Find drag targets" that integrates many programs performing the "upstream analysis" algorithm under default parameters.

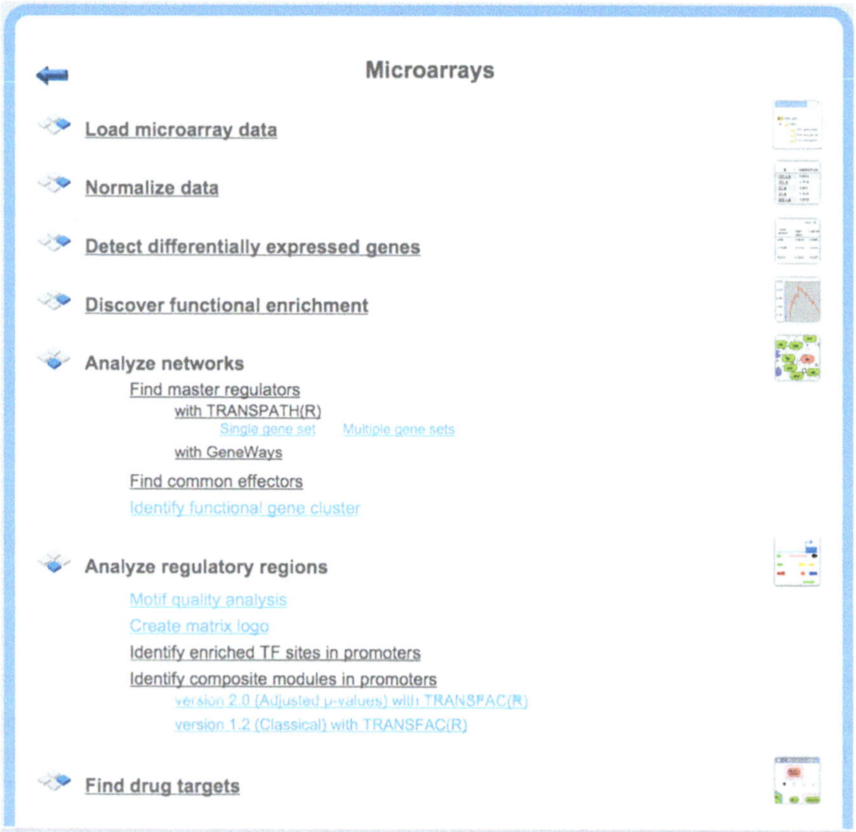

Fig. 2 List of workflows for analysis of microarray data. The workflows are put under several categories according to the consecutive steps of the data analysis. It starts by data loading and normalization of ends with most complex workflows for identification of drag targets

The GUI of the geneXplain platform is user-friendly and intuitive. In Fig. 3, we show an interface provided to the user to start a workflow of promoter analysis and search for so-called composite regulatory modules. This workflow opens after clicking on the respective link in the list of workflows (*see* Fig. 2). The user can see the full schema of the workflow in the Operation Field that consists of individual programs (blue boxes) performing necessary steps of the analysis connected between each other by arrows and input and output data files (green and yellow small boxes). In the Work Space the user sees the main window with the parameters of the workflow that the user has to set to initiate the analysis, such as "Input Yes gene set," "Input No gene set," and others. Under Yes set we define here a set of genes from our experiment whose promoters are going to be analyzed, for instance a set of up-regulated genes. Under No set we define here any control set, for instance a set of genes that did not change their expression in the experiment. The user can specify the respective data sets with Yes and No gene sets by "drag-and-drop," dragging the files from the

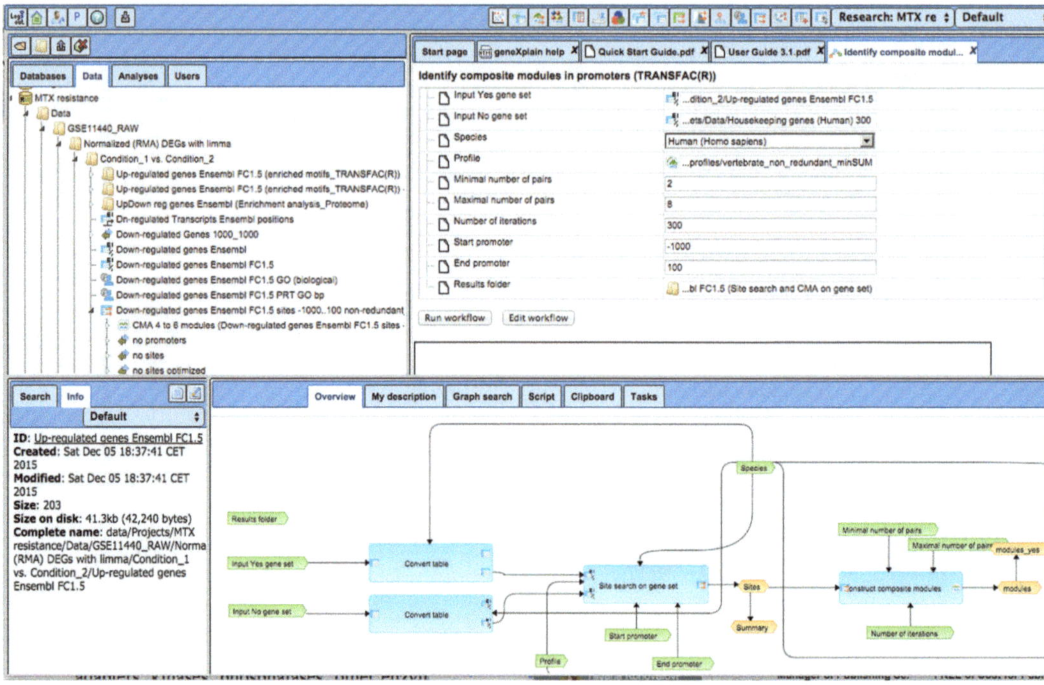

Fig. 3 User interface to start a workflow (as an example we use a workflow for the analysis of promoters of a gene set). The front window gives the possibility to set the input data and other parameters of the analysis. The window below shows the schema of the workflow that connects several programs (*blue boxes*) to each other to determine their running order

Tree Area into the necessary fields of the workflow. By clicking on the "Run workflow" button the user starts the workflow, which is performed automatically till the end of all calculations. The results are automatically displayed in new tabs in the Work Space and are stored as result files in the Tree Area in the user folder, so that they are accessible for the user at any time. The results are represented as tables, network diagrams, as well as various graphics that are available for export in many different formats.

4 Results

As was described above, our strategy of multi-omics "Upstream Analysis" of regulatory genomic regions comprises two main steps: (1) a systematic and comprehensive promoter and enhancer analysis on the basis of transcriptomics (differentially regulated genes) and epigenomic data (locations of regions of active chromatin) to identify transcription factors (TFs) involved in the regulation of the cellular process under study, and (2) an analysis of the topology of the signal transduction network upstream of transcription factors to identify master regulators, which are signaling proteins in the cell (receptors, their ligands, adapters, kinases, phosphatases, other enzymes involved in signal transduction) that may regulate the activity of transcription factors found in the previous step of the analysis. Applying this concept in previous studies has successfully revealed EGF and IGF2 as regulators during liver tumor development [16]. We also had analyzed a dataset of TNFα-induced genes in human endothelial cells [33], showing that our approach detects TNFα as a master regulator and explains the activity of other molecules from the TNFα pathway [14, 17].

4.1 Identification of Up- and Down-Regulated Genes

Identification of up- and down-regulated genes in gene expression data was done with the help of workflow "Compute differentially expressed genes using Limma." The workflow takes a table that summarizes all preprocessed gene expression values for all samples and applies Limma (Linear Models for Microarray Data) with a Benjamini-Hochberg adjusted p-value cutoff (≤ 0.05) to retrieve differentially expressed genes (DEG) between different conditions (cell lines) that are defined by the user before the start of the workflow. The table with expression values can be either directly uploaded from the results of experiments or prepared inside geneXplain platform with the help of a number of data preprocessing workflows that are working with raw microarray data supporting a big number of various microarray platforms, or preprocessing raw RNA-seq data converting them into read counts or FPKM values.

In this work, first, we analyzed the RNA-seq data provided in a form of read counts for the breast cancer cell lines in triplicates: normal-like, MCF10A cell line, and malignant MCF7 cell line. Comparison of the data for the MCF7 cell line versus the MCF10A cell line with the help of DEG analysis workflow resulted in the identification of 2066 upregulated (with higher expression values in MCF7 cell line compared to MCF10A cell line) and 2199 downregulated Ensembl genes, respectively. The availability of the read count data gave us an additional gene filtering possibility. We filtered out genes with read counts lower than 200 in all the samples from both cell lines. With this filtering we focused our attention on those genes that not only differ in their relative expression significantly between the two cell lines but also showed the high absolute level of expression and therefore were unlikely to bear any significant experimental noise. After such filtering, we obtained 764 upregulated and 1085 downregulated genes.

The second data set that we analyzed with the help of the workflow "Compute differentially expressed genes using Limma" was the microarray transcriptomics data on comparison between MTX-sensitive and resistant derivatives of the colon cancer cell line HT29. The resistant cell line in this case was a result of a long-term treatment of the HT29 cell line by MTX and selection of a stable MTX-resistant cell line. Similarly, as in the first data set, we identified up- and down-regulated genes from the comparison of transcriptomics data of resistant versus sensitive cells from raw microarray data available in GEO [28]. As a result, we identified 1951 up-regulated and 2185 down-regulated genes.

In both of these data sets, we were interested in understanding the nature of molecular switches that turn a normal-like cell line or a chemotherapy-sensitive cancer cell line into the cells with aggressive carcinogenic behavior. Understanding such switches can help us to find effective approaches for identification of promising drug targets.

In order to understand the basic functional changes in the cancer cells during such switches, we performed a classical GO and pathway mapping using the workflow "Mapping to GO categories and signaling pathways" of the geneXplain platform. The analysis of the up-regulated genes of the first data set (breast cancer cell lines) revealed the following major GO categories: anatomical structure development, cell-cell signaling, cell adhesion, ion transport; pathways (TRANSPATH®, REACTOME): beta-catenin network, Interferon alpha/beta signaling. In turn, the analysis of down-regulated genes revealed the GO categories: angiogenesis, developmental process, response to external stimulus, response to wounding, regulation of cell adhesion, cell migration; pathways (TRANSPATH®, REACTOME): metabolism of DAG, metabolism of eicosanoids, formation and degradation of triacylglycerols, IL-8 pathway, Extracellular matrix organization, Dissolution of Fibrin Clot.

For the second data set of the MTX-sensitive and resistant colon cancer cells, the GO and pathways mapping resulted in rather different profiles. The up-regulated genes are enriched by the following GO categories: oxidation-reduction process, lipid metabolic process, purine deoxyribonucleotide metabolic process, dephosphorylation, negative regulation of cell adhesion, cell migration; pathways (TRANSPATH®, REACTOME): serotonin degradation, cholesterol metabolism, release of active TGFbeta, metabolism of estrogens, regulation of lipid metabolism by peroxisome proliferator-activated receptor alpha (PPARalpha), extracellular matrix organization. The down-regulated genes are, in turn, enriched by the following GO categories: cell cycle, apoptosis, response to virus, protein phosphorylation, organelle fission, response to interferon-alpha, M phase, response to stress; pathways (TRANSPATH®, REACTOME): Aurora-B cell cycle regulation, E2F network, cyclosome regulatory network, interferon signaling.

Comparing results of GO and pathway mapping for these two data sets shows both high similarity and profound differences in the processes going on in these two types of cancer cells. We can see that up-regulation of such GO categories, as regulation of cell adhesion, organ morphogenesis, and regulation of epithelial cell proliferation is found common for both data sets. However, upregulation of cell-cell signaling and transmembrane transport is clearly specific for the breast cancer cells, although upregulation of oxidation-reduction process and lipid metabolic process is specific for the MTX-resistant colon cancer cells. As for the down-regulated genes, we found common such processes as regulation of biosynthetic process, regulation of endopeptidase activity, regulation of NF-kappaB cascade, and, in general, regulation of innate immune response and cell death. And we also found down-regulated processes such as anatomical structure development and extracellular matrix organization specific for the breast cancer cells and such processes as mitosis and regulation of cell cycle process specific for the MTX-resistant colon cancer cells.

Such a GO and pathway analysis gives a general idea of the global processes that changed their activity in the carcinogenic transitions that we study here. For instance, they coincided quite well with the existing knowledge about the mechanisms of MTX-resistance in cancer cells. According to the results of multiple studies, the most important resistance mechanisms to MTX were found to be connected with an increase of expression of the MTX primary target—enzyme DHFR [24, 25]. It is known that recovery of activity of this enzyme takes place as a result of amplification [34] and enhanced expression [35] of its gene. DHFR plays a central role in the synthesis of nucleic acid precursors, which are essential for cell proliferation and cell growth. As we can see from the GO analysis, many genes of cell cycle and proliferation are clearly down-regulated in the MTX-resistant cells, which might be explained by the influence of long-term inhibition of DHFR by MTX. But still those cells seem to be able to recover their proliferation, disregarding further treatment by MTX through some additional mechanisms that potentially are reflected by upregulation of the oxidation-reduction process and lipid metabolic process specific for these cells. It is clear that a more detailed analysis of molecular pathways is needed to fully understand the mechanisms of emergent resistance and carcinogenic transformation. To answer this question, we applied our concept of "upstream analysis" to the data on breast cancer and colon cancer cell lines.

4.2 Analysis of TF Site Frequency in Promoters and Enhancers

In order to identify transcription factors that may be activated during the carcinogenic transformations of breast cancer and colon cancer cell lines, we analyzed several important genomic regions of the genes that were differentially regulated during this process. For this, we identified the up- and down-regulated genes using a logFC cutoff (logarithm of the fold change to base 2) higher than 1.5 for up-regulated genes or lower than −1.5 for down-regulated genes ("Yes" sets of genes). As control we used genes expression of which did not change significantly in this experiment ("No" set of genes). From all these genes we extracted the promoter regions from −1000 to +1000 bp around the TSS (transcription start site). Next, we intersected the promoter regions of these genes with the ChIP-seq peaks identified by MACS algorithm (*see* Subheading 3). In both data sets, the ChIP-seq peaks are marking regions of active chromatin. In the breast cancer cell lines, the peaks correspond to the H3K4me3 signal that is one of the commonly accepted marks of active chromatin. In the colon cancer cell line, the respective peaks correspond to signal from binding of CDK8 kinase, which is associated with the mediator complex, a central integrator of transcription proven as a marker of active transcription regulatory regions in colorectal cancer cells (for the HT29 cell line) [29]. The central role of the CDK8 kinase complex in the Wnt pathway, which is very often dysregulated in colorectal cancers and contributes to their growth, invasion, and survival [36], renders it a suitable marker for active enhancers in colon cancer cells. The intersection procedure of promoter regions with ChIP-seq peaks was done with the help of "Intersect" function in the Galaxy section of the geneXplain platform. With such an intersection we extracted those regions of the genome that with the highest probability contain the active promoter and enhancer regions regulating the activity of the genes in the considered states of the cells. For further analysis, we considered only those regions that have got the length equal or above 400 bp.

We applied the F-Match algorithm to these regions of active promoters and enhancers. The F-Match algorithm searches and compares the frequency of TF-bindingsites in the Yes and No sets of sequences applying the nonredundant set of PWMs from the TRANSFAC® library. This program is able to find those PWMs and corresponding transcription factors whose sites are overrepresented in the Yes set compared to the No set (*see* Subheading 3). We applied this method separately for the up- and down-regulated genes to identify those specific transcription factors that are potentially involved in activation or inhibition of the expression of these sets of genes. The result of the analysis of up-regulated genes is presented in Table 1. Also, in Fig. 4, we show a map of predicted TF-binding sites in the enhancer region of *TFF1* gene located in the first intron, first noncoding exon and proximal promoter region that overlaps with the ChIP-seq peak of H3K4me3. This gene is one of the most highly upregulated genes in the MCF7 cell line when compared to MCF10A. The more distal promoter of this gene is well studied and a number of experimentally detected TF-binding sites are reported in the TRANSFAC database (*see* Fig. 4) that were found functionally active in the promoter of this gene in various cellular conditions. For instance, the NF-kappaB transcription factor-binding site marked in Fig. 4 which was also found in our analysis was previously experimentally identified as acting in gastric epithelial cells activated by TNF-alpha [37]. One can see that our method allows identifying not only previously known sites

Table 1
List of PWMs for transcription factors identified by site frequency search in regions of open chromatin in promoters of up-regulated genes in MCF7 cells and in MTX-resistant cells

ID	Yes-No ratio MCF7_vs_MCF 10A_UP	*p*-value MCF7_vs_ MCF10A_UP	Yes-No ratio_MTX-resistant UP	*p*-value MTXresis tant_UP
V$MTF1_Q5	1.697	1.78E-07	0.787	1.49E-01
V$REST_01	1.650	9.25E-04	0.599	8.12E-02
V$AHR_Q6	1.633	2.89E-08	0.994	4.76E-01
V$E2A_Q6_01	1.515	1.52E-11	0.974	4.25E-01
V$INSM1_01	1.492	9.17E-03	0.805	2.66E-01
V$NFY_Q3	1.481	1.15E-06	0.962	4.28E-01
V$GCM2_01	1.446	1.51E-10	0.812	5.94E-02
V$COE1_Q6	1.415	9.42E-06	0.751	3.55E-02
V$P53_04	1.375	7.34E-80	0.976	2.48E-01
V$RREB1_01	1.268	1.50E-13	0.898	4.22E-02
V$MYOGEN IN_Q6_01	1.252	6.35E-06	0.920	1.69E-01
V$HIF1A_Q5	1.245	2.82E-06	0.761	2.07E-03
V$EBOX_Q6_01	1.238	9.24E-06	0.933	2.24E-01
V$RFX1_01	1.237	1.48E-03	0.986	4.82E-01
V$DR4_Q2	1.127	8.43E-06	0.973	3.03E-01
V$ZIC1_05	1.277	1.59E-110	1.024	1.35E-01
V$AP2ALPHA_03	1.470	1.84E-222	1.049	3.47E-02
V$HES1_Q6	1.421	1.38E-13	1.053	3.14E-01
V$NF1A_Q6_01	1.114	3.65E-06	1.064	7.46E-02
V$MUSCLEINI_B	1.365	1.85E-27	1.070	1.41E-01
V$IK_Q5_01	1.189	9.18E-30	1.102	6.28E-04
V$MZF1_Q5	1.311	2.77E-11	1.103	1.20E-01
V$RNF96_01	1.575	4.75E-35	1.129	9.14E-02
V$GKLF_Q4	1.306	1.95E-80	1.155	1.41E-07
V$BEN_01	1.574	1.65E-156	1.159	6.48E-05
V$RELA_Q6	1.288	1.88E-05	1.196	6.96E-02
V$SP100_04	1.374	6.01E-34	1.208	1.73E-03
V$CPBP_Q6	1.490	1.09E-85	1.236	4.75E-07
V$CHCH_01	1.528	9.52E-169	1.239	2.98E-09

(continued)

Table 1
(continued)

ID	Yes-No ratio MCF7_vs_MCF 10A_UP	*p*-value MCF7_vs_ MCF10A_UP	Yes-No ratio_MTX-resistant UP	*p*-value MTXresis tant_UP
V$FPM315_01	1.498	2.22E-10	1.271	2.75E-02
V$GLI_Q3	1.304	3.35E-42	1.282	2.24E-10
V$E2F_Q6_01	1.517	6.60E-38	1.382	4.48E-05
V$ZFP161_04	1.650	0.00E + 00	1.389	5.38E-24
V$EGR1_Q6	1.640	2.67E-26	1.457	2.42E-04
V$MAZ_Q6_01	1.480	1.76E-30	1.615	8.96E-12
V$MAF_Q4	1.021	4.09E-01	1.628	4.46E-05
V$SP1_Q6_01	1.427	1.13E-31	1.717	8.48E-17
V$MAZR_01	1.374	1.42E-04	1.747	1.51E-03
V$EKLF_Q5_01	1.483	3.59E-03	2.120	3.02E-03
V$MECP2_02	1.404	1.94E-05	2.616	8.66E-06
V$CTCF_01	1.803	6.42E-03	19.183	3.53E-16
V$HMX1_02	0.645	6.04E-218	1.058	2.98E-03
V$PAX_Q6	0.958	2.11E-02	1.156	2.71E-05
V$FREAC3_01	0.639	4.52E-29	1.215	4.94E-04
V$RUSH1A_02	0.859	2.39E-03	1.236	7.76E-03
V$GATA_Q6	0.583	1.04E-17	1.239	7.59E-03
V$SRY_Q6	0.631	4.93E-19	1.284	4.65E-04
V$BBX_04	0.678	1.97E-21	1.291	9.17E-06
V$LEF1_Q5_01	0.865	6.70E-03	1.327	6.13E-04
V$CEBPA_Q6	0.720	9.54E-07	1.365	1.22E-03
V$PLZF_02	0.388	5.59E-20	1.376	7.92E-03
V$ERALPHA_Q6_01	0.860	1.61E-02	1.384	3.57E-03
V$CRX_Q4_01	0.645	4.13E-07	1.393	4.06E-03
V$ETS_Q6	0.982	2.97E-01	1.424	5.91E-11
V$IPF1_Q5	0.606	1.27E-10	1.469	1.56E-04
V$SOX2_Q3_01	**0.839**	**6.50E-02**	**1.530**	**6.04E-03**
V$HOXC13_01	**0.637**	**2.88E-07**	**1.554**	**3.20E-04**
V$CREBP1_01	**0.779**	**1.87E-03**	**1.581**	**9.32E-04**
V$AP1_Q6_02	**0.782**	**3.10E-09**	**1.586**	**1.93E-15**

(continued)

Table 1
(continued)

ID	Yes-No ratio MCF7_vs_MCF 10A_UP	*p*-value MCF7_vs_ MCF10A_UP	Yes-No ratio_MTX-resistant UP	*p*-value MTXresis tant_UP
V$HNF3B_Q6	0.712	7.00E-10	1.592	5.00E-09
V$STAT1_Q6	0.911	2.09E-01	1.918	8.48E-04
V$HOXD12_01	0.692	5.74E-04	2.020	7.56E-05
V$ZFX_01	0.998	4.73E-01	2.242	9.28E-03
V$DBP_Q6	0.835	1.44E-01	2.297	4.68E-04
V$NF1_Q6	0.901	2.81E-01	2.396	2.13E-04
V$DLX3_02	0.819	1.86E-01	2.670	3.33E-03
V$TEF1_Q6_04	0.645	4.77E-02	3.433	6.64E-04
V$HNF4A_Q3	0.580	8.54E-02	8.221	1.87E-04
V$ZFP206_01	0.772	5.61E-02	20.869	1.31E-06

PWMs are the identifiers from TRANSFAC database. Yes-No ratio is the ratio between frequencies of TF sites in Yes sequences (regions of open chromatin in promoters of genes up-regulated in MCF7 cells or MTX-resistance cells respectively) and No sequences (corresponding controls in those experiments). PWMs specific for the MCF7 cells are shown in *red* in the top of the table. PWMs specific for the MTX-resistant cells are shown in *blue* in the bottom of the table

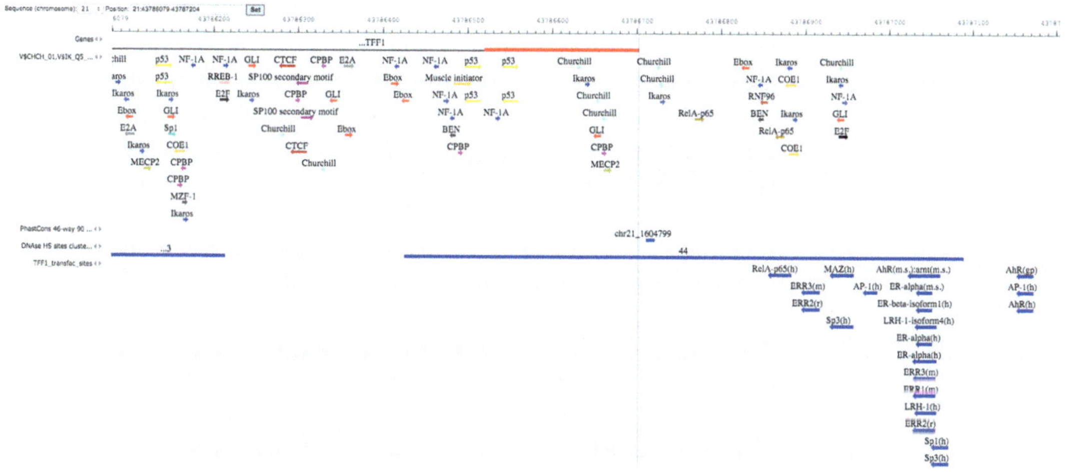

Fig. 4 A map of predicted TF-binding sites in the enhancer region of the *TFF1* gene located in the first intron, first noncoding exon and proximal promoter region that overlaps with the ChIP-seq peak of H3K4me3. Exons are represented by *red thick lines*, introns by thin *black lines*. The dotted vertical line indicates the TSS (transcription start site) for the *TFF1* gene. Colored arrows show positions of TF-binding sites (each color corresponds to one PWM). Regions of DNAse hypersensitivity (from ENCODE) are shown in a separate track which also indicates the area of open chromatin in *Hela* cells. In the last track, we show TF-binding sites previously identified in the promoter of this gene (from the TRANSFAC® database) acting in different types of cells. The site for RelA identified in our analysis coincides with the previously known RelA binding site. It can be seen that the region of active chromatin identified in the MCF7 cell line in this gene only partially overlaps with this area in other types of cells previously analyzed for this promoter

but also novel sites that were previously not detected in the regulatory regions of the genes. Such novel sites seemed to be responsible for the activation of these genes in the particular cellular conditions when during carcinogenic switch a new region of chromatin becomes active, which was not active in the normal-like cells of MCF10A cell line.

Results of analysis of regulatory genomic regions in the second data set are also shown in Table 1. Comparison of the overrepresentation of transcriptionfactor-binding sites in our two sets shows that there are few common transcription factors, of such families as E2F, GLI, ERG1, SP1, CTCF, and several others, whose sites are overrepresented in the transcriptionally active regions in promoters of up-regulated genes in both carcinogenic states. But there are also clear differences. Overrepresentation of sites for such TFs as AHR, E2F NFY, P53, and RELA is clearly specific for the MCF7 breast cancer cell lines, whereas overrepresentation of the sites for HNF4A, ZFP206, TEF1, STAT1, HNF3B, CREBP1, and AP1 is specific for the MTX-resistant colon cancer cells.

In Fig. 5, we show a map of identified TF-binding sites in the upstream aria of *DHFR* gene highly upregulated in the MTX-resistant cell line. Upregulation of this gene is known to be one of the most common mechanisms of the development of MTX resistance [35]. The promoter of this gene has been extensively studied and it was found that expression of the *DHFR* gene is tightly regulated during cell cycle through binding sites for transcription factor E2F [38]. Moreover, it was shown that at least one E2F site is located near an Sp1 site forming a composite element and that E2F and SP1 transcription factors act synergistically activating *DHFR* transcription [39, 40]. It was proposed earlier that the activation of the *DHFR* gene during development of MTX resistance is done through this E2F site [35]. Our site frequency analysis indeed revealed sites for E2F and Sp1 factors as overrepresented in the regions of open chromatin in the upstream areas of up-regulated genes (*see* Table 1). We also correctly identified the known E2F and Sp1 sites in the studied upstream region of the *DHFR* gene and even found a number of clusters of several E2F and Sp1 sites together with sites for the other important transcription factors. These site clusters colocalize with ChIP-seq peaks of the CDK8 mediator complex as well as with regions of DNase I hypersensitive sites (Fig. 5). Also, we found that the region of high homology between 46 mammalian genomes (PhastCons 46-way 50) is also located in the area near the detected site clusters (Fig. 5), which gives additional evidence about the functional importance of this regulatory area of the genome.

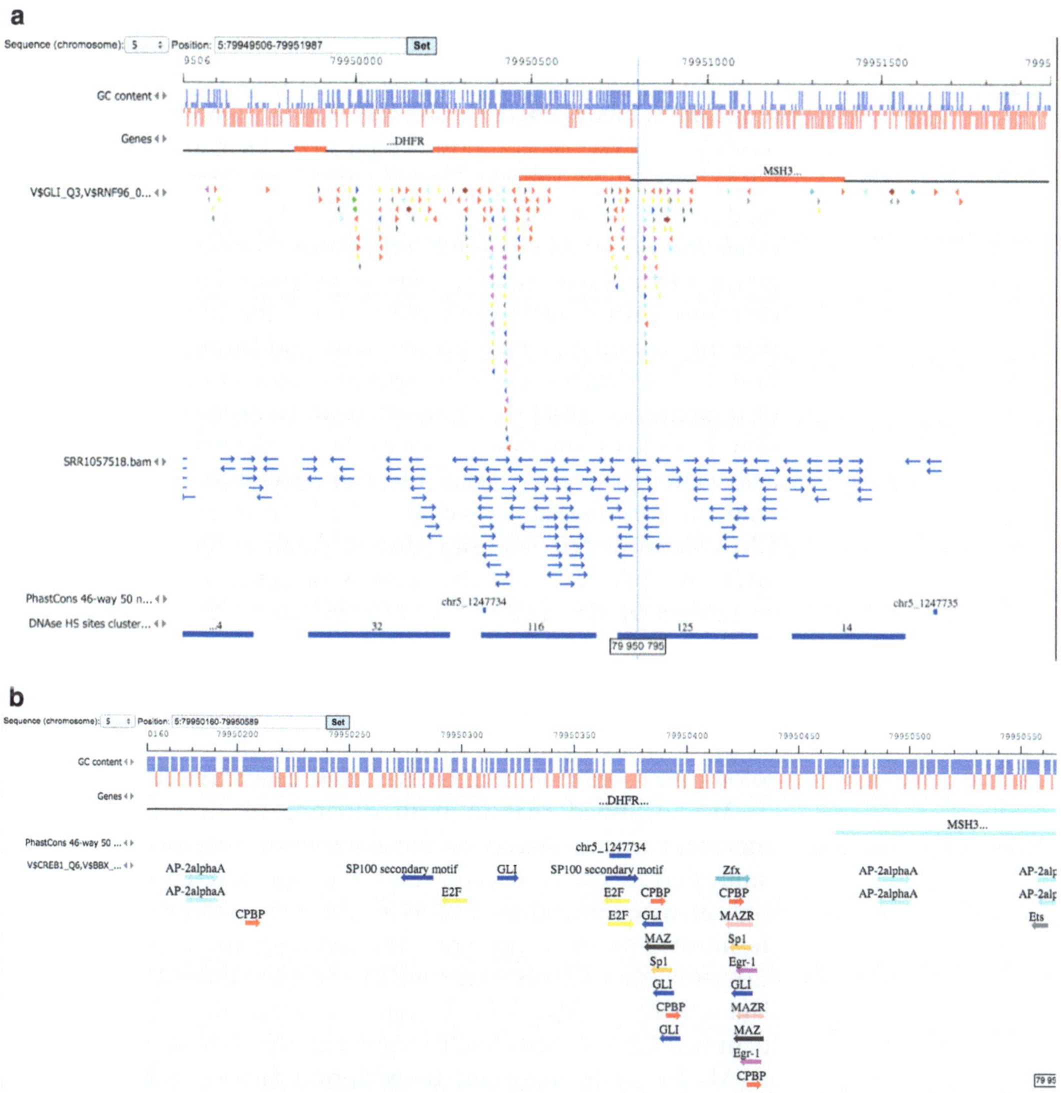

Fig. 5 Results of TF-bindingsites prediction in the overlapping promoters of *DHFR* and *MSH3*. (**a**) Low-resolution map of gene structures. Exons are represented by *red thick lines*, introns by *thin black lines*. (One can see that the first introns of *DHFR* and *MSH3* genes actually overlap). The dotted vertical line indicates the TSS (transcription start site) for the DHFR gene. Colored triangles show positions of TF-binding sites (each color corresponds to one PWM). Clusters of sites can be recognized as peaks of overlapping triangles. The track with *blue arrows* corresponds to the ChIP-seq reads from CDK8 experiment mapped to this genome region. The peak of the reads indicates the region of high regulatory transcription activity. Similar indicators of the open chromatin are the locations of the DNAse hypersensitivity (from ENCODE) shown in the bottom-most track. Two conserved regions (for 46-way 50% conservation between mammalian genomes) indicate potentially very important regulatory areas in these promoters. (**b**) High-resolution map. Each predicted TF-binding site is shown as an arrow with the name of PWM (from TRANSFAC®) on the top of it. The intensity of the *blue color* corresponds to the score of the binding site. The direction of the arrow shows at which DNA strand the site was recognized by the respective PWM. Known sites for E2F and Sp1 are shown in the center. One can see that predicted TF sites often overlap with each other indicating very complex potential regulatory switches

4.3 Identification of Composite Regulatory Modules

It is known that regulation of gene expression is controlled not only by single transcription factors but rather by their functionally connected combinations. Such combinatorial regulation of transcription is maintained through so-called composite regulatory modules CRMs (or composite regulatory elements) [20, 40]. A CRM describes specific combinations of transcriptionfactor-binding sites, sometimes with their specific arrangement to each other that often can be found in regulatory regions of co-regulated genes. Often such combinations of different TF sites located in close proximity in DNA can serve as targets for transcription factors that interact on a protein-protein level and through such interaction work in synergistic or antagonistic manner during regulation of transcription [40]. It is important to understand such interactions between transcription factors during their regulation of specific gene activity to reveal the causative mechanisms of gene regulation during carcinogenesis. We have therefore applied the CMA algorithm (Composite Module Analyst) for searching composite modules [20] in the regions of active chromatin of the promoters of up- and down-regulated genes in both data sets. The core of CMA is a genetic algorithm that identifies stable combinations of TF sites that are colocalized at a certain distance to each other in the analyzed regulatory sequences. In Figs. 6 and 7, we present the results of such an analysis for the MTX-resistant colon cancer cell lines and for the breast cancer cell line.

We identified that in both carcinogenic transitions a very important role is played by the transcription factors of the E2F family that are essential factors for regulation of cell division. CRMs for both data sets contain PWMs for the E2F-binding sites as one of the most important elements. Interestingly, the arrangements of E2F with other factors were found to be quite different in different data sets. In the MCF7 breast cancer cell line, the E2F sites in the identified CRM controlling upregulated genes are accompanied by PWMs for such important transcription factors as RELA, NFY, ITF1, E2A, EGR, and GLI. In MTX-resistant colon cancer cell line the CRM of upregulated genes arranges E2F sites with sites for such factors as: TCF4, SP1, HNF3B, AP1, SRY, ETS, GLI, and CTCF.

It is interesting that the TFs identified in this promoter analysis are highly relevant for the carcinogenic processes analyzed. For instance, the factors of the TCF/LEF family which were identified in the MTX-resistant colon cancer cell line are involved in the Wnt signaling pathway that is the most frequently deregulated pathway in colorectal cancers. AP-1 and Egr1, a known immediate-early response TFs, are activated by extracellular signals and mediating mitogenic responses [41]. It is very interesting also to see the presence of the RELA transcription factor (the NF-kB family member) in the CRM regulating the transformation of the breast cancer cells. The role of NF-kB factors in cancer is frequently discussed in

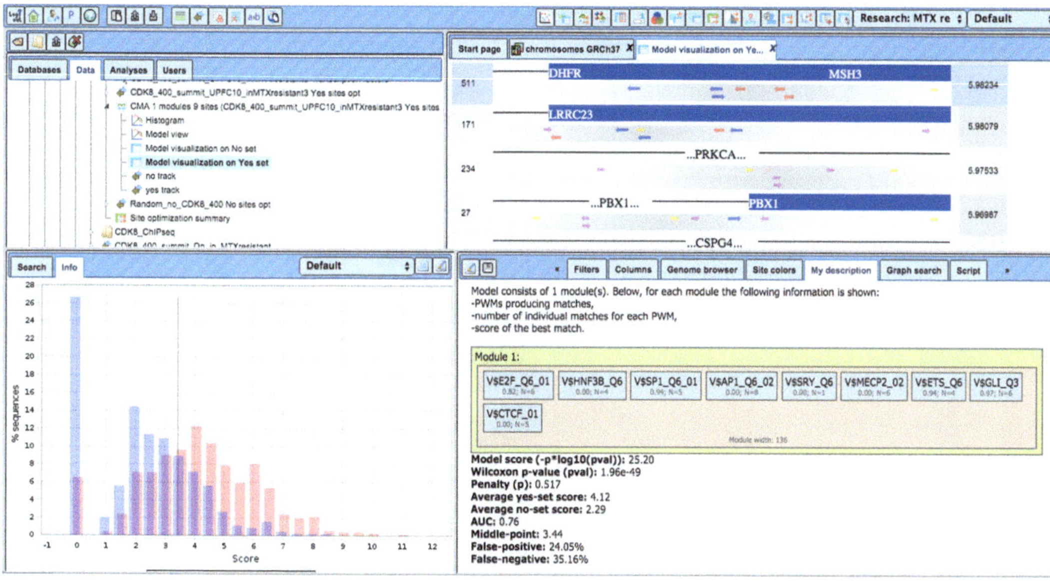

Fig. 6 Screenshot of geneXplain platform visualization of results of CMA analysis of genomic regions of activated chromatin in promoters of up-regulated genes of MTX-resistant colon cancer cell line. In the upper left window, it shows the user data and result repository in the platform with the highlighted result entry that is shown in the other three windows. The *bottom right window* shows the composition of the composite model consisting of nine PWMs (modules) with defined parameters of their cutoffs, maximal number of considered sites of the corresponding matrix (N), and preferable distance between the sites of the pair (Module width). The *bottom left window* shows two distributions of the values of the Composite Module Score (described in [20]; briefly, it is the sum of scores of all sites of the model found in the promoter) for the Yes promoters (*red*) and for the No promoters (*blue*). In the right window, it also shows the significance (Wilcoxon *p*-value) of the differences between these two distributions. The *upper right window* shows the map of the TF sites of the composite module found in the selected promoters. Here, we show promoters of the *DHFR* gene that is up-regulated in the MTX-resistant cells

recent literature (e.g., [42]) as factors that interact with other transcription factors such as p53, ETS, and EGR to regulate gene in various cancers and to link cancer and inflammation. It is especially interesting that in our analysis we propose the cooperative action of RELA factors with the E2F transcription factor. Moreover, such a cooperation between these two factors was reported earlier [43]. Also, the cooperation of E2F factors with GLI factors found in both our data sets was reported recently playing a central role in melanoma [44]. Evidence also suggests that HNF3B (FOXA1/2) is a tumor suppressor in certain cancers, including pancreatic and other cancers [44]. In summary, we can say that the CRMs identified by our genetic algorithm in the regions of open chromatin in the upstream regions of the upregulated genes in both our data sets provide a very reasonable hypothesis about transcription factors acting as key regulators in the processes of neoplastic transformation in our two systems under study.

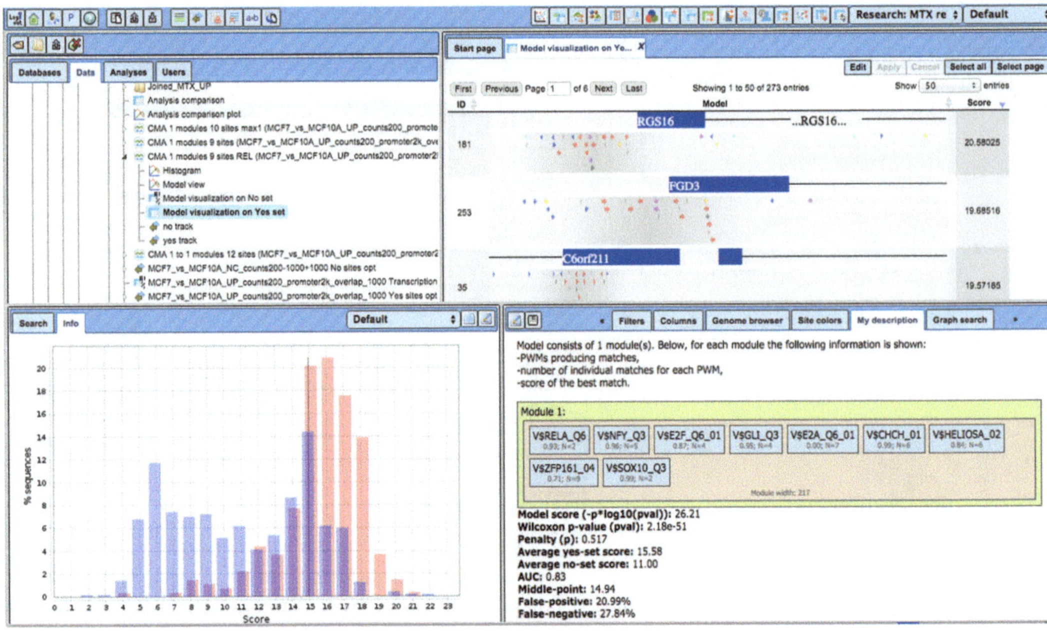

Fig. 7 Screenshot of geneXplain platform visualization of results of CMA analysis of genomic regions of activated chromatin in promoters of up-regulated genes of MCF7 breast cancer cell line. In the *upper left window*, it shows the user data and result repository in the platform with the highlighted result entry that is shown in the other three windows. The *bottom right window* shows the composition of the composite model consisting of nine PWMs (modules) with defined parameters of their cutoffs, maximal number of considered sites of the corresponding matrix (N), and preferable distance between the sites of the pair (Module width). The bottom left window shows two distributions of the values of the Composite Module Score (described in [20]; briefly, it is the sum of scores of all sites of the model found in the promoter) for the Yes promoters (*red*) and for the No promoters (*blue*). In the right window, it also shows the significance (Wilcoxon *p*-value) of the differences between these two distributions. The *upper right window* shows the map of the TF sites of the composite module found in the selected promoters

4.4 Find Master Regulators in Networks

The next step of the upstream analysis is the search for potential master regulators that can regulate the activity of the transcription factors identified in the previous step. In both data sets the master regulator search was done from the list of transcription factors fund by the CMA analysis. The sets of upregulated genes were used as sets of context nodes during the master-regulator search. To enable that, we converted the lists of upregulated genes into proteins from the TRANSPATH® database. During this conversion only genes encoding proteins involved in gene regulation and signal transduction in human cells were taken into account. As a result, we identified 1839 and 2462 TRANSPATH® proteins in MCF7 cells and MTX-resistant HT29 cells, respectively, encoded from the upregulated gene (including alternative protein isoforms) participating in various signal transduction and metabolic reactions according to the knowledge stored in this database.

In the current work, we set the maximal distance of the search for master regulator equal to ten steps, which gives a good chance to find regulators that are quite distant in the network and still can be responsible for the coordinated change of expression of the genes in the studied systems. When the search reaches the level of transmembrane receptors, or extracellular signaling molecules, the identified nodes in such a search can be considered promising drug targets.

Finally, after performing such a search for potential master regulators we checked which of them are actually up-regulated in the initial experimental data. We considered the fold change of the genes expressing the proteins that were found by the algorithm as potential master regulators. We require these genes to be statistically significantly up-regulated in MCF7 breast cancer cells or in MTX-resistant cells, respectively.

We hypothesized that the observed pathological switches from the non-malignant states into the MCF7 carcinogenic or MTX-resistant states of cells might be supported by the presence of **positive feedback loops.** We can observe such loops in the network when the genes expressing master-regulator proteins are working under the control of the transcription factors that receive activating signals through the signaling cascade starting from the proteins expressed by these genes (master regulators). Therefore, the up-regulation of the genes encoding master regulators in this analysis indicates the presence of such feedback loops. We believe that such positive feedback loops can contribute not only to the transition of one cellular state to another, but rather are necessary for the stabilization of the malignant state, since they maintain constant activation of a certain set of genes through the autoactivation loop. Therefore, we introduced into the algorithm an important requirement that the genes encoding selected master regulators should be up-regulated, which reflects the presence of such positive feedback loop in the system.

For MCF7 cells we identified 145 genes with LogFC > 2.0 that encode potential master regulators with a master regulator score > 0.3. For MTX-resistant cells we identified 29 such genes.

In Figs. 8 and 9, we show the networks of the top ten potential master regulators found by the algorithm in the MCF7 and MTX-resistant data sets. Genes encoding those ten master-regulator proteins, as well as some more intermediary proteins in the reconstructed signal transduction network, were also significantly up-regulated in the MCF7 and MTX-resistant cells, respectively, as it is indicated in the figures by the red circles decorating corresponding nodes on the diagram. The intermediary up-regulated nodes played the role of the "context nodes" in our algorithm. One can see that these context nodes often connect the identified master regulators with several transcription factors,

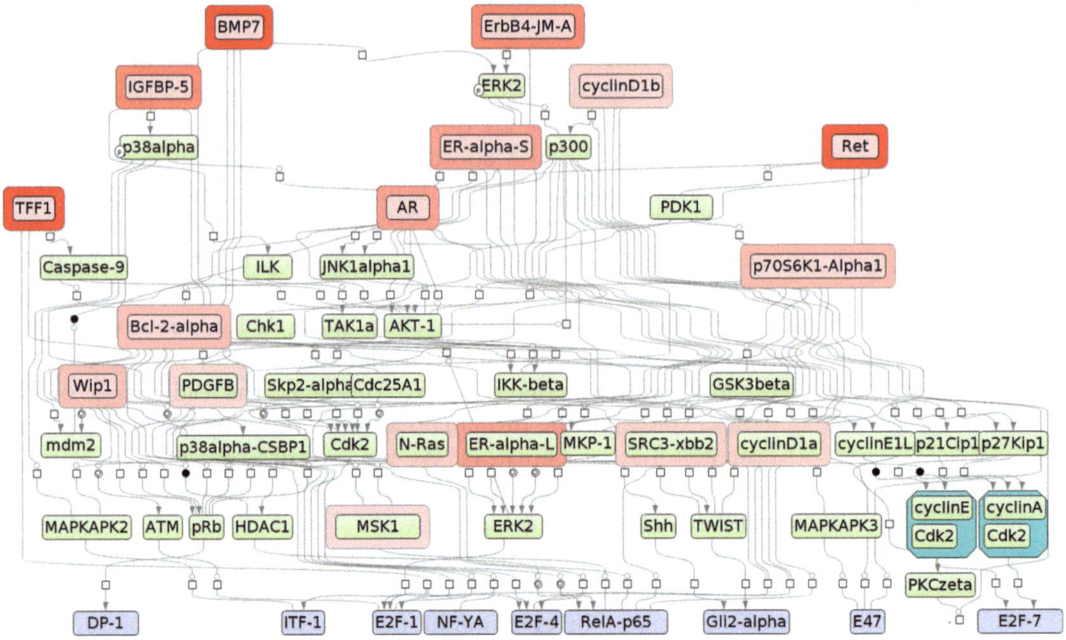

Fig. 8 A diagram of the signal transduction network of MCF7 breast cancer cells reconstructed with the help of the master-regulator search algorithm implemented in the geneXplain platform. Transcription factors (*blue*) are shown at the bottom of the diagram. Potential master regulators (*pink*) are shown at the top of the diagram. The direction of signal flow is from *top* to *bottom*. Intermediary molecules are *green*. *Red circles* around several nodes show those signaling proteins that are encoded by genes up-regulated in MCF7 cells. Such up-regulated nodes at the top of the diagram indicate the presence of positive feedback loops in the system since they transduce the activation signal to TFs that, in turn, activate transcription of the genes encoding these signaling proteins

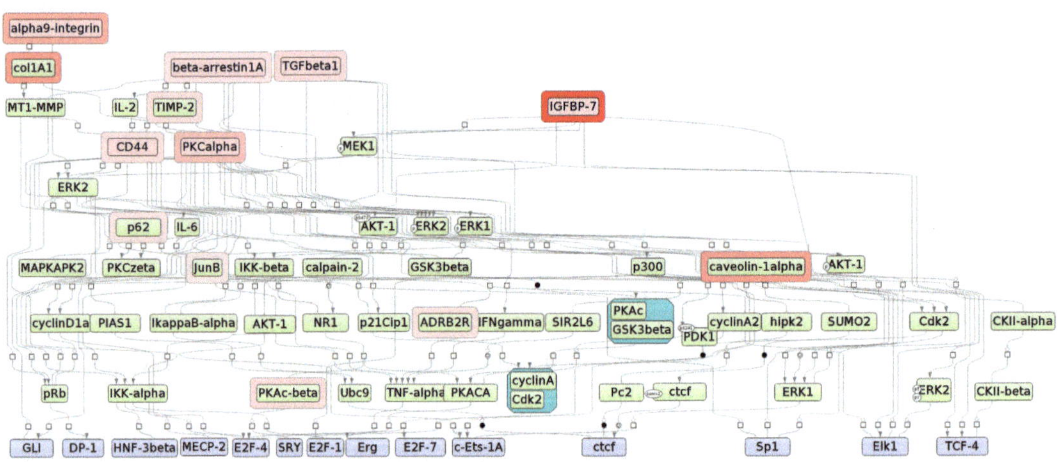

Fig. 9 A diagram of the signal transduction network of MTX-resistant colorectal cancer cells reconstructed with the help of the master-regulator search algorithm implemented in the geneXplain platform. Transcription factors (*blue*) are shown at the bottom of the diagram. Potential master regulators (*pink*) are shown at the top of the diagram. The direction of signal flow is from *top* to *bottom*. Intermediary molecules are *green*. *Red circles* around several nodes show those signaling proteins that are encoded by genes up-regulated in MTX-resistant cells. Such up-regulated nodes at the top of the diagram indicate the presence of positive feedback loops in the system since they transduce the activation signal to TFs that, in turn, activate transcription of the genes encoding these signaling proteins

therefore playing an important role in transducing the signal from the master regulators to these transcription factors, which in turn regulate their target genes upon receiving such a signal.

When focusing on the MTX-resistance data set, altogether, we noticed that many of the suggested master regulators are very important proteins that are known to be involved in regulating such process as cell cycle, apoptosis, cell adhesion, and metabolism of nucleotides. All those processes were detected as changed in MTX-resistant cells in our GO analysis above. Also, there are many lines of evidence showing the potential role of some of these proteins in sensitization of anti-cancer drug resistance mechanisms. For instance, TGF-beta, which is found in our master-regulator search and which is one of the most highly up-regulated proteins in MTX-resistant cells, has been found potentially responsible for acquired drug resistance in squamous cell carcinoma stem cells [45]. It was also shown that integrin alpha9 (ITGA9), which facilitates accelerated cell migration and regulates cancer cell proliferation and migration, is a target of epigenetic regulation; its overexpression leads to acquired resistance against 5-aza-dC treatment in human breast tumors [46]. Recently, it was shown that inhibition of insulin-like growth factor 1 receptor (IGF1R) leads to sensitization of head and neck cancer cells to cetuximab and methotrexate [47]. Therefore, it is extremely interesting that we identified the IGFBP7 protein as a potential master regulator, since this protein is a very potent modulator of IGF binding to its receptors.

All these facts show that the list of targets selected by the master regulator search algorithm has a very high potential. We can propose revealed master-regulators as promising drug targets for possible treatment of the MCF7-like breast cancer subtypes and for potential re-sensitization of MTX-resistant colon cancer cells toward action of MTX. Another possible application of such master regulators is in the field of biomarker identification. Our analysis can serve to reveal so-called causative biomarkers that during the evolution of a disease have got key positions in the combined signal transduction and gene regulatory networks acting in the disease and controlling the activity of a large number of genes; thus genetic or any other variations influencing activity of such master regulators should be a very good biomarker of disease subclasses which should be robust toward cohort variability.

5 Conclusions

In this paper, we have further developed our approach of "upstream analysis," [14, 17] which is an extension of the recently introduced idea of searching for "master regulators" in cellular networks [48]. We have introduced a new concept of "walking pathways," which emphasizes the plastic behavior of molecular networks in cells in the transition period between different states such as malignant and non-malignant states of tumors. We support the point of view that these different states are characterized by structurally different signaling networks acting in these states that are largely determined by the specific subsets of transcription factors cooperatively binding to open regions of chromatin that carry specific combinations of TF-bindingsites. Differences in the open and closed regions of the chromatin in different states of the cells lead to a different rewiring of respective signaling networks upstream of transcription factors. We also consider a very important role of positive (and potentially negative) feedback loops in such networks that have the ability to stabilize the rewired pathways and lead to the stable maintenance of a particular gene expression profile and respective cellular state (e.g., malignant state). This "upstream analysis" approach and the concept of "walking pathways" have been implemented in the software tool BioUML/geneXplain platform. An important novel part of the approach is the "Context Algorithm" of the master-regulator search, which is described in this paper. In this algorithm the sets of "context nodes" are defined using gene expression data or, if available, proteomics data, or even by defining any set of proteins that are known to be expressed in the cell types or tissues under study. These sets provide a specific "context" for the master-regulator search algorithm that searches through the signal transduction network, and they help to find most relevant components of the network in the given "context." We also introduced a novel way of integrating transcriptomics and epigenomic data, when peaks of active chromatin identified by ChIP-seq experiments are intersected with long 5′ upstream regions of differentially expressed genes to detect the locations of the most important "enhancers" of genes driving the pathway rewiring (pathway walking), and providing the state transformation.

We applied the developed tools to analyze two examples of cell transition from one state to another. In both cases, we have got multi-omics data that include transcriptomics (microarrays and RNA-seq) and epigenomics (ChIP-seq) data which helps us to do the analysis in a highly precise way. Frequency analysis of TFBS and analysis of composite regulatory modules in "enhancer" regions determined by the ChIP-seq data allow identifying transcription factors involved in the mechanism under study. Our approach gives us a nice possibility to integrate such different types of data, helping to achieve the goal of identifying drug targets with true potential. A considerable part of this analysis has been done with the help of automatic workflows in the BioUML/geneXplain platform, and therefore can be easily reproduced and can be applied to the analysis of other similar tasks. As a result, we identified a number of very promising drug targets, such as PKC-alpha,TGF-beta,insulin-like growth factor-binding protein 7, alpha9-integrin, and several others, and reconstructed a potential signal transduction network connecting these targets with the transcription factors triggering activity of the MTX-resistance genes. Many of these proteins are already known as important targets for anti-cancer drug therapy and our results suggest them for the use as anti-resistance targets. All these results demonstrate the validity of the presented approach of upstream analysis strategy with the concept of walking pathways.

Acknowledgments

This work was supported by a grant of the Federal Targeted Program "Research and development on priority directions of science and technology in Russia, 2014–2020," grant number: 14.604.21.0101 to the Institute of Chemical Biology and Fundamental Medicine, Novosibirsk, Russia. This work was also supported by the following grants of the EU FP7 program: "SysMedIBD," "RESOLVE," and "MIMOMICS." We are also very grateful to my colleague at the former Biobase GmbH, Niko Voss, for ideas on the algorithm on pathway analysis; my colleagues from Biosoft.ru: Dr. Tagir Valeev and Dr. Fedor Kolpakov for development of the BioUML framework; my colleagues at geneXplain: Dr. Holger Michael for the critical reading of the manuscript, Philip Stegmaier for development of the algorithms of TF site analysis, Dr. Jeannette Koschmann for creation workflows, Dr. Olga Kel-Margoulis and Prof. Edgar Wingender for the fruitful discussions of the work described here.

Conflicts of Interest: AK is an employee of geneXplain GmbH, which maintains and distributes the BioUML/geneXplain platform used in this study.

References

1. Sanyal AJ, Yoon SK, Lencioni R (2010) The etiology of hepatocellular carcinoma and consequences for treatment. Oncologist 15(Suppl. 4):14–22
2. Colussi D, Brandi G, Bazzoli F, Ricciardiello L (2013) Molecular pathways involved in colorectal cancer: implications for disease behavior and prevention. Int J Mol Sci 14:16365–16385
3. Hanahan D, Weinberg RA (2011) Hallmarks of cancer: the next generation. Cell 144:646–674
4. Guinney J, Dienstmann R et al (2015) The consensus molecular subtypes of colorectal cancer. Nat Med 21:1350–1356
5. Carro MS, Lim WK, Alvarez MJ et al (2010) The transcriptional network for mesenchymal transformation of brain tumours. Nature 463:318–325
6. Kolesnikov N, Hastings E, Keays M et al (2015) ArrayExpress update—simplifying data submissions. Nucleic Acids Res 43: D1113–D1116
7. Barrett T, Wilhite SE, Ledoux P et al (2013) NCBI GEO: archive for functional genomics data sets–update. Nucleic Acids Res 41: D991–D995
8. Petryszak R, Burdett T, Fiorelli B et al (2014) Expression atlas update—a database of gene

and transcript expression from microarray- and sequencing-based functional genomics experiments. Nucleic Acids Res 42:D926–D932
9. Smith CM, Finger JH, Hayamizu TF et al (2014) The mouse Gene expression database (GXD): 2014 update. Nucleic Acids Res 42: D818–D824
10. Fu J, Allen W, Xia A, Ma Z, Qi X (2014) Identification of biomarkers in breast cancer by Gene expression profiling using human tissues. Genom Data 2:299–301
11. de Gramont A, Watson S, Ellis LM et al (2015) Pragmatic issues in biomarker evaluation for targeted therapies in cancer. Nat Rev Clin Oncol 12(4):197–212. doi:10.1038/nrclinonc.2014.202
12. Subramanian A, Tamayo P, Mootha VK et al (2005) Gene set enrichment analysis: a knowledge-based approach for interpreting genome-wide expression profiles. Proc Natl Acad Sci U S A 102:15545–15550
13. Kanehisa M, Goto S, Sato Y et al (2012) KEGG for integration and interpretation of large-scale molecular data sets. Nucleic Acids Res 40: D109–D114
14. Kel A, Voss N, Jauregui R, Kel-Margoulis O, Wingender E (2006) Beyond microarrays: find key transcription factors controlling signal transduction pathways. BMC Bioinformatics 7:S13
15. Michael H, Hogan J, Kel A, Kel-Margoulis O, Schacherer F, Voss N, Wingender E (2008) Building a knowledge base for systems pathology. Brief Bioinform 9:518–531
16. Stegmaier P, Voss N, Meier T, Kel A, Wingender E, Borlak J (2011) Advanced computational biology methods identify molecular switches for malignancy in an EGF mouse model of liver cancer. PLoS One 6:e17738
17. Koschmann J, Bhar A, Stegmaier P, Kel AE, Wingender E (2015) "Upstream analysis": an integrated promoter-pathway analysis approach to causal interpretation of microarray data. Microarrays 4:270–286. doi:10.3390/microarrays4020270
18. Wingender E (2008) The TRANSFAC project as an example of framework technology that supports the analysis of genomic regulation. Brief Bioinform 9:326–332
19. Kel AE, Gössling E, Reuter I, Cheremushkin E, Kel-Margoulis OV, Wingender E (2003) MATCH: a tool for searching transcription factor binding sites in DNA sequences. Nucleic Acids Res 31:3576–3579
20. Waleev T, Shtokalo D, Konovalova T, Voss N, Cheremushkin E, Stegmaier P, Kel-Margoulis O, Wingender E, Kel A (2006) Composite module analyst: identification of transcription factor binding site combinations using genetic algorithm. Nucleic Acids Res 34(Web Server issue):W541–W545
21. Krull M, Pistor S, Voss N, Kel A, Reuter I, Kronenberg D, Michael H, Schwarzer K, Potapov A, Choi C, Kel-Margoulis O, Wingender E (2006) TRANSPATH: an information resource for storing and visualizing signaling pathways and their pathological aberrations. Nucleic Acids Res 34:D546–D551
22. Kel A, Stegmaier P, Valeev T, Koschmann J, Kel-Margoulis O, Wingender E (2016) Multiomics "upstream analysis" of regulatory genomic regions helps identifying targets against methotrexate resistance of colon cancer. EuPA Open Proteomics 13:1–13
23. Vogelstein B, Papadopoulos N, Velculescu VE, Zhou S, Diaz LA Jr, Kinzler KW (2013) Cancer genome landscapes. Science 339 (6127):1546–1558. doi:10.1126/science.1235122
24. Osborn MJ, Freeman M, Huennekens FM (1958) Inhibition of dihydrofolic reductase by aminopterin and amethopterin. Proc Soc Exp Blot Med 97:429
25. Morales C, Ribas M, Aiza G, Peinado MA (2005) Genetic determinants of methotrexate responsiveness and resistance in colon cancer cells. Oncogene 24(45):6842–6847
26. Messier T, Jonathan G, Boyd J, Tye C, Browne G, Stein J, Lian J, Stein G (2016) Histone H3 lysine 4 acetylation and methylation dynamics define breast cancer subtypes. Oncotarget 7 (5):5094–5109. doi:10.18632/oncotarget.6922
27. Smyth GK (2005) Limma: linear models for microarray data. In: Gentleman R, Carey V, Dudoit S, Irizarry R, Huber W (eds) Bioinformatics and computational biology solutions using R and Bioconductor. Springer, New York, pp 397–420
28. Selga E, Morales C, Noé V, Peinado MA et al (2008) Role of caveolin 1, E-cadherin, Enolase 2 and PKCalpha on resistance to methotrexate in human HT29 colon cancer cells. BMC Med Genet 1:35
29. Allen BL, Taatjes DJ (2015) The mediator complex: a central integrator of transcription. Nat Rev Mol Cell Biol 16:155–166
30. Langmead B, Trapnell C, Pop M, Salzberg SL (2009) Ultrafast and memory-efficient alignment of short DNA sequences to the human genome. Genome Biol 10:R25
31. Zhang Y, Liu T, Meyer CA et al (2008) Model-based analysis of ChIP-seq (MACS). Genome Biol 9(9):R137. doi:10.1186/gb-2008-9-9-r137

32. Dijkstra EW (1959) A note on two problems in connexion with graphs, vol 1. Numerische Mathematik, Mathematisch Centrum, Amsterdam, pp 269–271
33. Viemann D, Goebeler M, Schmid S et al (2004) Transcriptional profiling of IKK2/NF-kappa B- and p38 MAP kinase-dependent gene expression in TNF-alpha-stimulated primary human endothelial cells. Blood 103:3365–3373
34. Schimke RT, Kaufman RS, Alt FW, Kellems RF (1978) Gene amplification and drug resistance in cultured murine cells. Science 202:1051
35. Bertino JR, Göker E, Gorlick R, Li WW, Banerjee D (1996) Resistance mechanisms to methotrexate in tumors. Oncologist 1(4):223–226
36. Firestein R, Bass AJ, Kim SY et al (2008) CDK8 is a colorectal cancer oncogene that regulates beta-catenin activity. Nature 455 (7212):547–551. doi:10.1038/nature07179
37. Koike T, Shimada T, Fujii Y et al (2007) Up-regulation of TFF1 (pS2) expression by TNF-alpha in gastric epithelial cells. J Gastroenterol Hepatol 22(6):936–942
38. Good L, Dimri GP, Campisi J, Chen KY (1996) Regulation of dihydrofolate reductase gene expression and E2F components in human diploid fibroblasts during growth and senescence. J Cell Physiol 168(3):580–588
39. Lin SY, Black AR, Kostic D, Pajovic S, Hoover CN, Azizkhan JC (1996) Cell cycle-regulated association of E2F1 and Sp1 is related to their functional interaction. Mol Cell Biol 16 (4):1668–1675
40. Kel-Margoulis OV, Kel AE, Reuter I, Deineko IV, Wingender E (2002) TRANSCompel: a database on composite regulatory elements in eukaryotic genes. Nucleic Acids Res 30 (1):332–334
41. Zwang Y, Oren M, Yarden Y (2012) Consistency test of the cell cycle: roles for p53 and EGR1. Cancer Res 72:1051–1054
42. Hoesel B, Schmid JA (2013) The complexity of NF-κB signaling in inflammation and cancer. Mol Cancer 12:86. doi:10.1186/1476-4598-12-86
43. Kundu M, Guermah M, Roeder RG, Amini S, Khalili K (1997) Interaction between cell cycle regulator, E2F-1, and NF-kappaB mediates repression of HIV-1 gene transcription. J Biol Chem 272(47):29468–29474
44. Pandolfi S, Montagnani V, Lapucci A, Stecca B (2015) HEDGEHOG/GLI-E2F1 axis modulates iASPP expression and function and regulates melanoma cell growth. Cell Death Differ 22 (12):2006–2019. doi:10.1038/cdd.2015.56
45. Oshimori N, Oristian D, Fuchs E (2015) TGF-β promotes heterogeneity and drug resistance in squamous cell carcinoma. Cell 160 (5):963–976. doi:10.1016/j.cell.2015.01.043
46. Mostovich LA, Prudnikova TY, Kondratov AG et al (2011) Integrin alpha9 (ITGA9) expression and epigenetic silencing in human breast tumors. Cell Adh Migr 5(5):395–401. doi:10.4161/cam.5.5.17949
47. Hatakeyama H, Parker J, Wheeler D, Harari P, Levy S, Chung CH (2009) Effect of insulin-like growth factor 1 receptor inhibitor on sensitization of head and neck cancer cells to cetuximab and methotrexate. J Clin Oncol ASCO Annual Meeting Proceedings (Post-Meeting Edition) 27(15S):6079
48. Gevaert O, Plevritis S (2013) Identifying master regulators of cancer and their downstream targets by integrating genomic and epigenomic features. In: Proceedings of Pacific Symposium Biocomputing, USA, pp 123–134

Chapter 9

Mathematical Modeling of Avidity Distribution and Estimating General Binding Properties of Transcription Factors from Genome-Wide Binding Profiles

Vladimir A. Kuznetsov

Abstract

The shape of the experimental frequency distributions (EFD) of diverse molecular interaction events quantifying genome-wide binding is often skewed to the rare but abundant quantities. Such distributions are systematically deviated from standard power-law functions proposed by scale-free network models suggesting that more explanatory and predictive probabilistic model(s) are needed. Identification of the mechanism-based data-driven statistical distributions that provide an estimation and prediction of binding properties of transcription factors from genome-wide binding profiles is the goal of this analytical survey. Here, we review and develop an analytical framework for modeling, analysis, and prediction of transcription factor (TF) DNA binding properties detected at the genome scale. We introduce a mixture probabilistic model of binding avidity function that includes nonspecific and specific binding events. A method for decomposition of specific and nonspecific TF–DNA binding events is proposed. We show that the Kolmogorov–Waring (KW) probability function (PF), modeling the steady state TF binding–dissociation stochastic process, fits well with the EFD for diverse TF–DNA binding datasets. Furthermore, this distribution predicts total number of TF–DNA binding sites (BSs), estimating specificity and sensitivity as well as other basic statistical features of DNA-TF binding when the experimental datasets are noise-rich and essentially incomplete. The KW distribution fits equally well to TF–DNA binding activity for different TFs including ERE, CREB, STAT1, Nanog, and Oct4. Our analysis reveals that the KW distribution and its generalized form provides the family of power-law-like distributions given in terms of hypergeometric series functions, including standard and generalized Pareto and Waring distributions, providing flexible and common skewed forms of the transcription factor binding site (TFBS) avidity distribution function. We suggest that the skewed binding events may be due to a wide range of evolutionary processes of creating weak avidity TFBS associated with random mutations, while the rare high-avidity binding sites (i.e., high-avidity evolutionarily conserved canonical e-boxes) rarely occurred. These, however, may be positively selected in microevolution.

Key words Transcription factor, Avidity, Binding site, Mixture probability, Birth–death process, Skewed distribution, Kolmogorov–Waring distribution, Kemp distribution, Hypergeomeric function, Scale-free, Specificity, Sensitivity, ChIP-PET, ChIP-Seq, Scale dependence, Sample size

Tatiana V. Tatarinova and Yuri Nikolsky (eds.), *Biological Networks and Pathway Analysis*, Methods in Molecular Biology, vol. 1613, DOI 10.1007/978-1-4939-7027-8_9, © Springer Science+Business Media LLC 2017

1 Introduction

1.1 Protein–DNA Interaction In Vivo and Its Relative Avidity at the Genome Scale

The mathematical modeling of genome-scale next-generation sequencing (NGS) datasets is used to discover and analyze significant events arising in biological system. However, typically, NGS datasets are derived from one or a few samples that have a wide dynamic range with a large number of detected NGS signals. Furthermore, they are noise-rich and essentially incomplete. Therefore, one of the most serious dilemmas with genome-wide data analysis and statistical modeling is how to extract reliable, predictive, and meaningful knowledge about studied biomolecular systems from the resulting large but typically unique, noise-rich, and incomplete datasets. Based on huge databases of such large and multiscale biomolecular systems, common statistical properties can be identified and used for analysis, classification, understanding, and prediction of the system's properties.

Identification of molecular interactions of gene regulatory elements is a central problem in many biological disciplines, including biochemistry, molecular cell biology, systems biology, and functional genomics. Among those interactions, the transcription factors (TFs) and their binding site (BS) interactions are considered to be the basic units of functional gene activity [1–4]. A transcription factor is a sequence-specific DNA-binding protein that binds to such specific segments of DNA (called binding site, BS) [5, 6]. About 10% of the proteins in complex multicellular organisms carry out TF functions in living cells. It has been estimated that there are about 2500 proteins that potentially function as TF in human cells; for 570 of these proteins, manually curated data was reported (http://www.tfcheckpoint.org). These DNA-binding proteins are part of the complex biological system that controls the transcription activity of Pol-II and transcription of genetic information from DNA to RNA, establishing the gene expression patterns that can determine cellular programs and specific functions [7–9]. One TF could physically bind and functionally control hundreds and thousands of specific BSs in cells of complex eukaryotic organisms. Moreover, some highly similar short DNA sequences (TF-binding DNA motifs) have the potential to serve as direct binding targets of TFs [10, 11]. Different combinations of motifs are often clustered in the proximal promoter (upstream and downstream 5′end) regions and these events could provide specific regulatory signals for target genes.

Experimental models of DNA binding of these TFs have been studied in detail in vitro and by use of fluorescence techniques in live bacteria and eukaryotic cells. However, estimation of the DNA-binding affinity of a native TF to a defined DNA sequence in vivo in eukaryotes remains a challenge since it requires the quantitative analysis of (1) the intranuclear TF concentration, (2) the concentration of specific DNA sites that are accessible for TF binding, and (3) the fraction of DNA sites bound by the TF. Furthermore, the eukaryotic genome is complex, and each cell and in its different states contains a large number of different DNA-binding sites that are expected to have different avidity for a given TF. Additionally, in metazoans, the organization of the DNA in chromatin is exploited to gain tissue-specific gene regulation from a common genome achieved by composite binding of different TFs to the chromatin landscape, including regulatory DNA segments (i.e., enhancers). TFs are also capable of physical competition and/or synergistic protein–protein interactions with each other and BSs and overall interact with many other regulatory proteins, RNAs, metabolites, providing very complex and dynamic regulatory networks [11–18].

In a cell nucleus, physical binding by specific TF molecule(s) occur, leading to the formation of the TF–DNA pairs(s). Such binding events allow a TF molecule to regulate the transcription of a gene or a few neighboring genes in the proximal vicinity of a TF–BS complex. In the case of a population of the given TF molecules and corresponding BSs, TFs create within a cell nucleus a virtual interaction network between the given TF molecules and their target DNA BSs.

Here, we consider BSs as a population of genome-specific DNA species, called TF-binding sites (BSs) distributed by their intensity (or binding avidity, BA) for a given TF molecule. On a genome scale, the BSs have different TF–DNA BA with regard to the intensity of interactions with a given TF molecule. In this work, the TF molecules are considered identical functional species. According to this simplification, a BS set and its corresponding TF molecule are considered a point on the graph characterizing a virtual TF-BS binding network. In this graph, a TF is the single hub, and the protein–DNA BA characteristics can be considered the binding avidity-defined weighted domain of a BS set.

At the genome-wide level, the population distribution function (DF) of relative binding avidity (RBA) for a given TF can reveal significant statistical and functional attributes of the TF BSs. However, at the level of single cells or homogeneous cell samples, the DF of RBA for any specific TF is mostly unknown, since many technical obstacles in directly counting specific protein molecules bound to DNA have not yet been overcome. Experimental evidence at this time indicates that association and dissociation of a given TF at a promoter site of a given gene are very complex and stochastic events. The proportion of cells exhibiting TF–DNA binding in each promoter region and/or its BS, rather than the quantitative level of expression in each cell, is regulated at the cell and/or tissue level. Furthermore, based on the growing body of experimental findings on single cells, TF binding and dissociation events in a given promoter region likely occur in short bursts with relatively long and highly variable periods of inactivity. Transcription regulation in such discrete events is envisaged as involving changes in the probability rather than the rate of transcription initiation events. A quantitative transcription factor association and dissociation probabilistic model based on integrative NGS experimental data may explain and specify the TF–DNA binding mechanisms and associated gene expression profiling and cellular regulation phenomena, which appear to involve stochastic behavior.

According to the ChIP-based NGS detection methods, the number of overlapping TF-bound DNA fragments in a given genome locus (averaged across the cell population) can help to statistically characterize the TF–DNA RBA [19, 20]. However, direct experimental detection of the specific TF–DNA RBA and the respective construction of the EFD of binding events is a great biotechnology and statistical bioinformatics challenge. For one, TF–DNA RBA of different genome loci for the same TF is not a constant function. It has been observed that binding activity can vary within a genome by several orders of magnitude [20] and it can depend on many known and unknown factors. For instance, it can vary in distinct genome sequence compositions of BSs (TF binding with different motifs), the location of BSs in the gene region or its vicinity, loci with different genome architecture complexity, cell type, cell differentiation, genetic and epigenetic backgrounds, physiologic conditions, and environmental factors. In fact, due to biological complexity, the stochastic nature of regulatory processes and their spatiotemporal dynamics, detection system limitations, biased information about RBA the total number of binding sites per transcription factor in a given cell is unknown. Information about experimentally detected TF–DNA binding event values and associated EFD at the genome level is important for understanding the genome, transcriptome, and interactome biology and pathobiology of cells and tissues. However, only highly specific TFBSs with relatively high binding avidity have been reliably identified and characterized. A description of computational methodologies for the identification of DNA regions of overrepresented motifs that characterized TF binding via genome-scale experiments was reported in [21]. Such techniques allow for the prediction of TF–DNA binding events and can improve peak calling in genome scale sequencing experiments. However, in biology, the available experimental techniques do not sample and denoise the whole population of potential BSs system, but only study the noise-prone finite fraction. BA EFD is often not considered in the high-throughput studies of TF–DNA binding events.

1.2 High-Throughput Methods for Determining DNA-Binding Events

TF–DNA interactions can be detected by chromatin immunoprecipitation (ChIP) and the power of this technique lies in its ability to analyze protein–DNA interactions in vivo [15–17, 22–26] (Fig. 1). In ChIP experiments, TFs are cross-linked with DNA while an immune reagent (antibody) specific to a DNA-binding factor is used to enrich target DNA fragments where the TF was bound in a living cell. The bound DNA fragments that overlapped and are enriched with TF BSs are then identified and mapped on reference genomes. They are then further quantified to produce additional computational results.

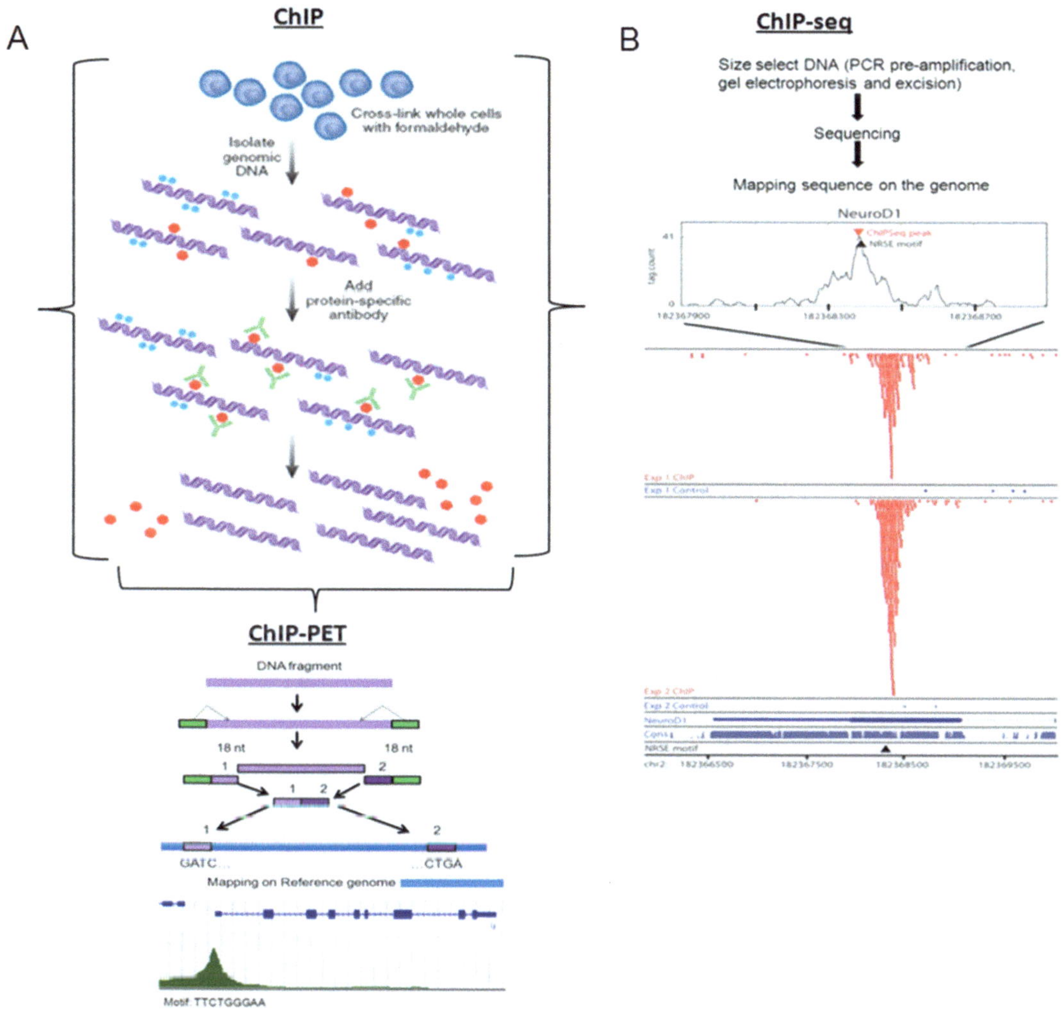

Fig. 1 Simplified work-flow of ChIP-based experiments for the genome-wide study of TF DNA binding sites in living cells: DNA and proteins are cross-linked and purified, then bound DNA fragments are isolated and amplified by massively parallel short-read sequencing methods. (**a**) ChIP-PET and (**b**) ChIP-seq-derived sequences are mapped onto the reference genome. This information is analyzed using statistical bioinformatics and computational genomics methods (*see* the next section)

Historically, the first technical platform to conduct wide-scale TF–DNA experiments was ChIP-on-chip DNA microarrays that tiled significant regions of the genome ([18], see also references in [2]). In ChIP-on-chip experiments, the copy number of DNA segments associated with TFs of interest is compared to a reference sample that is either genomic DNA or any DNA that might be immunoprecipitated with a negative control antibody. Starting with a biological question, a ChIP-on-chip experiment can be divided into a few major steps: (1) set up and design the experiment by selecting the appropriate array and probe type, (2) conduct experimental detection, and (3) carry out bioinformatics and computational analysis. The probabilistic models and statistical tests also are used for the identification of significant ChIP-on-chip signals. Although, over the years, ChIP-chip approaches have significantly improved and have greatly expanded our understanding of genome-wide TF–DNA interactions, it seems difficult to make ChIP-on-chip analyses affordable, reliable, and highly sensitive at the complex genome-wide scale [22–26]. In particular, several technological drawbacks with this method include complications in array hybridization and probe design, low resolution of the BS location, and experimental standardization.

Another way to achieve genome-wide identification of protein–DNA interactions is to adapt high-throughput DNA tag sequencing for analysis of chromosome mapping of ChIP DNA. Serial analysis gene expression (SAGE) is a short sequence tag mapping method, which was originally developed to analyze transcriptome profiles [2]. Several groups have modified the original SAGE protocol to isolate sequence tags from ChIP DNA and construct libraries of DNA tags for large-scale tag sequencing [16, 27]. For example, SACO [16] combines ChIP with a modification of SAGE. This method has the potential to semiquantitatively interrogate an entire metazoan genome by combining ChIP with a modification of long serial analysis of gene expression (Long-SAGE), a method normally used for transcriptome analysis [2, 16]. By sequencing many thousands of concatemerized 21 bp genomic signature tags (GSTs) generated from anti-TF ChIP sequences, a genome map of TF-binding sites can be identified and quantified. These and next-generation short-tag sequencing technologies [22–24] used to analyze protein–DNA fragments released after ChIP-on-chip have distinct advantages over standard microarray hybridization approaches. In particular, chromatin immunoprecipitation paired-end ditag (ChIP-PET) [4, 9] and ChIP-sequencing (ChIP-seq) methods [22–24] entail the possibility of a highly efficient process with a potentially unbiased coverage of the mammalian genome for large-scale identification of regulatory elements (promoters, enhancers, hypermethylated regions, etc.) mediated by DNA–protein interactions (Fig. 1). Current ChIP-PET technologies are capable of producing up to 100 or more millions of sequence reads during each instrument run [4, 9, 28].

The advantage of using PET over single tags is that the PETs mark the start and end of each ChIP fragment. When PET fragments are mapped to the reference genome, the identity of each ChIP fragment can be inferred by the PET mapping location, and binding sites can be accurately defined by the common regions within clusters of overlapping PETs. Furthermore, duplicate PET fragments arising from fragment amplification events during cloning can easily be distinguished and removed by treating these multiple PETs that map to identical locations as a single fragment. It has been demonstrated that the ChIP-PET method provides the most powerful short-tag sequencing technique for accurate localization of the physically specific mammalian TF-binding regions at a resolution of up to a few base pairs [23, 24, 28].

The ChIP-PET, ChIP-seq, and other IP-based sequence tag experiments have revolutionized genome-wide mapping, profiling, and interpretation of transcription factor-binding events. Although maturing sequencing technologies allow these experiments to be carried out with short (22–50 bps), long (75–100 bps), single-end, or paired-end reads, the impact of these read parameters on the downstream data analysis is still not well understood nor optimal [28]. Detecting TF–DNA interactions using ChIP-based sequence tag methods remains fraught with difficulties because it involves multiple and nonlinear experimental steps, sampling procedures, and unique data analysis methods. Our knowledge about optimization of the relationship between the specific and noisy binding events and sampling errors defined by ChIP technologies are still limited. Difficulty in discerning a successful experiment from a failed one and in choosing appropriate data analysis methods often presents a challenge.

1.3 Importance of Statistical and Computational Bioinformatics Analyses of Protein–DNA Interaction Events on the Genome-Wide Scale SAGE-Couple ChIP Assays

In many cases, it is desirable to quickly estimate the quality of the sequencing data mapped on the genome and know the specificity and sensitivity of genome-wide measurements of transcription factor-DNA-binding events. If one has prior knowledge of a set of all TF-binding sites (TFBSs) and the BSs that are not bound by the transcription factor, then conventional calculation of specificity and sensitivity of genome-wide TF-binding events is straightforward. However, in the absence of such prior knowledge, one must rely on statistical analysis, data-driven biophysical models, and computational predictions using currently available highly noisy and essentially incomplete DNA fragment samples.

For example, one can rank the identified target TFBSs based on the number of observed tags in a cluster of DNA fragments or in the cluster overlap, split the genomic regions into nonoverlapping blocks and count the frequency of events considered to be above binding. After ranking the values of such "protein–DNA binding events," this frequency information can be presented in the form of an empirical frequency distribution function (FDF) of a protein–DNA binding event, which is an essential starting point for any further statistical analysis and the planning of validation studies [19, 20]. This function has been used to identify the adequate statistical models required to perform appropriate statistical analysis catered to different types of genome-wide sequence datasets or prediction of specific TF binding regions [4, 12, 20]. This results in an ability to estimate the sensitivity of genome-wide transcription binding events using technically limited samples [19].

In 2005–2006, analyzing TF–DNA binding datasets detected using the ChIP-PET method, we recognized that TF–DNA binding events exhibit skewed frequency distributions [4, 12]. These binding events were detected by DNA sequence overlap and their corresponding peak height values were detected using the ChIP-PET method [12, 19]. This skewed frequency distribution shape and slope were changed in a predictable manner depending on sample size (the number of DNA fragment reads in the ChIP-PET library): the fat tail of the function and the proportion of specific detected TF–DNA binding sites grew when the number of sequence reads mapped on the genome increased. The EFD included two different functions related to technical noise and specific TF–DNA binding events. The EFD can be approximated by the mixture probabilistic model(s), where the EFD of the number of specific binding events followed the generalized discrete Pareto distribution (GPD) [19, 29].

The GPD exhibited the sample size (number of TF–DNA BSs) relevant properties and specific TF context dependence [12, 19, 30]. The statistical predictions of the TFBS regions and their relative avidity have been experimentally validated [4, 12] and successfully used in experimental studies [4, 11, 12, 20, 31, 32]. These findings have been used for the quality control of different high-throughput TF–DNA binding data analyses. Our probabilistic model approximation of the EFD of CHIP-PET binding signals allowed us to evaluate (1) the specificity of an Ig in an immunoprecipitation reaction, (2) the sequencing method depth, (3) sequence library data saturation level, and (4) critical cutoff values that separate specific and noise-rich TF–DNA binding sites. It also provided for (5) the quantitative comparison of the genome mapped TFs according to their relative binding avidity. In addition, adding motif-finding analysis, this approach allowed us to predict many novel putative target genes for these TFs and link variation in the TFBS binding activity to target gene expression levels. In the last 10 years, several useful mathematical models have been developed to quantify TFBS-binding avidity level and stratification of biological functions of specific genes at the genome scale [11, 13, 33–36]. Software for analyzing large-scale biomolecular systems and their networks has also been developed [11, 13, 19, 36, 37].

The importance of the mathematical analysis of empirical frequency distributions of TF–DNA BA events has been recognized, and some of the theoretical predictions have been experimentally validated [11–13, 20, 35, 36, 38, 39]. The characterization of the empirical frequency distributions of the biomolecular events is important for our understanding of this evolving and complex system's function. For instance, the genes directly repressed by c-myc showed low avidity of c-myc-DNA binding in the proximal regions of genes [11]. This result was a direct result of analysis using a mathematical model that assumes two classes of c-myc TFBSs, a high avidity class of BSs with consensus E-boxes and a second class of BS containing nonspecific DNA-binding signals. These findings and assumptions are consistent with our TF–DNA binding model and the results of the analysis of the frequency distribution of TF–DNA binding events, E-boxes, and expression patterns of the related target genes [34]. Both studies suggest that DNA binding itself (even in the vicinity of transcription sites) without additional quantitative and qualitative characterization of the BS cannot account for the functional activity of the TF–DNA binding event. Novel and potentially useful mathematical models were developed in [36, 38]. These models may help to identify the TF–DNA sites with low- and high-avidity potential and link these distinct avidity classes to different TF regulatory functions.

The objective of this work is to develop a basic probabilistic model and common statistical bioinformatics strategy to analyze different types of genome-wide ChIP-based TF–DNA binding experiments. We specifically propose a mixture probabilistic model of nonspecific and specific TF–DNA association–dissociation DFs. Our model estimates the basic statistical characteristics of the BA DF. We also summarize the findings of a newly developed procedure which can be used to estimate specificity and sensitivity of genome-wide tag-coupled ChIP assays (SACO, ChIP-PET, and ChIP-seq). Via parameterization of the model, we quantify the effect of denoising and sampling on the macroscopic characteristics of BA DF. We develop a uniform approach for quantitative analysis of such experiments which can (1) identify confidence subsets of TFBS, (2) reconstruct the low avidity part of the specific TF–DNA binding DF, (3) use such data to predict the total number of specific BSs for a given TF in mammalian organism genomes (e.g., rat and human), and (4) compare different ChIP-based tag sequencing approaches by uniform statistical parameters. In the end, we validate our results using TFBS motif search/prediction algorithms and microarray expression data. Using diverse experimental data, we investigated not only the effects of data undersampling but also proposed a method of estimating the number of nonobserved TFBSs in a noise-rich fraction with low avidity and the number of nondetected TFBSs. We then discuss how our study results allow us to better understand the functional significance of observed quantitative differences in the TF–DNA binding activity in low, moderate and high TF–DNA BSs in gene promoter regions, and to link TF-BS activity with gene expression profiling.

2 Data, Definitions, Empirical Models, and Methods

2.1 DNA Fragment Cluster, Cluster Overlap, Overlap Cluster Regions, Cluster Peak, and TFBS Avidity

Serial Analysis of Chromatin Occupancy (SACO) is a DNA sequence tag (GST) generation technology to identify genomic locations of ChIP-isolated DNA fragments. SACO is a method that has provided a conceptual basis for the development of next-generation sequencing (NGS) strategies currently utilized for conducting full-genome surveys of DNA-binding protein-binding sites.

"Paired-End ditagging" (PET) analysis revolves around the concept of extracting 18-base "signatures" or "tags" from each of the 5′ and 3′ termini of any contiguous DNA sequence and ligating them into Paired-End ditags (PETs) that are concatenated for enhanced sequencing efficiency (Fig. 1a). Each PET can subsequently be mapped onto the appropriate genome assembly (mapping 36 bp of the genome) to accurately define the location of the original fragment from which it was derived.

ChIP-PET is a ChIP-based method that uses SAGE-like PETs [14]. ChIP-PET analysis uses the principle of paired-end ditagging, which can be applied to the efficient mapping of TFBS identified by the ChIP method. It randomly shears genomic DNA fragments that are first enriched for the TFBS of interest by ChIP, inserted into a specific cloning vector (pGIS3) and then subjected to the same ditagging approach. ChIP-PET signals (probable TFBS) are indicated by the overlapping of multiple distinct PET sequences on defined chromosomal loci.

In ChIP-seq, a ChIP-enriched SAGE-like tag is represented by either a single internal 21 bp tag sequence (SACO), by a single 27 bp tag sequence (ChIP-Seq), or by a ~36 bp paired-end ditag sequence (ChIP-PET, with a ditag from the 18 bp 5′ and 3′ signature sequences extracted from each end of the ChIP DNA fragment) (Fig. 1b). Thus, SACO and ChIP-Seq demarcate a single end of the sonicated ChIP DNA fragment, while ChIP-PET, the full length of the sonicated ChIP DNA fragment. The protein–DNA BSs are then deduced based on the frequency with which tags in a given genome locus are extracted from ChIP DNA fragments using the background computational expectation or background control data.

A distinctive feature of the binding event defined by any large-scale ChIP-based technology is the DNA fragment cluster. With sequencing depth increases, the sample size grows in the ChIP-seq library, and the library includes more putative TF–DNA binding signals with less binding avidity potential. After DNA fragment mapping, the TFBS-associated DNA fragments are preferentially derived from the vicinities of TFBS and then accurately mapped to a definite chromosomal region of the reference genome, forming DNA fragment clusters. Such clusters can then be characterized as a putative TFBS for a given protein, and appropriately counted.

However, due to the fundamental differences in the properties of current high-throughput technologies, an identification of span of ChIP DNA sequence clusters and the DNA sequence aggregation procedures into clusters are not standardized. Hence, it may be technology-specific. For example, the SACO method requires 21 bp DNA fragments for the mapping of a specific region and forms clusters that identify and quantify the TFBS by incorporating GSTs in the "SACO cluster... that are within 2 kb of each other" [16]. Additionally, most SACO loci are confirmed by identification of chromosome location of that putative TFBS near or within "transcriptional open regions." A very different definition for fragment (tag) DNA cluster and corresponding TF–DNA binding event are used by the ChIP-seq method [22, 23].

ChIP-seq is a technique to specifically identify DNA sequences bound by the protein of interest (such as transcription factor, cofactor, or other chromatin protein of interest) (Fig. 1b). ChIP-seq combines the ChIP assay with largely parallel DNA sequencing to identify the protein–DNA binding sites on a genome. The methodology of ChIP-seq comprises six step process: (1) Protein-binding DNA is fixed in place with a cross-linking agent; (2) using immunoprecipitation, DNA is sheared and protein–DNA complexes with targeted antibodies are isolated; (3) DNA fragments are reverse cross-linked and isolated; (4) next a DNA fragment library is constructed with subsequent sequencing; (5) genome mapping is next; and (6) a Peak calling software is used to identify regions and intensity of protein–DNA interaction.

For ChiP-seq TF–DNA binding data, called ChIP-seq single-end tags (SETs), the authors define the "extended and overlapped" DNA fragment clusters formed by distinct DNA fragments as overlapping if (1) they are overlapped resulting in computational extension of the original 27 nt sequence into a 174 nt [22, 23] extended SET (eSET) and (2) they share common loci (at least 4 bp). After denoising the DNA fragment datasets, the overlapping clusters (observed as local peak heights on a genome coordinate) can provide genome mapping and the number of TF binding events in the entire reference genome, data that are usually reported in the processed ChIP-seq library dataset. Moreover, the height of the peak cluster region can be used as an observed measure of relative TFBS avidity. In essence, the number of transcription factor binding sites, reproducibility, and reliable identification of ChiP-seq binding events depends on the avidity of specific antibodies and the size of the sequence read library.

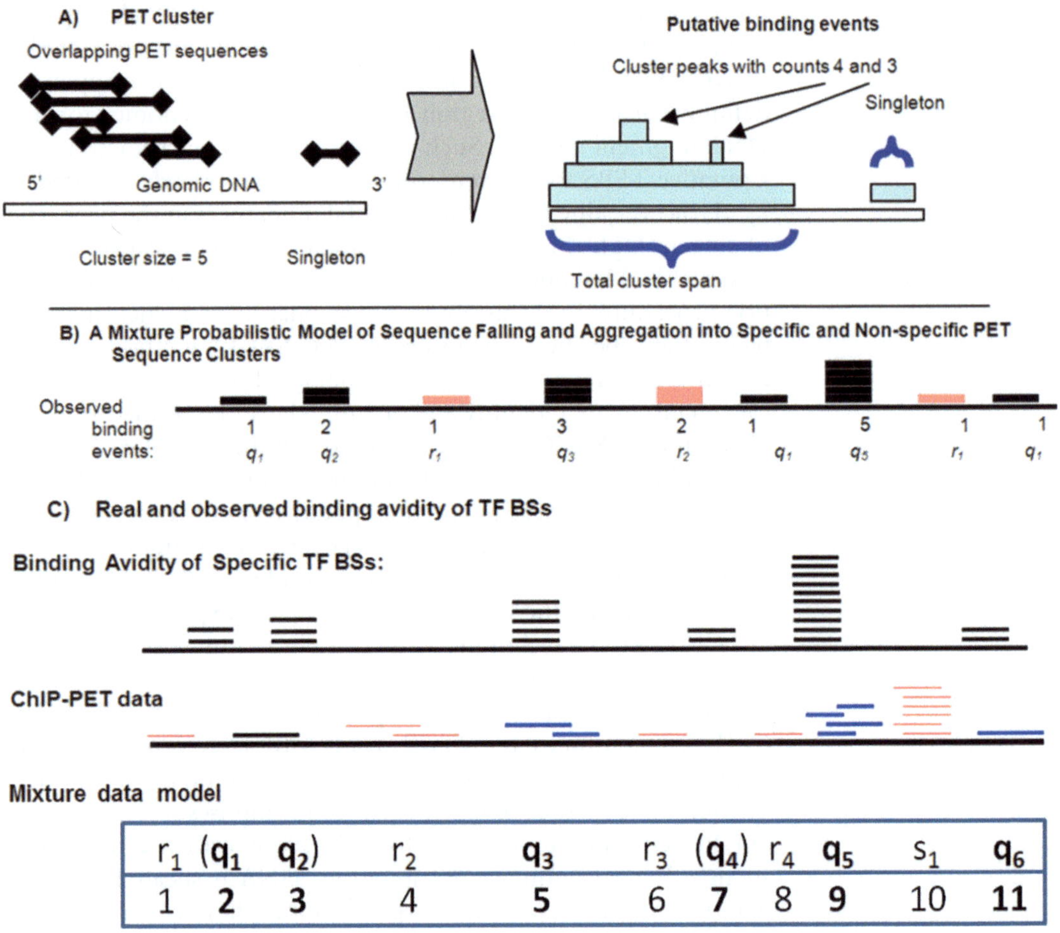

Fig. 2 Mapping, processing, and characterization of TF–DNA binding sites defined after ChIP-PET DNA fragment mapping onto genome and quantification of binding avidity. (**a**) Schematic example of the ChIP-PET DNA fragment cluster and singleton (*left panel*), processed data, providing information about genome location, total DNA fragment cluster span and peak height value of the local overlapped DNA fragment density for a cluster (*right panel*) . (b) Illustration of DNA fragment undersampling from a mixture of two frequency distributions where 1, 2, 3, and 5 (*black rectangles*) indicate the genome region with the intensities of true bindings q_1, q_2, q_3, q_5. (**c**) *Top panel* shows region locations and avidity values for true TFBS regions. *Bottom panel* illustrates ChIP-PET overlap with false positive data

According to ChIP-PET [4, 9], when PET DNA sequences share the same locus (4+ common nt) in the same chromosome region, they are recognized as a cluster and overlapping PET DNA sequences, called "cluster overlap" (Fig. 2). The chromosome locus of the cluster overlap can contain a TFBS region, and the number of overlapped PET DNA sequences in that cluster overlap region can represent a semiquantitative measure of relative avidity of a BS (*see* below). If more than one statistically confident overlapped PET DNA region is included in the PET DNA sequence cluster, the cluster overlap region containing the largest number of DNA sequences (largest peak) is counted as the most confident "binding event" associated with the cluster. The peak height in the cluster overlap region, together with span of the overlap region, are considered important experimental features of the sequencing experiment and are used for computational preprocessing of these data and their statistical analysis. Figure 2a shows schematically the DNA fragment cluster regions mapped on the genome and their quantitative characteristics used for quality control, prefiltering, assay optimization, and statistical analysis. Figure 2b illustrates our concept of specific DNA fragment undersampling, assuming random sampling from a mixture of frequency distributions, including high-avidity (specific) and low-avidity (noise) DNA fragments. Figure 2c shows the results of the ChIP-seq experimental mapping data including typical errors. In this work, we show how these errors can be related to real/(expected) genome data mapping precision, to the overlap peak heights (specificity limitation, related to variation of DNA–TF binding avidity), and to the incompleteness of ChiP-seq data (sensitivity limitation).

In the figure above, the right panel demonstrates a cluster size count of 5 (PET-5), and the cluster peak height value (binding event intensity) is 4. If a total DNA fragment cluster overlap span is longer and the cluster includes more than one peak, then such a cluster may be split into two or more independent clusters. For instance, using strict criteria, we may define two DNA fragment clusters related to each distinct pea) with peak height values of 4 and 3, respectively. This data processing analysis allows improving the accuracy of peak region identification. It was commonly used in our analysis.

In Fig. 2b, q_1, q_2, q_3, and q_5 quantify the binding signal intensity values detected in specific BS regions. The numbers 1 and 2 are associated with red rectangles r_1 and n_d which are quantified binding signal intensity values detected in low-avidity and/or nonspecific BS regions in a genome. Figure 2c, the top panel represents region locations and avidity values of true TFBS regions. The binding avidity value is described by the number of identical line segments in a given BS region. Comparing panel (b), this panel indicates that technical limitations and errors in the detection system can provide a proximal location of the true binding sites defined in a real experiment. Incomplete or bias sampling may lead to the missing of some low- or moderate-avidity BSs (i.e. binding site 6). The bottom panel of Fig. 2c illustrates that ChIP-PET (and ChIP-seq) mapping data can include two types of false-positive (noise) DNA fragment cluster overlap regions: the regions with relatively low binding avidity (r_1, r_2, r_3, r_4) and the regions with relatively high intensity binding events (s_1). The former TFBS region subset is usually random where results should are not likely reproducible/correlated over location and the binding intensity signal variation across biological samples. On the other hand, the last case (s_1) can be a false finding due to systematic errors of technology and/or intrinsic features of the biological system. This second type of error may be reproducible across different experimental systems (e.g., cell types or treatment conditions).

True TFBS regions defined in the experimental sample are represented by the DNA clusters with the binding intensities q_1, q_2, q_3, q_4, q_5, q_6. Among the TFBSs represented by this binding activity, some neighboring BS regions (q_1, q_2) could not be distinguished as the distinct TFBS-positive cluster. Furthermore, some of the true TFBSs may not be detected due to sampling and random error issues (true TFBS in position 7; q_4). Commonly, binding avidity of nonspecific BSs on average is lower than binding avidity for a true BS. Notice that ChIP-based sequencing methods provide samples that could be incomplete due to limited depth of sequencing reads and technical or biological noise reads. DNA sequence overlap regions can provide chromosome location of the highest-avidity BS, but due to limited sample size and noise signals, the moderate- and low-avidity specific BSs could not be reliably detected [20].

Incompleteness of the noise-depleted ChIP-seq signals and sample size dependence of the TF–DNA binding EFD are important features of this data. These statistical characteristics can be modeled and estimated by the mixture probability function, which will be considered in the next sections. Several basic quantitative characteristics of the ChIP-seq experiments are summarized in Table 1. In this table, the statistical characteristics of two STAT1–DNA binding ChIP-seq DNA fragment datasets are presented. These datasets (or sequence libraries) represent the samples of STAT1–DNA fragments extracted from INF-γ-stimulated and unstimulated human HeLa S3 cells. Note that the characteristics shown in Table 1 consist of the features important for the quality control and adequate quantification data analysis for ChIP-seq and other next-generation sequencing (NGS) methods. We present detailed biological and statistical characteristics of the STAT1–DNA binding profiles in Subheadings 4.4 and 5.

2.2 Modeling of the TF–DNA Binding Frequency Distribution

The number of clustered/overlapped DNA fragment sequences covering specific genome loci detected with a ChIP-based NGS method should roughly reflect the binding site avidity of the given TF–DNA interactome (Fig. 2b). We assume that when the number increases, the avidity of a locus becomes higher. Let us consider the number of occurrences of the DNA fragment sequence clusters/overlaps as a realization of a random process of TF–DNA binding. In this study, we consider the TF–DNA binding events in terms of the probability functions derived from a random continuous-time Markov jump process. The forward Kolmogorov differential equations for the birth–death nonhomogeneous in-time stochastic jump process are used in this work. We analyze the

Table 1
Characteristics of updated ChIP-seq libraries [23] (after our reprocessing)

Parameter	Stimulated	Un-stimulated
Reads		
Total sequenced (10^{-6})	24.1	22.7
Total, uniquely mapped onto genome (10^{-6})	15.1	12.9
In peaks (10^{-6})	2.71 (17.9%)	0.54 (4.2%)
Peak coverage (Mb)	34.5 (1.12%)	9.7 (0.31%)
Peaks		
Median width (bp)	473	519
Peak height at revised threshold	10	9
# peaks	63,309	16,470
• Average height	19.9	12.9
Median height	13	12
# peaks in problematic clusters (due to systematic errors)	853 (1.3%)	727 (4.4%)
# peaks after filtering (used in our work)	62,456	15,743
Average height of clusters after filtering	20	12.6
Average height of clusters in problematic clusters	17.3	18
Peak coverage after filtering (Mb)	34.1	9.3
Peak coverage of problematic clusters (Mb)	0.45	0.4
Median width (bp)	474	522
Median width of problematic clusters (bp)	414	427
Number of sequences on peaks	1,246,120	198,566
Number of overlapping loci	14,874 (23.8%)	14,303 (90.8%)
Number of non-overlapping loci	47,582 (76.2%)	1440 (9.2%)

properties of a distribution function derived from a global steady-state solution of Kolmogorov differential-difference equations for the birth–death nonhomogeneous continuous-time jump process [40], assuming that the transition rates between discrete states are non zero only for nearest neighboring states and at least linear transition functions of state's values are realized [30, 34]. In this manner, the Kolmogorov equations allow for finding the limiting probability function and the conditions for positive and stable recurrence of the transition rates of the birth (association event) and the death (dissociation event) processes. We have called this function the Kolmogorov–Waring (KW) distribution function. The dynamic model describing nonstochastic time-dependent trajectories and asymptotic values of mean and variance are proposed and analyzed. Next, we generalize our probabilistic model of binding events introducing the analysis of the global steady-state solution of the Kolmogorov differential equations, using the Kemp generalized hypergeometric probability distribution functions [40].

We use our global optimization method for goodness of fit analysis, summarizing the weighted discrepancy between observed values and the values expected under the model in question. We identify the parameters of the KW probabilistic function which describe how well its specific subfamilies fit a set of observations. We then specify mechanisms and quantify parameters of the transition rate law driving jumps between states in TF–DNA binding events.

2.3 Outcome, Event, Random Variable, and Probability Function

Let S be the sample space of an experiment. An event A is a set of outcomes in an experiment, a subset of the sample space S, to which a probability is assigned. A single outcome may be an element of many different events, and different events in an experiment are not equally likely since they may include very different groups of outcomes. By Kolmogorov's axioms, we assume that $P(A) \geq 0$, $P(S) = 1$ and A and B are mutually exclusive events; then $P(A \cup B) = P(A) + P(B)$. Any function of P that satisfies the axioms of probability is called a probability function. Any random variable X is a function from a sample space S into the nonnegative real numbers, with the probability that for every outcome there is an associated probability $\Pr(X = x)$ that exists for all values of x. Random variables will be denoted throughout this work by uppercase letters. Realized values of the random variable will be denoted by corresponding lowercase letter.

The random variable X takes on a finite or countably infinite number of possible event values. We determined $P(X = x_i) = p_i$ for all of the possible values of X and called it the discrete probability function (PF), which is a step frequency function with only an enumerable number of steps. Its height of the step at x_i is p_i; then $P(X = x_i) = p_i$. We say its support is the set$\{x_i\}$. The cumulative distribution function (CDF) of X is defined as $P(X \leq x_i)$. By definition $P(X \leq x_i)$ is a nondecreasing function of x, and $0 \geq P(X \leq x_i) \leq 1$. If $\lim_{x \to \infty} P(X \leq x) = 1$ then the distribution is called proper. Thus,

$$\sum_i p_i = 1$$

2.4 Experimental Frequency Distribution

In the ChIP-based NGS data, the TF–DNA binding events can be defined by the number of DNA fragment sequences belonging to their cluster/overlapped region mapped to a genome locus.

We define a ChIP-enriched DNA fragment library as a list of ChIP-derived DNA fragments which uniquely map to the genome and contain distinct tags (ditag in the case of the ChIP-PET method). The size of a library, M, is the total number of distinct DNA fragments observed in the library and uniquely mapped to the genome. Let $n(m,M)$ denote the number of TDB events in which ChIP-enriched DNA fragment sequences in a cluster overlap have peak height m; m is a count of distinct DNA sequence fragments within a cluster overlap in a given genomic locus in the library of size M. Let J denote the maximum observed TF–DNA binding events (e.g., peak height in a given locus) in the sequence library. Let N denote the number of specific binding events:

$$N = \sum_{m=1}^{J} n(m, M). \tag{1}$$

Then, we can also call M the "DNA sequence mass"

$$M = \sum_{m=1}^{J} mn(m, M). \tag{2}$$

The histogram or the frequency distribution of the number of DNA fragments in a given locus within a library

$$\bar{P}(X = m) = \bar{p}_m = n(m, M)/N$$

might be considered the empirical EFD of TF-binding activity.

Our analysis of the empirical EFDs in all studied datasets suggests that a binding event can be represented by two (or more) distinct random binding processes generated by different TF binding mechanisms. The binding events shown in Fig. 2c can be described by the following empirical mixture EFD function:

$$\bar{P}(X = m) = \alpha\bar{P}_s(X = m) + (1 - \alpha)\bar{P}_{ns}(X = m), \tag{3}$$

where $\bar{P}_s$ is the frequency distribution (FD) function of the specific TF–DNA binding event that occurred exactly m times in each genome-wide experiment. $\bar{P}_{ns}$ is the FD function of nonspecific binding; m – $\{0,1,2,..., m_{max}\}$ is the number of bindings in a given genome-wide experiment; $m_{max} = J$ denotes the maximum value of m. The parameter α is the fraction of specific DNA fragments of the experiment observed in a total population of DNA fragments mapped onto a genome in any location.

$0 < \alpha < 1$.

For a given TF-bound DNA fragment library, we assume that the corresponding probability function (PF) of the TF–DNA binding is defined as a sum of PFs of specific and/or nonspecific TF–DNA binding events:

$$P(X = m) = \alpha P_{sp}(X = m) + (1 - \alpha)P_{ns1}(X = m), \quad (4)$$

where $P(X = m)$ is the PF of occurrence of specific and nonspecific bindings, X is the random number of bindings in the genome, m {0,1,2,. . .} is the number of bindings, P_{sp} is the PF of specific bindings, $0 < \alpha < 1$ is the fraction of specific bindings, and P_{ns1} is the PF of nonspecific bindings. Note that P_{ns1} represents the experimental "background" noise and/or truly "low-avidity" binding events.

Using ChIP-PET, ChIP-seq, and SACO datasets, we construct an EFD of the binding events in a given experimental library. P_{ns} describes low-specific and/or nonspecific bindings, which are mostly represented by singleton DNA fragment sequences and low-height peak DNA fragment forming clusters. We found that P_{ns} can be well approximated by an exponential function. We also model the noise part of the EFD using a Monte-Carlo simulation. This method provides random sampling of DNA sequence fragments from a given library and random mapping of corresponding fragment spans onto a reference genome. DNA sequence fragment mapping was carried out onto "available" regions of a reference human genome by sampling the DNA fragments from a uniform distribution. We called this method "mapping at random without sequence specificity." After such mapping, random DNA fragment overlap clusters were identified and counted. As a result, the frequency distribution of the "random" clusters was constructed [4, 20]. We observed that both methods often provide similar FDs for the same library.

We described P_{sp} using the so-called generalized discrete Pareto distribution (GPD) function [20]. This probabilistic model constitutes an approximation for many skewed empirical frequency distributions with a long right tail in genome-scale biomolecule datasets. Such properties are often observed in evolving biological systems and in samples from diverse big-size bimolecular datasets.

In this work, we provide the parameter estimates of the functions P_{ns} and P_{sp} and the relative weight parameter α using the algorithm published in [30] and briefly described in Subheadings 2.5 and 3.8 of this chapter. The attributes of nonspecific or low-specific and relatively high-specific components of the mixture probability binding model in Eq. 1 are presented in Subheading 3.

Note that some nonspecific DNA fragment clusters can be found among highly abundant and experimentally reproducible clusters and/or cluster overlaps with large enough heights. The DNA sequence fragment clusters with the "problematic" or "unlikely" TFBS locations such as mitochondrial DNA, centromeric regions, locations where no gene is found within a 100 kb vicinity of a putative BS, that near a genome gap, or repeat element regions should be considered a source of systematic errors, reproducible across different datasets and experimental conditions. Such clusters can be eliminated from analysis by several empirical rules and excluded at the prefiltering stage of data processing. This type of systematic error can be identified computationally via additional analysis of suspicious genome regions.

In this case, the extended mixture probability model of TF–DNA binding can be considered:

$$P(X=m)=\alpha P_{\mathrm{sp}}(X=m)+\beta P_{\mathrm{ns1}}(X=m)+(1-\alpha-\beta)\times P_{\mathrm{ns2}}(X=m), \tag{5}$$

where the probability functions $P_{\mathrm{ns1}}(X=m)$ and $P_{\mathrm{ns2}}(X=m)$ are the probability of nonspecific random errors and of nonspecific systematic errors, respectively. α, β are unknown weight parameters. In our analyses, P_{ns2} binding events can be defined in experimental data as the false-positive DNA fragment clusters. Our computational analysis provides an advanced genome region annotation of the "problematic" regions, evolution conservation regions [41], and specific sites (e.g., TFBS). A relatively small fraction of DNA fragment data consists of systematic errors, about 3–7% of all ChIP-based library sequences mapped onto a genome.

Because sequence read sampling is an experimental parameter, the EFD Eq. 4 is a function of sample size. In this case, EFD is, in general, considered the sample size and a scale-dependent function [15, 19, 20, 42]. Therefore, when the sample size M increases, the shape of the EFD changes in accordance with a sequence library size M. This important experimental fact is closely associated with the skew form of the mixture EFD of binding events defined in ChIP-based experiments, and all other NGS experiments as well. This basic property of the EFD of TF–DNA binding events of NGS experiments and their mathematical modeling of this property is discussed in the next section.

2.5 Critical Cutoff Values, Specificity, and Sensitivity

If one has prior knowledge of all TFBS and ChIP-seq or ChIP-PET sequences not bound by a given TF, then conventional calculation of specificity and sensitivity is straightforward. However, in the absence of such knowledge, one needs to rely on statistical analysis of data-driven mathematical models and computational estimates using available highly noisy and incomplete DNA fragment samples [12, 28, 31, 34, 43, 44]. A significant amount of nonspecific genomic DNA fragments (background or technical noise) is always present in the immunoprecipitated DNA material of any ChIP-derived dataset [12, 28, 31, 34]. Some nonspecific DNA might easily be filtered out after computer mapping of the DNA fragments onto the genome. In a larger library, the number of TF-specific loci is of course increased. Furthermore, background or technical noise genomic DNA fragments are randomly located in the genome and thus false clusters occur in any region of the EFD.

The following basic statistical tasks are becoming imperative: (1) specificity of the library that identifies statistically significant TFBSs with a confidence level t; (2) power of the library which identifies "true" specific binding events in a noise-enriched subset of relatively low read counts ($0 < m < t$); and (3) sensitivity of detection of "lost" BSs that are available for TF binding in given cells under given conditions but were not detected due to a limitation of the TF library size or the technical implementation. We analyze these problems via probabilistic modeling, goodness-of-fit analysis, and computational modeling of nonspecific and specific TF–DNA binding event loci for a given TF in the ChIP-Seq library.

For a given TF-bound DNA fragment library mapped on a reference genome, let N denote the sum of two subsets of TF–DNA binding events:

$$N = N_1 + N_2 = \sum_{m=1}^{t-1} n(m, M) + \sum_{m=t}^{J} n(m, M), \qquad (6)$$

where N_1 is the number of observed "noise-rich" TF–DNA binding loci with relatively low/moderate TF -binding avidity potential; N_2 is the number of observed "specific-rich" TF–DNA binding loci with relatively high TF binding avidity potential; t is the TF–DNA binding specificity threshold value, indicating the critical cutoff for the true and false TF–DNA fragments forming clusters. At $m = t$ or $m > t$ the, loci of DNA fragment clusters will be called "confidence clusters."

To quantify specific and nonspecific TF–DNA binding events, we separate the uniquely mapped ChIP-seq or ChIP-PET DNA fragments into two subsets:

$$M = M_1 + M_2 = \sum_{m=1}^{t-1} n(m, M)m + \sum_{m=t}^{J} n(m, M)m, \qquad (7)$$

where M_1 is the number of ChIP-seq DNA sequences in the subset of "noise-rich" and nonreliable TF–DNA loci; M_2 is the number of ChIP-seq DNA fragments in the subset of reliable specific TF–DNA loci.

For a given ChIP-seq TF–DNA fragment library, let $\bar{N}$ denote the total number of specific TF–DNA binding loci in the ChIP-seq DNA fragment library. Then, a set of specific TF–DNA binding loci is split into two subsets as follows:

$$\bar{N}_s = \bar{N}_{s1} + \bar{N}_{s2} = \sum_{m=1}^{t-1} \bar{n}_s(m, \bar{M}) + \sum_{m=t}^{J} \bar{n}_s(m, \bar{M}), \quad (8)$$

where $\bar{n}_s(m, \bar{M})$ is the estimated number of specific binding events at value m; $\bar{N}_{s2}$ is the estimated number of loci in the subset of observed specific TF–DNA loci; $\bar{N}_{s1}$ is the estimated number of specific TF–DNA loci in the subset of unreliable low-/moderate avidity TF–DNA loci.

To quantify avidity-specific TF–DNA binding events and estimate parameter α in Eq. 3, we can estimate the number of ChIP-Seq DNA fragments in high confidence loci $\bar{M}$ and split this into two values:

$$\bar{M}_s = \bar{M}_{s1} + \bar{M}_{s2} = \sum_{m=1}^{t-1} \bar{n}_s(m, \bar{M})m + \sum_{m=l}^{J} \bar{n}_s(m, \bar{M})m, \quad (9)$$

Using Eq. 9, the weight parameter α in Eq. 3 can be estimated by the following:

$$a = \bar{\bar{M}}_s / M. \quad (10)$$

Parameter t is an unknown threshold value of a random variable X domain separating the domain on two sub-domains a binding specificity level defined by the following:

$$S_p = \bar{P}_s(X \geq t)/\bar{P}(X \geq t) \times 100\%, \quad (11)$$

where $\bar{P}_s(X \geq t)$ and $\bar{P}(X \geq t)$ are as defined in Eq. 3; $\bar{N}_{tot}$ denotes an estimate of the total number of binding events in the entire genome in a given cell population under given experimental conditions as follows:

$$\bar{N}_{tot} = \bar{N}_0 + \bar{N}_{s1} + \bar{N}_{s2}, \quad (12)$$

where $\bar{N}_0$ is the number of undetected TFBSs. Then, the sensitivity of the ChIP-seq assay is estimated by the following:

$$S_e = \bar{N}_s / \bar{N}_{tot} \times 100\%, \quad (13)$$

where $\bar{N}_s$ is the estimate of the number of true TF DNA binding events within the observed domain of library $(1, 2, 3, \ldots, J)$.

2.6 Models of Avidity Distribution Function of Specific Binding Events

2.6.1 Standard Pareto PF

Having big and diverse NGS datasets, it is possible to extract some common statistical properties that are robust and reproducible across many biological systems, and their networks, at the genome scale. The EFDs of the key biological parameters can be useful in such a discovery strategy. As mentioned earlier, the following has been observed: (1) the skewed nature of the EFD to the right and (2) the robustness of such relative to small changes in environment, cell type, or organism [15, 20, 30, 34].

One widely used skewed PF is the standard discrete Pareto distribution (SPD) [40, 45]

$$P(X = m) = p_m := f(m) = \frac{\zeta(k)}{m^{k+1}} \tag{14}$$

where X is the nonnegative random variable which has a power-law distribution if $m = 1, 2, \ldots ; +\infty > k > 1$, and

$$\varsigma(k) = 1/\left(\sum_{m \geq 1} m^{-(k+1)}\right) \tag{15}$$

is the Riemann zeta function. This PF is characterized by a single constant, which is the positive value exponent parameter k. Equations 14 and 15 are used for statistical characterization of many different calculations of bimolecular systems such as the frequency of specific DNA sequence subsets in the genome, protein domains in proteins of a proteome, paralogous gene family, and the number of links in network models.

The SPD function is a very useful for the approximation of the skewed EFDs. In double-logarithmic coordinates, the log p_m as a function of log(m) is a straight line with a negative slope [14] has scale-invariant property

$$f(n) = (\varsigma(k))^{-1} f(m) f(l)$$

for positive integers ≥ 1 , $l \geq 1$, $n = m \cdot l$. Equation 14 has "power-law" behavior, which is considered a basic feature in the so-called scale-free network [46], self-organized growing network and scaling theories [47, 48]. The similar scaling concept of self-similarity is in fractal and scaling polymer theories, where the probability function of set quantities is statistically self-similar at every scale.

Currently, a scale-free (SF) random network is commonly found in the literature today in random growing network models. The concept of SF consists of many low degree nodes and a few very large degree nodes. The number of degrees of nodal connectivity m in the SF random network, the number of connections the node has, is distributed according to the SPD, which has positive constant exponent parameter k [46]. All SF networks are unlikely stable outside the $(0 < k < 2.5)$ region [49, 50]. Thus, with these constraints, the SF model assumes an independence of the function shape (and exponent parameter $(k + 1)$) from scale (e.g., size of the network). This property suggests independence of the statistical characteristics of the network nodes (objects) and their links (node pair interactions) regardless of changes in the sample size of a network. Note that the SF network concept postulates that the SPD function, as a realization of a stochastic process of network growth, is fitted to the empirical data on the right tail of the function at $m \to +\infty$ and $t \to +\infty$.

The only known random mechanism of a growing network into a SF network is preferential linking [46, 49] where at each time step a new node (site) is added. It connects to old nodes via a fixed number of links. The probability of an old site getting a new link is proportional to the total number of connections to this site. Introduction of additional biologically or environmentally reasonable mechanisms such as aging or death of sites in the SF network or any other factor leading in increased power of the fat tail of the distribution function changes the value of parameter k and breaks the SF property even though $m \to +\infty$ and $t \to +\infty$ [30, 49, 50–53, 67]. These and many other theoretical results assume that the SF model is unlikely to fit to actual biological data.

2.7 Implementation of the Scale Free Random Network Model Is Unlikely Possible

Challenge 1. Identification of a complete skewed frequency distribution by its tail.

In a single cell or a cell population, TFBS avidity can range widely, often within 4–5 orders of magnitude. It has also been demonstrated that a large fraction of binding events in cells has very low or moderate binding avidity (*see* below).

The standard Pareto distribution can be asymptotically derived from many probability distribution functions at $m \to \infty$ and $t \to \infty$ [18, 19, 29, 40, 44, 49, 51–61], which consists of a few percent of the observed events in the right tail. For these reasons, many alternative mathematical and mechanistic models can provide similar asymptotic behavior and cannot be differentiated due to theoretical uncertainty, incomplete sample size, and experimental errors.

Thus, the goodness-of-fit of the parameter k of the right fat tail of the EFD for NGS data (or equivalent FD tail characteristic) cannot provide an identification of actual PF and the random mechanisms leading to an empirical skewed PDF.

Challenge 2. Actual sample size of observation datasets available for the analysis of biomolecular systems is often unknown and may be essentially incomplete [19, 20, 31, 34, 43, 44, 60, 62–66].

Thus, the goodness-of-fit of the parameter k of the right fat tail of the EFD for NGS data is a function of sample size and cannot be used for identification of actual PF and the random mechanisms leading to a skewed EFD.

Challenge 3. Genome-wide NGS information includes noise-rich quantities [4, 12, 14, 19, 20, 28, 34, 43, 68].

Due to sample preparation biases and nonlinear amplification of sequences, the right tail of an EFD could be overloaded with missing frequency data and false-positive events, reducing the reliability of the parameter estimation of the DF right fat tail (*see* Subheading 2.4).

Basically, analyzing only a relatively small part of an EFD for the identification of the probabilistic mechanisms leads to mathematical and experimental challenges. These great challenges are commonplace for any experimental information collected in big biomolecular datasets. Theoretical and simulation studies of the effects of false negatives in the detection of links and/or nodes (i.e., bond and/or node undersampling) on network topologies focused on the relation between true and observed degree distributions found that undersampling changes qualitatively the shape of the degree distribution. In other words, the shape of the best-fitted PF and the parameter k are dependent on sample size even at large values of m [19, 34, 43, 60, 63, 66].

Curiously, the SF network distribution models have been tested using correlation or linear regression models and estimation of k in the SPD using small sample size biological databases [46]. This results in the use of the intermediate part of the EFD (not the right hand tail) where noise-rich data and non-regular distribution of missing data are present. Alternative models have not been statistically compared and/or validated. Our visual inspection of the studied Pareto-like EFD in the log–log plots and statistically based goodness-of-fit model analysis criteria applied to the entire and independent datasets have demonstrated systematic deviations of the most empirical FDs from the SPD [19, 30, 31, 34, 60, 62, 66]. Thus, statistical analysis reveals a reproducible disagreement between the statistical properties of the SF model and the EFDs constructed based on different types of biomolecular datasets.

Summarizing the SF model statements, we conclude the following:

1. Space of the events may not be defined. The SF statements defined such that probability estimates may not satisfy the Kolmogorov axioms. Excluding low- and moderate-frequency event values, comprising a vast majority of the experimental data, presents a great challenge for identification and unbiased fitting, statistical analysis, and interpretation of the power law-like form of the right tail EFD.
2. The standard power law Eq. 14, with or without normalization factor [15], is systematically deviated from the EFD of ChIP-Seq and other NGS data.
3. Exponent parameter k is frequently a function of sample size and sample preparing procedure: the slope of the right tail of the standard power law DF decreases with increased sample size.
4. Technical and natural biological noise may lead to biased parameterization of the testing DFs.

These factors could lead to incorrect interpretation of statistical properties of biomolecular datasets.

It has been suggested that self-organization models predicting the power-law behavior may play a role in the course of natural selection in evolutionary biology, namely, population dynamics, molecular evolution, and morphogenesis [48]. However, self-organization is not a model of natural selection and molecular evolution. The reactions of biological systems to external factors and diversity of internal interactions in any cell, tissue, or multicellular organism are not simply self-organizing, as they are thermodynamically open and evolutionarily derived complex systems relying on continuous input of energy and fitness to the environment.

Thus, more mechanistic, exploratory, and predictive probabilistic models of the biomolecular systems and their network quantifying at the genome scale should be developed and validated using appropriate datasets and statistical methods.

2.8 Generalized Discrete Pareto Distribution

Power law-like distributions with a skewed form and heavy tails are common features of many large-scale evolving complex systems such as the frequency distribution of earthquake or solar flare size, the duration of neuronal avalanches in the brain, protein domains in the proteomes or TF–DNA binding avidity in genomes. In probability theory, a family of skewed DFs is very abundant with many dozens of distinct functions [29, 30, 40, 51, 56–58, 67, 68] and the family includes many well-studied DFs extensively used in the analysis of biological data [30, 33, 34, 36, 44, 69, 70], suggesting different stochastic processes occurring in these DFs.

Previously, we observed that several classes of skewed probability functions (Poisson, exponential, standard power law, lognormal, logistic functions [29, 40]) are available to fit the empirically defined frequency distribution functions FDFs observed in genomic and transcriptomic datasets. One such distribution is the FD of the TF–DNA binding events. After performing goodness-of-fit analysis using the method presented in [30, 34], we observed that the best fit for the most EFD empirical frequency distributions was obtained using the GPD PF [19, 30, 34]. Therefore, we propose to describe the EFD of the TF–DNA binding events using the (up-value) truncated GPD function [19, 30, 34, 45]:

$$f(X = m) = {z_J}^{-1} \frac{1}{(m + \beta)^{(k+1)}} \tag{16}$$

where in our case the random variable X is a positive discrete variable of TF–DNA binding avidity in a given TF–DNA BS; m can be 1, 2, ... J; in the EFD, the random variable X (in the m domain) is the peak height of a given DNA fragment cluster of a genome; $f(X = m)$ is the probability that a randomly chosen specific BS has a TF–DNA binding avidity value m; function f is characterized by two explicate parameters, k and β, where $k > 0$ and $\beta > -1$; and the normalization factor z is the generalized Riemann zeta-function value [29].

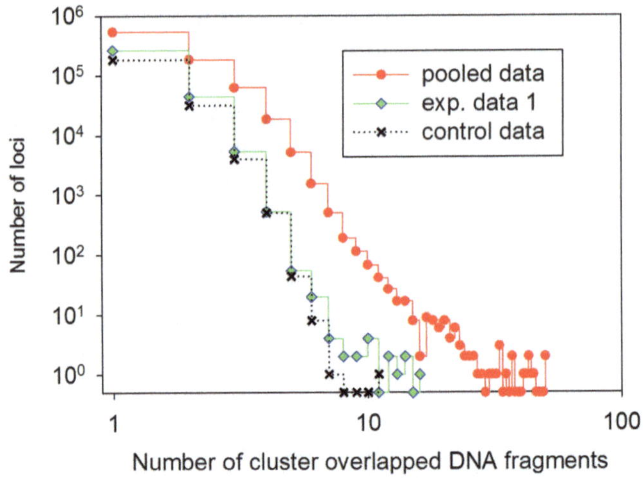

Fig. 3 Sample size-defined EFD of TF–DNA binding events for INF-gamma-induced STAT1 TF. ChIP-PET binding events for studied ChIP-PET library dha01, pooled (dha01, dha02, sha01), and negative control library are shown. Binding event is defined by the DNA fragment cluster size for ChIP-PET data using a strict cluster size definition

$$z_J = \sum_{m=1}^{J} 1 \Big/ (m+\beta)^{(k+1)}, \tag{17}$$

where k characterizes the skewedness of the probability function; β characterizes the deviation of the PF from a simple power law.

Figure 3 shows the sample size-define EFD for the INF-gamma-induced STAT1 TF-binding events detected in ChIP-PET binding experiments [15, 20]. ChIP-PET library dha01, the pooled (dha01, dha02, sha01) or virtual sequence library, and negative control library data (nonspecific binding, sequences pooled down at nonspecific immunoprecipitation) are shown. Binding events were defined by the DNA fragment cluster size. A strict cluster size definition was used for discrimination of binding events [15, 20].

Figure 4 shows the results of our goodness-of-fit analysis of the SPD and GPD distributions. The EFD has been reconstructed after data noise subtraction [34] using ChIP-Seq data for the Nanog protein bound to specific BSs in the genome of a mouse embryonic stem cell [13]. In a log–log plot, the EFD shows essentially different properties in comparison to SPD. An asymptotical part selected by visual inspection of the right tail of the EFD follows the SF concept; using a cutoff value of 39, the truncated EFD of the fat tail is approximated well by the SPD (at $k = 1.6$; Fig. 4a). However, the left part of the dynamical range represented by the majority of

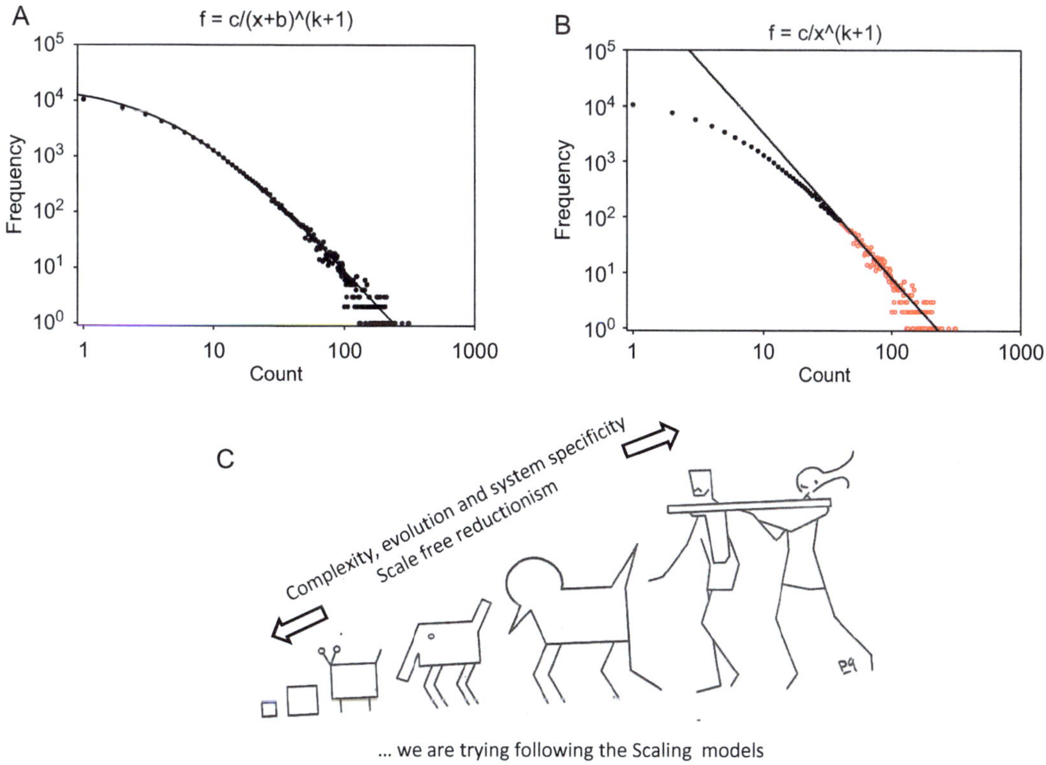

Fig. 4 Results of goodness-of-fit analysis of the standard Pareto DF and the GPD PFs. The EFD has been reconstructed after data noise subtraction [34], using ChIP-Seq data for the Nanog protein bound to specific BSs in the genome of a mouse embryonic stem cell [13]. (**a**) Truncated part of the EFD fat tail is well approximated by the standard Discrete Pareto PF (SPD) (at $k = 1.6$); however, it cannot be used for back-extrapolation. (**b**) Best-fit GDP PF at $b = 5.57 \pm 0.210$ (at t-test value 25.92 and $p < 0.0001$) and $k = 1.6 \pm 0.026$ (at t-test value 60.9 and $p < 0.0001$) using the cutoff value $m = 8$, and provides for accurate fit of the studied data within the whole dynamical range, including the left part of the EFD. Goodness-of-fit analysis was supported with the Fisher test ($F = 32{,}710$; $p < 0.0001$) and at a very small error (3.63). Parameters were estimated by SigmaPlot-11software. (**c**) Illustration of a track of research of the statistical properties of a complex system followed by the scaling concept. This figure was adapted by the author (V.K.) from *Scaling Concepts in Polymer Physics* by Dr. P.G. de Gennes [47] with minor changes

events cannot be back-extrapolated. Using the cutoff value $m = 8$, the best-fit GPD fits well the studied events for the whole dynamical range, including the left part of the distribution function on the interval from 1 to 8 (Fig. 4b). Goodness-of-fit analysis was strongly supported by the Fisher test ($F = 32{,}710$; $p < 0.0001$) with a small error (3.63), estimated using SigmaPlot-11 software. Based on these high-confidence results, supported by the appropriate statistical tests, we could select GPD as an appropriate model for future analysis of ChIP-Seq databases. Figure 4c illustrates a track of research of statistical properties of a complex system followed by the scaling concept.

2.9 Properties of the Scale-Dependent Pareto-Like Distribution

In Eqs. 16–17, the sample size dependence is modeled after the random parameter J. This parameter can be experimentally defined as the value of the most abundant event. $J = \max\{m\}$; $m = 1, 2, \ldots, J$. In Eqs. 16 and 17, the parameter J is a function of sample size M. In the context of NGS datasets, M is the number of DNA fragments in the NGS sequence library mapped onto the genome. When M increases, the right hand tail of the probability function at $\beta > -1$ and $k > 0$ becomes longer, and the shape of the probability function changes and gradually deviates from the SPD. If $\beta = 0$ and $J \to \infty$, then Eq. 16 converges to the SPD.

Let us describe several important statistical properties of the GPD distribution in more detail.

First, we show that when M is increased, the number of TFBSs (N) becomes larger. We assume that intrinsic parameters of the function (k, b, a) are constant.

For convenience, let us first find the relationship between N and M for the continued form of a double-truncated generalized Pareto PF:

$$f(m) = ks\frac{(m_0 + b)^k}{(m + b)^{k+1}}, \tag{18}$$

where $m_0 \geq m \geq aM$, $s = 1/\left(1 - \left(\frac{m_0+b}{aM+b}\right)^k\right)$, m_0, k, and b are constants, and $m_0 > 0$; $k > 0$, $b > -1$; $a_0 < a < 1$, $a > 0$.

Function f could be estimated as follows:

$$f = n_m/N, \tag{19}$$

where n_m is the observed number of distinct events (overlapped DNA sequence fragments, peak heights) which have the occurrence m in a given sample of size M.

Then, using estimation

$$M = \int_{m_0}^{aM} n_m m dm, \tag{20}$$

we can derive the formula

$$N = M \cdot A(M); \tag{21}$$

where

$$A(M) = \left(ks(m_0 + b)^k \int_{m_0}^{aM} \frac{m}{(m + b)^{k+1}} dm\right)^{-1}. \tag{22}$$

Taking the integral, we find that if $M \to \infty$, then

$$N \sim M^k$$

at $0 < k < 1$; and

$$N \sim M/\log M$$

at $k = 1$; $N \sim M$ at $k > 1$.

Thus, in all cases, for parameter k, we have $N \to \infty$ when $M \to \infty$.

Using a discrete form of the GPD function along with Eqs. 18, 19, and 22,

$$M = \sum_{m=1}^{aM} n_m m = N \sum_{m=1}^{aM} n_m m/(m+b)^{k+1}, \tag{23}$$

where

$$N = \sum_{m=1}^{aM} n_m = \sum_{m=1}^{aM} n_m/(m+b)^{k+1} \tag{24}$$

and $k > 0$; $b > -1$, $a_0 < a < 1$, $a_0 > 0$, we arrive at

$$N = MA_d(M), \tag{25}$$

where

$$A_\mathrm{d}(M) = \sum_{m=1}^{aM} \frac{1}{(m+b)^{k+1}} / \sum_{m=1}^{aM} \frac{m}{(m+b)^{k+1}}. \tag{26}$$

In the specific case of the Lotka–Zipf law ($b = 0$; $k = 1$), we have

$$A_\mathrm{d}(M) = \sum_{m=1}^{aM} \frac{1}{m^2} / \sum_{m=1}^{aM} \frac{1}{m} \approx \pi^2/6\,(\ln(aM) + \gamma)^{-1}, \tag{27}$$

where $\gamma = 0.5772\ldots$ is the Euler constant. Then,

$$\lim_{M \to \infty} N \approx \pi^2/6\,(M/\ln(aM)) \to \infty \tag{28}$$

In other specific case $b = 0$, $k = 2$, we have

$$\lim_{M \to \infty} N \approx \left(1 + \pi^2/15\right) M \to \infty. \tag{29}$$

One could conclude that Eq. 24 has a similar growth to infinity for N as we have shown for Eq. 20, when $M \to \infty$.

Thus, when the sample size M increases, the right hand tail of Eq. 18 becomes longer, and the shape of the distribution function is gradually deviated from the direct line with a constant slop (SF network assumption) and SPD ($J \to \infty$).

Notice that the truncated GPD function can be used for parameterization of EDFs, estimation of specificity and sensitivity of the experimental data, and comparative analysis of the statistical characteristics of the EFDs under different conditions and in differing cellular contexts. According to our goodness-of-fit analysis of diverse ChIP-seq/NGS datasets and our computational simulations of the sample size changes of the modeled frequency distribution, the left and/or right truncated GPD can often be extrapolated from the distribution tails and used for statistical inferences (see next sections).

This PGF can be used even when the PF does not have an asymptotical steady state at $M \to \infty$ and the tail exhibits regular (power law) reduction behavior at $m \to \infty$, $t \to \infty$. Examples of such practical applications can be seen in [34]. In the next section, we demonstrate that the scale dependence of the TF–DNA binding EFDs is a general attribute of such experimental data sampling.

2.10 Waring Distribution Function

There are many families of skewed distribution functions that have power law-like shape and vary regularly at infinity with exponent term [19, 40, 52, 62]. One of the useful skewed DFs is the Waring distribution [34, 40, 51]:

$$p_m = p_0 \frac{B(b+1, m)}{B(a, m)} \tag{30}$$

$$p_0 = \left(1 - \frac{a}{b}\right) \tag{31}$$

$m = 1, 2, \ldots$, $B(x)$ is the Beta function [40] and $+\infty > b > a > 0$.

This function was introduced by Irwin in 1963 [70]. It is a skewed DF that includes two positive parameters, which can be defined based on EFD data. The function has a power law-like shape and varies regularly [52] at infinity with exponent term $m^{-(1+b-a)}$ [52, 62]. Additionally, this probabilistic model allows us to predict p_0, which is the probability of unobserved events in a data sample. This parameter can easily be estimated using goodness-of-fit analysis of the EFD and estimation of parameters a and b. Importantly, the Waring distribution can be an explanatory model, generated with the help of a specific Markov [30, 65] stochastic process to model TF–DNA binding avidity.

Our previous studies have shown that this model fits well to many empirical frequency distributions in bimolecular systems analyzed at the genome scale [19, 34, 62]. Furthermore, the estimation of parameters a and b leads to the prediction of the total number of distinct species, or rather TF–DNA binding sites, present in the studied biological system [34]. Notice that the Waring distribution function consists of a subfamily in the three-parametric distribution function family, called the Kolmogorov–Waring (KW) distribution function, which we consider and use in this work.

3 Explanatory Relative TF–DNA Binding Avidity Model: The Kolmogorov–Waring Function, Its Properties, and Generalization

3.1 The Birth–Death Kolmogorov Random Process Model

Let us denote the KW probability function by $P_{KW}(X = m)$, where $m = 0, 1, 2, \ldots$ [30]. The KW probability function has been derived as a result of the modeling of fundamental biological phenomena—birth and death in a gradually evolving population. This function has been used in the analysis of different types of events at genome, transcriptome, proteome, and interactome scales [20, 30, 31]. In particular, we have shown that P_{KW} can be used as a possible exploratory model for stochastic "binding–dissociation" of TFs on specific DNA BSs at the genome scale for different TFs in differing cell types [34]. For statistical analysis of ChIP-based experiments, we consider a population of the TF molecules in which each TF molecule can bind specifically and/or nonspecifically to a given DNA locus with probable TFBS.

The mathematical part of this work focuses on relations between asymptotic solutions to the time nonhomogeneous birth–death Markov random process described using Kolmogorov differential equations [62], skewed distribution function families [34, 64, 65] and the Gaussian hypergeometric functions [40, 62].

Let us briefly describe the Markov process with continuous time and countable number of states 0, 1, 2, ... We characterize the N distinct BS loci for given TF molecules as a population of DNA target sequences of a genome of an organism. Let $d_1, d_2, \ldots, d_N$ denote these N BSs. Let m_i denote the number of countable TF–DNA binding events (discrete occupancy values) of the BS d_i presented in the list of distinct BSs of a given genome. $m_i = 0, 1, 2, \ldots, J$, where J is the occupancy value of most avidity (most abundant cluster of the DNA fragments in a given library) in a genome. Let n_m denote the number of distinct TFs which can bind to a BS exactly m times in a genome detected in a DNA fragment library. Then a likelihood estimate of the probability of such event p_m is n_m/N. Note $N = \sum_{m=1}^{J} n_m$ and J is the parameter, that may be dependent from a sample size of DNA fragments library. We model the continuous time evolution of TF–DNA binding events in a genome as the stochastic biding-dissociation events in the frequency of these events. Many distinct TFs may evolve in this way.

Let the random variable $D_t(d, P)$ be the number of TF binding events of the distinct BS d in the genome occurring at time t in the stochastic trajectory P of BS d. $D_t(d, P)$ is a realization of a continuous time stochastic process $D = \{D_t, t \geq 0\}$. We assume that time parameter t is a continuous parameter in $\{D_t, t \geq 0\}$. The set of possible state values of the process is denoted by the set of nonnegative integers $\{0, 1, 2, \ldots\}$. If $D_t = i$, then the process is said to be in state i at time t. We propose that conditional transition probability between states does not depend on states. We suppose that whenever the process is in state i at time t, there is a fixed probability

$P_{i,j}(t, s)$ that it will next be in state j at time s, where $\sum_{j \neq i} P_{ij} = 1$. This process with additional standard assumptions [53, 55, 56, 62] (see details below) can be considered a continuous-time Markov chain random process with a countable number of states.

Evolutionary progress along this process, in general, results in the appearance of which can never be other than −1, 0, or +1 and where binding events are "born" and "die" during the process. A birth indicates an increase in the number of binding events when counting the random fixed BS d, and death indicates a decrease in this number. We assume that the total number of BSs and the total number of TFs associated with any bond is constant over long stretches of time.

Let $p_i(t) = P(D_t = i)$ denote the probability function associated with the random process $\{D_t, t \geq 0\}$. Let transition probability $P_{i,j}(t) = P(D_{t+s} = j|D_s = i)$, assuming that for any $i = 0,1,2, \ldots$, $j = 0,1,2, \ldots$ the transition probability does not depend on s, and for $t \to 0$, we assume $P_{i,j}(t) = o(t)$ for $1 < |i - j| < \infty$, $P_{i,i+1}(t) = \lambda_i(t)t + o(t)$, $P_{i+1,i}(t) = \mu_{i+1}(t)t + o(t)$. These assumptions suggest that $P_{i,i}(t) = 1 - (\lambda_i(t) + \mu_i(t))t + o(t)$ when $t \to 0$. Then the rate of the probability functions p_m ($m = 0, 1, 2, \ldots$) at moment $t \geq 0$ can be described by the Kolmogorov differential-difference equations [34, 62]:

$$\mathrm{d}p_0(t)/\mathrm{d}t = -\lambda_0(t)p_0(t) + \mu_1(t)p_1(t) \tag{32}$$

$$\mathrm{d}p_m(t)/\mathrm{d}t = -(\lambda_m(t) + \mu_m(t))p_m(t) + \lambda_{m-1}(t)p_{m-1}(t) + \mu_{m+1}(t)p_{m+1}(t), \tag{33}$$

where $m = 1, 2, \ldots$. The initial probabilities $p_m(0) \geq 0$; ($m = 0, 1, 2, \ldots$) follow the condition $\sum_{m \geq 0} p_m = 1$. If $t \to +\infty$, the random birth and death processes describing TF–DNA binding events may be kept near the equilibrium. This equilibrium solution can be written explicitly by stating $dp_m/dt = 0$; $m = 0, 1, \ldots$ in Eqs. 32 and 33 as

$$\widehat{p}_m = p_0 \prod_{i=1}^{m} \frac{\lambda_{i-1}}{\mu_i}, \tag{34}$$

$$p_0 = \left(1 + \sum_{m=1}^{\infty} \prod_{i=1}^{m} \frac{\lambda_{i-1}}{\mu_i}\right)^{-1}. \tag{35}$$

In this case, a necessary and sufficient condition for the existence of the nontrivial stationary solution of Eqs. 34 and 35 is provided by convergence of the series

$$Q = \sum_{m=1}^{\infty} \prod_{i=1}^{m} \eta_i, \tag{36}$$

where $\eta_i = \frac{\lambda_{i-1}}{\mu_i} \leq v < 1$. This condition exists when starting from some $i = i_c$ the condition $\eta_i \leq v < 1$ takes place for all $i \geq i_c$ (i.e., on the right tail of the probability function).

In the case of the probability function, this inequality implies that starting from some large enough size i_c of high-avidity protein–DNA binding events, the intensity of the birth (binding) process must be less than or equal to the intensity of the death process (dissociation) for all consequent DNA–TF binding events characterized by i events for which $i \geq i_c$.

Using Eqs. 34 and 35, we can obtain the nonzero limiting probability function for the random process $D_{t \to \infty}$:

$$P(D_t = m) = p^*{}_m = \lim_{t \to \infty} p_m(t). \tag{37}$$

If we assume that the limiting probability distribution $p^*{}_m$ obtain m exists, then all dp_m/dt $(m = 0, 1, 2, \ldots)$ would necessarily converge to 0 as $t \to \infty$ and we can obtain

$$p^*{}_m = \widehat{p}_m = p_0 \prod_{s=1}^{m} \eta_s. \tag{38}$$

In general, the birth and death intensity rates between states can be approximated by the polynomial or rational functions of integer argument m.

Then Eq. 34 (or Eq. 35) can be defined by the product of the rational functions $\eta_m = \frac{\lambda_{m-1}}{\mu_m} = \theta \frac{\left(m^i + a_{i-1} m^{i-1} + \cdots + a_1 m + a_0\right)}{\left(m^j + b_j m^{j-1} + \cdots + b_1 m + b_0\right)}$ of argument m and constants θ, a_i, b_i $(i = 0,1,\ldots; j = 0,1,\ldots)$, multiplied by the positive constant p_0. Using a major theorem of algebra, we can present the probability Eq. 34 (or Eq. 38) as the m-th term of the generalized hypergeometric series ${}_pF_q$ [40, 62] (see below).

According to the previous considerations, we specified two binding transition probabilities: (1) "preferential binding" due to preferential attachment mechanism of a TF to specific DNA regions on the chromosome and (2) "nonpreferential" binding driven by the Poisson process. We assume two similar types of processes for TF–DNA detachment transition events. However, the preferential and nonpreferential dissociation processes could be realized with different intensities.

For our application purposes, we consider a Markov birth–-death random process such that the intensity rates are the linear functions of m [62]:

$$\lambda_m = \lambda_1^* + \lambda_2^* m \tag{39}$$

and

$$\mu_m = \mu_1^* + \mu_2^* m, \tag{40}$$

where $m = 0, 1, 2, \ldots$ and constants $\lambda_1^* > 0, \lambda_2^* > 0$, $\mu_1^* > 0, \mu_2^* > 0$. Hence, during an interval $(t, t + h)$ where h is small, we assume that there are four independent processes in the TFBSs: the spontaneous "birth" and "death" of DNA–TF pairs, with

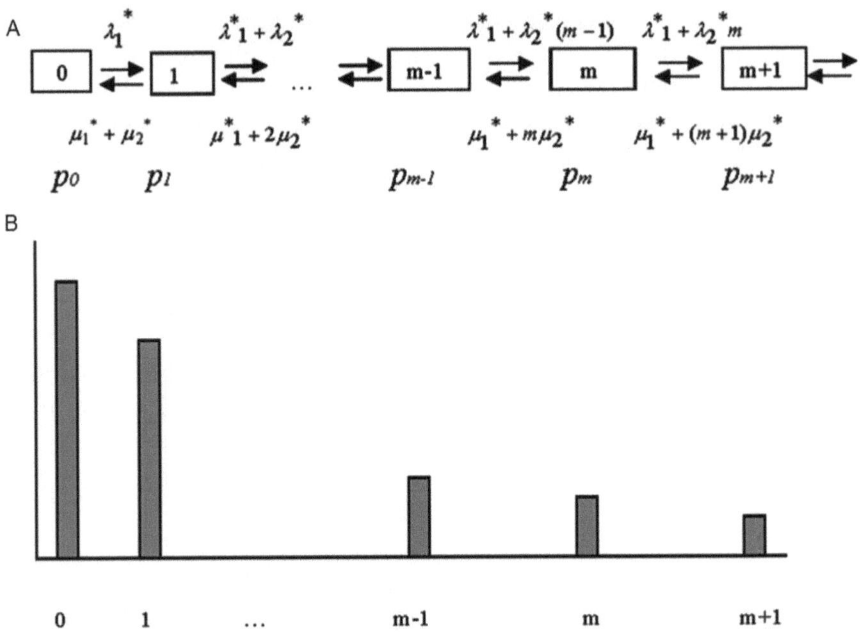

Fig. 5 DNA-TF binding–dissociation model by Kolmogorov birth–death process. (**a**) Directed graph of the process specified in our model. (**b**) Schematic presentation of the probability function of binding events under steady-state conditions.

constant intensities λ_1^* and μ_1^*, respectively, and the "flows" of the binding events with intensities proportional to the number of DNA–TF binding events which have already occurred in given TFBS $\mu_1^* m$ and $\mu_2^* m$. Figure 5 shows the random directed graph and schematic presentation of the skewed PF of the steady-state binding-dissociation process with intensities being the linear functions of m.

Note that, the intensities λ_1^* and $\mu_1^*$are the intensities of Poisson processes. During a time interval $(t, t + h)$, where h is small, the intensity λ_1^* is proportional to a transitional (birth) probability of sporadic increase of the number of DNA–protein binding event. During the same time interval $(t, t + h)$, the intensity μ_1^* is proportional to a transitional probability of a spontaneous DNA-- protein dissociation event.

At steady-state conditions, three parameters of the process can be used for characterization of the probability function. Let

$$a = \lambda_1^* / \lambda_2^*, \theta = \lambda_2^* / \mu_2^*, b = \mu_1^* / \mu_2^*.$$

Let us also denote factorial power $z^{[m]} = z(z + 1) \ldots (z + m - 1)$, where $m = 0, 1, 2, \ldots$; $z^{[0]} = 1$. Using Eqs. 34, 35, 38, 39, and 40, we can obtain the limiting (non-zero steady state solutions) probability function for the process in Eqs. 32 and 33 with the intensities given by Eqs. 39 and 40:

$$p^*{}_m = bp_0 \frac{a^{[m]}}{b^{[m+1]}} \theta^m = p_0 \frac{\Gamma(a+m)\Gamma(b+1)}{\Gamma(a)\Gamma(b+m+1)} \theta^m = p_0 \frac{B(b+1,m)}{B(a,m)} \theta^m \tag{41}$$

$$p_0 = \left(1 + \sum_{m=1}^{\infty} \left(\prod_{i=1}^{m} \frac{a-1+i}{b+i} \theta \right) \right)^{-1}, \tag{42}$$

where $m = 0, 1, 2, \ldots$, $\Gamma(x)$ is the Gamma function, and $B(x)$ is the Beta function [29, 40]. In the specific case $b > a > 0$ and $p_0 = \left(1 - \frac{a}{b}\right)$ distribution Eqs. 41 and 42 will be the well-known Waring distribution [40]. We called Eqs. 41 and 42 the Kolmogorov–Waring (KW) probability function [34, 62]. A series Σp_i is called the hypergeometric if the ratio p_{i+1}/p_i is a rational function of i ($i = 0, 1, 2, \ldots$).

The sum of probabilities Eq. 41 can be written in a form of hypergeometric series:

$$1 = p_0 \left(1 + \frac{a}{b+1} \theta + \frac{a(a+1)}{(b+1)(b+2)} \theta^2 + \ldots + \frac{a(a+1)\ldots(a+m-1)}{(b+1)(b+2)\ldots(b+m)} \theta^m + \ldots \right) \tag{43}$$

or

$$1 = p_0 \cdot {}_2F_1(a, 1; b+1; \theta), \tag{44}$$

where ${}_2F_1(a, 1; b+1; \theta)$ is the hypergeometric Gauss series [29, 40]:

$${}_2F_1(\alpha, \beta; \gamma; \theta) = \sum_{m=0}^{\infty} \frac{\alpha^{[m]} \beta^{[m]}}{\gamma^{[m]}} \frac{\theta^m}{m!} \tag{45}$$

at $\alpha = a$, $\beta = 1$ and $\gamma = b + 1$. The probability function in Eq. 41 has the probability generating function [62]

$$g_{\mathrm{KW}}(z) = \frac{{}_2F_1(a, 1; b+1; \theta z)}{{}_2F_1(a, 1; b+1; \theta)}.$$

A probability distribution may be characterized by its moments. If the points represent probability distribution then the zeroth moment is the total probability (i.e. one), the first moment is the mean, the second central moment is the variance, the third central moment is the skewness. Studying discrete distributions, it is often advantageous to use the factorial moments $\mu'_{[r]}$ of order r [57]. For the probability function Eq. 41, the factorial moment of order r is given by

$$\mu'_{[r]} = r! \theta^r \left(\frac{\alpha^{[r]}}{\gamma^{[r]}} \right) p_0 \cdot {}_2F_1(\alpha + r, 1 + r; \gamma + r; \theta).$$

Using Eqs. 44 and 45, we can obtain the following results [62]:

$$\text{If}\,{}_2F_1(a,1;b+1;\theta) < \infty, \text{then } p_0 = \frac{1}{{}_2F_1(a,1;b+1;\theta)} > 0; \tag{46}$$

$$\text{If } \theta < 1; a > 0; b > 0, \text{then } p_0 = \frac{1}{b\int\limits_0^1 (1-s)^{b-1}(1-s\theta)^{-a}ds}; \tag{47}$$

$$\text{If } b > a > 0 \text{ and } \theta \to 1-0, \text{then } \lim_{\theta\to 1-0} p_0 = \left(1-\frac{a}{b}\right). \tag{48}$$

Note that Eqs. 46–48 provide a theoretically justified way to estimate p_0 only if $\theta \leq 1$.

Using Eqs. 41, 42, and 46 and the integral presentation of the Beta and hypergeometric Gauss functions, we can prove:

Statement 1 [62]: *If $b + 1 > a > 0$ and $\theta \leq 1$, then the limiting probability function is*

$$p_m^* = p_0 \frac{B(a+m,b+1-a)}{B(a,b+1-a)}\theta^m = \frac{\int\limits_0^1 s^{b-a}(1-s)^{a+m-1}ds}{{}_2F_1(a,1;b+1;\theta)B(a,b-1+a)}\theta^m \tag{49}$$

and the probability is

$$P(X \geq m) = \sum_{s=m}^{\infty} p^*_s = p_m{}^* \cdot {}_2F_1(a+m,1;b+m+1;\theta) = p^*{}_m / p_{0,m}, \tag{50}$$

where $p_{0,m} = \left[(b+m)\int\limits_0^1 (1-s)^{b+m-1}(1-\theta s)^{-(a+m)}ds\right]^{-1}$ *and $m = 0$, 1, 2,...; $p_{0,0} \equiv p_0$, where p_0 is defined by Eq. 42 and $p_{0,m}$ is the hazard function of the KW distribution.*

3.2 Existence of the Nonzero Limiting Distribution [62]

In Eqs. 41 and 42, the nonzero limiting probability p^*_m $(m = 1, 2, \ldots)$ exists when $p_0 > 0$ is at $t \to \infty$. Therefore, it is important to define the conditions that provide a convergence of the series

$$Q = \sum_{m=1}^{\infty} \left(\prod_{i=1}^{m} \Psi(i)\theta\right), \tag{51}$$

where $\Psi(i) = (a - 1 + i)/(b + \text{i})$. Let us define the convergence conditions of this series.

Corollary: *The series Eq. 45 is converged at $\theta < 1$ or at $(\theta = 1; b > a)$.*

Proof: Let $\lambda_1^* > 0, \lambda_2^* > \theta, \mu_1^* > 0, \mu_2^* > 0$. Formula Eq. 51 can be presented as follows:

$$Q = \sum_{m=1}^{\infty} \exp\left\{\log\left[\prod_{i=1}^{m} \Psi(i)\theta\right]\right\} = \sum_{m=1}^{\infty} \theta^m \exp\left\{\sum_{i=1}^{m} \log\Psi(i)\right\}.$$

Let $\lim_{i\to\infty}\left[\frac{|O(i-k)|}{i-k}\right] = \text{const} > 0$. Using MacLaurin decomposition, we obtain $\Psi(i) = 1 + (a - b - 1)/i + O(i^{-2})$. Using $\sum_{i=k}^{m}\frac{1}{i} = \log(m/k) + O(1/k)$ and $\sum_{i=k}^{m} O\left(\frac{1}{i^2}\right) < \text{const} \sum_{i=k}^{\infty}\frac{1}{i^2} = O(1/k)$ we obtain

$$Q_k := \sum_{m=k}^{\infty} \theta^m (m/k)^{(a-b-1)} \exp(O(1/k)). \tag{52}$$

According to the comparison test of convergence, this series converges at $\theta < 1$ and diverges at $\theta > 1$. At the critical point $\theta = 1$ (or $\lambda_2^* = \mu_2^*$) the power series $\sum_{m=k}^{\infty}(m/k)^{(a-b-1)}$ converges when $b > a$ (i.e., $\lambda_1^* < \mu_1^*$) and diverges when $b \le a$ (i.e., at $\lambda_1^* \ge \mu_1^*$).

Statement 2: *For the intensity functions defined by Eqs. 39 and 40, the non-zero limiting solution of Eqs. 32 and 33 exists if*

$$\lambda_2^* < \mu_2^* \tag{53}$$

or

$$\left(\lambda_2^* = \mu_2^* \text{ and } \lambda_1^* < \mu_1^*\right). \tag{54}$$

Thus, conditions in Eqs. 53 and 54 provide the existence for the nonzero steady-state probability function in Eqs. 41 and 42.

If $\mu'_{[r]} < 0$ or $\mu'_{[r]} = \infty$ of the random variable X, then the factorial moments of order r and all higher orders are said not to exist. If $\mu'_{[r]} > 0$ for $i < r - 1$ and $\mu'_{[r]} = 0$, then all factorial moments exist but are said to have the value 0 for order r and all higher orders. According to the Statement 2 above, the results concerning the existence of the first and second factorial moments are as follows:

1. If $\theta < 1$, $a < 0$, $b > 0$, then

$$\mu'_{[1]} = p_0\left(\frac{a}{b+1}\right) \cdot {}_2F_1(a+1, 2; b+2; \theta)\theta$$

$$\mu'_{[2]} = p_0\left(\frac{a(a+1)}{(b+1)(b+2)}\right) \cdot {}_2F_1(a+2, 3; b+3; \theta)\theta^2,$$

where p_0 given by Eq. 47.

2. If $\theta \to 1 - 0$, $0 < a < b$, then

$$\mu'_{[1]} = \left(\frac{a}{b-a-1}\right), \text{at } b > a+1$$

$$\mu'_{[2]} = \left(\frac{a(a+b)}{(b-a-1)(b-a-2)}\right) \text{ at } b > a+2.$$

Having found the factorial moments, we can easily find the moments about the origin as well as the moments about the mean. In particular, $m_X^* = \mu'_{[1]}$, $D_X^* = \sigma^2 = \mu'_{[1]} + \mu'_{[2]} - \left(\mu'_{[1]}\right)^2$. Using this formula, the mean, variance and high order moments of the random variable X of a given probability function, represented by the hypergeometric series can be derived. For instance, if constrains $\lambda_1^* = \mu_2^*$ (or $\theta = 1$) and $\lambda_1^* < \mu_1^*$ (or $a\theta > b$), then we have $m_X^* = a/(b - a - 1)$ and $D_X^* = \frac{a(b-1)(b-a)}{(b-a-1)^2(b-a-2)}$. The derived formulas show that the mean value for the random variable X of the stationary distribution exists not only if $a > 0, b > 0, \theta > 0$ (or $\lambda_1^* > 0$, $\lambda_2^* > 0$, $\mu_1^* > 0$, $\mu_2^* > 0$), $\theta = 1$ (or $\mu_2^* = \lambda_2^*$) and $b > a$ ($\mu_1^* > \lambda_1^*$), as required by Statement 3, but it can also include $b > a + 1$ (or $\mu_1^* > \lambda_1^* + \lambda_2^*$). The second moment exists if more stringent constraint $b > a + 2$ (or $\mu_1^* > \lambda_1^* + 2\lambda_2^*$) is given.

Notice that the moments analysis is one of traditional analytical strategy in probability theory for characterization of essential properties of theoretical distribution(s) and also uses the parameter estimation of EFDs. Analytical identification of the moments can also provide a selection of adequate explanatory and predictive probabilistic model(s) for the parameterization and comparison of the skewed Pareto law-like empirical frequency distributions often occurring in bioinformatics, large-scale evolving biosystems, and network biology. The existence of moments is one of the central problems in the moment analysis, because its solution allows for the quantifying of the basic characteristics of a given distribution function and identifying of the distribution function subfamilies and provides their classification. The existence of moments and their analytical identification is important for selection of the more adequate explanatory and predictive probabilistic model for analyzing the empirical frequency distributions occurring in bioinformatics, large-scale evolving biosystems, network biology, etc. in the applications using advanced statistical methods and mathematical modeling.

3.3 Dynamics of Mean Value and Variance

Let us to analyze the features of TF–DNA binding as a birth–death continuous stochastic process. We are interested in the probabilistic characteristics (mean, variance) over a period of time and the steady state of the process when the intensities of association and dissociation rates of a given state m are independent linear functions of this state (Eqs. 39 and 40).

Here, to simplify analysis, we assume that the state transition rates are nonrandom real numbers and $\lambda > 0, \lambda > 0, \mu > 0, \mu > 0$.

Let us denote *mean value* $m_X(t) = E[X(t)] = \sum_{i \geq 0} ip_i(t)$ and variance $D_X(t) = \sum_{i \geq 0}(i - m_X(t))^2 p_i(t)$. Then, for Eqs. 32 and 33, we can derive differential equations for $m_X(t)$ and $D_X(t)$ (and also for the higher moments) of random variable $X(t)$. In particular, in Eqs. 39 and 40, we obtain [62]

$$\frac{\mathrm{d}m_X(t)}{\mathrm{d}t} = (\lambda_1^* - \mu_1^*) - (\mu_2^* - \lambda_2^*)m_X(t) + \mu_1^* p_0(t), \qquad (55)$$

$$\frac{\mathrm{d}D_X(t)}{\mathrm{d}t} = (\lambda_1^* + \mu_1^*) + (\mu_2^* + \lambda_2^*)m_X(t) - \mu_1^* p_0(t)(1 + 2m_X(t)) - 2(\mu_2^* - \lambda_2^*)D_X(t) \qquad (56)$$

defined at time t_0 by the nonnegative initial conditions $m_X(t_0) = m^0{}_X$, $D_X(t_0) = D^0{}_X$. Notice that it could be a challenge, then, to find a general solution for differential Eqs. 55 and 56 [57]. However, the simplifications of Eqs. 55 and 56 allow for specific but useful analytical solutions, kinetic and statistical characteristics of the stochastic process.

Combining the results presented in Eqs. 37, 39, 40, 41, 44, 47, 48, 53–56 and the results concerning the existence of the first and second factorial moments, the next statement immediately follows.

Statement 3: *If* $\lambda_2^* < \mu_2^*$ *(or* $\theta < 1$*) and* $\lambda_1^* > \mu_1^*(1 - p_0(t))$*, where at* $t \to \infty$, $p_0(t)$ *is defined by Eq. 47, then for any initial value* $m_X(0) \geq 0$*, the Kolmogorov birth and death stochastic process exponentially approaches a steady state at characteristic time* $1/(\mu_2^* - \lambda_2^*)$ *giving*

$$\lim_{t\to\infty} m_X(t) = (\lambda_1^* - \mu_1^*(1 - p_0))/(\mu_2^* - \lambda_2^*) \text{ and}$$
$$\lim_{t\to\infty} D_X(t) = (\lambda_2^* m_X^* + \mu_1^*(1 - p_0))/(\mu_2^* - \lambda_2^*) \qquad (57)$$

and all moments of the steady state process of Eq. 57 exist.

If $\lambda_2{}^* = \mu_2{}^*$ *(or* $\theta = 1$*) and* $\lambda_1^* < \mu_1^*$ *or* $a < b$*, then* $\lim_{t\to\infty} m_X(t) = \lambda_1^*/\mu_1^* - \lambda_1^* - \lambda_2^*)$ *(or* $a/(b - a - 1)$*) exists if* $\mu_1^* > \lambda_1^* + \lambda_2^*$ *(or* $b > 1 + a$*),* $E[p_0(t)] = \frac{1}{t}\int_0^t p_0(s)\, ds = (1 - \lambda_1^*/\mu_1^*)$ *(or* $E[p_0(t)] = (1 - \frac{a}{b})$*) and if* $\mu^*{}_1 > \lambda_1^* + 2\lambda_2^*$ *then*

$$\lim_{t\to\infty} D_X(t) = a(b-1)(b-a)/\left((b-a-1)^2(b-a-2)\right) \qquad (58)$$

and exists if $\mu_1^* > \lambda_1^* + 2\lambda_2^*$ *(or* $b > a + 2$*).*

The conditions $\lambda_2^* = \mu_2^*$, $\lambda_1^* < \mu_1^*$ for Eqs. 57 and 58 imply the process of the Waring distribution at steady state. We can see that the Waring distribution corresponds to the critical steady state of a linear Kolmogorov birth and death stochastic process, when (1) for a given discrete state, the rate of specific TF–DNA association transition from the m-th to the m + 1-th state equals the rate of specific transition from the m th to the $m - 1$-th state ($\mu_2^* = \lambda_2^*$) and (2) the rate of the sporadic dissociation of a TF from its BS is greater than the rate of sporadic association of a TF and its DNA fragment target ($\mu_1^* > \lambda_1^*$).

Moreover a mean for the non-zero steady state exists if the sporadic dissociation rate exceeds both the association rates

($\mu_1^* > \lambda_1^* + \lambda_2^*$), and the variance exists if the sporadic dissociation rate exceeds the sporadic association rate plus twice the specific association rate ($\mu_1^* > \lambda_1^* + 2\lambda_2^*$). These predictions have been tested via parametrization of the Eqs. 41 and 42 in goodness-of fit analysis of the skewed EFDs, identified in several molecular mechanisms and biological process [30, 34, 62, 65].

In several specific (and practically interesting) cases, the exact analytical solutions for Eqs. 55 and 56 can be derived and the kinetic properties of the mean value and variance of the stochastic process can be analytically studied.

For instance, at $\lambda^*{}_2 = \mu^*{}_1 = 0$ and $\lambda^*{}_1 = \mu^*{}_2 > 0$, the random variable $X(t)$ at any time point follows the Poisson process with the probability function $p_m(t) = (m_X(t)^{\mathrm{m}}/m!)\ \exp(-m_X(t))$, where $m_X(t) = (\lambda^*{}_2(t)/\mu^*{}_1(\mathrm{t}))(1 - \exp(-\mu^*{}_1(t))$ (at $m_X(0) = 0$) and $p(t) = \exp(-m_X(t))$.

Next, let us consider the analytical solutions in a more complex case.

Let us assume that, in the KW process, a probability for the unobserved events approaches zero in time. $p_0(t) \to +\ 0$ at $t > t_s > 0$. (For instance, if $p_0(t) \approx e^{-m_X(t)}$. Under such circumstances, Eqs. 57 and 58 can be simplified to

$$\frac{dm_X(t)}{dt} = (\lambda_1^* - \mu_1^*) - (\mu_2^* - \lambda_2^*)m_X(t),$$

$$\frac{dD_X(t)}{dt} = (\lambda_1^* + \mu_1^*) + (\mu_2^* + \lambda_2^*)m_X(t) - 2(\mu_2^* - \lambda_2^*)D_X(t).$$

At positive or zero initial conditions, this system of linear differential equations has an exact analytical solution and, thus, allows us to calculate the mean, variance and standard deviation of the random variable $X(t)$ for the stochastic process $\{D_t, t \geq t_s\}$. It also provides explicit and simple analytical forms of the mean, variance and standard deviation values under the steady-state condition of the process. For instance, at $\mu_2^* > \lambda_2^*\,(\theta < 1), \lambda_1^* > \mu_1^*(a > 1)$ and the initial conditions $m_X(0) = m_X^0$, $D_X(0) = D_X^0$ we have

$$m_X(t) = m_X^*(1 - \exp(-(\mu_2^* - \lambda_2^*)t) + m_X^0 \exp(-(\mu_2^* - \lambda_2^*)t),$$

where

$$m_X^* = \frac{\lambda_1^* - \mu_1^*}{\mu_2^* - \lambda_2^*} > 0;$$

$$\begin{aligned}D_X(t) &= D_X^* + D_X^* \exp(-2(\mu_2^* - \lambda_2^*)t) + \\ &m_X^0((\lambda_2^* + \mu_2^*)/(\mu_2^* - \lambda_2^*))(\exp(-(\mu_2^* - \lambda_2^*)t) - \exp(-2(\mu_2^* - \lambda_2^*)t)) - \\ &m_X^*((\lambda_2^* + \mu_2^*)/(\mu_2^* - \lambda_2^*))\exp(-(\mu_2^* - \lambda_2^*)t) + \\ &1/2\Big(((\lambda_2^* + \mu_2^*)(\lambda_1^* - \mu_1^*) - (\lambda_1^* + \mu_1^*)(\mu_2^* - \lambda_2^*))/(\mu_2^* - \lambda_2^*)^2\Big)\exp(-2(\mu_2^* - \lambda_2^*)t),\end{aligned}$$

where

$$D_X^* = \left(\lambda_1^* \mu_2^* - \lambda_2^* \mu_1^*\right) / \left(\mu_2^* - \lambda_2^*\right)^2.$$

Thus, this analytical solution of the differential equations of the mean and variance provides detailed kinetic and steady state characteristics of the Kolmogorov stochastic process as a function of time. At steady state, our analytical forms of the mean and variance parameters can be defined by the three parameters: $a = \lambda_1^*/\lambda_2^*$, $b = \mu_1^*/\mu_2^*$ and $\theta = \lambda_2^*/\mu_2^*$. Using these notations, we derive the expressions: $m_X^* = (a\theta - \mathrm{b})/(1 - \theta)$, $D_X^* = (a - b)\theta/(1 - \theta)^2$.

These simple expressions show that the mean and variance values of the stationary stochastic process exist at $\theta < 1$ ($\lambda_2^* < \mu_2^*$) and $a > b$ ($\mu_1^*/\mu_2^* > \lambda_1^*/\lambda_2^*$). These results are consistent with Statements 2 and 3. Notice that in the case $\mu_1^* = 0$, the stable point of our differential equations on the phase plot is defined by coordinates $m_X^* = a\theta/(1 - \theta)$ and $D_X^* = a\theta/(1 - \theta)^2$. Interestingly, the same expressions have been obtained for the stationary distribution of the Kolmogorov process via factorial moments analysis of a probability function presented in the form of the hypergeometric series [58].

Because currently the sample size of the sequences in an NGS experiment tends to be constantly growing, the frequency of non-occurring events in the NGS datasets becomes smaller. Thus, the estimates obtained based on this simplified model may actually be quite accurate.

Thus, this section introduces simple deterministic equations which allow one to estimate the kinetics of the mean values and variances of the stochastic process (Eqs. 32 and 33) and also provides explicit formulas for the steady state (limiting) values of the KW distribution parameters $m^*{}_X$ and $D^*{}_X$. The results can then be used to (1) identify the underlying stochastic mechanisms driving TF–DNA association and dissociation processes, (2) estimate the basic quantitative characteristics (best-fit KW parameters of the skewed EFD) of TF–DNA binding datasets, (3) provide an unbiased statistical comparison of the datasets, and (4) evaluate the incompleteness (by estimating p_0) of a sequence library [20, 34].

3.4 Family of Skewed Frequency Distributions

We can present Eq. 41 with the following recursive formula:

$$\eta_m = p^*{}_{m+1}/p^*{}_m = \theta \frac{(a+m)}{b+m+1}, \tag{59}$$

where $m = 0, 1, \ldots$. Using Eqs. 41, 42, and 59 and properties of the gamma function, beta function, and hypergeometric series [40], we can obtain several important degenerate forms (distribution subfamilies) of the KW distribution.

Statement 4: *If* $\theta \to 1$ *and* $b > 0$*, then we obtain the zero-truncated Waring distribution, and as* m *approaches infinity*

$$p^*{}_m \sim (b-a)\frac{\Gamma(b)}{\Gamma(a)}\frac{1}{m^{b+1}}\theta^m, \tag{60}$$

i.e., distribution Eq. 41 approximates the Champernowne distribution at $\theta \to 1$ *[71] and the SPD at* $\theta = 1$ *[71, 72] in the right tail. If* $p_0 > 0$*, and we can define the zero-truncated limiting distribution* $P(X = m|X = 0) = p_m^{\{0\}*}$ *as*

$$p_m^*\{0\} = p^*{}_m/(1-p_0) = \frac{B(b-a+1, a+m)}{B(b-a, a+1)}\theta^m, \tag{61}$$

where m = *1, 2, . . . and then prove it.*

Statement 5 [62]: 1. *If, in Eq. 41,* a → *0 +;* θ → *1; b > 0, then*

$$\lim_{a\to 0+,\theta\to 1} p^{\{0\}*}{}_m = \lim_{a\to 0+}\frac{(b-a)\Gamma(a+m)\Gamma(b+1)}{\Gamma(a+1)\Gamma(b+m+1)}\lim_{\theta\to 1}\theta^m$$
$$= bB(b+1, m), \tag{62}$$

where $bB(b + 1, m)$ *is the Beta function [29, 40].*

2. *If in Eq. 41* $a \to 0$, $b \to 0$ *and* $\theta < 1$, *then Eq. 41 approaches the Fisher logarithmic series distribution* $f_m = f_1\, \theta^{m-1}/m)$, *where* $f_1 = -\theta/log_e(1-\theta)$, $m = 1, 2, \ldots$
3. *If, in Eq. 41,* $a \to \infty$, $b \to \infty$ *but* $\theta a/b \to const = q < 1$, *then Eq. 41 approximates the geometrical distribution* $f_m = q(1-q)^m$, *where* $m = 0, 1, 2, \ldots$.

For case 1 of Statment 5, probability function Eq. 62 depends on one positive parameter, $b(b = \mu_1^*/\mu_2^*)$, and this function approximates the Yule distribution [40, 62] at $\lambda_1{}^*/\lambda_2{}^* \to 0; \lambda_2{}^*/\mu_2{}^* \to 1$ and, thus, $\lim_{a\to 0+,\theta\to 1} p_0 = 1$. Notice that the Yule distribution assumes $b < 1$. Simon constructed a birth random process model of word distribution in the text [72]. The limiting distribution of this model approaches the Yule distribution [31] at $b > 1$ and in this case the KW distribution characterizes GPD [34, 73]. Cases 2 and 3 can be simply tested using Eq. 59. Interestingly, Tripathi and Gurland [74] proposed the extend Katz family of discrete distributions [29, 40] with hypergeometric probabilities of which several are shared with the KW family.

Note that the critical case $\theta = 1$ was not included in the Tripathi and Gurland model [74]. Additionally, Shubert and Glänzel [73] reported the same relationship between the zero-truncated Waring and the Yule distributions [75]. Irwin [70] noted that the Yule distribution is a specific case of the Waring distribution. However, he made an error claiming that it is true at $a = 1$. These results demonstrate that the hypergeometric probability function, derived from a steady-state (limiting) solution of the KW differential equations with the state transition rates defined by the linear functions of state values, provides a large family of well-known and practically used distribution functions. However, the diversity of the distribution functions and their potential applications in data analysis could increase if more general functional dependencies of the state transition rates will be introduced and studied.

3.5 Generalized Hypergeometric Distributions of Stationary Birth and Death Kolmogorov–Waring Processes

In this section, we consider the generalization of the KW process assuming that the intensity parameters ν_m and μ_m $(m = 0, 1, \ldots)$ are the ratios of two polynomials of m with all roots being real. In this case, the steady-state solution of Eqs. 32 and 33 can be described using the special functions of the generalized hypergeometric functions (GHFs) theory [40, 56, 71]. ${}_pF_q(a_1, a_2,\ldots,a_p;\ b_1, b_2, \ldots b_q;\ \theta)$, where a_i, b_i are called the numerator and denominator parameters and θ is called the variable and p and q are arbitrary numbers of the numerator and denominator parameters.

${}_pF_q$ is symmetric in its numerator parameters and likewise in its denominator parameters. These functions often occur in the context of practical problems in the fields of physics, physical chemistry, engineering, and applied statistics [29, 34, 40, 51, 57–59, 62, 71]. However, practical applications of GHFs in genome and system biology are novel [30, 34, 51, 62]. The GHF function consists of the series representation $\sum_{m=0}^{\infty} f_m$ with f_{m+1}/f_m being a rational function of m. If the numerator and denominator parameters are different and can be factored, the ratio is usually written as

$$\frac{f_{m+1}}{f_m} = \frac{(m + a_1)(m + a_2)\ldots(m + a_p)}{(m + b_1)(m + b_2)\ldots(m + b_q)(m + 1)}\theta. \tag{63}$$

Parameter θ occurs because the polynomial that consists of coefficients and a single variable may not be monic, with the leading coefficient equal to one. The leading coefficient is found in the term that contains the variable with the largest exponent. Equation 63 is a rational function of m $(m = 0, 1, 2,\ldots)$. The factor $(m + 1)$ is added for convenience to introduce $m!$ into the hypergeometric series Σp_m. Here, b_i $(i = 1, 2, \ldots, q)$ are positive integers to avoid making the denominator zero. If f_0 is defined (see below), Eq. 63 can be solved for f_m as

$$f_m = \frac{a_1^{[m]} a_2^{[m]} \ldots a_p^{[m]}}{b_1^{[m]} b_2^{[m]} \ldots b_a^{[m]} 1^{[m]}} \theta^m \tag{64}$$

where $x^{[m]} = x(x + 1) \ldots (x + m - 1) = \Gamma(x + m)/\Gamma(m)$; $x^{[0]} = 1$; $\Gamma(m + 1) - m! = 1^{[m]}$, and

$${}_pF_q(a_1, a_2, \ldots, a_p; b_1, b_2, \ldots b_q; \theta) = \sum_{m=0}^{\infty} \frac{a_1^{[m]} a_2^{[m]} \ldots a_p^{[m]}}{b_1^{[m]} b_2^{[m]} \ldots b_q^{[m]} 1^{[m]}} \theta^m \tag{65}$$

is a standard notation for GHF. There are p numerator parameters and q denominator parameters. The series is well-defined as long as the lower parameters b_1, b_2, …, b_q, are not negative integers or zero; $i = 1, 2, \ldots, q$.

The simplest generalized hypergeometric series is ${}_0F_0(-; - ;\theta) = 1 + \theta + \theta^2/2! + \cdots = e^\theta$. A blank indicates the absence of a parameter. The series terminates if any of the upper parameters $a_1, a_2, \ldots, a_p$ are a nonpositive integer, otherwise it is nonterminating and therefore an infinite series. m is the summation index. Thus, if one of the numerator parameters a_i, $i = 1, 2, \ldots, p$ is a negative integer, $a_i = -n$ (n is positive integer) say, the series is terminated. If the series is well-defined and nonterminating, then questions of convergence or divergence become relevant.

Let us consider Eqs. 32 and 33, where $\lambda_m = P_1(m)/Q_1(m)$ and $\mu_m = P_2(m)/Q_2(m)$. The real functions $P_1(m)$, $Q_1(m)$, $P_2(m)$, $Q_2(m)$ are the polynomials of nonnegative integer m such that $0 < \lambda_m < \infty$, $0 < \mu_m < \infty$. In particular, the $P_1(m)$ and $Q_2(m)$ have all real roots such that λ_m or μ_m does not become infinite and $P_2(m)$ and $Q_1(m)$ have all real roots such that λ_m or μ_m does not become zero. Let us denote by k_1, s_1 the degrees of the polynomials $P_1(m)$, $Q_1(m)$, respectively. Let us denote with k_2, s_2 the degrees of the polynomials $P_2(m)$, $Q_2(m)$, respectively. Let α_0 , β_0 , γ_0, δ_0 denote the highest power coefficients of $P_1(m)$,$Q_1(m)$, $P_2(m)$, $Q_2(m)$, respectively. Then the limiting probability function of the Kolmogorov equations, if it exists, is given by

$$p^*{}_{m+1}/p^*{}_m = \frac{\lambda_m}{\mu_{m+1}} = \frac{P_1(m)Q_2(m+1)}{Q_1(m)P_2(m+1)}\theta$$

or

$$p^*{}_{m+1}/p^*{}_m = \frac{(m+\alpha_1)\ldots(m+\alpha_{k_1})(m+1+\delta_1)\ldots(m+1+\delta_{s_2})}{(m+\beta_1)\ldots(m+\beta_{s_1})(m+1+\gamma_1)\ldots(m+1+\gamma_{k_2})}\theta, \tag{66}$$

where α_i, β_i, δ_i, γ_i and θ are the parameters of the polynomials $P_1(m)$, $Q_1(m)$, $P_2(m)$, $Q_2(m)$ and $\theta = (\alpha_0\gamma_0)/(\beta_0\gamma_0) > 0$.

Using Eqs. 63 and 65, we can write Eq. 66 as

$$P(x = m) := p^*_m = p_0 \frac{\alpha_1^{[m]} \cdots \alpha_{k_1}^{[m]} (\delta_1 + 1)^{[m]} \cdots (\delta_{s_2} + 1)^{[m]}}{\beta_1^{[m]} \ldots \beta_{s_1}^{[m]} (\gamma_1 + 1) \ldots (\gamma_{k_2} + 1)^{[m]}} \theta^m, \tag{67}$$

where

$$p_0 = 1/{}_{p+1}F_q(1, \alpha_1, \ldots, \alpha_{k_1}, \delta_1 + 1, \ldots, \delta_{s_2} + 1; \beta_1, \ldots, \beta_{s_1}, \gamma_1 + 1, \ldots, \gamma_{k_2} + 1; \theta),$$

and probability generating function

$$G(z) = \frac{{}_{p+1}F_q(1,\alpha_1,\ldots,\alpha_{k_1},\delta_1+1,\ldots,\delta_{s_2}+1;\beta_1,\ldots,\beta_{s_1},\gamma_1+1,\ldots,\gamma_{k_2}+1;\theta z)}{{}_{p+1}F_q(1,\alpha_1,\ldots,\alpha_{k_1},\delta_1+1,\ldots,\delta_{s_2}+1;\beta_1,\ldots,\beta_{s_1},\gamma_1+1,\ldots,\gamma_{k_2}+1;\theta)}, \tag{68}$$

where $p = k_1 + s_2$; $q = k_2 + s_1$ and $z > 0$ [44, 58]. If a numerator and a denominator parameter coalesce, then omit the parameter, whence the ${}_{p+1}F_q$ becomes a ${}_pF_{q-1}$. The ${}_{p+1}F_q$ series terminates and, therefore, is a polynomial if a numerator parameters is a negative or zero. The denominator parameters are also non negative integer or zero, as that would make the denominator zero.

In the case of $k_1 = k_2 = k$, $s_1 = s_2 = s$, we convert Eq. 67 into a simple and useful form of the Beta function products:

$$p_m^* = p_0 \prod_{i=1}^{k} \frac{B(\gamma_i+1, m)}{B(\alpha_i, m)} \prod_{i=1}^{s} \frac{B(\beta_i, m)}{B(\delta_i + 1_i, m)} \theta^m. \tag{69}$$

In the specific case $Q_1(m) = Q_2(m) = 1$ and $k = 1$, Eq. 68 is reduced to Eq. 41. We call Eq. 69 the generalized power-Beta function (GBF). The existence of limiting probability distribution Eq. 67, where p_0 is defined by Eq. 67, is determined by the convergence of the ${}_{p+1}F_q$ series. Let us also suppose that neither the numerator nor the denominator parameters a_i, b_j of the GHF

$${}_{p+1}F_q(1, a_1, a_2, \ldots, a_p; b_1, b_2, \ldots, b_q; \theta) = \sum_{m=0}^{\infty} \frac{a_1^{[m]} a_2^{[m]} \ldots a_p^{[m]}}{b_1^{[m]} b_2^{[m]} \ldots b_p^{[m]}} \theta^m \tag{70}$$

(as well as PGF Eq. 68) are a negative integers or zero. Then,

$$p_{m+1}/p_m = \frac{(m+a_1)(m+a_2)\ldots(m+a_p)}{(m+b_1)(m+b_2)\ldots(m+b_q)}\theta = \theta_m^{p-q}\{1 + A/m + O(m^{-2})\}, \tag{71}$$

where $A = \sum_{i=1}^{p} a_i - \sum_{i=1}^{q} b_i$ [40, 62]. Using the ratio test, it has been shown that series Eq. 71 converges absolutely for all finite θ if $p < q$ and for $\theta < 1$ if $p = q$, and it diverges for all non-zero θ if $p > q$ and the series does not terminate. For $p = q$ and $\theta = 1$, the ${}_{q+1}F_q$ series (as well as PGF Eq. 68) is absolutely convergent if $A < 0$ and all polynomial coefficients are positive [40, 62].

Kapur [57] analyzed the global steady state solution of the time-homogeneous Kolmogorov differential Eqs. 32 and 33 under the assumptions that the intensity parameters are time-independent, $P_1(m)$ and $P_2(m)$ are the polynomial functions of m with real roots such that they do not become infinite for $P_1(m)$ and zero for $P_2(m)$. Of note, Kemp is a specific case when $Q_1(m) = Q_2(m) = 1$.

Thus, to arrive at the steady state solution for the time-inhomogeneous process defined by Eqs. 32 and 33, we use $dp_i/dt = 0$ for all m and solve the system of algebraic equations with the transition functions defined by

$$P_1(m) = \lambda_m = (b_1 + a_1 m)(b_2 + a_2 m) \ldots (b_p + a_p m)$$

and

$$P_2(m) = \mu_m = (d_1 + c_1 m)(d_2 + c_2 m) \ldots (d_p + c_p m)$$

$(m = 0, 1, 2, \ldots)$, where $P_1(m)$ and $P_2(m)$ are real functions of nonnegative integer m such that $0 < \lambda_m < \infty$, $0 < \mu_m < \infty$.

Let $\alpha_i = b_i / a_i (i = 1, 2, \ldots, p)$ and $\beta_j = d_j / c_j + 1$ $(j = 1, 2, \ldots, q)$ and $\theta = (\alpha_1 \alpha_2 \ldots \alpha_p)/(\beta_1 \beta_2 \ldots \beta_p)$ and assume that a_i and b_i $(i = 1, 2, \ldots, p)$, c_j and d_j $(j = 1, 2, \ldots, p)$ are not zeros. Then Eq. 67 yields

$$p_m^* = p_0 \left(\left(\alpha_1^{[m]} \cdots \alpha_p^{[m]} \right) / \left(\beta_1^{[m]} \ldots \beta_q^{[m]} \right) \right) \theta^m,$$

where

$$p_0 = 1/{}_{p+1}F_q \left(1, \alpha_1, \ldots, \alpha_p; \beta_1, \ldots, \beta_q; \theta \right).$$

The generalized hypergeometric series converges for all finite positive θ if $p < q$ and diverges for all non-zero θ if $p > q$. If $p = q$, the series converges for $\theta < 1$ and also when $\theta = 1$ according to Eq. 71. Interestingly, that Kemp's and Kapur's generalized hypergeometric functions have derived the analytical forms for dozens probability generating functions, PFs and their characteristics [40, 57, 58, 71]. In fact, most of these PGFs have not been associated with the stochastic birth-death in biological systems and never used for analysis of the statistical distributions in biology and omics studies.

Kapur has defined the factorial moments of all orders for the steady-state probability function referring to the Kolmogorov birth–death process with polynomial rate functions [57, 58]. These expressions could be simply generalized for the transition rates presented by rational functions. Indeed, let us combine the parameter sets $\{\alpha_1, \ldots, \alpha_{k1}\}$ of $P_1(m)$, and $\{\delta_1 + 1, \ldots, \delta_{s2} + 1\}$ of $Q_2(m)$, shown in the Eq. 67, into the joint set $\{\alpha_1, \ldots, \alpha_{k1}; \delta_1 + 1, \ldots, \delta_{s2} + 1\}$ and re-name the set elements by the following $\{\alpha_1, \ldots, \alpha_{k1}; \alpha_{k1} + 1, \ldots, \alpha_p\}$. The total number of elements in the set equals $p = k_1 + s_2$. Let us also combine the parameter sets $\{\beta_1, \ldots, \beta_{s1}\}$ of $Q_1(m)$ and $\{\gamma_1 + 1, \ldots, \gamma_{k2} + 1\}$ of $P_2(m)$, that shown in the Eq. 67, into the joint set $\{\beta_1, \ldots, \beta_{s1}; \gamma_1 + 1, \ldots, \gamma_{k2} + 1\}$ and re-name the set elements by the following $\{\beta_1, \ldots, \beta_{s1}; \beta_{s1} + 1, \ldots, \beta_q\}$. The total number of elements in the set equals $q = s_1 + k_2$.

Let β_j $(j = 1, 2, \ldots, q)$ and α_j $(j = 1, 2, \ldots, p)$ be the non-zero parameters of joint polynomial functions representing the birth (in the numerator) and death (in denominator) transition rate functions in Eq. 66, respectively. The factorial moment of order r of the probability function by differentiating the PGF Eq. 68 and $z = 1$ substitution is defined as follows

$$\mu'_{[r]} = p_0 r! \left(\alpha_1{}^{[r]} \alpha_2{}^{[r]} \ldots \alpha_p{}^{[r]}\right) / \left(\beta_1{}^{[r]} \beta_2{}^{[r]} \ldots \beta_q{}^{[r]}\right) \theta^r.$$
$$_{p+1}F_q\left(1 + r, \alpha_1 + r; \alpha_2 + r, \ldots, a_p + r; \beta_1 + r, \beta_2 + r, \ldots, \beta_q + r; \theta\right).$$

Using this formula, we can generalize Statement 2 as follows:

If $p < q$ or if $p = q$ and $0<\theta<1$ then the time inhomogeneous Kolmogorov process has a global non-zero steady state solution and the generalized KW distribution moments of all orders exist.

If $p = q$ and $\theta = 1$ the moments of the r-th order exist, if $-A > r + 1$, where A is the negative score, defined by Eq. 71, and r is the order of the moment, $r + 1$ ($r = 1, 2, \ldots$), and the moments do not exist in the opposite case.

Probability generating function Eq. 68 belongs to a broad family of Kemp's generalized hypergeometric probability generating functions [40, 56]. Among others, this family includes many subfamilies of skewed distributions such as Waring, Yule, hyper-Poisson, extended Katz, and many other useful distribution functions [29, 34, 40, 54, 55, 57, 58, 69, 73]. The multiparametric families of the distribution functions, derived from Eqs. 66–69, can also be useful in future theoretical studies and for diverse applications. Starting with Kemp' publications [56, 40], the Kolmogorov birth–death equations have been used to develop other types of generalized hypergeometric distributions [40, 53, 57, 58, 62]. The applications of the recently developed so-called regularly varying generalized hypergeometric distributions to bimolecular data were considered in [51–53].

3.6 Practical Implementation of the KW Distribution Function

We limit the analysis to steady-state solution of such a stochastic binding–dissociation process in which the proportions of the different states ($m = 0, 1, 2, \ldots$) become stable.

The exact steady-state solution of such a stochastic binding-–dissociation process model can be described by the KW distribution function, which is calculated via the following simple recursive formula from Eq. 65:

$$p_{m+1}/p_m = \theta \frac{(a + m)}{b + m + 1} \quad (72)$$

where $m = 0, 1, 2, \ldots$ and the other three parameters a, b, and θ are unknown. Importantly, the KW distribution function allows us to estimate the value p_0, which gives the fraction of lost (undetected) events in a given TF–DNA binding experiment.

$$p_0 = \frac{1}{{}_2F_1(a, 1; b + 1; \theta)} > 0, \quad (73)$$

where ${}_2F_1$ is the hypergeometric Gauss series [40]. In this case,

$$p_0 = \left(1 + \sum_{m=1}^{\infty}\left(\prod_{i=1}^{m}\frac{(a-1+i)}{(b+i)}\theta\right)\right)^{-1} \tag{74}$$

Specifically, if $b > a > 0$ and $\theta \to 1\text{–}0$, then

$$\lim_{\theta \to 1-0} p_0 = \left(1 - \frac{a}{b}\right) \tag{75}$$

In this important equation, estimating the parameters is simple. By following a recursive formula, we can estimate the TFBSs at each peak height intensity as the following:

$$p_1 = p_0\frac{a}{b+1}\theta, \ldots, p_{m+1} = p_m\frac{a+m}{b+m+1}\theta. \tag{76}$$

Note that the GPD function can be a fairly accurate approximation of the KW function throughout the entire dynamic range of random variable X ($m = 1, 2, \ldots$). We use this attribute of the KW probability function for (1) goodness-of-fit analysis of the model, (2) estimation of specificity and sensitivity of the ChIP-based dataset, and, finally, (3) estimation of the total number of real specific BSs for a given TF in a given genome.

It is important to note that in the specific case when $\theta = 1$, the model is described by the well-known Waring probability function ($P_W(X = m)$ [40, 62]). At $m \to \infty$ and $t \to \infty$, this function has the asymptotic properties of the generalized Pareto-like probability function [62].

3.7 Truncated Distribution

The robust and accurate estimation of the hypergeometric and GPD-like probability function parameters based on real data is a great challenge due to noise and overparameterization issues. The last problem becomes much more difficult when the number of unknown parameters increases. However, the three-, two-, or one-parameter family distributions can be fitted well to the empirical distributions.

Due to incompleteness of NGS samples, p_o can be large and may have to be estimated. On the other hand, week signals on the right side of the skewed EFD are commonly overloaded by false-positive noise signals, and data from this part of the dynamic range should be discarded from parameterization of the function. The fraction of distinct signals on the right side of the EFD is minor and may not cover the real dynamic range of the TF–DNA binding avidity distribution.

In order to apply KW distribution Eqs. 72 and 73 to such datasets, let us assume that the random variable X is doubly truncated. For instance, let us consider the random variable X in the restricted 1, 2, …, J; $J < \infty$. Using Eqs. 50 and 61, the probability function of the resulting truncated probability function is written as

$$p_m^{*^T} = p_m^* / \left(\sum_{s=1}^{s=J} p^*{}_s \right) = \frac{p^*{}_m}{1 - p_0 - p^*{}_{J+1}/p_{0,J+1}} \quad (77)$$

This probability function corresponds to a typical situation in which the values 0 and $J + 1, J + 2, \ldots, \infty$ are not observed. At $\lambda_2^*/\mu_2^* \to 1 - 0$, Eq. 77 is transformed to the expression

$$p_m^{*^T} = \frac{p^*{}_m}{a/b - (b + J + 1)p^*{}_{J+1}/(b - a)}. \quad (78)$$

which at $J \to \infty$ could be simplified to yield $p_m^{*^T} = bp_m^*/a$.

3.8 Methods for Parameter Estimates [34, 62]

Given an empirical histogram of the number of TF binding events in the DNA fragments sample of size M, and given the function $P(X = m) = f(m;\ a,\ b,\ \theta)$, where $m = 1, 2, \ldots$ and a, b, and θ are the unknown parameters of the PF model, we estimated these parameters based on multiple uniform partitioning of an initial rectangular area for two parameters $(a,\ b)$ at several chosen values of the third parameter (i.e., $\theta\{\theta_1,\ \theta_2,\ \ldots,\ \}$). Let data points $(m,\ n_m(M)/N)$ for $m = 1, 2, \ldots, J$, where J is the number of occurrences of the more abundant event value, form the histogram corresponding to the EFD $g_M(m)$.

Recall that $\sum_{m=1}^{J} g_M(m) = 1$ and that some of the n_m values may be 0 corresponding to missing data. Let $\{(a,\ b)|a \in (A_l,\ A_r),\ b \in (B_l,\ B_r)\}$ denote the initial rectangle area, where A_l, A_r and B_l, B_r are the left and the right boundaries of the parameter domain for a and b, respectively. To adjust the unknown parameters in the functions $f(m;\ a,\ b,\ \theta_i)$ $(i = 1, 2, \ldots)$ to fit the histogram points $(m,\ g_M(m))$, we minimized the sum of squared deviations, $D(f(m, \bar{a}, \bar{b}, \theta_i), g(m))$, between the theoretical distribution and the histogram in points $m = 1, 2, \ldots, J$ for each chosen value of parameter θ_i and estimated values $\bar{a}$ and $\bar{b}$.

Let S denote the number of uniform intervals taken for each parameter at each step of the optimization algorithm. Let s denote the number of subareas in which a value of the criterion D is smallest. $s = 0.1S$. Let δ_1, δ_2 denote the minimal intervals of the partitioning domain for a and b, respectively. The parameters δ_1, δ_2 provide the exactness of our estimates.

Step 1. The initial rectangle area is partitioned into S^2 smaller rectangles of the same size. The value of function f is calculated for the central points of each smaller rectangle. The central point (a_i^c, b_i^c) of the rectangle $\{(a,\ b)|a \in (a_i,\ a_{i+1}),\ b \in (b_i,\ b_{i+1})\}$ is defined as $(a_i^c = (a_{i+1} + a_i)/2, b_i^c = (b_{i+1} + b_i)/2)$, where $i = 1, 2,$

..., $S - 1$. These selected subareas become the input rectangle subareas for the second step of the algorithm. If $(A_r - A_l)/S < \delta_1$ and $(B_r - B_l)/S < \delta_2$, then the central point (a_i^{c*}, b_i^{c*}) corresponding to the minimum value of criterion D is taken as the final point estimate. Otherwise, the next step is executed.

Step k: Suppose we selected the "best" rectangle subareas (s^{k-1}) at step $k - 1$. The first step is then executed for each of these subareas. So, the "best" smaller rectangle subareas s are selected for each of the rectangles of the k – 1th step of the algorithm and, in total, we obtain s_k ($s_k = s^k$) "best-fit" smaller rectangles. If $(A_r - A_l)/S^{k+1} < \delta_1$ and $(B_r - B_l)/S^{k+1} < \delta_2$, then the central point $\left(a_i^{c^*}, b_i^{c^*}\right)$ corresponding to the minimum value criterion D is taken as a final point estimate. Otherwise, the next step is executed.

Parameters in the models were also estimated using MLAB mathematical modeling software (Civilized Software, Inc., Silver Spring, MD, www.civilized.com). For goodness-of-fit analysis, we also used the Model Selection Criteria (MSC) and the reverse cumulative function [30, 62].

3.9 Prediction Method of the Total Number of Binding Events (BEs)

Below, we fit the truncated empirical distribution of binding events BEs (e.g., ChIP-Seq peak height value), starting the fitting of the specific part of the empirical distribution using Eqs. 72–78 and estimating the "empirical" threshold which excludes the high-noise component of the distribution using Eqs. 3, 6–8 [30, 34]. Then, after parameterization of specific and nonspecific probability terms in the mixture probability function of the distribution with Eqs. 77 and 78, and estimating weight parameter α (using Eq. 9 and 10), we extrapolate the best-fit probability function of specific events into a noise-enriched event region of the empirical distribution to predict the entire specific frequency distribution of specific BSs in the given ChIP-based experiments. Using parametrized Eqs. 77 and 78, we estimate the total number of BSs in the entire genome, sensitivity, specificity and several other practically important quantitative characteristics of the parametrized KW function.

4 Data Preparation and Initial Data Analysis: SACO, ChIP-PET, ChIP-seq Methods and Characterization of Datasets

4.1 SACO: Detection of Transcription Factor DNA–CREB BSs in the Rat Genome

In this section, we present brief descriptions of three TF–DNA methods, SACO, ChIP-PET, and ChIP-seq, and provide characterization of datasets used in the validation of these methods.

TF CREB binds to the cAMP-response element (CRE), a sequence identified in the promoters of many inducible genes [16]. CREB has since been found to mediate calcium, neurotrophin, and cytokine signals as well as a variety of cellular stresses. The SACO library was prepared using ChIP DNA obtained from ~10^8 rat PC12 cells. Chromatin occupancy in DNA was isolated from

PC12 cells that were stimulated by forskolin to increase intracellular cAMP 15 min prior to DNA extraction. Forskolin-treated PC12 cells were subjected to a CREB-DNA binding assay using an anti-CREB antibody. For SACO experiments, sonicated CREB ChIP DNA was blunted (protruding 3′ and 5′ ends were made flush) and ligated to adapters for limited amplification. The resulting DNA fragments in the assay, averaging around 700 bp in length, were represented in a SACO library by 21 nt SAGE tags. These chromatin DNA fragments were digested by NlaIII, which cleaves genomic DNA at approximately every 120 bp, and a modified SAGE procedure was used to create concatemerized 21 bp genomic signature tags (GSTs). Approximately 5000 plasmids were used to obtain the sequence of ~3 × 10^6 GSTs. The resulting distinct 21 bp SACO GSTs were matched to genomic sites. GSTs with exact matches or matches with one substitution error that were uniquely assignable to a genomic location were considered positives. GSTs without a genomic match (SACO Tag0) or with multiple matches were not considered. GSTs within 2 kb of each other were taken to be associated with the same SACO locus and formed "SACO clusters." GSTs were counted in clusters and thus can be used for semiquantitative profiling of BSs of a given TF. For additional details of SACO loci, *see* [16].

After genomic mapping and noise sequence filtration, the authors selected ~41,000 GSTs that identified a single region in the rat genome. The authors considered at least two SACO overlapped tags as a SATO tag cluster (called Tag-2) and are regarded as high-specific clusters; clusters with sizes 2, 3,. . ., 94 were analyzed. We will call all SACO clusters a Tag-2 + SACO tag set. We used 6269 for Tag-2, represented by 24,082 GST sequences (http://genome.bnl.gov/SACO/). Additional information about observed frequency distribution of intensity of DNA–TF binding is presented in Table 2.

4.2 ChIP-PET: Estrogen Receptor Element (ERE)–DNA Binding Sites and STAT1–DNA Binding Sites

Estrogen receptor alpha (ER-α) is a member of the nuclear hormone family of intracellular receptors which is activated by the hormone 17β-estradiol (estrogen) [14]. The main function of ER-α is its role as a DNA-binding transcription factor that regulates gene expression. ER-α interacts either directly with genomic targets encoded by ER elements (EREs) (5′GGTCAnnnTGACC3′), or indirectly by tethering to nuclear proteins such as AP-1, Sp-1, or NF-κb that are bound to DNA at their cognate regulatory sites [14].

Briefly, hormone-deprived breast cancer cell line cells (MCF-7) were treated with 10 nM 17β-estradiol for 45 min and then DNA-bound receptor complexes were isolated through chromatin immunoprecipitation (ChIP) using anti-ERα antibodies. PET sequences were extracted from the raw reads and mapped to the human genome sequence assembly (hg18) . Ninety-five ChIP DNA fragments ranged from 100 bp to 2 kb. The distribution of the

Table 2
Observed and predicted frequency distributions of binding events in CREB BS (SACO) ERE BS (ChIP-PET), INF-gamma STAT1 BS (ChIP-PET)

(1) SACO, cluster size

CREB BSs	1	2	3	4[a]	5	6	7	8	9	10	C_{11} +	C_{+4}	Spec. C_4 +	C_{3+}	Spec. C_3 +
Observed loci	17,509	3588	1065	**518**	274	208	126	77	69	50	328	1650	1	2715	1
Specific loci (GPD)	**3036**	**1390**	**742**	**439**	**279**	**188**	**132**	**96**	**72**	**55**	**271**	1532	0.929	2274	0.838
Specific (K-W)	3037	1389	743	441	282	191	134	98	74	56	291	1568	0.950	2311	0.851
Nonspecific loci (NSL)	14,473	2228	343	**52.8**	8.1	1.3	0.2	0	0	0	0	62.4	0.04	405	0.15
Fraction of NSL	0.827	0.621	0.322	**0.102**	0.030	0.006	0.002	0	0	0	0	Non	Non	Non	Non

(2) ChiP-PET, cluster overlap

ERE BSs	1	2	3	4[a]	5	6	7	8	9	10	C_{11} +	C_{+4}	Spec. C_4 +	C_{3+}	Spec. C_3 +
Observed loci	117,024	6245	771	279	133	93	70	40	22	24	55	716	1	1487	1
Specific loci (GPD)	3693	1202	512	258	145	88	57	39	27	20	79	713	0.996	1225	0.824
Specific (K-W)	**3693**	**1200**	**513**	**257**	**144**	**87**	**56**	**37**	**26**	**19**	**68**	693	0.969	1206	0.811
Nonspecific loci (NSL)	113,331	5046	225	10	0.4	0.0	0.0	0.0	0.0	0.0	0.0	10	0.01	235	0.16
Fraction of NSL	0.9684	0.8081	0.291	0.036	0.003	0	0	0	0	0	0		Non	Non	Non

(3) ChiP-PET, overlap height															
INF-g STAT1 BSs	**1**	**2**	**3**	**4**	**5**	**6[a]**	**7**	**8**	**9**	**10**	c_{11+}	c_{6+}	**Spec.** c_{6+}	c_{5+}	**Spec.** c_{5+}
Observed loci	212,447	39,025	7103	1444	400	157	113	66	39	38	121	534	1	934	1
Specific loci (GPD)	**2763**	**1368**	**734**	**420**	**253**	**159**	**104**	**70**	**48**	**34**	**110**	526	0.985	779	0.834
Specific (K-W)	2764	1368	735	422	255	161	105	71	49	35	111	531	0.995	786	0.841
Nonspecific loci (NSL)	209,697	37,516	6712	1201	215	38	7	0	0	0	0	45	0.085	260	0.279
Fraction of NSL	0.99	0.96	0.94	0.83	0.54	0.24	0.06	0.00	0	0	0		non	non	non

GPD best-fitted (truncated) generalized discrete Pareto probability function, *K-W* best-fitted Kolmogorov-Warning probability function, *c* occupancy level of BS locus

[a]Cut-off value at selected specificity (Spec)

sequence span of these DNA fragments followed the log-normal function with a span average of 674 bp, median of 458 bp, and mode of 277 bp.

To find relationships between relative binding avidity of ERE BSs and expression level of putative direct ERE TF gene targets, we used U133A&B expression profiling of transcripts of human ER-positive MCF7 cells defined before and after stimulation with 10 nM 17β-estradiol [14]. In these experiments, the RNA was extracted at 12, 24, and 48 h and hybridizations were performed in microarray triplicates according to the manufacturer's protocol.

The 480,042 original ChIP-PET sequencing reads of INF-γ-stimulated STAT1–DNA binding (library shc016 in T2G DB; HeLa S3 cells [15]) were mapped to single loci in the human genome assembly (hg17), and 327,838 distinct (nonredundant) PETs (68%) were identified. Of these unique fragments, the PET tags whose DNA fragment spans 5′- and 3′-ends were <6 kb apart were selected. Then, the PET DNA fragments mapped to the regions of unlikely locations (mitochondrial DNA, Y chromosome, centromeric loci, long gene-free chromosome regions, etc. (see above)) were excluded from that PET DNA fragment dataset. We then selected 324,523 distinct sequences in the ChIP-PET library shc016, representing 260,953 DNA fragment cluster overlaps (including singleton DNA fragment genome hits). From the untreated control dataset (library shc019 in T2G DB; HeLa S3 cells), 507,828 original ChIP-PET sequencing reads were mapped to a single location in the human genome assembly (hg17) and 263,901 distinct PETs (52%) were identified. Of these unique fragments, the PET tags whose DNA fragment spans 5′- and 3′-ends were <6 kb apart were selected. Finally, we selected 254,233 ((254,233/507,828) × 100% = 50%) distinct sequences of the ChIP-PET library shc019, representing 212,982 DNA fragment cluster overlaps, for further analysis. Additional information about the observed frequency distribution of DNA–TF binding events is presented in Table 2.

4.3 ChIP-PET: STAT1–DNA Binding in INF-γ-Stimulated and -Unstimulated HeLa S3 Cells

Interferons (INFs) are well-known cytokines that play pivotal roles in antiviral, antibacterial, cell proliferation and differentiation, and antitumor responses primarily by modulating gene expression via the JAK/STAT pathway [3, 23]. STAT1 regulates proliferation by promoting growth arrest and apoptosis in response to INF signals. These effects have given the INFs their major therapeutic value in the treatment of hepatitis, melanoma, leukemia, lymphoma, and multiple sclerosis, although their mechanism of action is still unclear [23, 27]. The regulation of STAT1 by INF-γ provides a useful system to understand how transcription factors select specific binding sites, and ChIP-PET has been used to study INF-γ-induced DNA–STAT1 binding. In our analysis, we used original ChIP-enriched DNA fragments of INF-γ-induced the

DNA–STAT1 ChIP-PET library (sch016) and noninduced DNA--STAT1 (sch019) library stored in the T2G DB [37]. Based on the ChIP-PET protocol [4], STAT1 ChIP-PET libraries were prepared from INF-γ-stimulated (sch016) and un-stimulated (sch019) HeLa S3 cells as described in [15]. Briefly, for each biological replicate, 12×10^8 cells were used in total. These cells were split into INF-γ-treated and untreated halves for STAT1 ChIPs. The ChIP-enriched DNA fragments were cloned into the cloning vector pGIS3 to generate the ChIP DNA fragment library. Additional information about the libraries and mapping of ChIP-PET DNA fragments onto the human genome can be found in the USCS browser track "GIS-PET" (http://genome.ucsc.edu/).

The T2G DB contains 327,838 and 263,901 distinct ChIP-PET DNA fragments obtained from INF-stimulated and unstimulated HeLa S3 cells, respectively. These sequences that uniquely mapped to the reference human genome (hg18) were clustered by the T2G DB algorithm into 247,846 and 212,982 clusters (including singletons), respectively. We also identified and excluded probable false-positive clusters (and cluster overlaps) having a very large size and located distantly (>100 kilobases (kbs)) from most neighboring genes. Sequences and clusters located in centromeric regions, Y and M chromosomes as well as alpha-satellites, and overlapping gap regions were also excluded from analysis. Finally, we used 324,523 and 259,759 ChIP-PET fragments obtained from INF-γ-stimulated and unstimulated HeLa S3 cells, respectively. We identified computationally these sequences in 246,644 and 211,619 clusters (including singletons) for INF-stimulated and unstimulated datasets, respectively. To identify specific binding loci, we also recalculated the number of PET cluster overlap peaks (see below); for INF-stimulated and -unstimulated HeLa S3 cells, we defined 260,953 and 218,440 PET cluster overlap peaks, respectively.

To study the relationships between relative TF-binding avidity of BSs and expression level of putative direct STAT1 gene targets, we used U133 plus2 expression profiling of transcripts of HeLa cells at 0, 2 and 4 h after stimulation with INF-γ in microarray triplicates under standard experimental conditions (S. Hartman's microarray dataset; [14]).

To validate TF–DNA BS predictions, we provide computational identification of the position weighted matrix (PWM) of canonical STAT1 motifs [6]. The PWM method of TF motifs is used to scan the locations of STAT1 canonical motifs in unstimulated and INF-γ-stimulated DNA sequence datasets derived via the ChIP-PET method. The canonical STAT1 motifs are scanned by the nmscan program of NestedMiCA tools using threshold-5 (Score Threshold-5.0) and dual strand search (http://www.sanger.ac.uk/Software/analysis/nmica/).

4.4 ChIP-seq: STAT1–DNA Binding in INF-γ-Stimulated and Unstimulated HeLa S3 Cells

ChIP-seq maps single-end 27 bp tags using the Illumina 1G system (1G), which provides for a 1.5–2-fold increase in the number of sequences compared to other systems (e.g., SACO). The method is based on deep 1G sequencing of short-read single-end tags (SETs), which are simpler to prepare than PETs. Using ChIP-seq, the authors in [23] compared STAT1–DNA binding in INF-γ-stimulated and unstimulated human HeLa S3 cells, by generating approximately 47 million reads from immunoprecipitated DNA fragments. For each dataset (or DNA fragment library), the authors calculated the number of "extended" overlapped SETs (composing the extended SETs or "XSETs") and the corresponding genomic locations into a DNA fragment overlap profile. They identified significantly overlapped SET cluster peaks by thresholding the profiles (with peaks $\geq$11 XSETs) at a height equivalent to an estimated false discovery rate (FDR) of $<$0.001. The global properties of the resulting two profiles were distinct and consistent with high ChIP enrichment.

4.5 ChIP-seq Nanog and Oct4-Binding Data

We used a similar method in our analysis of ChIP-Seq TFBS datasets for two ChIP-seq libraries (GEO ID: GSE 11431) derived from mouse embryonic E14 cell lines [13].

5 Computational Preprocessing of Raw Sequence Data, Sequence Mapping, Aggregation, and Preliminary Data Analysis

5.1 Long PET DNA Fragments Presented in the ChIP-Based Generated Library Can Form False Clusters and Produce Bias in the Count of Real Clusters and Their Sizes

The long tails of the frequency distribution of a span of DNA fragments in ChIP-PET experiments can produce errors in mapping, counting, and modeling of background (noise) events, since longer DNA fragments have a greater chance to form random (false-positive) clusters while shorter fragments can be a source of multipeak clusters. For example, Fig. 6a shows the frequency distribution of a span (nt) of PET DNA fragments found in an INF-γ-induced STAT1 ChIP-PET binding event dataset. Goodness-of-fit analysis of the empirical frequency distribution showed that this distribution is a mixture of the following distributions: relatively short DNA fragments ($<$700 nt) and very long DNA fragments (up to several thousands of nucleotides). Figure 6b demonstrates that the lengths of DNA fragment clusters are often longer than that of PET DNA fragments in PET fragment singletons. Data analysis suggests that the longer PET DNA fragments can form false-positive clusters and inappropriately mapped regions, thus producing bias in the count of PET clusters, incorrect peak heights, and erroneous locations for BS loci. It seems that prefiltering very long ChIP DNA fragments before mapping them onto genomes is helpful reducing the size of false DNA clusters and counts.

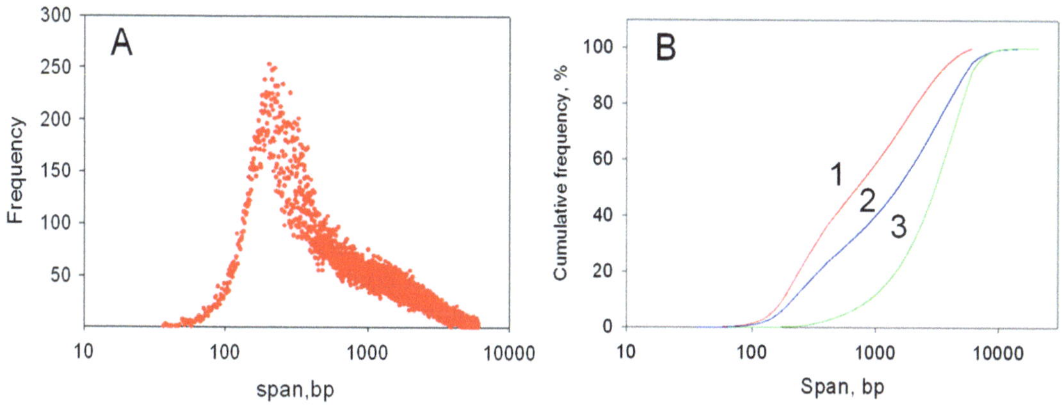

Fig. 6 False-positive "significant" clusters in ChIP-PET datasets occurred due to suboptimal ultrasound generation of DNA fragments with very long and very short PET DNA fragment spans. (**a**) An example of the frequency distribution of a span of PET DNA fragments found in an INF-γ-induced STAT1 ChIP-PET binding event dataset. This data contains a large fraction of long DNA fragments. (**b**) *Red line*: the cumulative functions of the length of PET DNA fragments in the total set of DNA fragments which uniquely mapped onto the reference genome (used in panel **a**); *blue line*: the spans of singleton PET DNA fragment sequences; *green line*: and the spans of PET DNA fragments forming clusters of size 2 or greater (PET-2+)

5.2 Sequence Prefiltering and the Compromise Between Specificity and Sensitivity in ChIP-seq Data Analysis in Identification of TF STAT1 BSs in INF-γ-Stimulated HeLa S3 Cells

An INF-γ-stimulated STAT1–DNA binding ChIP-seq assay produces millions of short (27 nt) unique sequence reads ($20–40 \times 10^6$), and in combination with the unstimulated (basal) STAT1–DNA binding ChIP-seq assay, this experiment can be used to estimate the specificity, sensitivity, and accuracy of the STAT1 TFBS mapping onto complex genomes [23]. After ChIP for STAT1 in INF-γ-stimulated HeLa S3 cells, the 1G system produced a total of 24.1 million 27-bp sequence reads (SETs [23]; Table 1). Of these, 15.1 million reads (63%) uniquely aligned to the non-repeat-masked regions of the human genome. For unstimulated HeLa S3 cells, the authors generated 22.7 million reads and uniquely mapped 12.9 million reads (57%).

A cluster peak height (maximum number of DNA fragments overlapped in a genome locus) for the XSET overlaps was the maximum number of overlapped XSETs in the cluster; peak height and FDR were inversely related. For both STAT1 sequence libraries, the authors estimated FDR = 0.001 at peak height = 11 XSETs. The resulting 41,582 overlapping peak regions containing 17.9% of mapped reads were experimentally defined as the probable STAT1 BS in INF-γ-Stimulated HeLa S3 cell DNA. In the INF-γ-Stimulated HeLa S3 cell DNA samples, only 11,004 DNA fragment overlap cluster peak regions and only 4.2% of mapped reads were experimentally defined and interpreted as the probable STAT1 BS ([23], Table 1). However, after the paper had been published [23], the authors revised their statistical criteria. In particular, new peak height critical cutoff values of 9 and 10 were used for genome mapping using DNA fragment clusters derived from the ChIP-seq libraries of the stimulated and unstimulated STAT1 binding tumor cells, respectively. The authors also identified and split the largest clusters, redefined the genome coordinates of the significant DNA fragment clusters and recalculated the number and the heights of the peaks in DNA fragment overlap regions. In our analyses, we used the revised sets as our raw data. With the new statistical criteria, we counted 63,309 INF-γ-"stimulated peaks" and 16,470 "unstimulated peaks" for HeLa S3 cells, respectively. We discarded all sequence reads that could not be uniquely mapped to the genome and did not follow our filtering criteria.

Shot-height DNA fragment clusters might be observed in multiple unusual positions within genic or nongenic regions, including 3′ends of the gene and downstream gene regions (Fig. 7). Such unlikely real clusters are often superlative in a given genome region across the different experiments and their number is very sensitive to the statistical method and criteria used to estimate specificity of the experiment (or FDR by [23], Table 1). Figure 7a shows an example of a massive multicluster DNA fragment mapping in the centromere region of chromosome 1. These clusters exhibit the same location and the same shape of distribution of overlapped DNA sequences in the clusters for unstimulated and INF-γ-

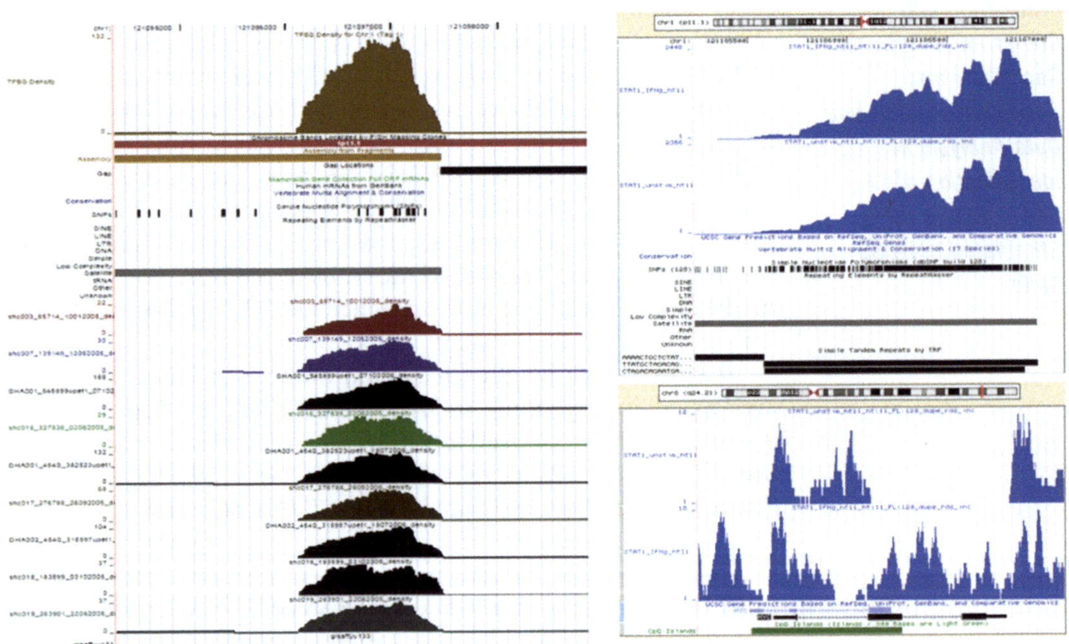

Fig. 7 Typical examples of nonspecific binding regions. (**a**) False-positive ChIP-PET clusters can often be located within or near centromeric regions. Large clusters occur near chromosome gap regions in many ChIP-PET libraries for different TFs (T2G data base). (**b**) False-positive ChIP-seq clusters can often be located within or near centromeric regions. STAT1-DNA binding in INF-γ stimulated and unstimulated HeLa S3 cell ChIP-PET datasets. (**c**) False-positive group(s) of shot-height multiple clusters in ChIP-seq-defined STAT1-DNA binding in the genome of INF-γ stimulated and unstimulated HeLa S3 cells

stimulated samples. Such clusters in the centromere regions and near chromosome gap regions were considered false-positive clusters (Fig. 7a, b). Clusters with relatively long spans, large sizes, and distantly located (>100 kb) from the closest genes (and their related cluster peaks) were also excluded from our analysis as false-positive events. False-positive group(s) of shot-height multiple clusters in ChIP-seq-defined STAT1–DNA binding in the genome of INF-γ stimulated and unstimulated HeLa S3 cells are shown in Fig. 7c.

Due to suboptimal experimental design, a small number of biological samples, and ignoring the genome sequence complexity information in peak calling and statistical background models of computational algorithms, ChIP-qPCR validation experiments can provide an optimistic estimate of specificity and sensitivity of a ChIP-seq experiment [32].

In total, we filtered out 1.3% (853/63,309 × 100%) and 4.4% (727/16,470 × 100%) problematic and false-positive stimulated and unstimulated peaks, respectively (Table 1).

The resulting 62,456 stimulated peaks contained 8.2% (1,246,120/15,100,000) of confidence-mapped reads, whereas the 16,470 unstimulated peaks contained 36.8% (198,566/540,000) of confidence-mapped reads (Table 1).

Finally, after all filters were applied, we selected 62,456 and 15,743 ChIP-seq fragments obtained from the libraries of INF-stimulated and unstimulated HeLa S3 cells, respectively. Table 1 summarizes the characteristics of the revisited ChIP-PET sequence libraries derived from the stimulated and unstimulated HeLa S3 cells.

5.3 ChIP-PET ERE Data: Sequence Filtering, Mapping Onto Reference Genome, and Sequence Clustering Are Essential Steps for Data Analysis

A significant amount of nonspecific (background) genomic DNA is always present in the inmmunoprecipitated DNA material of any ChIP-based DNA sequence library. In fact, about 35–80% of original ChIP-based tag sequences are nongenomic/noise sequences. Fortunately, these DNA sequences can easily be filtered out after mapping of the DNA fragments onto the genome. For example, in one ERE–DNA binding study [14] (library shc07; http://t2g.bii.a-star.edu.sg), 635,371 PETs were sequenced, of which 361,241 (~56.86%) were mapped unambiguously to unique loci in the reference genome [14]. We filtered and reanalyzed ChIP-PET DNA fragment sequence data of ERE-binding sequences using our algorithm of filtering, mapping, and clustering [20]. Due to PET sequence duplication, the number of unique PET sequences mapped uniquely to the genome was reduced to ~137,500 distinct PET sequences. After additional filtering of the sequences mapped to Y chromosomes, centromeres, gaps, and alpha satellite regions, we finally selected 136,348 ((136,348/635,371) × 100% = 21.5%) distinct sequences of the ChIP-PET library representing 124,756 DNA fragment cluster overlaps (including singleton DNA fragments).

Additional information about the observed frequency distribution of the intensity of DNA–TF binding is presented in Tables 2 and 3.

5.4 Specific TF Binding Events in ChIP-PET and SACO Datasets Are Following the Common Skewed Statistics

The EFDs of the binding events for all ChIP-based datasets exhibit monotonically skewed shapes with a greater abundance of rare binding events, and more gaps and irregularities among the higher-confidence binding events (Fig. 8). The forms of the distributions of binding events for different methods and TF–DNA binding events and cell types are very similar to each other (*see* also [4, 19, 20, 30, 31]).

Figure 8a–d demonstrate that the frequency distribution shapes of the BEs are similar for both the ChIP PET and SACO methods. Figure 8a–c show the histograms specifying the binding events frequency in the ChIP-PET experiments: (panel a) INF-γ-stimulated and unstimulated STAT1 DNA binding data; (panel b)the same INF-γ-stimulated STAT1 DNA binding data; and (panel c) estradiol-induced ER-DNA binding data. Comparing Fig. 8d to a–c, the histogram specifying the frequency of CREB binding events in the SACO experiment is similar to the histograms for the ERE TF and STAT1 TF ChIP-PET experiments. *See* Table 3 for additional information and details. Thus, the shapes of the EFDs in different experimental systems are robust across the methods. Furthermore, our mixture distribution function (Eq. 4) provides for a quantifiable description of these distributions. Next, the KW and exponential functions characterizing the specific and nonspecific TF–DNA binding events allow accurate analysis of the EFDs for different mammalian organisms, cell types and TF–DNA inducing pathways.

5.5 Specificity and Critical Cutoff Values of Binding Events in ChIP-PET and SACO Experiments

Using our curve-fitting analysis method for parameterization of the GPD for the experimental data presented in Fig. 8a–d, we decomposed the noise and the specific components of binding events in the ChIP-PET and SACO datasets and estimated a cutoff value for binding events at a reasonable specificity. Table 2 contains detailed information about the number of binding events (changed from 1 to 11) for the CREB SACO library and two ChIP-PET libraries and their distributions. The numerical results of goodness-off-fit analysis of the truncated GPD and KW functions are also presented in Table 2 and both functions show an accurate approximation of the available data. Based on the results, we found optimal cutoff values in each experiment with acceptable p-values: $p = 0.038$ SREB BS, SACO method; $p = 0.014$ ERE BSs, ChIP-PET method; $p = 0.084$ INF-γ-stimulated STAT1 BSs, ChIP-PET. There were also acceptable specificity estimates for detection in each experiment: 96.2% for SREB BS, 98.6% for ERE BSs, and 91.6% for INF-γ-stimulated STAT1 BSs. A detailed description and numerical values of these and other important statistical characteristics of the libraries are presented in Table 3. Importantly, the parameterization of the mixture distribution function (Eq. 4) allowed us to

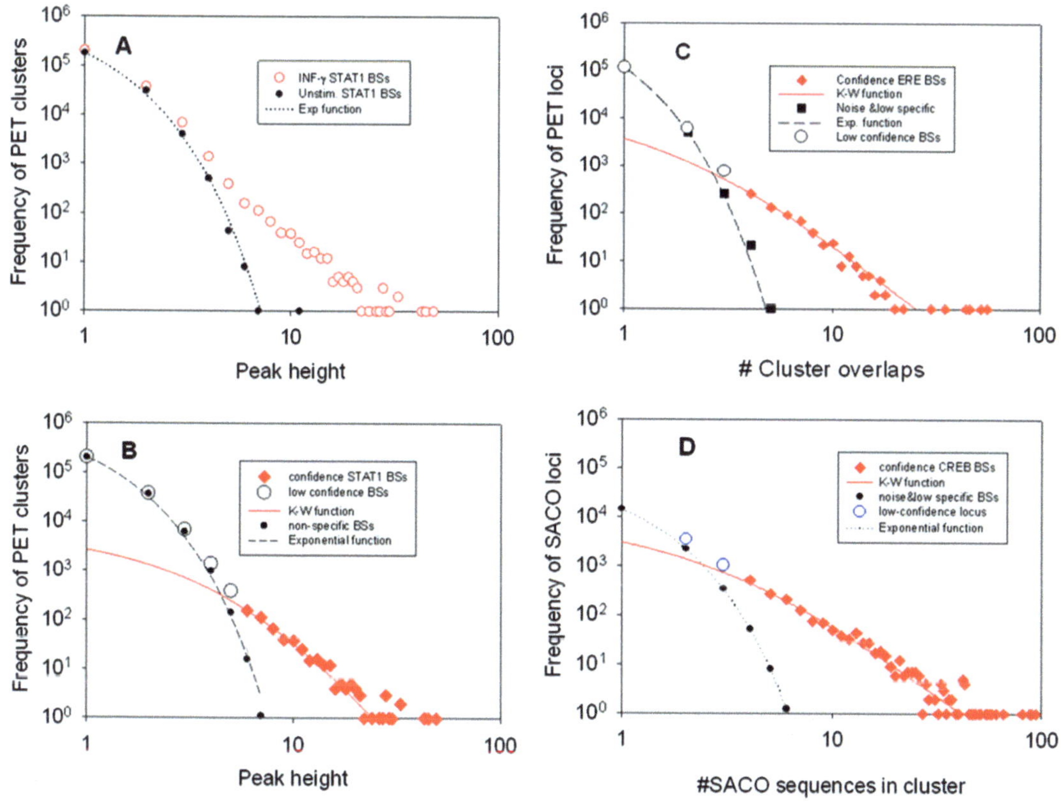

Fig. 8 Common statistical properties of binding events and goodness-of-fit analysis of the distribution of binding events across ChIP-based methods (ChIP-PET and SACO) and different TF–DNA binding systems: (**a**) STAT1-DNA binding events in INF-γ-stimulated (*open circle*) a n d unstimulated (control—*filled circle*) cell samples where a binding event was defined by the highest peak in overlap of cluster ChIP-PET DNA sequences; *dotted line*: best-fit exponent function at power coefficient $s = 1.95 \pm 0.136$; $t = 14.4$; $p < 0.0001$. (**b**) A decomposition and estimation of the parameters of our mixture probability distribution model: (*open diamond*) best-fit GPD to specific (right) tail of the empirical distribution; (*solid circle*) model-predicted nonspecific binding events; (*open circle*) total number of specific and nonspecific events in the overlapped region of the mixture distribution. The exponential function with a power coefficient of $s = 1.86 \pm 0.072$ ($t = 25.96$; $p < 0.0001$) fits well to the predicted frequency distribution and simultaneously fits to the observed frequency distribution of nonstimulated HeLa S3 cells (panel **a**). The GPD function [3] fits well to the truncated statistics of specific binding events with a cutoff value of 6 and at $k = 4.44 \pm 0.0806$ ($t = 56.9$; $p < 0.0001$); $b = 6.258 \pm 0.1260$ ($t = 51.9$; $p < 0.0001$). (**c**) Frequency distribution of the estradiol-induced ERE-DNA binding events in the genome of human breast cancer MCF7 cells, detected by via the ChIP-PET method and the results of decomposition of this frequency distribution based on our mixture distribution model (Eq. 4). (**d**) Frequency distribution of the CREB-DNA binding events in the genome of forskolin-treated rat PC12 cells, detected by via the SACO method and the results of decomposition of this frequency distribution based on our mixture distribution model (Eq. 4). Exponential and KW functions are used for the modeling and parameterization of the frequency distributions of the nonspecific and specific TF–DNA binding events, respectively. See Table 3 for additional information and details

Table 3
Statistical characteristics of the TF–DNA datasets and the frequency distributions of binding events generated by SACO, ChIP-PET and ChIP-seq platforms

Definition	CREB (SACO)	ERE (ChIP-PET (1)	STAT1 (ChIP-PET (2))	STAT1 (ChIP-seq (1))[a]	STAT1 (ChIP-seq (2))[b]
Number of mapped genome loci, in dataset N	23,812	124,756	260,953	62,456 at $c_+ = 9$	15,743 at $c_+ = 10$
Number of DNA fragments in N loci, M	41,702	136,348	324,523	1,246,120 at $c_+ = 9$	198,566 at $c_+ = 10$
Mean of binding event (e.g. height of peak value), M/N	1.75	1.09	1.25	19.95 at $c_+ = 9$	12.61 at $c_+ = 10$
Frequency of DNA fragment singletons, p_1	0.74	0.94	0.81	N.A.	N.A.
Estimated number of specific BS occurred in dataset, N_{sp}	6737	6099	6074	16,872	4694
Number of DNA fragments in N_{sp}, M_{sp}	20,962	12,104	15,599	493,992	25,794
Mean value of specific binding events, M_{sp}/N_{sp}	3.11	1.98	2.57	29.28	5.5
Estimated frequency of specific BS out of dataset, p_o	0.563	0.77	0.5	0.04	0.22
Predicted number of specific BS in genome, N_{tot} (by KW)	15,438	26,350	12,252	17,661	5985
Critical cut-off value of specific binding (e.g. peak height), c	4	3	6	61	31
Number of observed BS for binding events $\geq c_+$, $N(c_+)$	1650	1487	534	2322	64
Number of DNA fragments in Nc_+, $M(c_+)$	13,822	6834	4941	271,521	2809
Number of specific BS in N_{c+}, $N_{sp}(c_+)$ (by KW)	1568	1206	526	2134	58
Number of DNA fragments in $N_{sp}(c_+)$, $M_{sp}(c_+)$	12,918	6011	4662	237,790	2448
N_{sp}/N * 100%	28.3	4.89	1.87	N.A.	N.A.
% Specific DNA fragments in dataset, α (%) = M_{sp}/M * 100%	50.3	8.8	4.54	N.A.	N.A.
% at $\geq c$, $N_{sp}(c_+)/N_{sp}$ * 100%	25.5	25.84	8.09	12.65	1.36
% Specific sequences in Nc_+, $M_{sp}(c_+)/M_{sp}$ * 100%	61.6	49.66	29.88	48.14	9.49
Specificity = $N_{sp}(c_+)/N(c_+)$ * 100%	95.0	81.10	91.52	91.9	90.7

(continued)

Table 3 (continued)

Definition	CREB (SACO)	ERE (ChIP-PET (1)	STAT1 (ChIP-PET (2))	STAT1 (ChIP-seq (1))[a]	STAT1 (ChIP-seq (2))[b]
Sensitivity $= N_{sp}(c_+)/N_{tot} * 100\%$	10.16	4.58	4.36	12.08	0.97
θ (KW)	0.999	0.987	0.997	0.996	0.969
a (KW)	1.712	0.806	4.341	37.318	7.633
b (KW)	3.924	3.481	8.757	39.064	9.737
k (GDP)	2.224	2.565	4.446	2.266	2.975
β (GDP)	2.649	1.701	6.258	43.347	10.29
s (by exponential distribution; non-specific binding events)	1.871	3.112	1.86	N.A.	N.A.

Column 1: Characteristic; Column 2: Forskolin-induced CREB binding (SACO clusters); Column 3: Estradiol-stimulated ERE-DNA binding; Column 4: INF-γ-stimulated specific STAT1 BS Type I (ChIP-PET peaks); Column 5: INF-γ-stimulated specific STAT1 BS Type I (ChIP-seq peaks); 6. Unstimulated specific STAT1 BS Type I (ChIP-seq peaks) θ, a, b: estimated parameters of KW function; k, β: estimated parameters of GPD function; s: estimated power parameter in exponential function

[a]Data for high-avidity STAT1 BS Type I

[b]Data for low-avidity STAT1 BS Type I

arrive at viable statistical estimates and effectively compare EFDs (Eq. 3) across different methods, organisms, cell types, TFs and their inducing pathways.

Figure 8a demonstrates that GPD fits well to the less noisy binding events in the right tail of the empirical distribution of binding events of TF STAT1in INF-γ-stimulated HeLa S3 cells, as the exponential function fits well to the distribution of the binding events in the unstimulated STAT1 HeLa S3 cells (Fig. 8b). Our colocalization analysis of spans of ChIP-PET-defined significant cluster peaks in the stimulated and unstimulated datasets showed that only a few BS regions were common in the stimulated and unstimulated HeLa S3 datasets. The canonical STAT1 BS (TTCCNGGAA) was also well-defined in a significant part of the empirical distribution of TF–DNA binding events (cluster peaks) of TF STAT1 in INF-γ-stimulated HeLa S3 cells, but rarely defined in unstimulated HeLa S3 cells (not presented).

These results suggest that TF–DNA binding events in unstimulated HeLa S3 cells cannot be reliably detected in the given ChIP-PET experiment under the given experimental conditions and design.

Figure 8c, d show that, for the SACO and ERE TF BEs analyses, the exponential distribution statistics fit well to the mostly false-positive low-abundance cluster overlap peaks, and that the

GPD and KW functions can approximate the tail of the empirical frequency distribution of specific binding events (after filtering out nonspecific clusters) efficiently. Fit parameters are represented in Table 3. The estimates presented in Table 3 suggest that only a relatively small fraction of real BSs in all these experiments can be identified within the experiments. Notice that, we [4, 12, 14, 20] and others [3, 8, 15, 22, 23] have shown that in combination with motif search procedures (e.g., [6]) and expression data analysis, many low-abundance ChIP-defined binding events can be considered the TF-specific BSs.

It is possible that there are many low-abundant and moderate-abundant binding events (BEs) that cannot be considered true binding sites due to the high level of noise in ChIP detection, as well as the nonuniform distribution of those BEs. Even for singleton PET data, the fraction of ChIP-PCR-confirmed specific BSs was 10–20% of all observed single sequences mapped on the genome [9, 12].

In general, larger clusters (or cluster peaks) tend to map more specific binding sites [4, 19, 20] (*see* also Fig. 8; Tables 2 and 3). This property was observed for p53, Statl, c-Myc ChIP-PET TF binding loci in combination with microarray expression data analysis and by a combination of direct qPCR measurements with motif search analyses [4, 12, 14, 15, 20]. Those combined data analyses, including validation of limited random subsets of PET clusters with ChIP-qPCR, show that specific loci for these TFs also can be determined in the corresponding PET datasets even in the smallest of clusters (PET-2 peaks) and in singletons [12]. Moderate-sized and large clusters can contain a relatively small fraction of nonspecific clusters. However, these clusters can be accurately validated in combination with other (mentioned above) independent methods [4, 9, 12].

5.6 Sensitivity and Estimates of the Total Number of TFBSs in ChIP-PET and SACO Experiments

To estimate the number of specific BSs, we begin with an estimation of the specific BS events in a highly noisy enriched region of the distribution function in Fig. 8. For instance, for INF-γ-induced STAT1 binding data, this region includes peak values 5, 4, 3, 2, 1 (Table 2; Fig. 8b). For such estimation, the best-fit GPD function can be "back extrapolated" to the highly noisy enriched region of the distribution. We then identify the dynamical subregion of the random variable which can be used for fitting via the KW function. According to this method, the entire GPD function was accurately fitted using the KW distribution function, Eqs. 72 and 73 (discontinued line; Fig. 8b), with model parameters $\theta = 0.997$, $a = 4.341$; $b = 8.757$. Finally, using the best-fitted Eq. 73, we can estimate the fraction of undetected specific BSs of the STAT1 TF $p_0 = (1 - a/b) = 0.50$. Empirical distribution and KW fitting results in ERE ChIP-PET experiments with $p_0 = 0.77$. Empirical distribution and GPD fitting results in SREB SACO experiments with $p_0 = 0.56$. Table 3 provides detailed information about the identified statistical distribution. For example, the total number of specific SACO-defined SREB BSs in the rat genome is 15,438, and the total number of specific PET-defined BSs for ERE and STAT1 in the human genome is 26,350 and 12,252, respectively. These estimates are consistent with those of previous publications [3, 13–15, 19, 20, 22, 23].

5.7 ChIP-seq DNA Fragments for Nanog and Oct4 TF Binding in Mouse Embryonic E14 Cells

Nanog and Oct4 are key regulatory genes involved in self-renewing and development of mouse and human embryonic stem cells [9, 13]. Identification of the binding avidity EFD for these TFs is carried out in this section.

The DNA fragments have been mapped onto the reference mouse chromosomes [13] and collected in the T2G libraries [37]. In our analysis of a ChIP-seq dataset, a random variable X represented the count of maximum DNA sequences overlapped in a given genome region and identified as the isolated DNA fragment density region with peak region height at $X = m > c$. We used the peak region height as a measure of the TF–DNA binding intensity (or avidity). We analyzed the peak region height EFD and quantified the peak region height of the 200-nt ChIP-seq DNA fragments for Nanog and Oct4 TF binding in the genome of mouse embryonic E14 cells.

Let c denote the critical cutoff value of the peak region height. This value provides the requested specificity (0.95%) of the biding intensity (quantified by the peak region height). According to previous publications, we selected $c = 11$ DNA fragments for Nanog and $c = 8$ DNA fragments for Oct4 binding quantity events [13]).

Let M_{c+} and N_{c+} denote the numbers of DNA fragments and the number of peak regions in the datasets in the NGS sequence library, respectively. At confidence cutoff values defined on the EFD, these numbers provide the subsets of most likely TF–DNA specific binding DNA fragments and the peak region height values above the critical cutoff, respectively. Thus, N is the subpopulation of preferentially true positive peak regions on the genome and M is the subpopulation of DNA fragments in these preferentially true positive peak regions of the genome. For Nanog TF dataset: $M = 468{,}156$; $N = 52{,}004$; Oct4 TF peak dataset: $M = 84{,}810$; $N = 19{,}160$.

Both datasets show a similar shape of the EFD of binding events (peak region heights). We used the KW distribution in Eqs. 72 and 73 to describe the distribution of specific binding events. Equations 77 and 78 allow us to fit the data well to the left-side and right-side truncated empirical distributions. In our goodness-of-fit analysis, we carried out back extrapolation of the best-fitted truncated KW function into the left-sided high-noise region of the EFD (to the left part of the distribution with height values between 1 and the cutoff value of specificity).

Moreover, using Eqs. 72 and 73, we can estimate what fraction of specific BSs has not been detected in the ChIP-seq experiments (p). Estimated parameters of Eqs. 72 and 73 for Nanog TF data are the following: $\theta = 0.997$; $\alpha = 5.870$; $\beta = 7.465$. Thus, by Eq. 73, $p = 0.22$. Estimated parameters for Oct4 data are $\theta = 0.998$; $\alpha = 5.681$; $\beta = 8.32$. Thus, by Eq. 73, $p = 0.32$. Finally, we estimated the total number of specific BSs in the mouse genome for a given TF. This estimate equals ~66.7 × 10 BSs for the Nanog TF and ~28.2 × 10 BSs for Oct4.

Figure 9 shows the results of our goodness-of-fit analysis for Nanog-DNA binding events in the genome of mouse embryonic E14 cells. Our mixture distribution function of TF–DNA binding events in high-throughput experiments [3] specifies three of the most important zones in the dynamical ranges of the events for data analysis. As for other datasets, the KW distribution accurately identifies the specific zone of the right tail of the distribution that comprises the most intense TF–DNA binding events. In total, our modeling and analysis leads to extension of the high-confidence

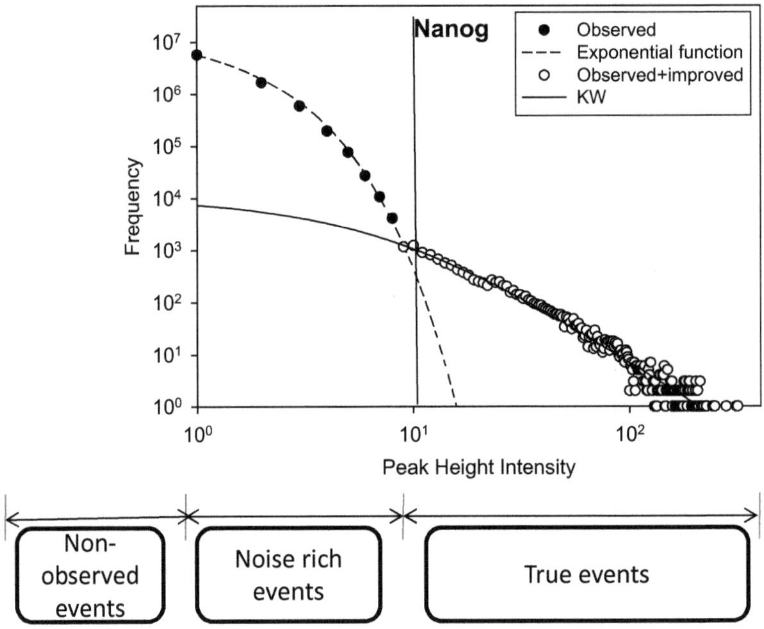

Fig. 9 Three zones in the EFD in NGS experiments. Our mixture PF (Eq. 3) specifies the three most important zones for data analysis. The KW distribution accurately identifies the specific statistical properties for Nanog-DNA binding events in the genome of mouse embryonic E14 cells. Goodness-of-fit analysis of the frequency distribution of binding events (# ChIP-seq peak region heights) fitted starting from the cutoff peak value defined in [13]. For these data, the KW distribution fits to the observed right tail (reliable part of the mixture frequency distribution), as well as the best-fit truncated GPD extrapolated to the smaller peak values, including value 1 (belonging to the highly noise-rich subset). The KW function fits to the best-fit estimated GPD function values on the entire dynamical range of peak values. Due to the high confidence of curve-fitting of the KW distribution, the parameters of the KW function were accurately estimated. Vertical line: threshold value $c = 11$

zone of TF–DNA binding events to the moderate and low-avidity zones. As result, we can provide identification of the entire EFD of specific binding events even in the noise-rich zone. This predictive ability of the KW distribution makes our method practically reasonable for the quantitative analysis of the binding profiles of Nanog, Oct4, and other DNA-binding proteins in the NGS experiments.

5.8 Sample-Size Dependence Analysis Can Define Rescaling in Binding Mechanisms of ChIP-Based Methods

ChIP-seq uses a much larger number of sequence reads than other ChIP-based methods. Therefore, the examination of ChIP-seq TF-binding sites might more adequately represent the true complexity and diversity of TF-binding patterns. We suggest that due to this advantage, the empirical distribution of the ChIP-seq dataset allows us to find more diverse/complex patterns of STAT1–DNA binding than ChIP-PET and other ChIP-based methods.

We hypothesized the existence of the "low-avidity" and "high-avidity" subsets of binding events. We suggested that two distinct specific distributions of STAT1 binding should be observed in a ChIP-seq experiment. Figure 10 shows that a mixture model of two KW functions fits well to the empirical frequency distribution of STAT1–DNA binding events. ChIP-seq binding events are represented by a number of XSETs. Detailed notations and estimated parameter values of the model are presented in Table 3. Table 4 shows that, in INF-γ-stimulated cells, the high-avidity STAT1 BSs in HeLa S3 cells are strongly associated with the location of canonical motifs (82% BS association), while small a number of relatively high-avidity binding sites (64 BSs) also exists in a population of unstimulated cells, and further that 47% ($30/64 \times 100\%$) of these BSs are supported by canonical motifs. A large number of stimulated and unstimulated HeLa S3 cells exhibit low STAT1-BS avidity.

We also compared ChIP-PET and ChIP-seq detection of high-avidity binding sites of the STAT1 TF in INF-γ-stimulated and unstimulated HeLa S3 cells (Fig. 8a). ChIP-PET binding events are presented by the peaks that correspond to the number of overlapping PET DNA sequences; ChIP-seq binding events are represented by XSETs. For stimulated HeLA S3 cells, the ChIP-seq and ChIP-PET data are consistent in chromosome location (common loci) and the relative value of the TF–DNA binding events. However, the ChIP-PET dataset (library size being much smaller than that of the ChIP-seq dataset) lost essential information about truly existing binding sites in the genome of the stimulated and unstimulated cells. The fraction and height values of specific BEs in the ChIP-seq dataset are expressed much more strongly and are often more statistically significant than in the ChIP-PET dataset. Using a 95% specificity cutoff value of 31 eSETs, we suggest that ChIP-seq data for unstimulated HeLa S3 cells can be used to detect 64 high-confidence specific STAT1 BSs in unstimulated HeLa S3 cells (Figs. 10 and 11).

We found 45 RefSeq genes predicted as possible direct targets for "basal" transcription regulation by STAT1 in INF-γ-unstimulated cells including interferon regulatory factor 9 (IRF9), ATP-dependent DNA helicase homolog (*Saccharomyces cerevisiae* (HFM1)), polyamine-modulated factor 1 (PMF1), WD repeat domain 74 (WDR74), polymerase (DNA directed), delta 1 catalytic subunit 125 kDa (POLD1), Huntingtin Associated Protein 1 (HAP1) encoding the protein that interacts with huntingtin, with two cytoskeletal proteins (dynactin and pericentriolar autoantigen protein 1), and with a hepatocyte growth factor regulated tyrosine kinase substrate playing a role in vesicular trafficking or organelle transport, substrate recognition component of a SCF (SKP1-CUL1-F-box protein) E3 ubiquitin-protein ligase complex (SKP2) which produces mediators to the ubiquitination and subsequent proteasomal degradation of target proteins involved in cell cycle progression, signal transduction, and transcription (Table 5). This table also includes poorly described and in silico predicted genes such as LOC284801.cApr07, kloypeybo.aApr07, RNU2P2.cApr07, LOC554226.bApr07. The listed genes provide the important resource for study of the role STAT1 plays in regulation of transcription and the functions of these and other genes of Table 5.

We would like to indicate that noisy BEs in ChIP-seq and ChIP-PET datasets are increased when data set sample size

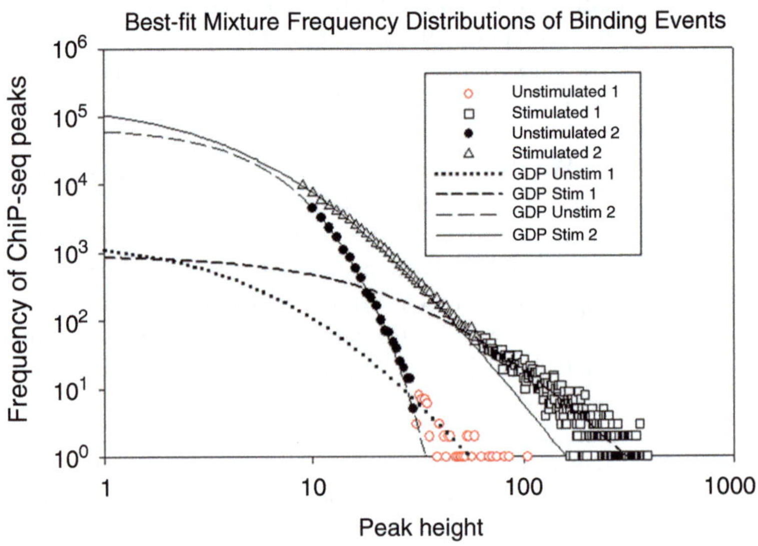

Fig. 10 The frequency distribution of INF-γ-stimulated and unstimulated STAT1-DNA binding events in the ChIP-seq experiment [23]. We consider the empirical distribution as a mixture of distributions of two distinct classes of binding events. Two KW functions fit well to the empirical frequency distribution of STAT1-DNA binding events due to a decomposition of the mixture of "low-avidity" and "high-avidity" binding events. ChIP-seq binding events are represented by XSETs. *See* detailed notations on the graphic and the estimated parameter values of the model in Tables 3 and 4

Table 4
STAT1 motif frequency in high- and low-avidity STAT1 BSs in the genome of the INF-γ-stimulated and nonstimulated HeLa cells

Subset of STAT1 BSs	Avidity	Subset of BS	# Loci	STAT1 motifs	Fraction of BSs
Stimulated	High	Peak height ≥ 61	2322	1898	0.82
Stimulated	Low	Peak height < 61	60,134	23,653	0.39
Un-stimulated	High	Peak height ≥ 31	64	30	0.47
Overlapped with stimulated BSs	Low	Peak height < 31	14,239	4452	0.31
Non-overlapped with stimul. BSs	Low	Peak height < 31	1440	282	0.20

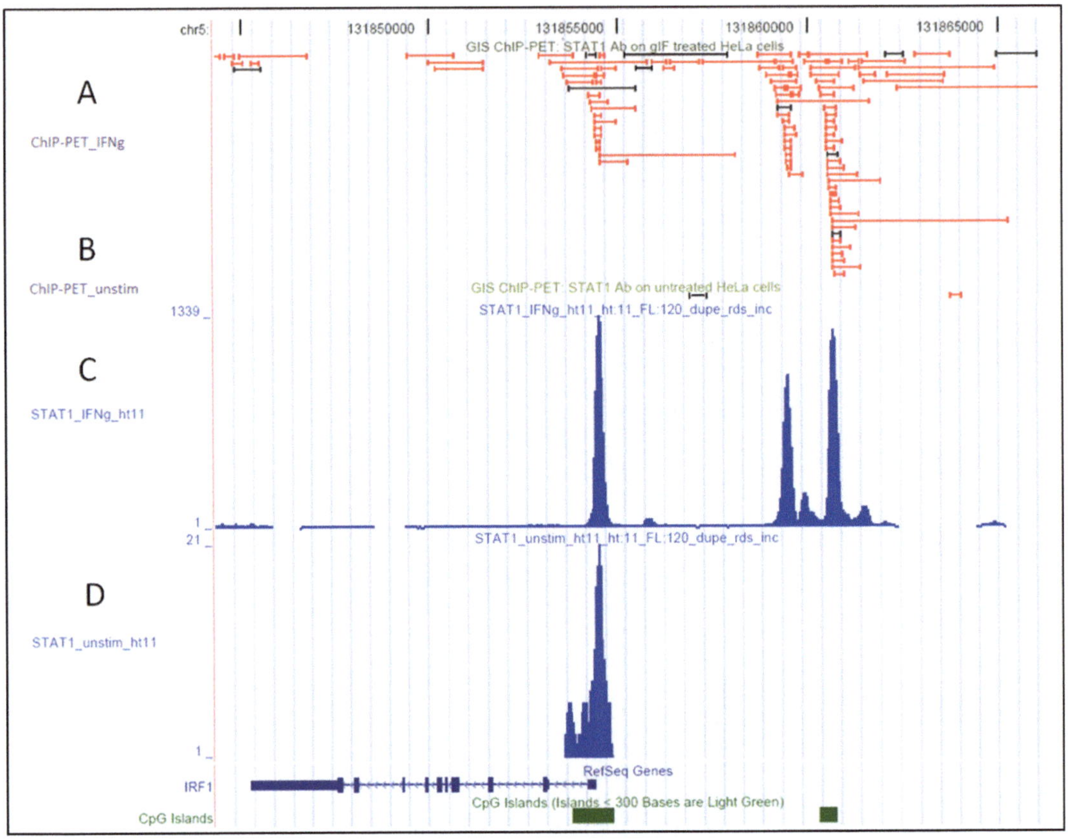

Fig. 11 Sample-size dependence of ChIP-based methods. Comparison of ChIP-PET [15] and ChIP-seq [23] detection of high avidity binding sites of the STAT1 TF in INF-γ-stimulated (**a**, **c**) and unstimulated HeLa S3 cells (**b**, **d**). For stimulated cells, the data provides strong consistency in chromosome location (common loci) and relative amplitude of the signals. However, the Chip-PET dataset lost essential information about truly existent binding sites in the genome of unstimulated cells. Chip-PET binding events are presented by the peaks corresponding to the number of overlapping ditags at a given genomic coordinate. ChIP-seq binding events are represented by XSETs

becomes larger. This common property of ChIP-based sequencing methods is still a complex challenge to resolve in order to achieve reliable detection of true low-avidity BS events even in a very large sample size.

Figure 12 provides useful information regarding the comparative sensitivity of ChIP-seq and ChIP-PET analyses. In our meta-analysis of colocalization of STAT1-binding sites, both methods provided a similar shape for binding site distributions near transcription start sites (TSSs) of a gene. However, spatial variation of the BSs in the ChIP-seq experiment is relatively smaller than in the ChIP-PET experiment (Fig. 12a, c). Consistency between frequencies of ChIP-PET and ChIP-seq binding events approaches 100% at values PET-6 (solid line, Fig. 12d) and above (PET6+).

Table 5
Chip-seq-defined high-avidity STAT1 TF BSs (peak heights >30 DNA fragments) in unstimulated HeLa S3 cells found in 50 kb upstream transcription start site (TSS) region and in downstream gene regions

Cluster	Max height in cluster (un-stimulated cells)	Chr coordinate	RefSeq	Gene symbol	Strand	Distance btw TSS of gene and start of cluter, nt
56	32	chr5:32,803,910–32,804,468	NM_000908	NPR3	+	−56,488
37	35	chr5:37,448,874–37,449,213	NM_018034	WDR70	+	−33,705
51	33	chr5:17,302,178–17,302,751	NM_006317	BASP1	+	−31,428
5	71	chr1:154,452,793–154,453,347	NM_014655	SLC25A44	+	−22,439
5	71	chr1:154,452,793–154,453,347	NM_001135672	SLC25A44	+	−22,439
31	39	chr5:134,237,616–134,288,360	NM_032151	PCBD2	+	−18,907
38	35	chr17:35,390,245–35,390,634	NM_178171	GSDMA	+	−17,493
1	105	chr14:23,700,051–23,700,428	NM_017999	RNF31	+	−13,552
52	33	chr11:47,556,913–47,557,379	NM_175732	PTPMT1	+	−13,203
44	34	chr12:52,131,714–52,132,136	NM_001005354	PRR13	+	−10,014
44	34	chr12:52,131,714–52,132,136	NM_018457	PRR13	+	−10,014
11	55	chr11:62,355,421–62,365,972	NM_003164	STX5	–	−9284
33	36	chr17:30,502,100–30,502,605	NM_001014445	NLE1	–	−3664
33	36	chr17:30,502,100–30,502,605	NM_018096	NLE1	–	−3664
5	71	chr1–:154,452,793–154,453,347	NM_007221	PMF1	+	−3385
33	36	chr17:30,502,100–30,502,605	NM_001033576	UNC45B	+	−3151
33	36	chr17:30,502,100–30,502,605	NM_173167	UNC45B	+	−3151
29	40	chr5:40,870,875–40,871,529	NR_002583	SNORD72	–	−2280
49	33	chr5:10,363,449–10,364,226	NM_138809	CMBL	–	−2280

(continued)

Table 5
(continued)

Cluster	Max height in cluster (un-stimulated cells)	Chr coordinate	RefSeq	Gene symbol	Strand	Distance btw TSS of gene and start of cluter, nt
11	55	chr11:62,365,421–62,365,972	NM_018093	WDR74	–	−1216
22	45	chr17:37,144,358–37,145,318	NM_001079871	HAP1	–	67
22	45	chr17:37,144,358–37,145,318	NM_177977	HAP1	–	67
22	45	chr17:37,144,358–37,145,318	NM_001079870	HAP1	–	67
53	33	chr15:47,235,030–47,235,368	NM_001143887	COPS2	–	117
53	33	chr15:47,235,030–47,235,368	NM_004236	COPS2	–	117
57	32	chr5:36,187,640–36,188,151	NM_001007527	LMBRD2	–	133
52	33	chr11:47,556,913–47,557,379	NM_016506	KBTBD4	–	185
1	105	chr14:23,700,051–23,700,428	NM_006084	IRF9	+	211
52	33	chr11:47,556,913–47,557,379	NM_018095	KBTBD4	–	231
52	33	chr11:47,556,913–47,557,379	NR_024222	KBTBD4	–	231
53	33	chr15:47,235,030–47,235,368	NM_001001556	GALK2	+	238
42	34	chr1:143,807,525–143,807,996	NM_004892	SEC22B	+	239
39	35	chr19:55,571,256–55,571,614	NM_007121	NR1H2	+	241
43	34	chr1:171,953,450–171,950,995	NM_014458	KLHL20	+	253
29	40	chr5:40,870,875–40,871,529	NM_000997	RPL37	–	270
52	33	chr11:47,556,913–47,557,379	NM_004551	NDUFS3	+	295
41	34	chr1:93,317,075–93,317,639	NM_007358	MTF2	+	305
57	32	chr5:36,187,640–36,188,151	NM_005983	SKP2	+	306
57	32	chr5:36,187,640–36,188,151	NM_032637	SKP2	+	306

(continued)

Table 5
(continued)

Cluster	Max height in cluster (un-stimulated cells)	Chr coordinate	RefSeq	Gene symbol	Strand	Distance btw TSS of gene and start of cluter, nt
48	33	chr1:161,558,037–161,558,717	NM_031423	NUF2	+	310
48	33	chr1:161,558,037–161,558,717	NM_145697	NUF2	+	310
38	35	chr17:35,390,245–35,390,634	NM_002809	PSMD3	+	341
34	35	chr2:74,535,334–74,535,335	NM_031288	INO80B	+	373
45	34	chr17:46,585,535–46,586,093	NM_000269	NME1	+	384
45	34	chr17:46,585,535–46,586,093	NM_198175	NME1	+	384
45	34	chr17:46,585,535–46,586,093	NM_001018136	NME1-NME2	+	384
44	34	chr12:52,131,714–52,132,186	NM_031989	PCBP2	+	439
44	34	chr12:52,131,714–52,132,186	NM_005016	PCBP2	+	439
44	34	chr12:52,131,714–52,132,186	NM_001098620	PCBP2	+	439
44	34	chr12:52,131,714–52,132,186	NM_001128914	PCBP2	+	439
44	34	chr12:52,131,714–52,132,186	NM_001128913	PCBP2	+	439
44	34	chr12:52,131,714–52,132,186	NM_001128912	PCBP2	+	439
44	34	chr12:52,131,714–52,132,186	NM_001128911	PCBP2	+	439
59	32	chr12:52,355,841–52,356,497	NM_005176	ATP5G2	–	536
24	44	chr5:324,159–324,820	NM_013232	PDCD6	+	567
59	32	chr12:52,355,841–52,356,497	NM_001002031	ATP5G2	–	939
54	32	chr5:5,474,861–5,475,820	NM_015325	KIAA0947	+	946
34	35	chr2:74,535,334–74,535,835	NM_012477	WBP1	+	3751
29	40	chr5 40,870,875–40,871,529	NM_032587	CARD6	+	6292
39	35	chr19:55,571,256–55,571,614	NM_002691	P0LD1	+	8149

(continued)

Table 5
(continued)

Cluster	Max height in cluster (un-stimulated cells)	Chr coordinate	RefSeq	Gene symbol	Strand	Distance btw TSS of gene and start of cluter, nt
34	35	chr2:74,535,334–74,535,835	NM_006302	GCS1	–	10,712
17	51	chr17:38,821,136–38,822,164	NM_001661	ARL4D	+	10,743
2	35	chr1:91,625,230–91,625,782	NM_001017975	HFM1	–	17,735
36	35	chr5:16,947,382–16,948,205	NM_012334	MYO10	–	42,004

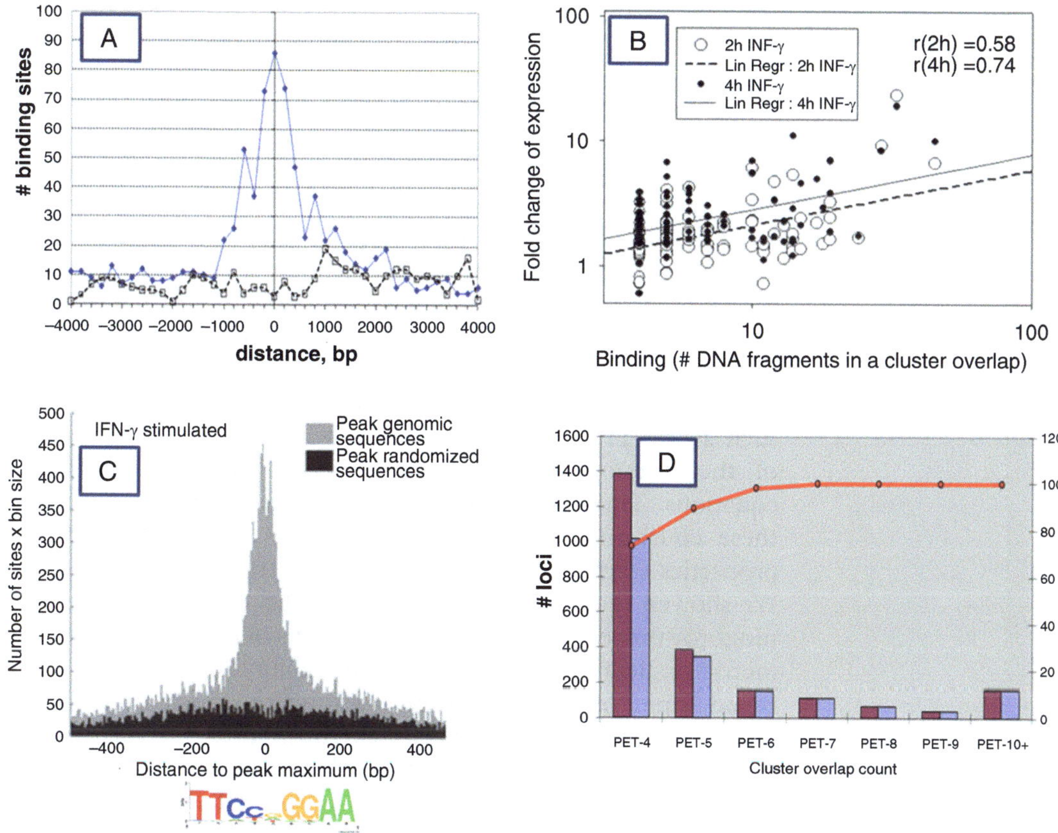

Fig. 12 Comparison analysis of the results of ChIP-seq and ChIP-PET datasets. (**a**) ChIP-PET binding site distributions near a transcription start site (TSS) of a gene. (**b**) Spearman correlation between relative avidity (peak height in the ChIP-PET experiment) and expression signal value (time-course expression microarray Hartmann's data [3]). (**c**) ChIP-PET and ChIP-seq binding site distributions near transcription start sites (TSSs) of a gene have a similar shape; however, variation of the distribution in the ChIP-seq experiment is better compared to ChIP-PET. (ChIP-seq data in [23]). (**d**) Consistency between frequencies of ChIP-PET and ChIP-seq binding events approaches 100% (*solid line*) at values PET-6 and above

5.9 The ChIP-Based Approach Allows Measuring a Relative Avidity Function

Interestingly, we determined a strong correlation between peak height in STAT1 ChIP-PET experiments and expression signal value (defined by microarray Hartmann's data; [76]) in putative gene targets located in the vicinity of 2 kb from the identified STAT1 BS. Figure 12c shows that this correlation, estimated by the Spearman coefficient correlation, presents in time-course microarray experimental design (0, 2, and 4 h). These observations suggest the probability of binding events in ChIP-PET clusters located in a canonical promoter region can reflect a relative avidity of STAT1 binding and increase the probability of regulation of the transcription of STAT1 direct gene targets.

6 Discussion and Conclusion

The stationary distributions of the birth–death stochastic processes, which are skewed to the right, can be used as explanatory models of commonly observed EFDs of diverse biological phenomena taking place in large-scale evolving systems [9, 12, 19, 20, 34, 51–53, 55, 59, 62, 64, 65, 69]. In this work, probabilistic models of the TF–DNA binding site events on the genome scale and an algorithm for the accurate identification of corresponding frequency distributions of TF binding events from IP-based sequence read clustering and mapping on the genome were developed. The probabilistic mechanisms of the dynamics of large-scale bimolecular systems such as the TFBS virtual network can be modeled with the help of the Kolmogorov birth–death process differential-difference equations. This is done by deriving the steady-state solutions to these equations with various assumptions regarding functional properties of the process coefficients and transition probabilities. We showed that the steady-state probability function of the Kolmogorov process can be described in terms of a series of hypergeometric or Beta functions and the probability function includes a broad family of skewed distribution functions, including but not limited to Waring, GPD, several well-known power law-like probability functions, and scale-free network models.

For a given biological data and NGS method, selection and fitting of appropriate probabilistic model of the biological phenomena is often a difficult problem. We have specified our probabilistic model in the context of analyzing different TF interactions with their specific binding loci, cell types, and conditions defined by different NGS methods. Our implementation of the KW process model to TF binding, based on simple but reasonable binding–dissociation assumptions, leads to a uniform explanation for experimental TF–DNA binding avidity distributions at mammalian genome scale, providing estimates of the total number of BSs for a given TF in contexts of cell types and conditions. Additionally, our model provides a simple, robust, quantitative, and visual method for estimating the sensitivity, specificity, and accuracy of any NGS-based experimental data in which a skewed form of the distribution of events is observed.

We further demonstrated that the empirical distributions for all the studied ChIP-based datasets are well fitted by a mixture model with specific components of BEs described by the skewed Pareto-like distribution function subfamilies (including Waring, Yule, Pareto, and other known skewed distributions), whose shape depends in a predictable manner on the sample size [19]. Such distributions can be generated as limiting/critical distributions of the KW random birth–death process, where the birth and death intensities are linear functions of (binding) events [30]. The power law for the analysis of ChIP-seq data was also used in [20, 32, 33, 42] and supports these findings.

Actually, the skew form of the EFD is consistent with the well-known fact that TFBSs of moderate and low predicted affinity have many sequence variants of similar quality (noncanonical c-Myc E-boxes), whereas the highest-affinity motifs have far fewer alternatives of similar quality (e.g., canonical c-Myc E-boxes) [34]. This means that there are many more ways to be a weak avidity-binding site than a strong binding site at the genome scale. This is further testable.

Our quantitative TF–DNA binding activity model [3], including the KW distribution, allows us not only to identify a common statistical law (Figs. 8–10) for specific TF–DNA binding, but also to estimate the number and fraction of specific BSs for a particular TF in different ChIP-based experiments even if the dataset is essentially incomplete and enriched with noisy events.

Note that, by looking at the experimentally resulting mixture distribution of peak region heights, a researcher can visually evaluate the overall success of the chromatin immunoprecipitation experiment and the depth of sequencing: the larger the difference between the two distributions, the more specific was the antibody and the DNA fragment numbers mapped onto a genome. Decomposing the mixture distribution function based on the visual analysis of the frequency distribution curvature slope change can provide the location of the proximal critical cutoff value, separating a fraction of the significant putative specific TFBS (represented by the peak height values) higher than the cutoff value.

The sensitivity and the specificity of our mixture model [3] are demonstrated by applying the model to analysis of different ChIP-seq datasets across different platforms. In this work, we developed the probabilistic model of TF–DNA binding and binding statistics of five biologically essential and well-characterized human transcription factors: ERE (estrogen receptor-α), CREB (cAMF-response element), Nanog (Nanog homeobox), Oct4 (POU class 5 homeobox 1), STAT1 (signal transducer and activator of transcription protein 1). By our estimates, the number of BSs in the genome is 66.7×10^3, 28.2×10^3, 15.5×10^3, and 26.4×10^3 for the Nanog (ChIP-seq), Oct4 (ChIP-seq), CREB (SACO), and ERE (ChIP-PET) TFs respectively. For STAT1, the number of BSs in the human genome are 12.25×10^3 (ChIP-PET), 17.66×10^3 (ChIP-seq; high-avidity BS), and 5.99×10^3 (ChIP-seq; low-avidity BS). Our goodness-of-fit analysis of the mixture probability function Eq. 3 leads to prediction of a novel subset of STAT1 BS in unstimulated cells.

These findings provide insight into the field of basal transcription machinery and predict new gene targets for direct STAT1 transcription control.

The frequency distribution of gene expression in human and many other complex cell types is skewed; the long right tail usually decreases when gene expression is increased. Most genes have very low expression levels, whereas a few genes have high expression levels. We found that all observed large-scale gene expression datasets follow a Pareto-like distribution model, skewed by many low-abundance transcripts [19, 30, 60, 66]. These findings in combination with our knowledge about the driving role of TFBS binding events in the transcription process and the KW distribution of the TF–DNA avidity suggest the predictable correlations between TFBS avidity and gene expression patterns. Our results of the correlation analysis for STAT1 TFBS avidity and the expression of their strong target genes support this hypothesis.

It is important to note that evolutionary enhancer sequence alignment analysis suggested that, although many predicted low-affinity sites are poorly conserved, overall, TF occupancy on an enhancer may be maintained despite significant sequence turnover. This may occur either through the rapid gain and loss of individual sites, or through the maintenance of relatively weak binding affinity at a site that is unstable at the level of the DNA sequence [77–79]. While this last idea requires further direct testing, it is consistent with the fact that Gli sites of moderate predicted affinity have many sequence variants of similar quality, whereas the highest-affinity motifs have far fewer alternatives of similar quality [65]. This has led the authors to conclude that low-affinity TF–DNA interactions, mediated by nonconsensus and often poorly conserved sequence motifs, could play important and widespread roles in developmental patterning and cis-regulatory evolution, and therefore can be tested.

We determined that the specificity of all the studied experiments was high (91–99%). These results are consistent with published estimates of the specificity of ChIP-based methods [4, 13, 16, 22, 23]. The BE sensitivity issue raises the issue of how many physical BSs, including low-avidity BSs, are present in the genome of a given cell under given environmental conditions. Sensitivity is difficult to assess experimentally because of the lack of reliable bench markers and methods for detection of low binding avidity. We used a computational approach to estimate the sensitivity of ChIP-based experiments. By our estimations, the sensitivity of all ChIP-based methods is low: 6.3%, 4.8%, 10.2%, 4.6% for Nanog (ChIP-seq), Oct4 (ChIP-seq), CREB (SACO), and ERE (ChIP-PET), respectively; and 4.36% (ChIP-PET), 12.1% (ChIP-seq, high-avidity BS); 0.97% (ChIP-seq; low-avidity BS) for the STAT1 TF (Table 4; Fig. 9). The surprisingly low sensitivity levels of the current ChIP-based sequencing methods for the identification of TF–DNA binding can be associated with (1) a large fraction of noisy sequences forming low- and moderate-avidity binding events and (2) missing specific ChIP-derived sequences (not-detect sequences) due to the limited sample size of the experimental dataset and/or suboptimal design of experiments. The efficiency of sequencing (the percentage-specific DNA sequences at a given specificity cutoff peak, Table 4), defined by qPCR-ChIP, was also low: 3.6%, 1.0%, 8.8%, 50.3%, and 0.54% for Nanog, Oct4, ERE, SREB, and STAT1 (ChIP-PET) TFs, respectively.

We can conclude that although ChIP-seq is a powerful technique, nevertheless, it still produces essentially incomplete and noise-rich data, underrepresenting the low- and moderate-avidity DNA–protein binding events of TFs in complex genomes. Ten years after Chip-based sequencing technology became available, it still remains largely unclear how the Paired-end ditag (PE) and single-end (SE) (in which only one end of the fragment is sequenced) designs and long and short reads influence alignment/mapping rates and accuracy, coverage of repetitive elements, sensitivity and specificity in peak calling, and allele-specific binding detection [34].

While a higher number of reads may increase sensitivity and resolution, it may also increase the fraction of noise sequence reads. In fact, subsampling in all ChIP-seq and ChIP-PET datasets showed that the noise component increases when the dataset sample size becomes larger. This common property of ChIP-based sequencing methods is a dilemma in the reliable detection of real BSs even with very large samples. In this case, other factors, such as the specificity of antibodies, optimal (shorter/homogeneous) length of ChIP DNA fragments, and better computational processing of raw data, have a direct impact on the sensitivity.

Similar to what was described in [12, 20, 42], we suggest that the distance of BS from gene transcription start site influences the distribution of relative avidity for binding (STAT1–DNA binding; Fig. 12a–c).

Thus, we would like to propose the importance of the following steps for the further optimization of ChIP-seq analysis: (1) an adequate experimental design (assuming biological replication), standardization and optimization (assuming a design of appropriate positive and negative controls); (2) automatic, high-quality control methods of ChIP-derived sequences; (3) nonredundant mapping of the sequences onto complex genomes; (4) specification of filtering and clustering procedures of ChIP-derived DNA sequences in different regions of the chromosomes (pericentromere, low-complexity, repeat regions, etc.); (5) adequate statistical methods to define a binding event (e.g., cluster peak height and cluster overlap span); (6) minimization of signal-to-noise ratio; (7)

adequate controls and statistical modeling of background (noise) signals; and (8) deep data sampling to improve the sensitivity of ChIP-based experiments, allowing the detection of low-affinity genomic binding events which are frequent and could be functionally relevant.

Genome-wide assay development for direct TF–DNA binding avidity is important. Recently, the apparent dissociation constant (K_d) for specific binding of glucocorticoid receptor (GR) and androgen receptor (AR) to DNA was determined in vivo in Xenopus oocytes [59]. The total nuclear receptor concentration was quantified as the amount of specifically retained [(3)H]-hormone in manually isolated oocyte nuclei. DNA was introduced by nuclear microinjection of single stranded phagemid DNA. Chromatin was then formed during second strand synthesis. The fraction of DNA sites occupied by the expressed receptor was determined by dimethylsulfate in vivo-footprinting and used for the calculation of receptor-DNA binding affinity. Furthermore, the forkhead transcription factor FoxA1 enhanced DNA binding by GR with an apparent K_d of $\sim$1 μM and dramatically stimulated DNA binding by AR with an apparent K_d of $\sim$0.13 μM at a composite androgen-responsive DNA element containing one FoxA1 binding site and one palindromic hormone receptor binding site known to bind one receptor homodimer. It was shown that FoxA1 exerted both weak constitutive- and strongly cooperative DNA binding with AR but had a less prominent effect on GR, the difference reflecting the licensing function of FoxA1 at this androgen-responsive DNA element. This study provides a new way to carry out more direct detection of TF–DNA avidity in a given cell and also includes in the analysis of the interactions between transcription factors and avidity in these dynamic and complex processes.

Note that, during developmental process, acute phase of cell stress response, severe medical conditions associated with genome instability, transcriptome alterations and cellular reprogramming processes, the form of the scale-dependent skewed frequency distribution function of the associated events can be changed significantly and provide important biologically significant predictions [19, 68]. In cell development or pathologic transformation, cellular properties and regulatory pathways change globally in time and the stochastic processes of TF–DNA binding and gene transcription may significantly deviate from initial steady-state conditions. Analysis of the time-dependent transformation of the statistical distributions of transcription machinery, including TF–DNA binding events, is of great interest and a challenge for future studies. In such studies, stochastic process models and wide classes of skewed distributions associated with these models could be useful.

The distribution functions derived based on birth–death stochastic process models have been developed and used for a long time [20, 34, 36, 51–53, 55, 59, 62, 68–70, 75]. In the last several years, a group of authors has developed new families of GHFs, generated from the Kolmogorov steady state birth–death stochastic process [51–53]. By limiting with several empirical statistical facts, they elaborated novel families of the generalized hypergeometric distributions. Specifically, a multiparametric family of stationary distributions of stochastic processes, called the Regularly Varying Generalized Hypergeometric Distribution of the Second Type (GHS), has been proposed and studied [53]. A subfamily of the GHS that varies regularly at infinity and exhibits asymptotically constant slowly varying component decreases, is log-downward convex and unimodal. Such features can be observed in diverse large biomolecular evolving system datasets. It is imperative to identify the members of these regularly varying distribution functions which fit to the EFDs of the TF–DNA binding events. Overwise, only few subfamilies of the skewed PFs have been associated with stochastic birth-death mechanisms and fitted appropriately used biological data and biomolecular processes genome-wide. Further works on estimation for the families/subfamilies of these explanatory-relevant models is clearly needed.

New stochastic process models of evolving biomolecular systems could provide a powerful and unbiased statistical basis for analysis of CHIP-seq and other NGS methods. An integrative genomics strategy can form critical links between transcription regulation or diverse gene expression patterns and cellular phenotypes or cell functions. However, due to the high complexity of biological systems in general, technologic limitations, and experimental biases, as well as different types of data noise, the goodness-of-fit analysis results and statistical conclusions are requiring independent experimental validations [19, 20, 34, 43].

On the other hand, accurate and reproducible mapping of the regulatory sequences combined with functional tests and stimuli response experiments at the level of homogenous cell populations and individual cells is of significant benefit for the identification of more mechanistic, informative, testable and predictive probabilistic and biological models of the TF–DNA binding process acting at the genome and cellular phenotype scales.

Acknowledgments

I would like to express my special thanks to Yuri Nikolsky and Tatiyana Tatarinova who encouraged me to develop and carry out this study. This work was supported by Bioinformatics Institute/A-STAR, Singapore.

References

1. Ren B, Robert F, Wyrick JJ, Aparicio O, Jennings EG, Simon I, Zeitlinger J, Schreiber J, Hannett N, Kanin E et al (2000) Genome-wide location and function of DNA binding proteins. Science 290:2306–2309
2. Kim TH, Ren B (2006) Genome-wide analysis of protein-DNA interactions. Annu Rev Genomics Hum Genet 7:81–102
3. Hartman SE, Bertone P, Nath AK, Royce TE, Gerstein M, Weissman S, Snyder M (2005) Global changes in STAT target selection and transcription regulation upon interferon treatments. Genes Dev 19:2953–2968
4. Wei CL, Wu Q, Vega VB, Chiu KP, Ng P, Zhang T, Shahab A, Yong HC, Fu Y, Weng Z et al (2006) A global map of p53 transcription-factor binding sites in the human genome. Cell 124:207–219
5. Stormo GD (2000) DNA binding sites: representation and discovery. Bioinformatics 16:16–23
6. Down TA, Hubbard TJ (2005) NestedMICA: sensitive inference of over-represented motifs in nucleic acid sequence. Nucleic Acids Res 33:1445–1453
7. Lovegrove FE, Pena-Castillo L, Mohammad N, Liles WC, Hughes TR, Kain KC (2006) Simultaneous host and parasite expression profiling identifies tissue-specific transcriptional programs associated with susceptibility or resistance to experimental cerebral malaria. BMC Genomics 7:295
8. Fernandez PC, Frank SR, Wang L, Schroeder M, Liu S, Greene J, Cocito A, Amati B (2003) Genomic targets of the human c-myc protein. Genes Dev 17:1115–1129
9. Loh YH, Wu Q, Chew JL, Vega VB, Zhang W, Chen X, Bourque G, George J, Leong B, Liu J et al (2006) The Oct4 and Nanog transcription network regulates pluripotency in mouse embryonic stem cells. Nat Genet 38:431–440
10. Boeva V, Lermine A, Barette C, Guillouf C, Barillot E (2012) Nebula—a web-server for advanced ChIP-seq data analysis. Bioinformatics 28:2517–2519
11. Lorenzin F, Benary U, Baluapuri A, Walz S, Jung LA, von Eyss B, Kisker C, Wolf J, Eilers M, Wolf E (2016) Different promoter affinities account for specificity in MYC-dependent gene regulation. Elife 5. pii:15611
12. Zeller KI, Zhao X, Lee CW, Chiu KP, Yao F, Yustein JT, Ooi HS, Orlov YL, Shahab A, Yong HC et al (2006) Global mapping of c-myc binding sites and target gene networks in human B cells. Proc Natl Acad Sci U S A 103:17834–17839
13. Chen X, Xu H, Yuan P, Fang F, Huss M, Vega VB, Wong E, Orlov YL, Zhang W, Jiang J et al (2008) Integration of external signaling pathways with the core transcriptional network in embryonic stem cells. Cell 133:1106–1117
14. Lin CY, Vega VB, Thomsen JS, Zhang T, Kong SL, Xie M, Chiu KP, Lipovich L, Barnett DH, Stossi F et al (2007) Whole-genome cartography of estrogen receptor alpha binding sites. PLoS Genet 3:e87
15. Euskirchen GM, Rozowsky JS, Wei CL, Lee WH, Zhang ZD, Hartman S, Emanuelsson O, Stolc V, Weissman S, Gerstein MB et al (2007) Mapping of transcription factor binding regions in mammalian cells by ChIP: comparison of array- and sequencing-based technologies. Genome Res 17:898–909
16. Impey S, McCorkle SR, Cha-Molstad H, Dwyer JM, Yochum GS, Boss JM, McWeeney S, Dunn JJ, Mandel G, Goodman RH (2004) Defining the CREB regulon: a genome-wide analysis of transcription factor regulatory regions. Cell 119:1041–1054
17. Ozsolak F, Song JS, Liu XS, Fisher DE (2007) High-throughput mapping of the chromatin structure of human promoters. Nat Biotechnol 25:244–248
18. Lieb JD, Liu X, Botstein D, Brown PO (2001) Promoter-specific binding of Rap1 revealed by genome-wide maps of protein-DNA association. Nat Genet 28:327–334
19. Kuznetsov VA (2002) Statistics of the number of transcripts and protein sequence encoded in the genome. In: Zhang W, Shmulevich I (eds) Computational and statistical approaches to genomics, 1st edn. Springer, Boston, MA, pp 125–171
20. Kuznetsov VA, Orlov YL, Ruan Y, Wei CL (2007) Computational analysis of genome-scale avidity distribution of TFBS in ChIP-PET experiments. Genome Inform 19:83–94
21. Boeva V (2016) Analysis of genomic sequence motifs for deciphering transcription factor binding and transcriptional regulation in eukaryotic cells. Front Genet 7:24
22. Johnson DS, Mortazavi A, Myers RM, Wold B (2007) Genome-wide mapping of in vivo protein-DNA interactions. Science 316:1497–1502
23. Robertson G, Hirst M, Bainbridge M, Bilenky M, Zhao Y, Zeng T, Euskirchen G, Bernier B, Varhol R, Delaney A et al (2007) Genome-wide profiles of STAT1 DNA association using chromatin immunoprecipitation and massively parallel sequencing. Nat Methods 4:651–657

24. Barski A, Cuddapah S, Cui K, Roh TY, Schones DE, Wang Z, Wei G, Chepelev I, Zhao K (2007) High-resolution profiling of histone methylations in the human genome. Cell 129:823–837
25. Massie CE, Mills IG (2008) ChIPping away at gene regulation. EMBO Rep 9:337–343
26. Mardis ER (2007) ChIP-seq: welcome to the new frontier. Nat Methods 4:613–614
27. Bhinge AA, Kim J, Euskirchen GM, Snyder M, Iyer VR (2007) Mapping the chromosomal targets of STAT1 by sequence tag analysis of genomic enrichment (STAGE). Genome Res 17:910–916
28. Zhang Q, Zeng X, Younkin S, Kawli T, Snyder MP, Keles S (2016) Systematic evaluation of the impact of ChIP-seq read designs on genome coverage, peak identification, and allele-specific binding detection. BMC Bioinformatics 17:96
29. Johnson NL, Kotz S, Balakrishnan N (1997) Discrete multivariate distributions. John Wiley & Sons, New York, NY
30. Kuznetsov VA (2003) Family of skewed distributions associated with the gene expression and proteome evolution. Signal Process 83: 889–910. Available online 14 December 2002
31. Kuznetsov VA (2006) Emergence of size-dependent networks on genome scale. In: Lecture series on computer and computational sciences, vol 7a. Brill Academic Publishers, Amsterdam, pp 754–757
32. Kuznetsov VA, Singh O, Ng HH, Wei CL (2008) Modelling and prediction of DNA-protein interaction events of transcription factors (TF) in ChIP-seq experiments. In: The sixth international conference on bioinformatics of genome regulation and structure (BGRS'2008). Institute of Cytology and Genetics SB RAS, Novosibirsk, Russia, p 131
33. Zhang ZD, Rozowsky J, Snyder M, Chang J, Gerstein M (2008) Modeling ChIP sequencing in silico with applications. PLoS Comput Biol 4:e1000158
34. Kuznetsov VA, Singh O, Jenjaroenpun P (2010) Statistics of protein-DNA binding and the total number of binding sites for a transcription factor in the mammalian genome. BMC Genomics 11(Suppl 1):S12
35. Walz S, Lorenzin F, Morton J, Wiese KE, von Eyss B, Herold S, Rycak L, Dumay-Odelot H, Karim S, Bartkuhn M et al (2014) Activation and repression by oncogenic MYC shape tumour-specific gene expression profiles. Nature 511:483–487
36. Chu D, Zabet NR, Mitavskiy B (2009) Models of transcription factor binding: sensitivity of activation functions to model assumptions. J Theor Biol 257:419–429
37. Chiu KP, Wong CH, Chen Q, Ariyaratne P, Ooi HS, Wei CL, Sung WK, Ruan Y (2006) PET-tool: a software suite for comprehensive processing and managing of paired-end diTag (PET) sequence data. BMC Bioinformatics 7:390
38. Zabet NR, Adryan B (2015) Estimating binding properties of transcription factors from genome-wide binding profiles. Nucleic Acids Res 43:84–94
39. Wang J, Lu J, Gu G, Liu Y (2011) In vitro DNA-binding profile of transcription factors: methods and new insights. J Endocrinol 210:15–27
40. Johnson NL, Kotz S, Kemp AW (1992) Univariate discrete distributions, 2nd edn. John Wiley, New York, NY
41. Tuch BB, Li H, Johnson AD (2008) Evolution of eukaryotic transcription circuits. Science 319:1797–1799
42. Jothi R, Cuddapah S, Barski A, Cui K, Zhao K (2008) Genome-wide identification of in vivo protein-DNA binding sites from ChIP-seq data. Nucleic Acids Res 36:5221–5231
43. de Silva E, Thorne T, Ingram P, Agrafioti I, Swire J, Wiuf C, Stumpf MP (2006) The effects of incomplete protein interaction data on structural and evolutionary inferences. BMC Biol 4:39
44. Marshall N, Timme NM, Bennett N, Ripp M, Lautzenhiser E, Beggs JM (2016) Analysis of power laws, shape collapses, and neural complexity: new techniques and MATLAB support via the NCC toolbox. Front Physiol 7:250
45. Pareto V (1896) Cours d'Économie Politique Professé a l'Université de Lausanne
46. Barabasi AL, Albert R (1999) Emergence of scaling in random networks. Science 286:509–512
47. de Gennes PG (1979) Scaling concepts in polymer physics. Cornell University Press, Ithaca, NY
48. Kauffman S (1993) The origins of order: self-organization and selection in evolution. Oxford University Press, New York, NY
49. Dorogovtsev SN, Mendes JF, Samukhin AN (2001) Size-dependent degree distribution of a scale-free growing network. Phys Rev E Stat Nonlinear Soft Matter Phys 63:062101
50. Timar G, Dorogovtsev SN, Mendes JF (2016) Scale-free networks with exponent one. Phys Rev E 94:022302
51. Astola J, Danielian EA, Arzumanyan SK (2010) Frequency distributions in bioinformatics: the

development. A review.Proceedings of Yerevan state university. Phys Math Sci 3:3–22
52. Astola J, Danielian EA (2007) Frequency distributions in biomolecular systems and growing networks. In: Tampere International Center for signal processing (TICSP), Tampere, Finland, vol 31
53. Danielian EA, Chitchyan R, Farbood D (2016) On a new regulatory varying generalized hypergeometric distribution of second type. Math Rep 18(68):217–232
54. Duerr HP, Dietz K (2000) Stochastic models for aggregation processes. Math Biosci 165:135–145
55. Novozhilov AS, Karev GP, Koonin EV (2006) Biological applications of the theory of birth-and-death processes. Brief Bioinform 7:70–85
56. Kemp AW (1968) A wide class of discrete distributions and the associated differential equations. Ind J Stat A 30:401–410
57. Kapur JN (1978) Application of generalized hypergeometric functions to generalized birth and death processes. Indian J Pure Appl Math 9:1059–1069
58. Kapur JN (1979) Probabilities of ultimate extinction for general birth and death process. Indian J Pure Appl Math 10:105–108
59. Crawford FW, Suchard MA (2012) Transition probabilities for general birth-death processes with applications in ecology, genetics, and evolution. J Math Biol 65:553–580
60. Kuznetsov VA (2001) Distribution associated with stochastic processes of gene expression in a single eukaryotic cell. EURASIP J Appl Signal Process 2001:285–296
61. McMullen PD, Morimoto RI, Amaral LA (2010) Physically grounded approach for estimating gene expression from microarray data. Proc Natl Acad Sci U S A 107:13690–13695
62. Kuznetsov VA (2003) Hypergeometric stochastic model of evolution of conserved protein coding sequence in the archaeal, bacterial and eukaryotic proteomes. Fluct Noise Lett 3:295–324
63. Annibale A, Coolen ACC (2011) What you see is not what you get: how sampling affects macroscopic features of biological networks. Interface Focus 1:836–856
64. Kuznetsov VA (2009) Relative avidity, specificity, and sensitivity of transcription factor-DNA binding in genome-scale experiments. Methods Mol Biol 563:15–50
65. Kuznetsov VA, Pickalov VV, Senko OV, Knott GD (2002) Analysis of the evolving proteomes: predictions of the number of protein domains in nature and the number of genes in eukaryotic organisms. J Biol Syst 10:381–407
66. Kuznetsov VA, Knott GD, Bonner RF (2002) General statistics of stochastic process of gene expression in eukaryotic cells. Genetics 161:1321–1332
67. Amaral LA, Scala A, Barthelemy M, Stanley HE (2000) Classes of small-world networks. Proc Natl Acad Sci U S A 97:11149–11152
68. Chua ALS, Ivshina AV, Kuznetsov VA (2006) Pareto-gamma statistic reveals global rescaling in transcriptomes of low and high aggressive breast cancer phenotypes. In: Ragapakese JC, Wong L, Acharya R (eds) Pattern recognition in bioinformatics (PRIB-2006), vol 4146. Springer, Berlin, pp 49–59
69. Karev GP, Wolf YI, Rzhetsky AY, Berezovskaya FS, Koonin EV (2002) Birth and death of protein domains: a simple model of evolution explains power law behavior. BMC Evol Biol 2:18
70. Irwin JO (1963) The place of mathematics in medical and biological statistics. J R Stat Soc 126:1–41
71. Kemp CD, Kemp AW (1956) Generalized hypergeometric distributions. J R Stat Soc Ser B Methodol 18:202–211
72. Simon HA (1955) On a class of skew distribution functions. Biometrika 42:425–440
73. Shubert A, Glanzel W (1984) A dynamical look at a class of skew distributions – a model with scientometric applications. Scientometrics 6:149–167
74. Tripathi RC, Gurland J (1977) A general family of discrete distributions with hypergeometric probabilities. J R Stat Soc B 39:349–356
75. Yule U (1925) A mathematical theory of evolution, based on the conclusions of Dr. J.C. Willis, F.R.S. Philos Trans R Soc Lond B 213:21–87
76. Wormald S, Hilton DJ, Smyth GK, Speed TP (2006) Proximal genomic localization of STAT1 binding and regulated transcriptional activity. BMC Genomics 7:254
77. Jaeger SA, Chan ET, Berger MF, Stottmann R, Hughes TR, Bulyk ML (2010) Conservation and regulatory associations of a wide affinity range of mouse transcription factor binding sites. Genomics 95:185–195
78. Ramos AI, Barolo S (2013) Low-affinity transcription factor binding sites shape morphogen responses and enhancer evolution. Philos Trans R Soc Lond Ser B Biol Sci 368:20130018
79. Belikov S, Berg OG, Wrange O (2016) Quantification of transcription factor-DNA binding affinity in a living cell. Nucleic Acids Res 44:3045–3058

Chapter 10

A Weighted SNP Correlation Network Method for Estimating Polygenic Risk Scores

Morgan E. Levine, Peter Langfelder, and Steve Horvath

Abstract

Polygenic scores are useful for examining the joint associations of genetic markers. However, because traditional methods involve summing weighted allele counts, they may fail to capture the complex nature of biology. Here we describe a network-based method, which we call weighted SNP correlation network analysis (WSCNA), and demonstrate how it could be used to generate meaningful polygenic scores. Using data on human height in a US population of non-Hispanic whites, we illustrate how this method can be used to identify SNP networks from GWAS data, create network-specific polygenic scores, examine network topology to identify hub SNPs, and gain biological insights into complex traits. In our example, we show that this method explains a larger proportion of the variance in human height than traditional polygenic score methods. We also identify hub genes and pathways that have previously been identified as influencing human height. In moving forward, this method may be useful for generating genetic susceptibility measures for other health related traits, examining genetic pleiotropy, identifying at-risk individuals, examining gene score by environmental effects, and gaining a deeper understanding of the underlying biology of complex traits.

Key words Polygenic score, Weighted network, GWAS, Height

1 Introduction

While genome-wide association studies (GWAS) have led to some ground-breaking discoveries [1], overall their success has been somewhat underwhelming. In general, GWAS have not been very effective in identifying the genetic contributions to complex traits that do not follow Mendelian laws of inheritance [2]. In general, many of the results coming out of GWAS fail to replicate, or for those markers that are independently validated, the majority only explain a very small proportion of the variance in a given trait [3, 4]. This is a valid concern as it impedes the ability to incorporate "personalized medicine" into disease prevention and treatment.

Tatiana V. Tatarinova and Yuri Nikolsky (eds.), *Biological Networks and Pathway Analysis*, Methods in Molecular Biology, vol. 1613, DOI 10.1007/978-1-4939-7027-8_10, © Springer Science+Business Media LLC 2017

GWAS relies on linkage to examine the association between loci and a given trait. Due to recombination during meiosis, the markers in a GWAS—single-nucleotide polymorphisms, or SNPs—are used as proxies for detecting nearby variants, which are potentially causal [5]. In a typical GWAS, the association between the trait of interest and a large number m of SNPs (often in the millions) is assessed using m regression models where for each model, the trait is regressed on a single SNP. Such approaches fail to capture the complex nature of biology, and suffer from a number of statistical limitations that impede our ability to identify replicable molecular mechanisms. For instance, because GWAS require testing of millions of hypotheses, these studies tend to lack the power needed to detect the very small individual effects observed for most SNPs [2]. Further, there is significant evidence suggesting that many complex traits are highly polygenic [6], implying multiple causal variants contribute simultaneously to the genetic susceptibility of a trait. Thus, examination of genetic scores, rather than individual SNPs, may lead to better insights when studying the genetic contributions to complex traits.

Polygenic methods that move beyond the one marker approach have the ability to aid in genetic association studies by (1) increasing statistical power to detect true effects via dimension reduction; (2) providing biological insight regarding important pathways; and (3) improving our ability to examine gene by environment interactions. In 2007, Wray et al. proposed a method for examining the aggregate influence of multiple genetic markers [7]. The method involved generating a Polygenic Risks Score (PRS) based on results from a GWAS. After running a GWAS on a discovery sample, SNPs are selected for inclusion in the PRS on the basis of their association with the phenotype. Using a validation sample, the PRS can be calculated as a sum of the phenotype-associated alleles (often weighted by the SNP-specific coefficients from the GWAS). Using this score, the joint association of multiple SNPs with the given trait can be evaluated. Overall, PRS techniques have become increasingly popular, facilitating genetic discoveries for complex traits [6, 8–11]. However, given that they are based on linear combinations of markers, traditional PRS may fail to capture nonlinearity between SNPs.

While PRSs often account for a larger proportion of the variance in a trait than individual SNPs, much of the heritability remains unaccounted for—a phenomenon known as "missing heritability" [12]. One hypothesis is that the surprisingly low proportion of heritability being explained may be due to the exclusion of gene–gene interactions—or genetic network structure [13, 14]. However, very few methods exist that generate PRSs by incorporating gene network topology.

Weighted gene correlation network analysis (WGCNA) has been used repeatedly for the successful identification of epigenetic and transcriptomic networks, which relate to a number of physical, behavioral, and disease traits [15–19]. In WGCNA, network modules are identified using unsupervised machine learning methods—hierarchical clustering based on topological overlap similarity measure—and then represented using a single synthetic profile referred to as the "eigengene" or more generally eigen-node, which can be used to examine the association between a module (network) and the trait of interest [20, 21]. However, the underlying linkage-based structure of GWA data prevents the use of SNPs in traditional WGCNA methods.

Here we present a WGCNA-based method that can be applied to SNP data, which we call the weighted SNP correlation network analysis (WSCNA). Aside from accounting for the influence of LD, this method also incorporates a semisupervised machine learning approach that will facilitate the detection of modules that are trait specific. We demonstrate this method using human height as the phenotype. Human height has been extensively studied using GWAS, PRS, and heritability analyses. It is also predicted to be approximately 80% heritable and highly polygenic.

2 Materials

In order to conduct WSCNA one either needs access to genotype data or published GWAS results from multiple studies/cohorts. For our analytic example, we used genotype data from 10,466 persons of European ancestry who were participants in the Health and Retirement Study (HRS), a nationally representative longitudinal study of health and aging in the US. Genotyping was done using the Illumina Human Omni-2.5 Quad beadchip, with coverage of approximately 2.5 million single nucleotide polymorphisms (SNPs). Depending on both sample size and the number of genotyped markers, the ability to carry out WSCNA will also likely require access to a multi-core, 64 GB computer. For our example we used both [1] the University of Southern California's high performance super computer (https://hpcc.usc.edu/), for GWA, clumping, PRS estimation, and [2] a 24-core desktop workstation with 64GB of memory, for WSCNA and validation.

3 Methods

3.1 Using Published GWAS Results

As mentioned previously, WSCNA can be run using published GWAS results or by generating new GWAS results. When using published results, many of the same criteria and concerns that go into constructing traditional PRS apply. Namely, one should be aware of strand ambiguous SNPs (A/T and C/G), linkage disequilibrium (LD), and overlap in availability of SNPs across datasets.

While using results from imputed data will help with the latter concern, when it comes to strand-ambiguous SNPs, likely the safest option is to drop them. To account for LD, conventional practice is to prune data prior to conducting analysis—clump SNPs based on R^2 (typically between 0.1 and 0.5) and physical distance (typically around 500 kb), and then select the most significant SNP to represent the given block. One issue in pruning for WSCNA is that unlike PRS, the analysis requires multiple sets of GWAS results in order to look at SNP-SNP correlations. For that reason, identifying "the most significant SNP" in an LD block is ambiguous, but a natural solution is to select SNPs based on meta-analysis P values.

3.2 Running GWAS for WSCNA

When conducting original GWAS for WSCNA it is essential to use a training sample that is completely independent from the sample that will be used to assess the predictive ability of the score/s. In our example, we randomly divided our samples into a training set (70%) and a test set (30%). Before conducting the GWAS, quality control filters must be applied, which in our case resulted in 1,224,285 SNPs retained for the analysis. Additionally, principal components were generated in accordance with the methods described by Patterson et al. [22] to use as covariates to adjust for population structure.

As mentioned before, SNPs need to be pruned according to LD. To do so, a GWAS should be carried out in the training set, and results should be used to clump SNPs according to linkage disequilibrium ($R^2 > 0.5$) and physical distance ($\leq$250 kb), such that only the most significant SNP is used to represent a given haplotype block. Once SNPs have been pruned and QC has been performed, one can now conduct the GWAS that will be used as input for WSCNA. Because the network structure in WSCNA is based on pairwise correlations of beta coefficients for individual SNPs, multiple GWAS have to be run using either different samples or different phenotypes. For our example, the training data was used to create 60 subsamples of 500 participants each (with replacement) and a GWAS for human height was run for each of the subsamples using only those SNPs selected from the clumping procedure, producing 60 GWAS results for each SNP.

3.3 Preparing Data for WSCNA

Once one has either (1) collected results from multiple published GWAS or (2) generated original results from multiple GWAS, inclusion criteria based on significance can be used to select SNPs for WSCNA. While it is possible to use all SNPs, this will likely be very computationally demanding. Therefore, as with traditional polygenic score estimation, we suggest significance criteria to select SNPs (e.g., consider all SNPs with $P < 0.05$) in the training data. For instance, in our example, we selected SNPs with $P < 0.05$ ($n = 32{,}284$). The *P*-values used for selecting SNPs can be the same as used for inclusion criteria when pruning. After SNPs of interest have been selected, the beta coefficients from each of the GWAS can be used to populate an $n \times m$ matrix, where n refers to the number of examined SNPs, and m refers to the number of GWAS from which results have been gathered. In the case of our example, we had a 32,284 × 60 matrix. Assuming all results files are placed in the current working directory, the following R code can be used to generate the appropriate matrices.

```
Mat=matrix(NA,nrow=32284,ncol=60)
for (i in 1:60){
 temp=read.csv(paste("Height_sample",i,".assoc.
linear", sep=""), sep="")
Mat[,i]=temp$BETA
}
 Data=as.data.frame(Mat)datSNP=as.data.frame(t
(Data[,]))
```

3.4 Module/Network Detection Using WSCNA

WSCNA (using the WGCNA package in R) is run much like WGCNA; however, instead of using levels of expression or methylation as inputs, it uses the SNP associations with the trait/s of interest, with the goal of identifying trait-specific SNP networks, also known as modules. Weighted network construction requires a user-specified soft-thresholding power, β, to which SNP-SNP relationships are raised to calculate adjacency. Adjacency, as shown in Eq. (1), implies that the weighted adjacency a_{ij} between two SNPs is proportional to their similarity on a logarithmic scale,

$$a_{ij} = s_{ij}^{\beta}, \tag{1}$$

As suggested by Zhang et al., the value for β could be selected so that the resulting network is approximately scale-free. The WGCNA package provides the functions *pickSoftThreshold* for evaluating scale-free topology as a function of β.

The next step in WSCNA is module detection. Modules represent clusters (or networks) of densely interconnected SNPs. Topological Overlap Matrix (TOM) is used to define a dissimilarity matrix that is then used as input to cluster SNPs into modules by applying hierarchical clustering. In order to define modules, one can select to implement either a constant- or variable-height tree cut—the latter is known as the Dynamic Tree Cut [23]. The constant-height tree cut allows the user to visually inspect the dendrogram and decide on a cut height that will be used to differentiate modules. However, in most cases, there is no single cut height that captures all prominent branches. For this reason, the Dynamic Tree Cut can be employed, in which branches below the cut height can be evaluated based on various branch shape measures and sufficiently "different" branches are called separate modules [23].

In addition to selecting the branch cutting methods and the β (thresholding power), as described above, a number of other network construction and module identification options can be specified, including the correlation function (Pearson correlation or the robust biweight midcorelation), signed vs. unsigned network, minimum module size and the sensitivity of Dynamic Tree Cut to branch splits (argument deepSplit). The deepSplit command specifies a value between 0 and 4 (lower values will produce larger, less finely split clusters). Signed vs. unsigned networks refer to whether negative SNP-SNP correlations are considered connected or not. In a signed correlation network, negative correlations are considered unconnected. Conversely, in unsigned correlation networks, network adjacency is based on the absolute value of correlation, such that strong negative correlations are treated as strong connections.

3.5 Network-Based Polygenic Scores

Using the network modules identified from WSCNA, we estimate the module eigen-nodes in our validation sample. Eigen-nodes are defined as the first singular vector of all SNP profiles in a given module and values can be interpreted as module-specific polygenic scores. The input data to calculate eigen-nodes for a validation subsample includes a matrix of m SNPs by n participants, where each cell represented a participant's minor allele count for that given SNP. This is similar to the information that would be summed to generate a traditional polygenic score. The module eigen-nodes can then be used to validate the modules in respect to the phenotype of interest.

3.6 Gaining Biological Insight from SNP Correlation Networks

Biological insights can be gained from network modules by examining their topology and relationships to known pathways, biological processes, and molecular functions. For instance, for relevant networks, intramodular connectivity can be calculated for each gene in the network. The function intramodularConnectivity in the WGCNA R package computes the whole network connectivity (kTotal), the within module connectivity (kWithin), kOut = kTotal-kWithin, and kDiff = kIn-kOut = 2*kIN-kTotal.

```
ADJ1=abs(cor(datSNP,use="p"))^8
 Alldegrees1=intramodularConnectivity(ADJ1, modu-
leColors)
```

The measure of within module connectivity can be used to identify hub SNPs within the module. Similarly, module membership values can be estimated to identify hubs, by examining how strongly SNPs relate to module eigen-nodes.

```
datKME=signedKME(daSNP, datME, outputColumnNa-
me="MM.")
```

Finally, SNPs within modules of interest can also be mapped to genes to be used as input in pathway enrichment and Gene Ontology (GO) analyses.

4 Results

4.1 Identification of SNP Networks Using WSCNA

Here we illustrate the use of WSCNA using height as a phenotype. For this example, we conducted original GWAS, rather than incorporating existing GWAS results. All GWAS were carried out on European populations and included adjustments for age, sex, and population stratification (using the first four principal components). SNPs with $P < 0.05$ were selected and pruned, resulting in 32,284 SNPs to perform WSCNA with. These SNPs were used to conduct 60 GWAS—one for each training subsample of 500 subjects—and the resulting beta coefficients were used to create a $32{,}284 \times 60$ matrix. Scale-free topology analysis of this network was used to select a soft-thresholding power of 8.

WSCNA was run using the following specifications to identify SNP modules: signed network, dynamic tree cut, a module detection cut-height of 0.998, soft-thresholding power of 8, minimum module size of 50, biweight midcorrelation, and a medium branch split sensitivity (deepSplit = 2). Fifty-five modules (excluding the grey module, which represents ungrouped SNPs) were identified (Fig. 1).

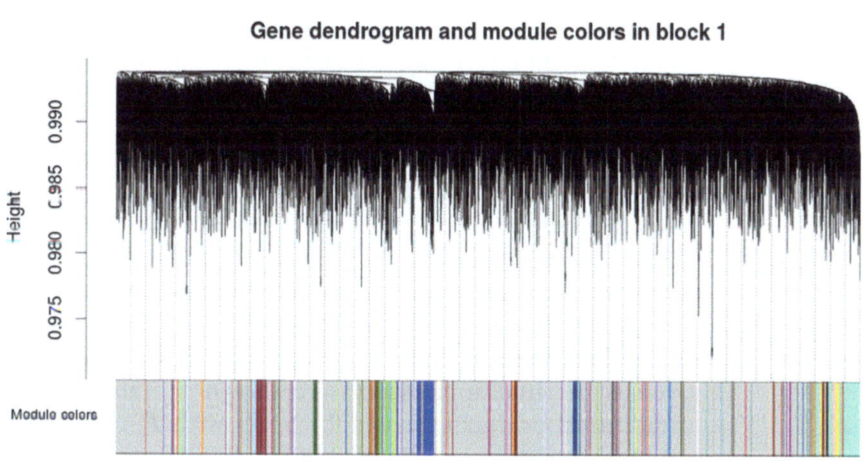

Fig. 1 WSCNA SNP clustering tree and modules. SNP clustering tree (*dendrograms*) obtained from hierarchical clustering of SNPs based on their WSCNA dissimilarity. Module assignment for each SNP is indicated in the *color row* below the *dendrograms*. Each module is represented by a single color, *grey* represents unassigned SNPs

Table 1
Significant modules identified using WSCNA

Modules	SNPs	SNPs that mapped to genes	Beta coefficient (*P*-value)
Light green	193	100	0.305 (3.84E−6)
Salmon	235	90	−0.236 (0.003)
Ivory	109	48	0.187 (0.015)
Navajo white2	63	37	0.162 (0.021)
Violet	131	66	−0.150 (0.028)
Thistle2	85	38	0.145 (0.030)
Dark orange2	93	49	−0.137 (0.043)

Beta coefficients and *P*-values come from a single model with the residual of height (adjusting for age, sex and PC1–4) as the dependent variable and all 55 modules identified in WSCNA as the independent variables

4.2 Validation for Associations between SNP Modules and Human Height

We used a single linear model to relate all module eigen-nodes, which represent network-specific polygenic models, to human height in the validation samples. Seven modules were found to be associated with height (Table 1). The most significant was the lightgreen module ($P = 3.84\text{E}{-6}$), which contains 193 SNPs. Next, we compared the amount of the variance in human height explained by WSCNA-based polygenic scores (module eigen-nodes) versus traditional polygenic scores. Three traditional polygenic scores were calculated using various significance based inclusion criteria after SNP pruning—SNPs with $P < 0.05$ in the original GWAS ($n = 32{,}284$; the same SNPs included in WSCNA), SNPs with $P < 0.005$ in the original GWAS ($n = 4318$), and SNPs with $P < 0.0005$ in the original GWAS ($n = 570$). Traditional polygenic scores were estimated in accordance with the methods outlined by Wray et al., such that the score was equal to the sum of the minor alleles, weighted by the beta coefficients from the GWAS. We examined correlations between the three polygenic scores and the seven module scores and found evidence that they were unique from one another (Fig. 2). Overall most correlations between the traditional polygenic scores and the WSCNA scores were less than $r = 0.20$.

The results for the comparison of the WSCNA polygenic scores versus the traditional scores are shown in Table 2. Overall, the traditional polygenic scores for SNPs with $P < 0.05$, 0.005, and 0.0005 had adjusted R^2 of 0.0093, 0.0082, and 0.0079, respectively. Conversely, the model with the seven significant modules of interest had an adjusted R^2 of 0.0139, while a model with just the

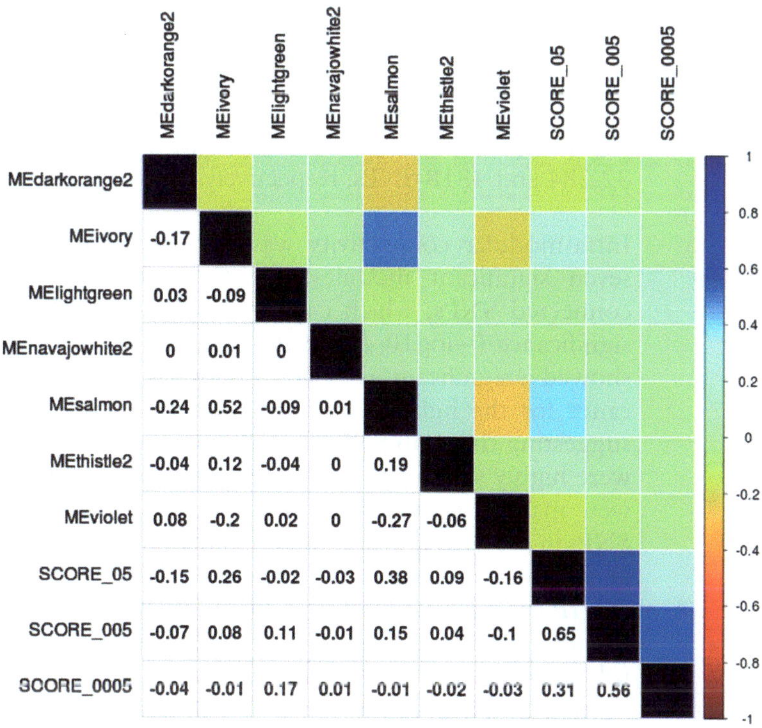

Fig. 2 Correlations between WSCNA and traditional polygenic scores. Pearson's correlations between the seven significant polygenic scores (eigen-nodes of WSCNA modules), labeled using module colors, and three traditional polygenic scores—SCORE_05 (included SNPs with $P < 0.05$), SCORE_005 (included SNPs with $P < 0.005$), SCORE_0005 (included SNPs with $P < 0.0005$), are displayed both numerically (*bottom*) and using a heatmap (*top*). In general, SNP scores were weakly correlated ($R < 0.20$). The strongest correlation between a WSCNA score and a traditional polygenic score is found between the score for the Salmon module and SCORE_05 ($r = 0.38$)

Table 2
Variance in human height explained by models containing different polygenic scores

Independent variable/s in each model	R^2	Adjusted R^2
PRS 0.05 ($n = 32{,}284$)	0.0096	0.0093
PRS 0.005 ($n = 4318$)	0.0084	0.0082
PRS 0.0005 ($n = 507$)	0.0083	0.0079
Light green module ($n = 193$)	0.007	0.0066
The seven significant WSCNA modules ($n = 909$)	0.0163	0.0139

n refers to the number of SNPs used to generate the polygenic score/s in each model light green module had an adjusted R^2 of 0.0066. This is noteworthy, given the adjusted R^2 for the model containing the significant WSCNA scores was 50–70% higher than the R^2 for the models containing the first two traditional polygenic scores, even though it was based on information from only 909 SNPs, compared to 32,284 and 4318 SNPs, respectively.

4.3 Hub Genes, Pathways, and Gene Ontology

Intra-modular connectivity was calculated for each SNP in the seven significant modules. We then examined whether more connected SNPs, which can be thought of as hubs, had higher significance (−log10(*P*)) in the training sample (Fig. 3). Results showed a significant association between connectivity and significance for the light green module ($r = -0.28$, $P = 8 \times 10^{-5}$), suggesting that hub SNPs for this module tended to be SNPs that were highly significant in our original training GWAS. Next SNPs were mapped to genes. We find that genes mapped from the hub SNPs in the light green module included *HHIP* (kWithin = 0.213), as well as its neighboring gene *ANAPC10* (kWithin = 0.605). *HHIP* has been implicated in both GWAS and microarray studies examining genes related to height [24] and has a well-established role in chondrogenesis [25].

Finally, pathway enrichment, GO, and protein interaction network module analysis were performed using WebGestalt (http://bioinfo.vanderbilt.edu/webgestalt/). When examining all the genes from the seven networks ($n = 428$), we find enrichment for "Signaling events mediated by the Hedgehog Family" (enrichment = 4.99, Bonferroni adjusted $P = 0.046$), "Calcium signaling" (enrichment = 3.73, adjusted $P = 0.001$), and "G alpha (*q*) signaling events" (enrichment = 3.76, adjusted $P = 0.046$). "Signaling events mediated by the Hedgehog Family" also was found to be enriched when only examining genes in the light green module (enrichment = 14.20, adjusted $P = 0.011$). This pathway has been repeatedly shown to influence height in genetic association studies [26–28].

We also found significant enrichment for GO biological processes involved in "anatomical structure development" (enrichment = 1.32, adjusted $P = 0.046$). Lastly, we identified a significantly enriched protein-protein interaction network using WebGestalt (enrichment ratio of 2.22, adjusted *P*-value = 0.0006), shown in Fig. 4, which was highly associated with the molecular function, "non-membrane spanning protein tyrosine kinase activity."

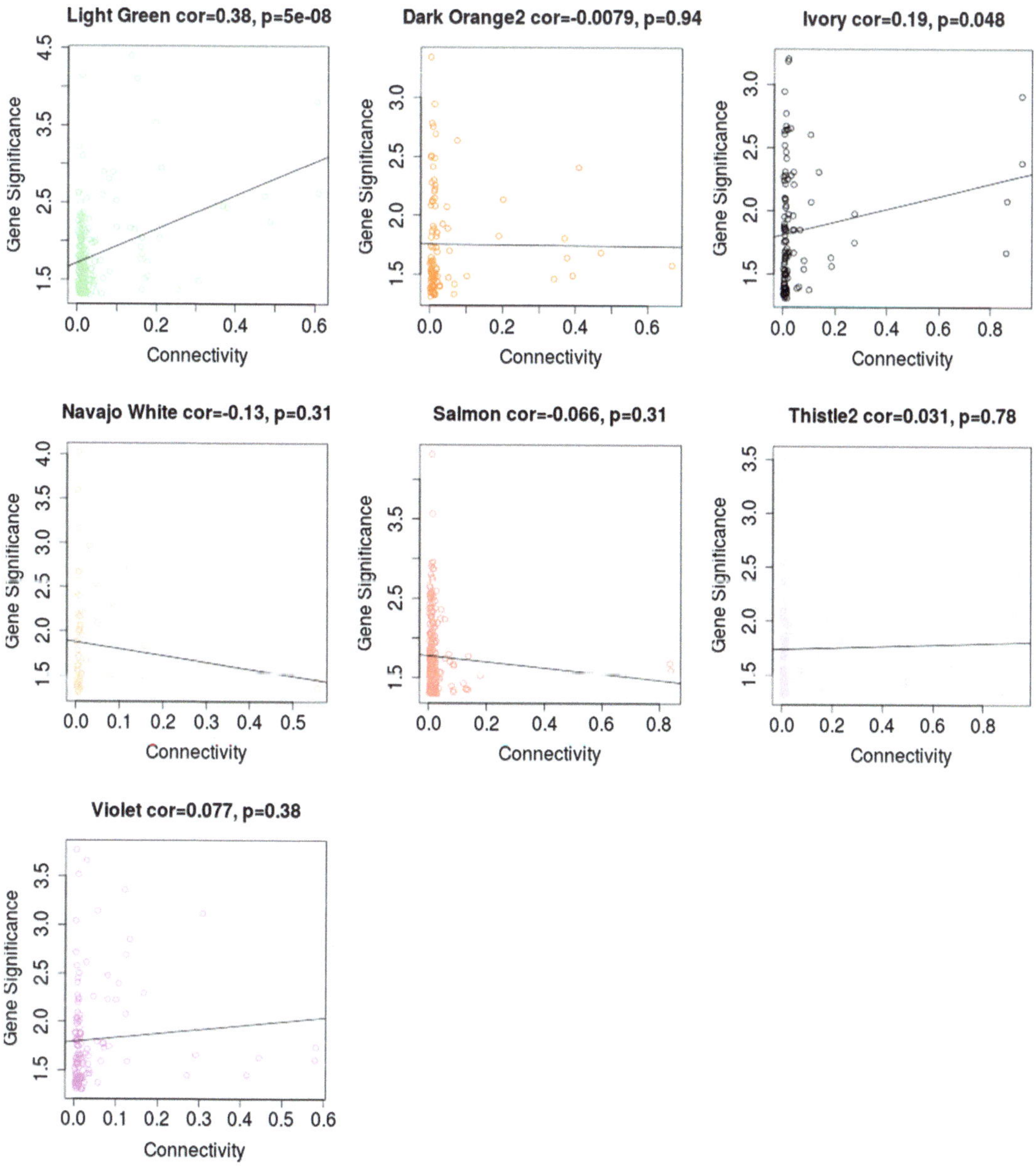

Fig. 3 Module connectivity and significance in the GWAS. Within module connectivity for each SNP in the seven significant modules is plotted against significance of that SNP in the original GWAS. Overall, we find a moderately strong correlation between connectivity and significance among SNPs in the light green module, suggesting that hub SNPs in this module tended to be those that had more significant relationships to height in our training sample

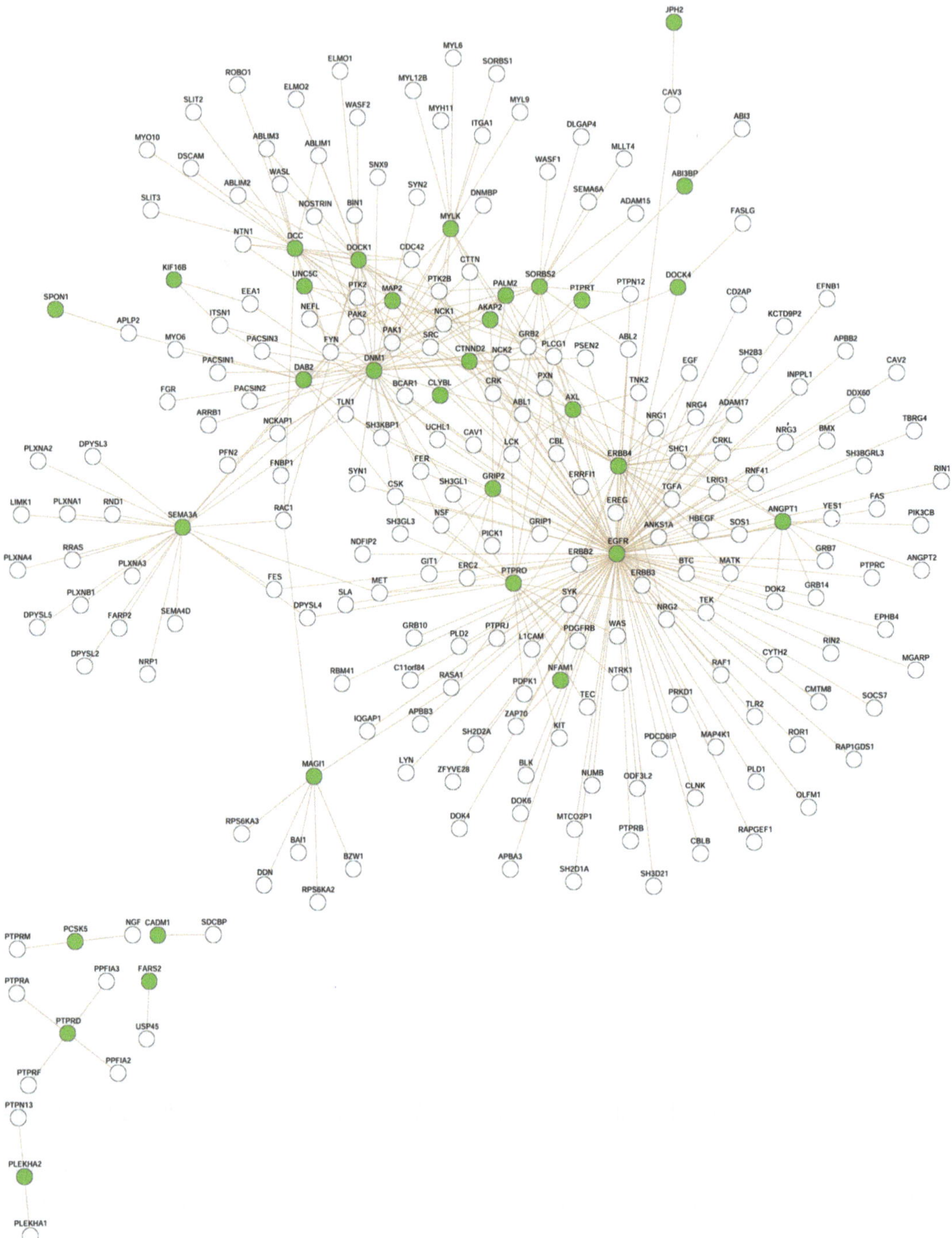

Fig. 4 Protein-protein interaction network enriched in genes mapped to significant SNP modules. This PPI network was significantly enriched in genes that mapped to SNPs in the seven significant modules. Of the 619 genes in this protein interaction network, 32 are present on our gene list (enrichment ratio of 2.22, Bonferroni adjusted P-value = 0.0006). Genes present in our WSCNA modules are shown in *green*

5 Conclusion

We illustrate here the methodology for performing WSCNA using results from GWAS. We also show that the incorporation of network structures in the analysis of large-scale genetic association data can be used to estimate genetic scores for specific traits, identify hub SNPs/genes, and lead to biological insight into the pathways involved. Our example also demonstrated that the scores generated from WSCNA can more closely relate to the phenotype of interest in validation analysis than traditional polygenic risk scores. We were also able to identify hub genes and pathways that are known to relate to human height. This is consistent with what has been found for co-expression analysis, for which the use of topological overlap matrix, coupled with a signed correlation network gives rise to biologically meaningful modules [29]. The ability to identify hub genes/SNPs is a significant advantage of this methodology, as it has been demonstrated that intramodular hub genes are often biologically meaningful and represent the module [30, 31].

The presented methodology also has important limitations. First, since WSCNA relies on results of multiple GWAS, care must be taken to ensure the underlying GWAS results are reliable. Thus, it may be more reliable to use previously published results or including as many studies as possible. Along those same lines, the number of participants needed to carry out such analyses may be quite large. In our example we used GWAS results from only 7,326 participants; however, sample sizes will need to increase in order to improve efficacy of gene scores, particularly for traits whose heritability is not as high as that of human height. Second, WSCNA is relatively computationally expensive and may require the user to have access to a multi-core workstation or a supercomputing cluster. Third, users have to specify various parameters for network construction and module identification.

The WSCNA method presented here offers a new and innovative way to incorporate networks—a dominant feature in biology and physiology—into genetic association studies of complex traits. In moving forward, this type of methodology may be useful for generating genetic susceptibility measures for other health related traits, examining genetic pleiotropy, identifying at-risk individuals, examining gene score by environmental effects, and gaining a deeper understanding of the underlying biology of complex traits.

References

1. McCarthy MI et al (2008) Genome-wide association studies for complex traits: consensus, uncertainty and challenges. Nat Rev Genet 9 (5):356–369
2. Risch NJ (2000) Searching for genetic determinants in the new millennium. Nature 405 (6788):847–856

3. Hindorff LA et al (2009) Potential etiologic and functional implications of genome-wide association loci for human diseases and traits. Proc Natl Acad Sci U S A 106(23):9362–9367
4. Manolio TA et al (2009) Finding the missing heritability of complex diseases. Nature 461 (7265):747–753
5. Hardy J, Singleton A (2009) Genomewide association studies and human disease. N Engl J Med 360(17):1759–1768
6. Dudbridge F (2013) Power and predictive accuracy of polygenic risk scores. PLoS Genet 9(3):e1003348
7. Wray NR, Goddard ME, Visscher PM (2007) Prediction of individual genetic risk to disease from genome-wide association studies. Genome Res 17(10):1520–1528
8. Levine ME, Crimmins EM (2015) A genetic network associated with stress resistance, longevity, and cancer in humans. J Gerontol A Biol Sci Med Sci 71(6):703–712
9. Peterson RE et al (2011) Genetic risk sum score comprised of common polygenic variation is associated with body mass index. Hum Genet 129(2):221–230
10. Purcell SM et al (2014) A polygenic burden of rare disruptive mutations in schizophrenia. Nature 506(7487):185–190
11. Wray NR, Goddard ME, Visscher PM (2008) Prediction of individual genetic risk of complex disease. Curr Opin Genet Dev 18(3):257–263
12. Eichler EE et al (2010) Missing heritability and strategies for finding the underlying causes of complex disease. Nat Rev Genet 11 (6):446–450
13. Cordell HJ (2009) Detecting gene-gene interactions that underlie human diseases. Nat Rev Genet 10(6):392–404
14. Hemani G, Knott S, Haley C (2013) An evolutionary perspective on epistasis and the missing heritability. PLoS Genet. 9(2):e1003295
15. Ghazalpour A et al (2006) Integrating genetic and network analysis to characterize genes related to mouse weight. PLoS Genet. 2(8): e130
16. Horvath S et al (2006) Analysis of oncogenic signaling networks in glioblastoma identifies ASPM as a molecular target. Proc Natl Acad Sci U S A 103(46):17402–17407
17. Langfelder P et al (2012) A systems genetic analysis of high density lipoprotein metabolism and network preservation across mouse models. Biochim Biophys Acta 1821(3):435–447
18. Oldham MC, Horvath S, Geschwind DH (2006) Conservation and evolution of gene coexpression networks in human and chimpanzee brains. Proc Natl Acad Sci U S A 103 (47):17973–17978
19. Oldham MC, Langfelder P, Horvath S (2012) Network methods for describing sample relationships in genomic datasets: application to Huntington's disease. BMC Syst Biol 6:63
20. Langfelder P, Horvath S (2008) WGCNA: an R package for weighted correlation network analysis. BMC Bioinformatics 9:559
21. Zhang B, Horvath S (2005) A general framework for weighted gene co-expression network analysis. Stat Appl Genet Mol Biol. 4: Article 17
22. Patterson N, Price AL, Reich D (2006) Population structure and eigenanalysis. PLoS Genet. 2(12):e190
23. Langfelder P, Zhang B, Horvath S (2008) Defining clusters from a hierarchical cluster tree: the dynamic tree cut package for R. Bioinformatics 24(5):719–720
24. Visscher PM (2008) Sizing up human height variation. Nat Genet 40(5):489–490
25. Lui JC et al (2012) Synthesizing genome-wide association studies and expression microarray reveals novel genes that act in the human growth plate to modulate height. Hum Mol Genet 21(23):5193–5201
26. Lango Allen H et al (2010) Hundreds of variants clustered in genomic loci and biological pathways affect human height. Nature 467 (7317):832–838
27. Liu JZ et al (2010) Genome-wide association study of height and body mass index in Australian twin families. Twin Res Hum Genet 13 (2):179–193
28. Wood AR et al (2014) Defining the role of common variation in the genomic and biological architecture of adult human height. Nat Genet 46(11):1173–1186
29. Song L, Langfelder P, Horvath S (2012) Comparison of co-expression measures: mutual information, correlation, and model based indices. BMC Bioinformatics 13:328
30. Horvath S, Dong J (2008) Geometric interpretation of gene coexpression network analysis. PLoS Comput Biol. 4(8):e1000117
31. Langfelder P, Mischel PS, Horvath S (2013) When is hub gene selection better than standard meta-analysis? PLoS One 8(4): e61505

Chapter 11

Analysis of *cis*-Regulatory Elements in Gene Co-expression Networks in Cancer

Martin Triska, Alexander Ivliev, Yuri Nikolsky, and Tatiana V. Tatarinova

Abstract

Analysis of gene co-expression networks is a powerful "data-driven" tool, invaluable for understanding cancer biology and mechanisms of tumor development. Yet, despite of completion of thousands of studies on cancer gene expression, there were few attempts to normalize and integrate co-expression data from scattered sources in a concise "meta-analysis" framework. Here we describe an integrated approach to cancer expression meta-analysis, which combines generation of "data-driven" co-expression networks with detailed statistical detection of promoter sequence motifs within the co-expression clusters. First, we applied Weighted Gene Co-Expression Network Analysis (WGCNA) workflow and Pearson's correlation to generate a comprehensive set of over 3000 co-expression clusters in 82 normalized microarray datasets from nine cancers of different origin. Next, we designed a genome-wide statistical approach to the detection of specific DNA sequence motifs based on similarities between the promoters of similarly expressed genes. The approach, realized as *cisExpress* software module, was specifically designed for analysis of very large data sets such as those generated by publicly accessible whole genome and transcriptome projects. *cisExpress* uses a task farming algorithm to exploit all available computational cores within a shared memory node.

We discovered that although co-expression modules are populated with different sets of genes, they share distinct stable patterns of co-regulation based on promoter sequence analysis. The number of motifs per co-expression cluster varies widely in accordance with cancer tissue of origin, with the largest number in colon (68 motifs) and the lowest in ovary (18 motifs). The top scored motifs are typically shared between several tissues; they define sets of target genes responsible for certain functionality of cancerogenesis. Both the co-expression modules and a database of precalculated motifs are publically available and accessible for further studies.

Key words Promoters, Motifs, Gene expression, Genome annotation, Co-expression clusters, Cancer

1 Introduction

In any living organism, proteins function in groups, complexes, biochemical and signaling pathways, and networks. Precise timing and spatial assembly of protein machinery is closely regulated at multiple levels, transcription being the major one. Genome-wide transcriptional profiling has been the most accessible "omics" method since the dawn of "genomics." The amount of raw data

Tatiana V. Tatarinova and Yuri Nikolsky (eds.), *Biological Networks and Pathway Analysis*, Methods in Molecular Biology, vol. 1613, DOI 10.1007/978-1-4939-7027-8_11, © Springer Science+Business Media LLC 2017

in private and publicly accessible transcriptional databases grows exponentially for over 20 years, both in a form of microarray datasets and, more recently, as DNA and RNA sequences. In order to be suitable for biomedical research and clinical practice, these datasets are analyzed by methods of statistical and functional analysis and annotated with phenotypic "metadata." Yet only a small fraction of accumulated expression data is now utilized in biomedicine, typically in a form of transcriptional biomarkers ("gene signatures"). Moreover, with a few exceptions, even these markers are not approved for clinical use, mostly due to high heterogeneity of expression patterns for any complex phenotype, such as disease or drug response. As it was clearly demonstrated in a massive FDA-led study MAQC/SEQC, predictive power of expression signatures depends mostly on the biology of end point, and there is no "silver bullet" statistical correlation method to deal with natural heterogeneity [1]. Functional analysis approach takes into account experimentally proven grouping of gene expression signals, such as protein interactions, linear pathways and networks (reviewed by Kristensen et al. [2]), as well as contextual modeling of expression patterns, using text mining methods [3]. A combination of these methods were proven to marginally improve predictive power and specificity of expression profiling in cancers [2] and other diseases [4]. However, in spite the progress in analytical methodology and technique, the bulk of expression data is not usable and is waiting for exploring.

Therefore, in this study we intended to combine two independent methods producing transcription regulation patterns: co-expression networks and precalculated transcription regulation motifs. Recently, we conducted a comprehensive study on global co-expression profiling of 82 independent expression datasets derived from nine major human cancers of different tissue origins [5]. We applied the Weighted Gene Co-Expression Network Analysis methodology (WGCNA) [6] and identified over 3000 distinct gene co-expression modules. Functional analysis revealed that the clusters cover a range of known tumor features, such as proliferation, extracellular matrix remodeling, hypoxia, inflammation, angiogenesis, specific biological pathways, various genomic alterations, tumor differentiation programs and biomarkers of individual tumor subtypes. Yet over 900 co-expression modules have shown no direct reference to organized knowledge of cancer biology. Exploration of these modules may shed light on yet unknown aspects of cancer mechanisms. Taken together, we generated a comprehensive, normalized, and well-documented collection of gene co-expression modules in a variety of cancers as a rich data resource to facilitate further progress in cancer research.

The key reason for the genes to be clustered in accordance with expression is their co-regulation (activation or inhibition) by transcription factors. Such regulation is orchestrated by *cis*-regulatory elements generally located in promoters of target genes. High confidence identification of potential DNA *cis*-regulatory elements was long associated with recovering putative genomic functionality. Knowledge of regulatory motifs bound by transcription factors can provide crucial insights into the mechanisms of transcriptional regulation [7]. Moreover, a specific pattern of transcription regulation may be a more robust expression biomarker for cancer phenotypes than "gene signatures" per se [8]. Regulatory motifs may be specific enough for tissues, cell types, pathological conditions, and other contexts. Therefore, a comprehensive landscaping of major regulatory motifs can contribute to understanding molecular mechanisms of many complex diseases. In the mouse genome, many in vitro-derived motifs performed similarly to motifs derived from in vivo data [9]. However, specificity of the binding of transcription factors to DNA motifs is difficult to access, due to lack of data and deficiencies in predictive models for motif detection.

The problem of motif detection can be formulated in several ways, but the most common is the following: given a set of co-expressed (and presumably, transcriptionally co-regulated) genes and their promoters, one needs to identify the motif appearing in all/most of the promoters. This approach is limited by ambiguity in the selection of any particular co-expressed set of genes and is, therefore, rather subjective. Alternatively, there are genome wide methods that search for statistically significant associations between "words" in both biological sequence and phenotypic data. Both approaches are commonly used and frequently are useful. From an algorithmic perspective, tools that implement these approaches can be classified as using either ***enumerative*** methods or ***alignment-based*** methods.

Enumerative methods typically list, or enumerate, all the words of defined length in the text (expression dataset in this case) and then assess statistical significance of each word. The most significant words are then suggested as sequence motifs. The computational complexity of these methods can be represented as $O(NmAeLe)$, where N is the number of sequences, m is their length, A is the alphabet size, and e is the number of errors allowed in a match to a catalog entry [7, 10]. Several commonly used motif-finding tools are based on this approach, including Motifer [11], REDUCE and MatrixREDUCE [12, 13], WordSpy [14], Vocabulon [15, 16], Allegro [17, 18], and *cisExpress* [19].

Some of the most commonly used alignment-based methods include AlignACE [20, 21], MEME [22] and Dialign [23]. This class of methods generally builds probabilistic models of the observed sequence data and then use optimization techniques to find the words common to all input sequences. Most frequently, researchers use expectation-maximization (EM) and Gibbs sampling [24, 25] as optimization methods. Most of the motif-finding algorithms do not take motif position information into account [26], although the position with respect to the transcription start site is important [27].

EM is a general technique used for maximizing the likelihood of a function with hidden variables. In the case of motif detection based on a probabilistic model of observed data, the hidden variables are the positions of motifs in the input sequences. An EM algorithm consists of two steps. The E-step calculates the expected likelihood of the observed sequence data, using the current parameter setting. The M-step updates the parameters to improve the expected likelihood (EM algorithms belong to the hill-climbing algorithm family). Like every hill-climbing algorithm, the EM is not guaranteed to converge to the global optimum and may become stuck in a local optimum. EM is, therefore, very sensitive to the input parameters and is typically run several times with different initialization conditions. This feature is advantageous for identification of biologically relevant motifs, which need not to be necessary corresponded to the global optimum [24].

Gibbs sampling uses a global undirected search over the parameterized distribution. In the context of motif discovery, Gibbs sampling typically initiates the hidden variables (the motif locations) by using random samples from the distribution. The parameters are then reestimated based on the random samples, and the sampling is repeated. Gibbs sampling usually requires many iterations to obtain adequate results, so its computation is time consuming [25].

An alternative approach consists of cataloguing known motifs, mapping them on promoters of interest and, using gene expression information, calculating motifs' influence on gene expression. Such a strategy was described, for example, by Jolma, Yan [26] as well as by other groups.

The existing methods for computational identification of regulatory patterns were benchmarked several times, with mixed reviews [11, 18, 28, 29]. Combination of various genomic features using machine-learning methods is a promising approach. It was demonstrated that no single feature provides sufficiently strong predictive power; however, sophisticated approaches, such as *TargetFinder*, can identify an optimal combination of features that is highly informative and achieves a low false discovery rate [30]. Independently, the authors noted that prediction of regulatory elements remains a challenge; more work is needed to optimize the algorithms and users are advised combining several motif-finding tools for the best results. As the issue seems to be unresolved, we developed an accurate tool, called *cisExpress,* designed for more effective analysis of large datasets in a manner that is both cost effective and highly robust in its predictive capacity. Here, we applied *cisExpress* for the comprehensive analysis of co-expressed modules from 49 cancer datasets.

2 Materials

2.1 Gene Expression Data

Following Ivliev, 't Hoen [5], 82 cancer-related datasets were downloaded from the GEO database [31–33], all healthy control samples excluded, and the resulting datasets normalized using MAS5 algorithm followed by the quantile normalization, as described in [34, 35]. The outliers were removed using the following procedure: first, Pearson's correlation coefficients were computed between all samples within every dataset; second, mean and standard deviation of the correlation coefficient were computed; and third, samples that were more than four standard deviations below the mean correlation coefficient were excluded [35, 36]. In total, we selected 6 tissues, resulting in 11 "Brain" datasets, 14 "Colon" datasets, 6 "Kidney" datasets, 12 "Lung" datasets, 7 "Ovary" datasets, and 4 "Prostate" datasets.

2.2 Promoter Sequences

Positions of transcription start sites were obtained from the DBTSS database [37] (ftp://ftp.hgc.jp/pub/hgc/db/dbtss/dbtss_ver9/hg38/TSSseq/tsc_data/Adult/). DBTSS includes transcription start site data for human adult and embryonic tissues; in total it contains 491 million TSS tag sequences collected from 20 tissues and seven cell cultures. The putative TSS were processed using the TSSer algorithm [11] and its more recent nonparametric version, NPEST [38], selecting one transcription start site per locus per tissue. The resulted collection contained tissue-specific promoters that have both TSS and gene expression support for six tissues: Brain (15,068 sequences), Colon (12,262 sequences), Kidney (12,976 sequences), Lung (12,438 sequences), Ovaries (10,825 sequences), and Prostate (12,037 sequences).

3 Methods

3.1 Weighted Gene Co-expression Network Analysis (WGCNA)

Gene co-expression networks were calculated as in Ivliev et al. [5] using the R package WGCNA [6]. For every gene expression dataset, the co-expression networks were constructed independently in the following manner. First, an adjacency matrix A [6, 39] was computed for all genes in a given dataset from a matrix of Pearson's correlation coefficients raised to a fixed power β (adjacency = correlation$^{\beta}$). The value of β, ranging from 7 to 15, was chosen separately for each dataset to penalize spurious weak correlations as an alternative to setting an ad hoc correlation coefficient cut-off [6, 39].

Distance matrix D was obtained from the adjacency matrix A: $D_{ij} = 1 - \frac{\sum_k A_{ik}A_{kj} + A_{ij}}{1+\min\{\sum_k A_{ik}, \sum_k A_{kj}\} - A_{ij}}$. This transformation reinforced consistent patterns and robustness of the network [6, 39]. Matrix D was used for hierarchical clustering of genes within each dataset, with clustering dendrogram's branches corresponding to gene co-expression modules [40]. Each module can be characterized by a representative expression trend, corresponding to the first principal component of the gene expression matrix, known as a module *eigengene* [6]. Modules with correlated eigengenes (Pearson's correlation >0.8) were combined [35].

As a result, each gene in a genome can be characterized by its connectivity to each of the co-expression modules, measuring how strongly a gene is connected with all the other genes in the module.

From the networking perspective, highly connected genes represent central genes in the module, while lowly connected ones can be interpreted as peripheral genes. For each gene and each module the expression connectivity is defined as a Pearson's correlation between the expression profile of a gene and the eigengene of the module [6].

3.2 cisExpress *Algorithm*

cisExpress algorithms is based on two important assumptions:

1. The function of promoter motifs is position-specific.
2. Microarray data provide reasonable measurements of transcript abundance and reflect promoter activity.

cisExpress divides the problem into two general stages:

1. *Finding "seed" motifs.* This part of the method outputs the motif in the form of a consensus sequence and includes its approximate position in promoter region.
2. *Optimizing the motifs obtained by the first part of the method.* This part inputs a motif that was detected in the first step and applies a Genetic Algorithm (GA) [41] to find the best possible motif model and motif position. The output consists of an *N*-by-4 motif matrix (where *N* is the length of the motif), representing the relative frequencies of nucleotides in the motif. For each position within the motif there is a probability that each base occurs at that position. The matrix structure also includes motif conservancy and position of the motif. This stage is unique to *cisExpress.*

Identification of the "seed" motifs consists of four stages: initial data processing, merging similar motifs, clustering, and selection of the best ancestor word.

Initial data processing. *cisExpress* takes as its input a set of promoter sequences for all genes in a genome of interest (promoter sequences are aligned by the position of Transcription Start Site (TSS)), and a set of gene expression values for an experimental condition of interest. The expression values can be taken from individual microarray experiments, or may be averaged across similar experimental conditions. The expression levels can be either log-fold change, number of ESTs, gene connectivity to a co-expression module (as described above) or any other expression measure. Then the promoter region is partitioned into overlapping windows, each of which is independently searched for the motifs. This decomposition into discrete tasks makes the algorithm naturally suited for implementation in a parallel computing environment.

Within each window k, and for each l-letter word w, a summary statistic d_{with}, d_{without} (resp. s_{with}, s_{without}) is computed across all N promoter sequences as follows:

$$d_{\text{with}}(w, k) = \sum_{i=1}^{N} e_i \delta(w, k, i) \tag{1}$$

$$d_{\text{without}}(w, k) = \sum_{i=1}^{N} e_i (1 - \delta(w, k, i)) \tag{2}$$

$$s_{\text{with}}(w, k) = \sum_{i=1}^{N} e_i^2 \delta(w, k, i) \tag{3}$$

$$s_{\text{without}}(w, k) = \sum_{i=1}^{N} e_i^2 (1 - \delta(w, k, i)) \tag{4}$$

where e_i is the expression value of the ith gene and $\delta(w,k,i)$ is the Kronecker delta symbol (equal to 1 if the word w is present in promoter i in the window k, and *0* otherwise). In every window, the significance of each word is determined by the corresponding Z-score:

$$n_{\text{with}}(w, k) = \sum_{i=0}^{N} \delta(w, k, i) \tag{5}$$

$$n_{\text{without}}(w, k) = \sum_{i=0}^{N} (1 - \delta(w, k, i)) = N - n_{\text{with}}(w, k) \tag{6}$$

$$\text{Stdev}_{\text{with}}(w, k) = \sqrt{\frac{s_{\text{with}}(w, k)}{n_{\text{with}}(w, k)} - \left(\frac{d_{\text{with}}(w, k)}{n_{\text{with}}(w, k)}\right)^2} \tag{7}$$

$$\text{Stdev}_{\text{without}}(w, k) = \sqrt{\frac{s_{\text{without}}(w, k)}{n_{\text{without}}(w, k)} - \left(\frac{d_{\text{without}}(w, k)}{n_{\text{without}}(w, k)}\right)^2} \tag{8}$$

$$Z_{\text{score}}(w, k) = \frac{\frac{d_{\text{with}}(w,k)}{n_{\text{with}}(w,k)} - \frac{d_{\text{without}}(\text{w},\text{k})}{n_{\text{without}}(\text{w},\text{k})}}{\sqrt{\frac{\text{Stdev}^2_{\text{with}}(w,k)}{n_{\text{with}}(w,k)} + \frac{\text{Stdev}^2_{\text{without}}(w,k)}{n_{\text{without}}(w,k)}}} \tag{9}$$

The algorithm displays a list of significant motifs with Z-scores above a user-defined threshold.

Merging similarmotifs. Deterministic approaches frequently report several highly similar motifs, which differ in only one nucleotide. In this case it is reasonable to assume that, in fact, just one motif has been found and that it can occur in several variations. *cisExpress*, is tailored to deal with these ambiguous nucleotides in transcription factor binding sites as described below.

Suppose that two motifs, ACTGA and ACTGC, are found in a single window, and have a *Z*-score above some predefined threshold. ACTGA and ACTGC may represent two instances of the same motif, occurring as ACTGA and ACTGC (this ambiguity can be written as ACTG[AC]. We show the ambiguous nucleotides in square brackets. For example, [AG] stands for either adenine or guanine). The *Z*-score of the composite motif is computed, and, if the composite motif is also significant, then the two original motifs can be replaced by the composite. The possibility of further merging of the resulting composite motif with other motifs in the same window is investigated in an iterative manner. For example, if there is a motif CCTGC in the same window as the ACTG[AC], the significance of the composite motif [AC]CTG[AC] is investigated. If the *Z*-score of the new composite motif is above a desired threshold, the merge is accepted. This process is continued until there are no more candidate motifs that result in significant composite motifs.

To make the computation as fast as possible, each character of the original DNA motif is treated as a bit-mask, where each base is represented by one bit. Bit-mask corresponds to the possibility of nucleotide occurrence in every position. For example, the binary mask 0001 indicates that there is Adenine, 0010—Cytosine, 0100—Guanine, 1000—Thymine. Binary mask 1101 indicates that there can be any nucleotide except Cytosine. Using binary masks allows usage of bitwise *AND* operations to find the intersections of two motifs. For example, for a given position in a motif: [ACT] (1011) AND [AGT] (1101) = [AT] (1001). The union of two motifs is found using the bitwise *OR* operator, e.g., *T* (1000) OR [AT] (1001) = [AT] (1001). Two words may be merged if the intersection of each corresponding pair of nucleotides (obtained by performing bitwise *AND* over their binary representations) is non-zero (0000). If the intersection is zero, then the investigated pair of nucleotides does not match. The results for all positions in the motif are added and compared to one (only one nucleotide difference is allowed between two motifs that are being merged). The pseudo-code for merging of words is displayed below:

```
Initialize: collect all words into the pool
do{
Find two words, which differ only in one nucleotide and
find their composite word (union)
if (Z-score (composite word)> cutoff) then
Replace the words by the composite
}While (words were merged in last cycle)
```

Next, it is necessary to define the merging relation $A \sim B$, indicating that words A and B can be merged (i.e., words A and B are *in relation*), and the merging operation $A \times B$ which results in the merged composite words. The relation and operation have the following properties:

1. Relation ~ is symmetric: if $A \sim B$ is true, then $B \sim A$ is also true (and likewise for false).
2. Operation × is commutative: $A \times B = B \times A$.
3. For all words A, B, C, if $A \sim C$, then $(A \times B) \sim C$.
4. All motifs that are in relation with word A are also in relation with word $(A \times B)$ as well, i.e., $A \sim C => (A \times B) \sim C$.
5. There are different ways to assemble a composite word from simple words: If $A \sim B$, $B \sim C$, $A \sim C$ then $(A \times B) \times C = A \times (B \times C) = (A \times C) \times B$.

Note that the merge operation is not associative. For example, consider the words A: ACCGT, B: CCCGT, and C: ACCAT. The word C is not in relation with word B (they differ in nucleotides 1 and 4), but it is in relation with $(A \times B)$ ($A \times B =$ [AC]CCGT, motif C differs only in the nucleotide 4). Thus, it is necessary for the algorithm to cycle through more than one iteration in order to account for all possible merges: in this example the merged word $(A \times B) \times C$ only becomes possible after A and B have been merged.

Clustering. Up to this point treatment of fixed-size motifs has been described. Usually the sizes of motifs are not known a priori. To accommodate arbitrary size words, one can cluster basic words of fixed size into larger composite motifs. For example, two 5-nucleotide words TACCT and ACCTG, can be parts of a composite motif TACCTG. We define the noncommutative clustering relation as $A + B$, indicating that word B can extend the word A in 3′ direction. Note, that $B + A$ being true does not necessarily imply that $A + B$ is true. It is also possible to define an associative clustering operator $A^\circ B$ that inputs two words of size n and outputs one motif of size $n + 1$. Operator ° is associative, so for any words A, B, C, $(A^\circ B)^\circ C = A^\circ(B^\circ C)$.Using the described operator ° shorter words can be clustered to longer ones. The clustering process is iterative, described by following pseudo-code:

```
    Initialize: collect all words after merging process
into the pool
     do{
     Find two words A,B, which are in relation A+B
     if (Z-score (A°B)> cutoff) then
     Replace the words A,B by the composite C=A°B
     }While (words were clustered in last cycle)
```

Selection of the best ancestor word. Using the approaches described above, the words in the dictionary can be merged and clustered while there are at least two words that are in a relation A ~ B or A + B (two words that can be merged by operator "×" or clustered by operator "°") and the *Z*-scores of the resulting composite words remains above some predefined threshold. After each merging/clustering operation the statistical properties are recomputed and the *Z*-scores are reevaluated. The danger of this approach is that important but short motifs maybe sacrificed in favor of long and spurious ones. To avoid this, the ancestor of each motif with the best *Z*-score ("the best ancestor") is also stored. When a new basic motif is identified, it becomes its own best ancestor. After a merger/clustering step, the *Z*-score of the composite motif is compared to the *Z*-scores of the best ancestors. If the new *Z*-score is greater than the ancestral *Z*-scores, the new word becomes its own "best ancestor." Otherwise it inherits the "best ancestor" from the best ancestors. The process is shown in the Fig. 1.

3.2.1 Parallelization

The method described in previous subsections independently examines promoters by windows (in respect to their distance from the TSS). Therefore, the algorithm can be parallelized by examining the various windows in parallel. A task-farming algorithm has been implemented to allow *cisExpress* to exploit multiple computational cores within a single shared memory computer system to perform the window search. Significant parallel scaling can be achieved using intra-node parallelism due to the current trend towards more cores per node in high-performance computing (HPC) systems.

The task-farming algorithm is illustrated by the flowchart in the Fig. 2. A given set of input data contains a fixed number *W* of windows. The search of each of these windows is considered to be a task and all of the tasks are added to a queue. A master process is launched to act as a queue manager. Note that the number *P* of cores assigned should be less than the number of windows present (since there are only as many tasks as there are windows, so additional cores will be redundant). The master process initially loads the input data into memory and forks *P* slave processes, each of which is then assigned one window search (such that *W* – *P* tasks then remain in the queue). The slave processes read the promoter and expression data via shared memory, and therefore do not

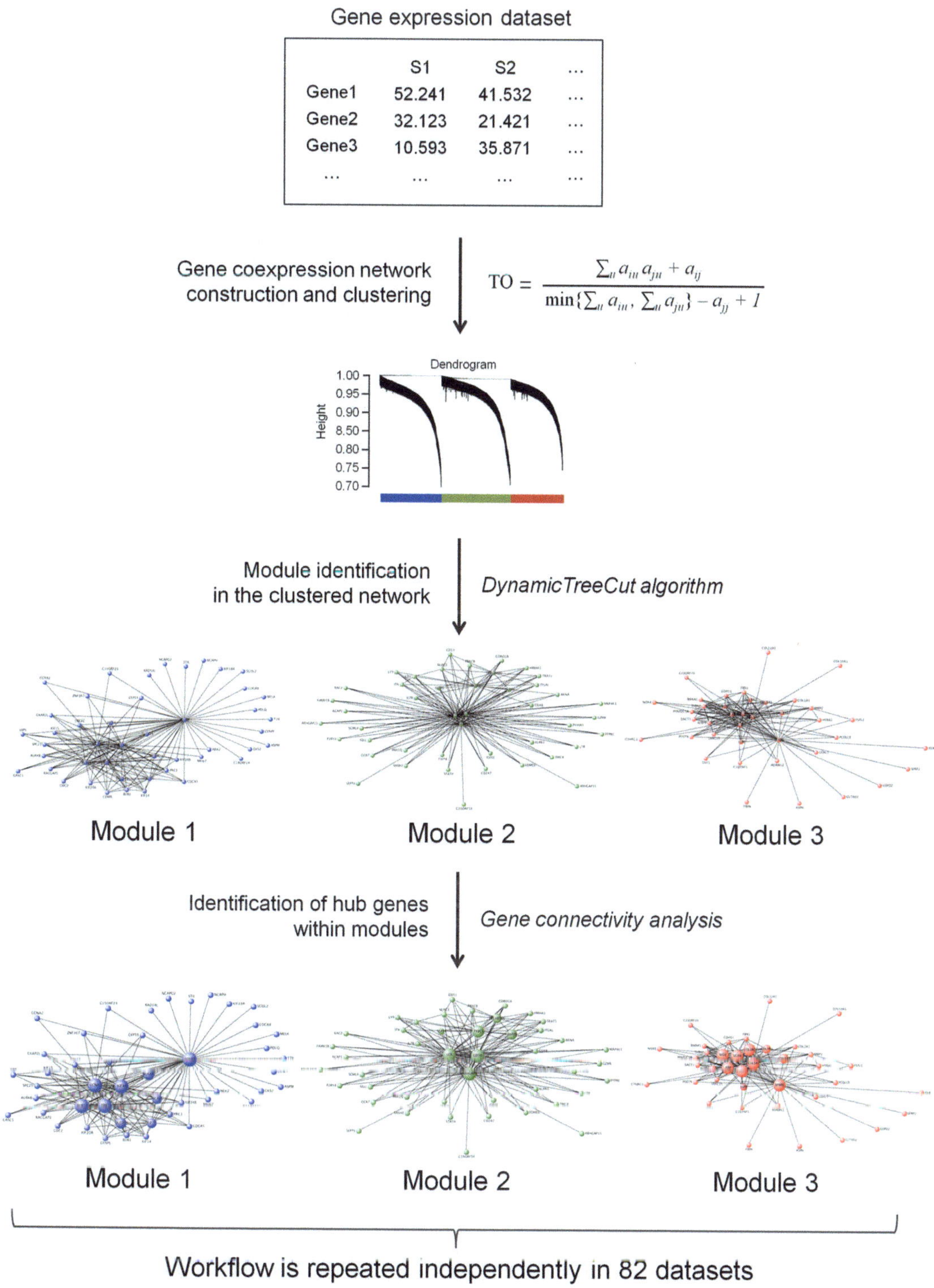

Fig. 1 Workflow overview. In each dataset, the following workflow was applied. (1) The dataset was used as a starting point to construct a gene co-expression network based on Topological Overlap between genes. TO determines similarity between gene expression profiles taking into account a systems level context. The

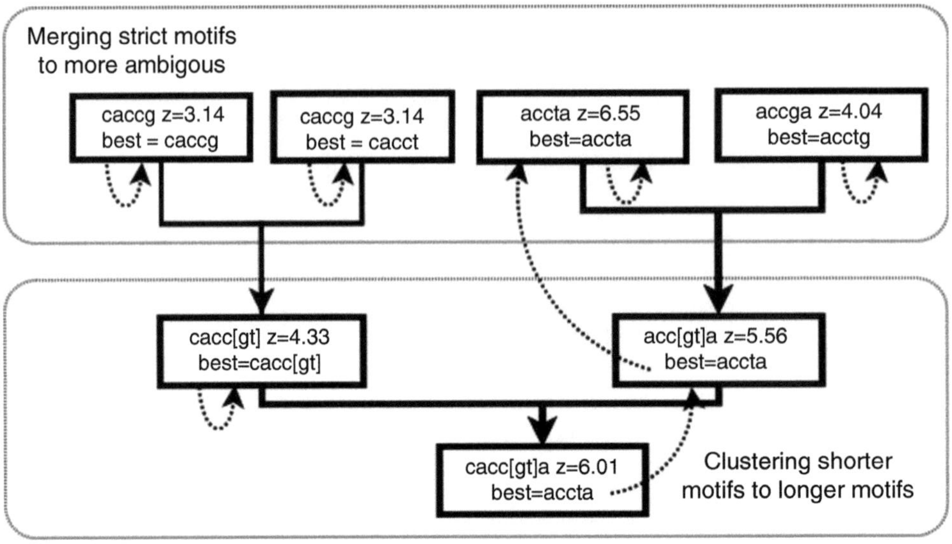

Fig. 2 Selection of the best ancestor motif

perform any *I/O*. The master process monitors the progress of the slaves and ensures that as tasks are completed, new ones are started, so that all CPU cores are maximally utilized at all times. Once all the window searches have completed, the master process post-processes the data for output.

3.2.2 Second Stage: Motif Optimization

After the approximate sequences and positions of motifs have been identified, the next step is to find the optimal sequence model and position. Since regulatory motifs are rarely conserved perfectly, the best possible results cannot be reached using only consensus representations. In the optimization step, the matrix that best describes its probabilistic model is being searched for. At the same time the best positional boundaries describing the position of the motif are being sought. The sequence of a motif is represented by the $N \times 4$ matrix (where N is the length of motif) and number *match threshold* which represents the motif conservancy. An example of a matrix representing motif is as follows:

Fig. 1 (continued) network was next hierarchically clustered, resulting in a cluster dendrogram. (2) Using DynamicTreeCut algorithm, branches were identified in the dendrogram, leading to identification of gene co-expression modules. (3) Genes in each module were further prioritized by intramodular connectivity, providing a distinction between lowly and highly connected genes. The entire workflow was repeated independently for 82 datasets, resulting in a set of gene co-expression modules in each of them

#A	C	G	T
1	0	0	0
0.26	0.49	0.09	0.14
0	0	1	0
0	0	0.15	0.84
0.60	0	0.28	0.11
0	0.21	0.03	0.75

from: 400
to: 426
match_threshold: 0.8

Each row represents one position in the motif, numbers in A, C, G, T columns represent the probability of given base being on this position.

The *cisExpress* optimization routines are based on the GA approach, and are the only nondeterministic part of *cisExpress* (due to the nondeterministic nature of GAs). The GA was implemented using the GAlib library [42]. The GA optimization method uses terms such a *genotype*, *genome*, *population*, *mutation* in different contexts to those used in the rest of this chapter. Therefore, in the following section we write terms in italic when referring to their GA meaning.

3.2.3 GA Specification

Representation of a *genotype*: From the optimization point of view a solution (*genome*) is a motif. One motif consists of the following properties: motif matrix, degree of conservation (*match_threshold*) and positional boundaries. The motif matrix is encoded as a one-dimensional array, where each cell of this array contains one row of the matrix. (In the context of GA, each cell in this array is called a *gene*. Operations such as *mutation* and *crossover* (described below) are therefore applied to entire rows rather than to separate numbers in the matrix.)

***Population* initiation:** Initially, the population is completely filled by a matrix representation of the motif to be optimized. The matrix is built from the motif as follows. For each row: if the motif allows only one base in this position, then the value in the corresponding column is 1.0 and all other values are zero; if the motif allows two possible bases in this position, then the values in each corresponding column are 0.5 and the remaining values are zero.

***Mutation* operator:** The *Mutation* operator is the tool of the GA responsible for bringing variability into a *population*. A *mutation* is divided into three steps: (1) Matrix mutation, (2) Positional boundaries mutation and (3) Motif conservancy mutation. For *Matrix mutation*, each row of the matrix undergoes the *mutation* with a certain probability, given by one of optimization parameters. If the row is selected for *mutation*, a random number given by a Gaussian distribution with mean value zero and variance given by the optimization parameter is added to each entry in the row. *Positional boundaries* and *Motif conservancy mutation*: each of these entries is mutated with a given probability. If a value is chosen for mutation, a random number given by Gaussian distribution with a mean value of zero and variance given by the optimization parameter is added to it. After the *mutation*, all mutated values are checked to be nonnegative and each row of the matrix is normalized.

Crossover operator: The *crossover operator* is the part of the GA responsible for information exchange between *individuals* in the *population*. A one-point *crossover* operator, implemented at the library level (Fig. 3), was used. From the GC point of view, *genes* are not numbers in the matrix but rather whole rows of the matrix. The *crossover* operator (Fig. 4) does not move entries between columns, but operates on entire rows of the matrix.

Fitness function: In GA terms, the *fitness function* is a function able to assess how "good" a *genome* is. The principle of evaluation of a motif represented by *genome* described above is similar to the

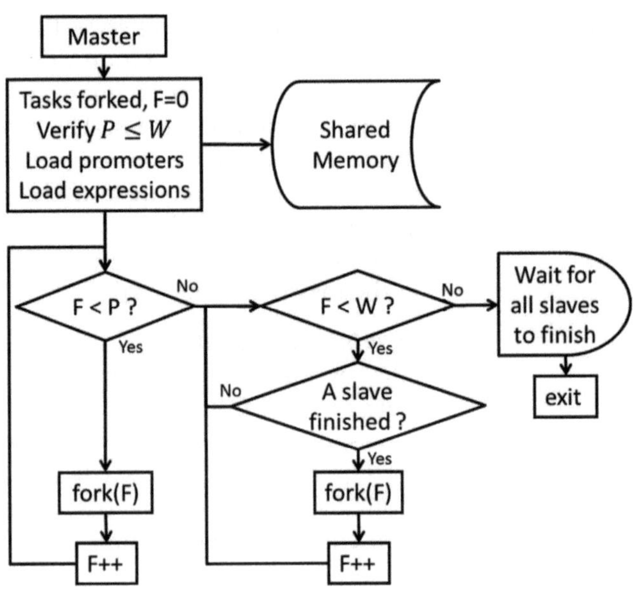

Fig. 3 Task-farming flowchart of cisExpress. A master process loads the promoters and expressions, then forks a number of slave processes equal to the number of cores available (which must be no more than the total number of windows) to perform the window search

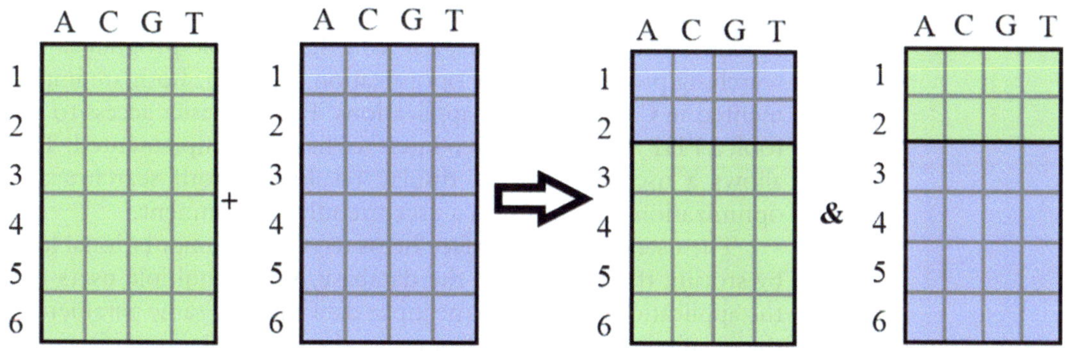

Fig. 4 The crossover operator is the tool of GA responsible for information exchange between individuals in population

evaluation described in the section on *Initial data processing*. The only difference is in the way of deciding whether or not the motif is present in the observed window of the promoter. In this case one must compute a match score (similarity of the observed promoter sequence to the motif matrix), which can be subsequently compared to the match threshold (motif conservation). If the match score of the matrix to the promoter is greater than or equal to the match threshold, then the motif is considered to be present in the promoter. If the match of the matrix to the promoter is lower than the match threshold, then the motif is considered not to be present. The match score is computed as follows: (1) compute the match score for each position in the motif—i.e., each row in the matrix; (2) the match score of a row is equal to the mean value of all columns that the promoter sequence allows on this position. This is usually only one column; however, sometimes there is uncertainty in promoter sequence; (3) the match score of the matrix to the promoter sequence is equal to the mean of the match scores of each row. After deciding whether or not the motif is present in the promoters, the evaluation of the motif is the same as described in *Initial Data Processing* and Eqs. (1–9) apply.

4 Notes

Availability, Installation, and Configuration of the Software

Latest version of the WGCNA R package can be downloaded from: https://labs.genetics.ucla.edu/horvath/CoexpressionNetwork/Rpackages/WGCNA/ or from the CRAN website. The package is installed in R using standard R package installation procedures. Detailed instructions are provided on the WGCNA home page (see the UCLA website above). Once the package is installed, it is ready for use in an R session.

cisExpress is available from http://chcb.saban-chla.usc.edu/*cisExpress/*home.php. *cisExpress* consists of two programs: the searching program and the optimization program. Both are implemented in C++ as console applications. For the easier access to the tool, a PHP/SQL based web interface has been implemented. This allows a user to run and display results of motif searches and optimization algorithms in a user friendly environment.

The user-friendly interface helps to reduce computational load by storing the results into the database. When multiple users run the application using the same input data with the same parameters, the results are returned from the database. The interface also includes tools to display graphical logos and statistical properties of discovered motifs.

cisExpress requires two input files: (1) a file containing promoter sequences aligned by the position of Transcription Start Site (in multi-fasta format; length of promoter is arbitrary) and (2) a file with expression values (in format [Sequence_Name<TAB> Expression_Value<EndLine>]).

Analysis of Identified Motifs

A comprehensive functional analysis of all found motifs and their possible role in cancerogenesis is beyond the scope of this chapter. Here, we only briefly describe the approach we took and several most relevant motifs. In order to access functionality of identified promoter sequence motifs, we have compiled lists of motifs common for different tissues and expression clusters. Consider motifs that are present across many expression clusters for a given tissue (Table 1).

Table 1
Most common motifs in promoters

Tissue	Motif	Fraction of clusters containing motif, %	Fraction of promoters containing motif, %
Brain	CGGAA	37	86
	TTCCG	33	65
Colon	CGGAA	37	75
	TTCCG	40	81
Kidney	CGGAA	16	49
	TTCCG	22	44
Lung	CGGAA	40	78
	TTCCG	41	81
Ovary	CGGAA	40	76
	TTCCG	41	81
Prostate	CGGAA	37	42
	TTCCG	50	78

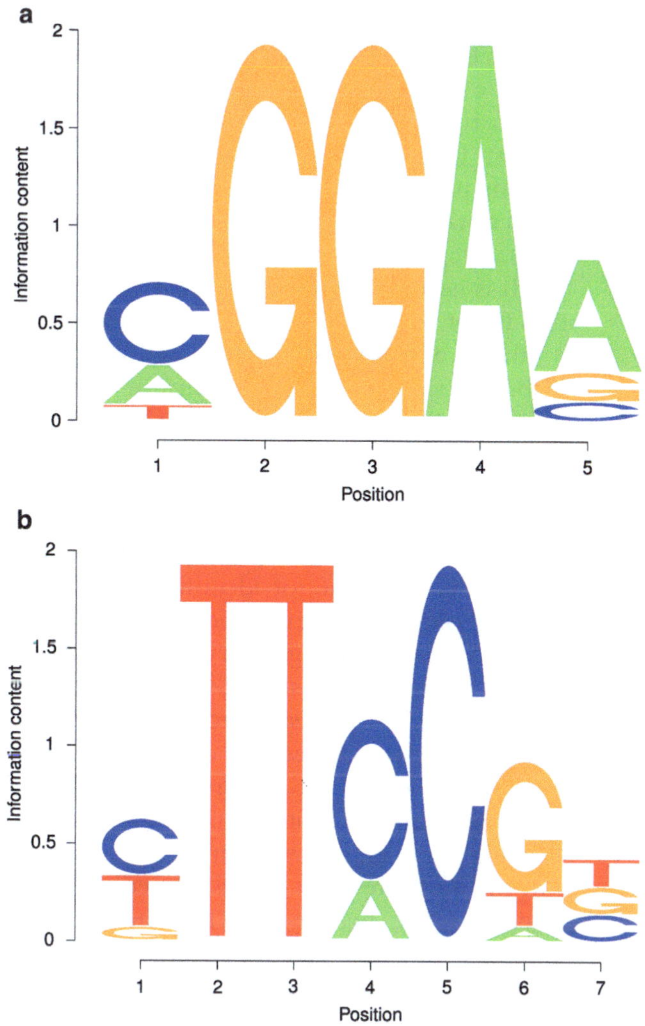

Fig. 5 Instances of motifs CGGAA (**a**) and TTCCG (**b**) in brain-specific promoters

TTCCG or its complement, CGGAA are the most common significant motifs (Z-score > 4) in promoters of genes expressed in five out of six tissues (Prostate, Ovary, Lung, Brain and Colon (Fig. 5)). This sequence is also present in the promoters of Kidney-related genes, although it is twice less prevalent (Table 1). TTCCG is a well-known target for E2f transcription factors family, the major regulators of initiation of DNA replication and G1/S transition in both human and mouse. E2f targets include cyclins, CDks, DNA repair proteins, such as major cancer-related tumor suppressor gene BRCA [43, 44]. Importantly, Rb/E2f pathway is deregulated in virtually all cancers [45], the phenomenon which contributes to cell proliferation and other cancerogenic phenotypes. A number of E2f-controlled genes are directly involved in cancerogenesis, such as Thymidylate synthase (TYMS). Inactivation of TTCCG motif in the promoter of the TYMS gene may lead to moderately decreased gene expression [46, 47].

The most tissue-specific motif is GGAAG, found at 11% in the "Brain tumor" promoters and only at 2% of promoters in other tissues. Tandem repeat GGAAG is the binding site for the GA-binding protein (GABP) [48]. GABP reactivates telomerase reverse transcriptase, which allows cancer cells to avoid apoptosis, a fundamental step in initiation of cancer development [49].

It is necessary to point out that *cisExpress* represents transcription factor binding sites using position weight matrices. Therefore, it inherits all known deficiencies of PWM-based methods, such as inability to take into account gaps of variable length, and dependencies between the residues in the binding site [9, 50]. Nevertheless, there are classes of transcription factors that can be efficiently modeled by PWMs [9]; *cisExpress* is a convenient tool for the PWM identification.

In conclusion, we developed a computational organism-independent pipeline for identification of co-expression modules from whole-genome expression profiles and a novel method of sequence analysis of regulatory elements corresponding to expression patterns of interest, *cisExpress*. We applied this pipeline to a diverse set of 82 expression profiling experiments in nine cancers of different origin. The resulted database is, arguably, the most comprehensive database of cancer-related co-expression modules and promoter sequence motifs. Functional analysis of modules revealed very high enrichment with cancer-related pathways and processes, and the most prevalent motifs proved to be highly cancer-related. Therefore, we believe that the modules and motifs with yet unknown function may be an important resource for follow-up studies on molecular mechanisms of cancerogenesis in tumors of different origin. The resource is freely available for further investigation.

References

1. Consortium, S.M.-I (2014) A comprehensive assessment of RNA-seq accuracy, reproducibility and information content by the sequencing quality control consortium. Nat Biotechnol 32 (9):903–914
2. Kristensen VN et al (2014) Principles and methods of integrative genomic analyses in cancer. Nat Rev Cancer 14(5):299–313
3. Zhao Z et al (2016) Drug drug interaction extraction from biomedical literature using syntax convolutional neural network. Bioinformatics 32(22):3444–3453
4. Zolotarenko A et al (2016) Integrated computational approach to the analysis of RNASeq data reveals new transcriptional regulators for psoriasis. Exp Mol Med 48(11):e268
5. Ivliev AE, 't Hoen PAC, Borisevich D, Nikolsky Y, Sergeeva MG (2016) Drug repositioning through systematic mining of gene coexpression networks in cancer. PLoS One 11(11):e0165059. doi:10.1371/journal.pone.0165059
6. Langfelder P, Horvath S (2008) WGCNA: an R package for weighted correlation network analysis. BMC Bioinformatics 9:559
7. MacIsaac KD, Fraenkel E (2006) Practical strategies for discovering regulatory DNA sequence motifs. PLoS Comput Biol 2(4):e36

8. Shi Q et al (2010) Biomarkers for drug-induced liver injury. Expert Rev Gastroenterol Hepatol 4(2):225–234
9. Weirauch MT et al (2013) Evaluation of methods for modeling transcription factor sequence specificity. Nat Biotechnol 31(2):126–134
10. Pavesi G, Mauri G, Pesole G (2001) An algorithm for finding signals of unknown length in DNA sequences. Bioinformatics 17(Suppl 1): S207–S214
11. Troukhan M et al (2009) Genome-wide discovery of cis-elements in promoter sequences using gene expression data. OMICS J Integr Biol 13(2)
12. Bussemaker HJ, Li H, Siggia ED (2000) Building a dictionary for genomes: identification of presumptive regulatory sites by statistical analysis. Proc Natl Acad Sci U S A 97 (18):10096–10100
13. Foat BC, Morozov AV, Bussemaker HJ (2006) Statistical mechanical modeling of genome-wide transcription factor occupancy data by MatrixREDUCE. Bioinformatics 22(14): e141–e149
14. Wang G, Yu T, Zhang W (2005) WordSpy: identifying transcription factor binding motifs by building a dictionary and learning a grammar. Nucleic Acids Res 33(Web Server issue): W412–W416
15. Sabatti C, Lange K (2002) Genomewide motif identification using a dictionary model. Proc IEEE 90(11):1803–1810
16. Sabatti C et al (2005) Vocabulon: a dictionary model approach for reconstruction and localization of transcription factor binding sites. Bioinformatics 21(7):922–931
17. Halperin Y et al (2009) Allegro: analyzing expression and sequence in concert to discover regulatory programs. Nucleic Acids Res 37 (5):1566–1579
18. Orenstein Y, Linhart C, Shamir R (2012) Assessment of algorithms for inferring positional weight matrix motifs of transcription factor binding sites using protein binding microarray data. PLoS One 7(9):e46145
19. Triska M et al (2013) cisExpress: motif detection in DNA sequences. Bioinformatics 29 (17):2203–2205
20. Hughes JD et al (2000) Computational identification of cis-regulatory elements associated with groups of functionally related genes in *Saccharomyces cerevisiae*. J Mol Biol 296 (5):1205–1214
21. Roth FP et al (1998) Finding DNA regulatory motifs within unaligned noncoding sequences clustered by whole-genome mRNA quantitation. Nat Biotechnol 16(10):939–945
22. Bailey TL et al (2006) MEME discovering and analyzing DNA and protein sequence motifs. Nucleic Acids Res 34(Web Server issue): W369–W373
23. Subramanian AR, Kaufmann M, Morgenstern B (2008) DIALIGN-TX: greedy and progressive approaches for segment-based multiple sequence alignment. Algorithms Mol Biol 3:6
24. Do CB, Batzoglou S (2008) What is the expectation maximization algorithm? Nat Biotechnol 26(8):897–899
25. Thompson W, Rouchka EC, Lawrence CE (2003) Gibbs recursive sampler: finding transcription factor binding sites. Nucleic Acids Res 31(13):3580–3585
26. Jolma A et al (2013) DNA-binding specificities of human transcription factors. Cell 152 (1–2):327–339
27. Tharakaraman K et al (2005) Alignments anchored on genomic landmarks can aid in the identification of regulatory elements. Bioinformatics 21(Suppl 1):i440–i448
28. Sandve GK, Drablos F (2006) A survey of motif discovery methods in an integrated framework. Biol Direct 1:11
29. Tompa M et al (2005) Assessing computational tools for the discovery of transcription factor binding sites. Nat Biotechnol 23(1): 137–144
30. Whalen S, Truty RM, Pollard KS (2016) Enhancer-promoter interactions are encoded by complex genomic signatures on looping chromatin. Nat Genet 48(5):488–496
31. Barrett T et al (2011) NCBI GEO: archive for functional genomics data sets—10 years on. Nucleic Acids Res 39(Database issue): D1005–D1010
32. Barrett T et al (2013) NCBI GEO: archive for functional genomics data sets—update. Nucleic Acids Res 41(Database issue): D991–D995
33. Edgar R, Domrachev M, Lash AE (2002) Gene expression omnibus: NCBI gene expression and hybridization array data repository. Nucleic Acids Res 30(1):207–210
34. Ivliev AE, Hoen PA, Sergeeva MG (2010) Coexpression network analysis identifies transcriptional modules related to proastrocytic differentiation and sprouty signaling in glioma. Cancer Res 70(24):10060–10070
35. Miller JA, Oldham MC, Geschwind DH (2008) A systems level analysis of transcriptional changes in Alzheimer's disease and normal aging. J Neurosci 28(6):1410–1420
36. Miller JA, Horvath S, Geschwind DH (2010) Divergence of human and mouse brain transcriptome highlights Alzheimer disease

pathways. Proc Natl Acad Sci U S A 107 (28):12698–12703
37. Suzuki A et al (2015) DBTSS as an integrative platform for transcriptome, epigenome and genome sequence variation data. Nucleic Acids Res 43(Database issue):D87–D91
38. Tatarinova T et al (2013) NPEST: a nonparametric method and a database for transcription start site prediction. Quant Biol 1(4):61–271
39. Zhang B, Horvath S (2005) A general framework for weighted gene co-expression network analysis. Stat Appl Genet Mol Biol 4:17
40. Langfelder P, Zhang B, Horvath S (2008) Defining clusters from a hierarchical cluster tree: the dynamic tree cut package for R. Bioinformatics 24(5):719–720
41. Whitley D (1994) A genetic algorithm tutorial. Stat Comput 4(2):65–85
42. Wall M (2007) GAlibA C++ library of genetic algorithm components. Massachusetts Institute of Technology, Cambridge, MA
43. Sonkin D et al (2013) Tumor suppressors status in cancer cell line encyclopedia. Mol Oncol 7(4):791–798
44. Thakur S et al (2003) Regulation of BRCA1 transcription by specific single-stranded DNA binding factors. Mol Cell Biol 23(11):3774–3787
45. Nevins JR (2001) The Rb/E2F pathway and cancer. Hum Mol Genet 10(7):699–703
46. Evangelou K, Havaki S, Kotsinas A (2014) E2F transcription factors and digestive system malignancies: how much do we know? World J Gastroenterol 20(29):10212–10216
47. Xanthoulis A, Tiniakos DG (2013) E2F transcription factors and digestive system malignancies: how much do we know? World J Gastroenterol 19(21):3189–3198
48. Sadasivan E, Cedeno MM, Rothenberg SP (1994) Characterization of the gene encoding a folate-binding protein expressed in human placenta. Identification of promoter activity in a G-rich SP1 site linked with the tandemly repeated GGAAG motif for the ets encoded GA-binding protein. J Biol Chem 269 (7):4725–4735
49. Bell RJ et al (2015) Cancer. The transcription factor GABP selectively binds and activates the mutant TERT promoter in cancer. Science 348 (6238):1036–1039
50. Siddharthan R (2010) Dinucleotide weight matrices for predicting transcription factor binding sites: generalizing the position weight matrix. PLoS One 5(3):e9722

Chapter 12

Rule Mining Techniques to Predict Prokaryotic Metabolic Pathways

Rabie Saidi*, Imane Boudellioua*, Maria J. Martin, and Victor Solovyev

Abstract

It is becoming more evident that computational methods are needed for the identification and the mapping of pathways in new genomes. We introduce an automatic annotation system (ARBA4Path Association Rule-Based Annotator for Pathways) that utilizes rule mining techniques to predict metabolic pathways across wide range of prokaryotes. It was demonstrated that specific combinations of protein domains (recorded in our rules) strongly determine pathways in which proteins are involved and thus provide information that let us very accurately assign pathway membership (with precision of 0.999 and recall of 0.966) to proteins of a given prokaryotic taxon. Our system can be used to enhance the quality of automatically generated annotations as well as annotating proteins with unknown function. The prediction models are represented in the form of human-readable rules, and they can be used effectively to add absent pathway information to many proteins in UniProtKB/TrEMBL database.

Key words Pathway prediction, Machine learning, Rule mining, Automatic annotation, Functional genomics, Proteomics

1 Introduction

The widening gap between the amount of known proteins and knowledge of their functions has encouraged the development of methods to automatically infer annotations. Functional annotation of proteins encoded in newly sequenced genomes is expected to meet the conflicting requirements of providing as much comprehensive information as possible while avoiding erroneous functional assignments. This trade-off imposes a great challenge in designing intelligent systems to tackle the problem of automatic protein annotation. Hence, the need for automated methods is urgent to help increase the annotation coverage, detect

*These authors contributed equally to this work.

Tatiana V. Tatarinova and Yuri Nikolsky (eds.), *Biological Networks and Pathway Analysis*, Methods in Molecular Biology, vol. 1613, DOI 10.1007/978-1-4939-7027-8_12, © Springer Science+Business Media LLC 2017

inconsistencies and provide seeds for manual curation. There are several approaches proposed in the literature for such a task. A quite promising approach is to apply knowledge discovery and data mining techniques to predict some protein features based on a set of known data. Such rule-based methods provide rich automatic functional annotations and aid in performing integrity checks. For instance, Kretschmann et al. [1] applied C4.5 data mining algorithm [2] to gain knowledge about the Keyword annotation from UniProtKB/Swiss-Prot [3]. Rule-base [4] is another semiautomatic annotation system run on UniProtKB/TrEMBL [3]. It uses the annotation of UniProtKB/Swiss-Prot entries that possess a set sequence signatures to annotate UniProtKB/TrEMBL entries that contain the same signature, fundamentally with keywords and comments. Other examples of automatic annotation systems that generate annotations of several protein features integrated into UniProtKB/TrEMBL are HAMAP-Rule [5], EDIT to UniProtKB/TrEMBL [6], and PIR [7].

One of the central research goals of systems biology is modeling various biological processes. Elucidation of chemical reactions and pathways is one of the challenging problems in this field. A biological pathway is formed by a series of chemical reactions catalyzed by enzymes within a cell. Some of the most common biological pathways are those associated with metabolism, regulation of gene expression and transmission of molecular signals. A metabolic pathway involves the step-by-step modification of an initial molecule to form another product. The resulting product can be stored by the cell, secreted, used immediately, or used to initiate another metabolic pathway. An example of a metabolic pathway is the cellular respiration equation where glucose is oxidized by oxygen to produce ATP, adenosine triphosphate [8]. Pathways play a key role in advanced studies of functional genomics. For instance, identifying pathways involved in a disease may lead to effective strategies for diagnosing, treating, and preventing diseases. Moreover, by comparing the behavior of certain pathways between a healthy person and a diseased person, researchers can discover the roots of the disorder and use the information gained from pathway analysis to develop new and better drugs [9–11]. It is increasingly clear that mapping dysregulated pathways associated with various diseases is crucial to fully understand these diseases [12]. In addition, pathways are often conserved, thus studying their interactions in model organisms may help elucidate cellular response mechanisms in other organisms.

One of the very first pathway prediction systems was PathFinder [13] which aims to identify signaling pathways in protein–protein interaction networks. It extracts the characteristics of known signal transduction pathways and their functional annotations in the form of association rules. There are also tools that predict biodegradation pathways such as META [14], CATABOL [15], and UM-PPS [16]. In addition, relative reasoning has been used in the prediction of mammalian detoxification pathways in order to limit combinatorial explosion [17]. The PathoLogic component of the Pathway Tools software [18] is the state of the art in pathway prediction. It performs prediction of metabolic pathways in sequenced and annotated genomes using MetaCyc [19] as the reference metabolic pathway database. One of the limitations of this system is extendibility due to the fact that its logic is hardcoded. That is because PathoLogic incorporates rules and heuristics developed using feedback from biologists to improve the accuracy of the predictions. Another limitation is becoming more apparent with the growth of MetaCyc size, resulting in PathoLogic suffering from more false-positive pathway predictions. In addition, the algorithm is limited to Boolean predictions with a coarse measure of prediction confidence making it difficult to filter the predictions with a probability cutoff. A comparative analysis was conducted [20] revealing that some machine learning approaches performed better than PathoLogic in pathway prediction.

The tremendously increasing growth of UniProtKB raises a double challenge for both high-quality and high-coverage annotations. Although literature-based manual annotation of pathways ensures incorporation of a valuable knowledge and the quality of the database, it is very far away from keeping up with the ever increasing amount of recently sequenced genomes. Therefore, we suggest that association rule mining could be used effectively as a computational method for pathway prediction. Association rule mining is a technique originated from the analysis of data on market baskets. The objective is to locate trends by means of association relationships and correlations within a dataset. Essentially, the aim of such analysis is to discover a set of useful rules that are shared by a percentage of the dataset. An association rule is an implication expression of the form $X \Rightarrow Y$, where X and Y are disjoint itemsets. Association rule mining was used in several applications of bioinformatics including mining gene expression data [21], analyzing microarray data [22], and identifying related GO terms [23]. Moreover, association rule mining was used to improve the quality of automatically generated annotations by detecting anomalies in annotation items [24]. In the context of automated protein annotation, we consider association rules in the form of many to one implications. If an annotation satisfies a rule with accepted quality of metric values, then we hypothesize that such a rule may reflect a biological regularity. An example of an association rule in a database of annotated proteins is: "Nuclear localization ⇒ Origin:eukaryota", which describes that every protein which is annotated as localized in nucleus has a eukaryotic origin [24].

In this chapter we describe ARBA4Path, an automatic annotation system, that was introduced in [25]. ARBA4Path utilizes rule mining techniques to predict metabolic pathways for UniProtKB/TrEMBL data. ARBA4Path can be used to enhance the quality of automatically generated annotations as well as annotating proteins with unknown function. The pathway prediction system utilizes proteins from UniProtKB/Swiss-Prot [3], which is a high quality manually annotated and nonredundant protein sequence database containing experimental results, computed features, and scientific conclusions. Specifically, the pathway prediction system uses InterPro [26] signatures and organism taxonomy attributes of UniProtKB/Swiss-Prot entries to predict metabolic pathways associated with each protein entry. The association algorithm, Apriori [27], is used in the learning phase to identify significant relationships between the attributes of UniProtKB/Swiss-Prot annotations. Furthermore, we use a filtering method, SkyRule [28, 29], to select the best rules based on a combination of several interestingness metrics. The resulting rules represent the prediction models that are in the form of human-readable rules, and they can be used effectively to add absent pathway information to many proteins in UniProtKB/TrEMBL database. We finally present an evaluation study on UniProtKB prokaryotic entries to demonstrate the performance, capability, and robustness of this approach.

2 Methods

ARBA4Path is designed to solve the following problem: given a set of UniProtKB/SwissProt entries, generate models for pathway prediction using rule mining techniques. As any machine learning system, the system has two major phases: the learning phase and the applying phase. The learning phase involves the training and testing on UniProtKB/Swiss-Prot input data to obtain the prediction models while the applying phase involves applying the prediction models on the respective UniProtKB/TrEMBL entries.

2.1 Creation of Itemsets

First, we extract the necessary information from the desired input entries of UniProtKB/SwissProt with metabolic pathways as targets, and InterPro signatures and organism taxonomic lineages as attributes. We chose InterPro signatures as an attribute since it covers 96.3% of UniProtKB/Swiss-Prot entries and 77.2% of UniProtKB/TrEMBl entries (as of November 2015). This high coverage will aid us in the learning process by using InterPro signatures identifiers as an attribute for the prediction models as well as in the annotation phase. The attributes and target representation in UniProtKB are described as follows:

2.1.1 Target: Metabolic Pathway Comment

Represented as a structured hierarchy of controlled vocabulary where each process is split up into superpathway, pathway, and/or subpathway. When known, the step number mediated by the protein within the pathway is also indicated. On the other hand, when the metabolic pathway is not fully known, only the superpathway and pathway labels are indicated. Moreover, a protein can participate in different pathways or in different steps of the same pathway. An example of a fully known pathway representation in UniProtKB for the protein Anthranilate synthase component 1 is: Amino-acid biosynthesis; L-tryptophan biosynthesis; L-tryptophan from chorismate: step 1/5. Also *see* **Note 1**.

2.1.2 Attribute: InterPro Signature ID

The InterPro signature IDs are cross-referenced from InterPro database, which is an integrated resource for protein families, domains, and functional sites. InterPro provides functional analysis of proteins by classifying them into families and domains. Protein signatures are combined from 11 member databases into a single searchable resource. A protein entry could be associated with one or more InterPro IDs. An example of InterPro IDs associated with the protein Anthranilate synthase component 1 is: IPR005801 (a domain), IPR019999 (a family), IPR006805 (a domain), IPR005256 (a family), and IPR015890 (a domain).

2.1.3 Attribute: Taxonomic Lineage

The taxonomy in UniProtKB is based on the NCBI taxonomy database and is organized in a tree structure that represents the taxonomic lineage. It contains the taxonomic hierarchical classification lineage of the source organism. It lists the nodes as they appear top-down in the taxonomic tree, with the more general grouping listed first. An example of taxonomic lineage representation for protein Anthranilate synthase component 1 is: Bacteria; Proteobacteria; Gammaproteobacteria; Pseudomonadales; Pseudomonadaceae; Pseudomonas.

The extracted list of attributes and targets for each loaded entry will be characterized in the form of an itemset. Table 1 describes some examples of the forms of itemsets that are associated with some UniProtKB/Swiss-Prot protein identifiers.

Table 1
Examples of itemsets corresponding to some UniProt/Swiss-Prot entries

Entry ID	Corresponding itemset
Q8TRZ4	PATHWAY: One-carbon metabolism; methanogenesis from acetate, TAXON:Archaea, TAXON: Euryarchaeota, TAXON: Methanomicrobia, TAXON:Methanosarcinales, TAXON: Methanosarcinaceae, TAXON: Methanosarcina, IPR: IPR017896, IPR: IPR017900, IPR: IPR004460, IPR:IPR004137, IPR: IPR009051, IPR: IPR011254, IPR: IPR016099
P18335	PATHWAY: Amino-acid biosynthesis; L-arginine biosynthesis; N(2)-acetyl L ornithine from L-glutamate: step 4/4, PATHWAY: Amino-acid biosynthesis; L-lysine biosynthesis via DAP pathway; LL-2,6-diaminopimelate from (S) tetrahydrodipicolinate (succinylase route): step 2/3, TAXON: Bacteria, TAXON: Proteobacteria, TAXON: Gammaproteobacteria, TAXON: Enterobacteriales, TAXON: Enterobacteriaceae, TAXON: Escherichia, IPR: IPR017652, IPR: IPR004636, IPR:IPR005814, IPR: IPR015424, IPR:IPR015421, IPR: IPR015422

2.2 Generation of Association Rules

The prepared itemsets form the input of Apriori algorithm proposed by Agarwal and Srikant [27]. Apriori, a bottom-up approach, is one of the well-known association rule mining techniques. Apriori aims to discover all significant association rules that represent trends in a large database of entries or transactions. It proceeds by identifying the frequent individual items in the database and extending them one item at a time as long as those itemsets appear sufficiently often in the database. The frequent itemsets determined by Apriori are used to generate association rules which highlight general trends in the database: We use the Apriori implementation developed by Borgelt [30]. This implementation uses a prefix tree to organize the support counters and a doubly recursive procedure to process the transaction to count the support of candidate itemsets. Apriori could be configured to provide different evaluating measures with each generated association rule. Each evaluation measure tries to quantify the dependency between the antecedent and the consequent of and association rule. The user has the freedom to select the threshold values for each evaluation measure based on his requirements. We use a combination of four measures to effectively minimize false positives and the number of rules generated out of pure randomness. The chosen metrics are:

2.2.1 Support

According to [31], the support of an association rule R = A AND B ⇒ C (noted supp(R)) is the support of the set S = A, B, C which is defined by the absolute or relative number of cases in which the rule is correct. In the prior example, it is the number of cases where the occurrence of item C follows from the occurrences of items A and B. However, this definition may cause some problems if multiple evaluation measures are used [30]. Hence, we adopt the definition proposed by [30, 32, 33] which describes the support of an association rule as the absolute or relative number of cases in which it is applicable, in other words, in which its antecedent part holds. Unlike the original definition, the support in this case provides a useful statistical meaning of the support of a rule and its confidence [30].

2.2.2 Confidence

Confidence metric is used to measure the quality of a particular association rule. More intuitively, it measures the reliability of the inference made by a rule. Introduced in [31], the confidence of an association rule $R = X \Rightarrow Y$ (noted conf(R)), where X and Y are itemsets, is calculated as the support of the set of all items that appear in the rule divided by the support of the antecedent set. More formally,

$$\mathrm{conf}(R) = \frac{\mathrm{supp}(X \cap Y)}{\mathrm{supp}(X)}$$

In other words, the confidence of a rule is the number of cases in which the rule is correct relative to the number of cases in which it is applicable. A high confidence ratio indicates that its associated rule has a high probability of correctness and thus makes correct predictions. It is worth mentioning that rules with high confidence may occur by chance. Determining whether the antecedent and the consequent are statistically independent is used to detect such spurious rules. One of the measures that could assist with this is the lift value.

2.2.3 Lift Value

The lift value, or confidence quotient, is basically the quotient of the posterior and the prior confidence of an association rule. Mathematically speaking, the lift of a rule $R = X \Rightarrow Y$ is:

$$\text{lift}(R) = \frac{\text{conf}(X \Rightarrow Y)}{\text{conf}(\varnothing \Rightarrow Y)}$$

where $\varnothing$ is the empty set and hence supp($\varnothing$) is the number of transactions (entries) in the database. Lift measures how far from independence the antecedent and consequent are. A lift value equals to one implies that the antecedent and consequent are independent and that the support of a rule is expected considering the supports of its components which renders such rule not interesting. If the resulting lift value is greater than one, this implies that the presence of the antecedent items raises the confidence. Likewise, if the lift value is less than one, then the presence of the antecedent items lowers the confidence.

2.2.4 p-Value

In statistics, the *p*-value is used to measure the statistical significance of a result. Several statistical tests have been used to calculate *p*-values of association rules [34, 35]. Here, we adopt the *p*-value computed from G-Statistic. Under independence, the G-statistic also has a χ^2-distribution. The chi-squared statistic can be used to calculate a *p*-value by comparing the value of the statistic to a χ^2 distribution. That is, the *p*-value is computed as the probability that the χ^2-value of an association rule can be observed by chance assuming that the antecedent and the consequent of the rule are independent. This measure does not assess the strength of correlation between antecedent and consequent. It only assists in deciding about the independence of the antecedent and the consequent in a rule. The *p*-value is used to infer how likely the occurrence of the rule is due to a systematic effect instead of pure random chance. If a rule has a low *p*-value, then this rule has a low chance to occur if its two sides are independent. Given that this rule is observed in the data, then its two sides are unlikely to be independent, and thus, the association between them is likely to be real. On the other hand, high *p*-value means that the rule has a high chance to occur even if there is no association between its two sides. Such rules should be discarded.

Some examples of rule representation along with their quality metrics are shown in the Table 2 (*see* **Notes 2–4**).

Table 2
Examples of rules generated by Apriori along with their evaluation measures from a set of UniProt/Swiss-Prot entries

Consequent	Antecedent	Support	Conf	Lift	*p*-value
PATHWAY:Cofactor biosynthesis; adenosylcobalamin biosynthesis	IPR:IPR003705	3.25E-04	1	90.5787	6.47E-63
PATHWAY:tRNA modification; archaeosine-tRNA biosynthesis	IPR:IPR004804 IPR:IPR002616 TAXON:Archaea	3.35E-04	1	2983.44	2.72E-127
PATHWAY:Amino-acid biosynthesis; L-leucine biosynthesis; L-leucine from 3-methyl-2-oxobutanoate: step 2/4	IPR:IPR004430 IPR:IPR018136 IPR:IPR001030 TAXON: Enterobacteriaceae TAXON: Proteobacteria TAXON:Bacteria	8.07E-04	1	94.6184	1.07E-155

2.3 Selection of Association Rules

Apriori typically generates a large number of rules especially for large databases (mining irrelevant rules etc.). The user is thus unable to determine the most interesting association rules and make decisions based on these rules. Hence, we need an efficient evaluation of rules to select those that are actually relevant. The generated list of rules will be analyzed by SkyRule software [28, 29] to selects the rules that are supposed to be the most interesting ones accounting several measures. In our case, the interestingness measures considered are support, confidence, lift, and *p*-value that were discussed in the previous subsection. SkyRule approach adopts the notion of dominance and comparability between association rules to discover interesting association rules without favoring or excluding any measure among the used ones. SkyRule also eliminates the need for the threshold value specification through the use of dominance relationship. The dominance relationship which is the cornerstone of the SkyRule operator is applied on rules and can be summarized as follows:

- A rule r is said to be dominated by another rule r', if for all used measures, r' has better measures than r.
- A rule $x = (A \Rightarrow B)$ is said to be comparable to rule $x' = (C \Rightarrow D)$ if $B = D$ AND $A \cap C = \emptyset$.

The dominance relationship describes the relevance of rules whereas the semantic comparability helps to decide if the considered association rules are semantically related (i.e., comparable). Comparability defines a kind of semantic relationship between rules in order to restrict the use of dominance. Concretely, the dominance between two rules must be applied only if a semantic relationship exists between them. SkyRule utilizes the concepts of dominance and comparability to select a family of inter-independent and statistically relevant rules, we term them representative rules. SkyRule works as follows:

1. Add all rules of the set of candidate rules.
2. For each rule, compute the Euclidean distance to the normalized ideal metrics (1.0) for all four quality metrics.
3. Sort the set of rules in a descending order by their associated distances.
4. Select a representative rule as the rule which has metrics closest to the normalized ideal metrics (smallest distance value).
5. Discard all rules comparable to the representative rule from the set of candidate rules.
6. Repeat **items 4** and **5** until no more rules in candidate set.

Essentially, SkyRule will filter out rules so that only undominated and incomparable rules are maintained. The set of representative rules will be used to construct our prediction models as described in the next subsection.

2.4 Construction of Prediction Models

The chosen rules by SkyRule will be aggregated to create a model for each pathway target. For example, if we have two rules of the form A $\Rightarrow$ C and B $\Rightarrow$ C, then we aggregate them to a single rule such that A OR B $\Rightarrow$ C. The set of the aggregated rules will build the final prediction models that are presented in a human readable format. Table 3 shows some examples of the aggregated rules presented in the form of prediction models. For each rule, the antecedent set is accompanied by its four evaluation measures and its Euclidean distance to normalized ideal metrics. These prediction models are applied to UniProtKB/TrEMBL entries to annotate them accordingly (*see* **Note 5**).

2.5 Annotation of UniProtKB/TrEMBL Entries

The final step in our pipeline is to apply the generated prediction models on the respective data set (or part of it) of those entries in UniProtKB/TrEMBL that belong to the same taxonomic group as those from UniProtKB/Swiss-Prot for the learning phase. That is, the target entries (with their InterPro signatures and taxonomic lineage attributes) will be annotated based on the satisfaction of one or more of the prediction models rule and consequently, annotated with one or more pathways (*see* **Notes 6** and **7**).

3 Materials

In this section, we discuss the choice of the dataset and present a case study of our system on prokaryotic data in UniProt which illustrates an application of our methods with evaluation and comparison to other existing automatic annotation systems.

3.1 Dataset Preparation

The current status in UniProtKB for prokaryotes is summarized in Table 4. Firstly, the system loads all prokaryotic protein entries from UniProtKB/Swiss-Prot. After that, we filter out the entries

Table 3
Examples of prediction models obtained in the form or aggregated rules along with their evaluation measures. Each rule is accompanied by its four evaluation measures and its Euclidean distance to normalized ideal metrics

[PREDICT] PATHWAY:Quinol/quinone metabolism; 1,4-dihydroxy-2-naphthoate biosynthesis; 1,4-dihydroxy-2-naphthoate from chorismate: step 7/7
[IF]
[IPR:IPR022829] 0.000332364–1.0–0.030303074366431242–1.0$\,\to\,$1.3927122854520324
OR
[IPR:IPR029069, TAXON:Cyanobacteria] 0.000332364–1.0–0.030303074366431242–1.0$\,\to\,$1.3927122854520324
[END]

[PREDICT] PATHWAY:Purine metabolism; IMP biosynthesis via de novo pathway; N(2)-formyl-N(1)-(5-phospho-D-ribosyl)glycinamide from N(1)-(5-phospho-D-ribosyl)glycinamide (formate route): step 1/1
[IF]
[IPR:IPR005862] 0.00232655–1.0–0.004464281262982967–1.0 → 1.4094125301401756
OR
[IPR:IPR001509, IPR:IPR011761] 0.000633569–1.0–0.004464281262982967–1.0 → 1.4106114385935296
OR
[IPR:IPR003135, IPR:IPR013815, TAXON:Enterobacteriaceae] 0.000436228–1.0–0.004464281262982967–1.0 → 1.4107512543237724
[END]

Table 4
Current status in UniProtKB for prokaryotes

	Swiss-Prot	TrEMBL
Total number of entries	351,649	34,356,770
Entries with pathway annotations	30.44%	5.22%
Entries with InterPro annotations	98.76%	76.17%

As of November 2015

that do not contain pathway functional annotation as an attribute. Moreover, in order to maintain data quality, the system only considers entries with manual assertion evidence. An evidence is described by a code from the Evidence Codes Ontology (ECO) [36]. ECO is a controlled vocabulary of terms that describe scientific evidence in the realm of biological research. ECO can be used to document both the evidence that supports a scientific conclusion and how that conclusion was recorded by a scientist. The evidence types that are used in UniProtKB for manual assertion are described in Table 5. At this stage, we ended up with a total of 96,280 entries that will form our itemsets (*see* **Note 8**).

Table 5
Considered evidences for pathway annotation in UniProtKB/Swiss-Prot

Evidence ID	Evidence label	Description
ECO:0000269	Experimental evidence	Manually curated information for which there is published experimental evidence.
ECO:0000303	Non-traceable author statement evidence	Manually curated information that is based on statements in scientific articles for which there is no experimental support.
ECO:0000305	Curator inference evidence	Manually curated information which has been inferred by a curator based on his/her scientific knowledge or on the scientific content of an article.
ECO:0000250	Sequence similarity evidence	Manually curated information which has been propagated from a related experimentally characterized protein.
ECO:0000255	Sequence model evidence	Manually curated information which has been generated by the UniProtKB automatic annotation system or by various sequence analysis programs that are used during the manual curation process and which has been verified by a curator.
ECO:0000244	Combinatorial evidence	Manually curated Information inferred from a combination of experimental and computational evidence.

Table 6
Apriori threshold values considered for the system

Parameter	Value
Minimum number of items per association rule	2
Minimum support of an itemset (absolute number of transactions)	20
Minimum confidence of a rule as a percentage	100%

3.2 Apriori and SkyRule

For our given scenario, Table 6 displays the suggested threshold values we considered for Apriori. Given our selected dataset and parameters, Apriori successfully generated 568,006 rules in total. Next, out of all the rules generated by Apriori, SkyRule selected 1347 rules as representative rules. These rules were aggregated to form 356 prediction models.

3.3 Annotation of UniProtKB/TrEMBL

In order to capture the performance of our system, we considered the reference proteome set of prokaryotic entries of UniProtKB/TrEMBL for the purpose of annotation using our prediction models. Reference proteomes are a subset of proteomes that have been selected either manually or algorithmically according to some criteria to provide a broad coverage of the tree of life and a representative cross-section of the taxonomic diversity found within

UniProtKB. It also covers the proteomes of well-studied model organisms and other species of interest for biomedical research. These reference proteomes are tagged with the keyword "Reference proteome." As of November 2015, the reference proteome set of UniProtKB/TrEMBL entries of prokaryotes represents a fraction of around 18% over all prokaryotic UniProtKB/TrEMBL entries available in UniProtKB. In details, there are 6,193,540 prokaryotic reference proteome entries in UniProtKB/TrEMBL out of 34,356,770 total prokaryotic UniProtKB/TrEMBL entries. The coverage of our automatic annotations over the set specified is illustrated in the next section.

3.4 System Evaluation

In order to evaluate the robustness of our system, we use the cross-validation technique with multiple runs. Cross-validation is a standard technique to give an insight on how the prediction models will generalize to an independent dataset. A single round of cross-validation involves partitioning data into complementary subsets and performing the analysis on one subset (called the training set), and validating the generated predictor on the other subset (called the validation set or testing set). For this experiment, we used the set of UniProtKB/Swiss-Prot prokaryotic entries containing pathway annotations with manual assertion evidence (96,280 entries in total as of November 2015). We define our positives and negatives relative to our reference set of pathways present in at least 20 protein entries of our target set. The positive class contains associations between each entry and its pathway annotation, which is present in the reference set of pathways. On the other hand, the negative class has associations between each entry and all pathways of the reference set which are not present in the entry annotations. Moreover, we define a true positive (TP) as to when we successfully predict a pathway present in the protein from the reference set. Likewise, a true negative (TN) case is when we do not predict a pathway annotated for the protein and present in the reference set of pathways. For example, if a protein has an annotation with two pathways (both present in the reference set) and the system predicted only one of two pathways, then we will count one True positive and one false negative. A more general example is; if we have x number of pathways in the reference set of pathways and n proteins where each protein is annotated with a unique pathway from the set of x, assuming we predicted them all correctly, then we will have n TP and $n(x - 1)$ TN.

Our validation results are averaged over two runs where at each run, a five-fold cross-validation is performed. There are five different evaluation metrics considered which are accuracy, precision, recall, F_1-measure, and area under the curve (AUC) defined as:

- Accuracy $= (TP + \frac{TN)}{(TP} + FP + TN + FN)$
- Precision $= \frac{TP}{TP} + FP$

- Recall $= \frac{TP}{(TP}+FN)$
- F_1-measure $= 2 \times \frac{\text{Precision} \times \text{Recall}}{\text{Precision}} + \text{Recall}$
- AUC = Area under the ROC curve. A receiver operating characteristic (ROC) is a plot that illustrates the performance of a binary score-based classifier. It depicts the trade-offs between the true positive rate and the false positive rate while varying the score threshold from best to worst values. The area under the ROC curve is a summary measure that essentially averages diagnostic accuracy across the spectrum of threshold values. Calculating the global evaluation metrics over all target pathways, our system achieved a very high accuracy of pathway identification with an F_1-measure of 0.982, a precision of 0.999, a recall of 0.966 and an AUC of 0.987.

3.5 Relevance of GO Annotations to Pathway Annotations

Recently, the use of similarity measures [37] for comparison between various biological ontologies or, by extension, between entities annotated with some concepts (functional annotations in our case) had increased rapidly. We aim to study the relevance of Gene Ontology (GO) [38] of the entries we annotated by our prediction models to those entries known to possess the same target pathway annotation. We considered, as a case study, the set of protein entries of *Escherichia coli* in UniProtKB/Swiss-Prot (NCBI Taxonomy Identifier: 83333) since the coverage of GO annotation on UniProtKB/TrEMBL is low. We applied our prediction models on those entries that lack pathway annotations (4749 entries in total). The prediction models provided 365 predictions touching 326 entries with 62 pathways that vary in their hierarchical representation. The set of those entries along with their pathway annotation are mapped to their corresponding entries of UniProtKB/Swiss-Prot with manual assertion evidence (171 entries) that share the same pathway annotation. This mapping is constructed in a form of pairs such that if protein P1 is known to participate in pathway P and protein P2 is predicted to participate in the same pathway P, then we form a pair in our mapping list as (P1 P2), and so on. We intend to test the hypothesis that the computed GO semantic similarity scores of these pairs will be significant compared to the GO semantic similarity scores computed for the rest of protein pairs.

We computed semantic similarity for all GO ontology annotations available for the set of UniProtKB/Swiss-Prot entries of *Escherichia coli* (taxid: 83333) (5394 entries in total). We used Semantic Measures Library SML [39] with Resnik measure [40] that is based on the information content of the most informative common ancestor. The GO scores of the resulting pairwise semantic similarity computation with best matching average are recorded. The GO scores that correspond to the computed mapping pairs will form the set of our positives for the Wilcoxon rank-sum test and the

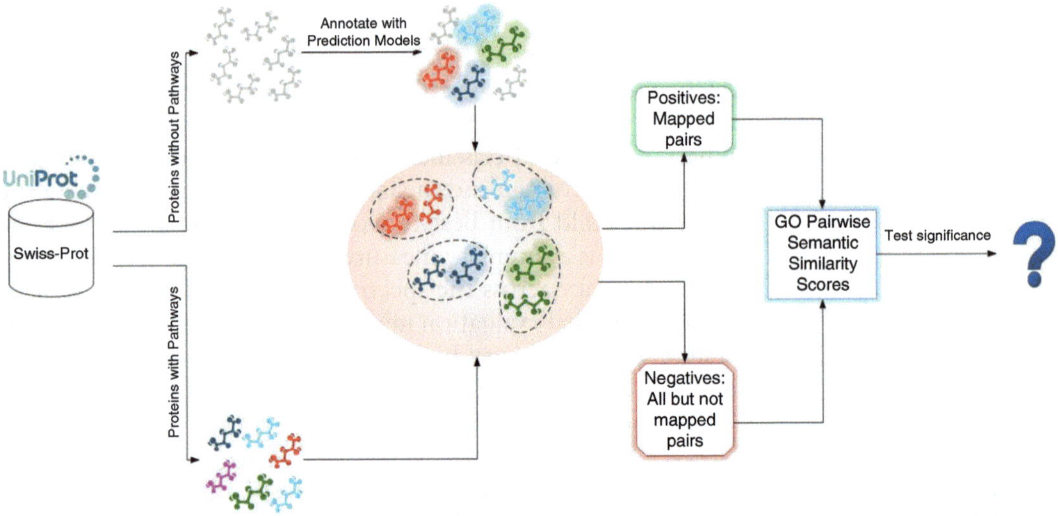

Fig. 1 Workflow for evaluating GO annotation relevance to our pathway annotations. A coloured molecular symbol represents a protein with a pathway annotation whereas a gray coloured molecular symbol represents a protein lacking any pathway annotation. Also, a shaded molecular symbol depicts a protein we annotated using our prediction models with a pathway and a non-shaded molecular symbol depicts a protein manually asserted with a pathway annotation. Different colours of the symbols illustrate different pathways associated with proteins

rest of the pairs are the negatives. We found that the p-values are less than 2.2e-16 which firmly indicates that there is a strong significance of the Go scores corresponding to the positive pairs. A summary of this workflow is illustrated in Fig. 1. Therefore, we conclude from this analysis that the current state of GO annotations of protein entries can provide information that can be used reliably to relate proteins that share the same pathway.

3.6 Distribution of Annotation Coverage

Here, we provide a comparison of our system annotation coverage over UniProtKB/TrEMBL with reference to all other automatic annotation systems run on UniProtKB/TrEMBL such as Rule-Base [4] and HAMAP-Rule [5]. Fig. 2 illustrates some statistics about the UniProtKB/TrEMBL entries annotated by our system as follows. Out of 6,193,540 prokaryotic reference proteomes entries in UniProtKB/TrEMBL, 663,724 were annotated using the prediction models built by our system. Interestingly, a considerably large set of 436,510 entries lacked any previous pathway annotations and is now annotated by our system. A total of 150,295 entries of those covered constitute the entries that had previous annotations by other systems in addition to the annotation proposed by our system. The remaining set of only 76,919 entries represents those that had been annotated by other systems and were not touched by our prediction models.

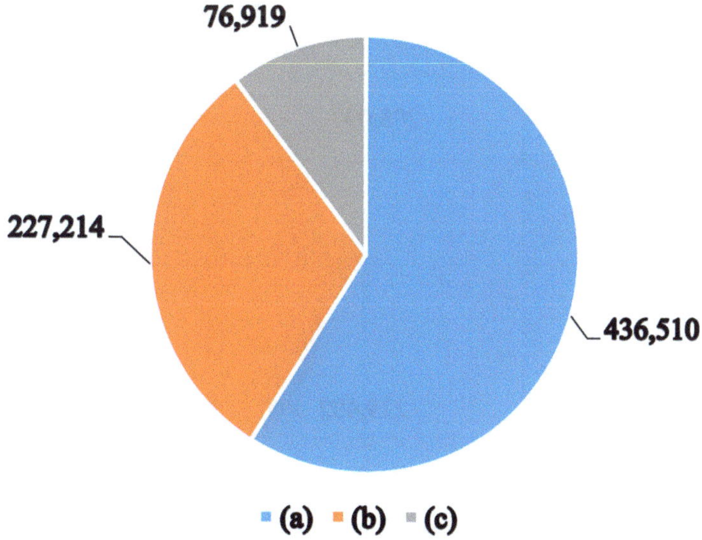

Fig. 2 Annotation coverage for UniProtKB/TrEMBL reference proteome prokaryotic entries. (**a**) represents entries we could cover which lack pathway annotation, (**b**) represents entries we could cover which already have pathway annotation, and (**c**) represents entries we could not cover which already have pathway annotation

3.7 Comparison of Annotation Coverage

In Fig. 3, we compare the coverage of entries of our system to three other main automatic annotation systems present in UniProtKB/TrEMBL, namely SAAS, HAMAP-Rule, and Rule Base. Our system significantly surpasses the other three systems in terms of the number of entries covered. It annotated 663,724 entries where the next best system was HAMAP-Rule with a coverage of only 229,402 entries. Rule-base touched the least number of entries of only 93,613 entries.

3.8 Comparison of Total Number of Prediction

In Fig. 4, we take a deeper look into the various predictions made by our system in comparison to those made by Rule-base, SAAS, and HAMAP-Rule. Note that an entry in UniProtKB/TrEMBL could gain multiple predictions and hence obtain multiple pathway annotations accordingly. Here we were able to make a total of 786,819 predictions by our system where the majority of these predictions, 516,042, touched entries that have no previous pathway annotation. Moreover, 237,784 predictions were found to be identical matches to the annotations proposed by other systems. We also found 20,901 of our annotations similar to those proposed by other systems either being more specific or more general in their pathway hierarchical representation. Finally, there were 12,092 predictions distinct from those already assigned by the other systems.

In order to better quantify the proportion of identical or similar predictions shared between our system and the other three main automatic annotation systems, Rule Base, HAMAP-Rule and

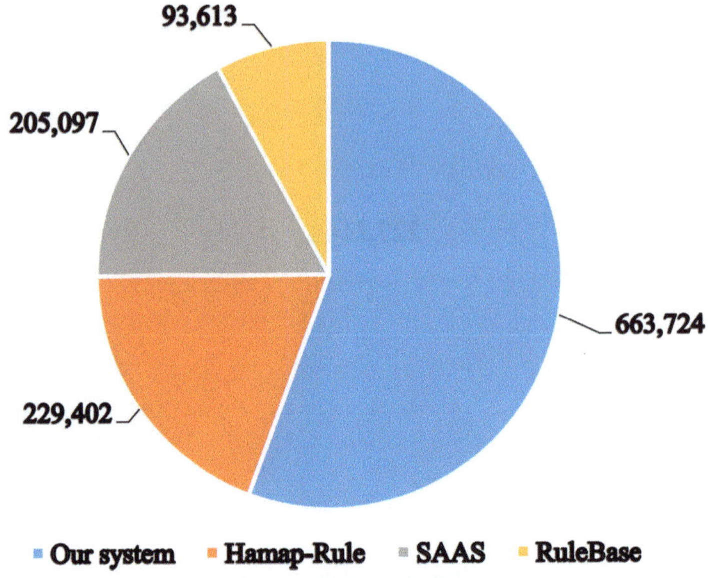

Fig. 3 Comparison of annotation coverage of UniProtKB/TrEMBL reference proteome prokaryotic entries with three main automatic annotation systems present in UniProtKB/TrEMBL which are SAAS, HAMAP-Rule, and Rule-base

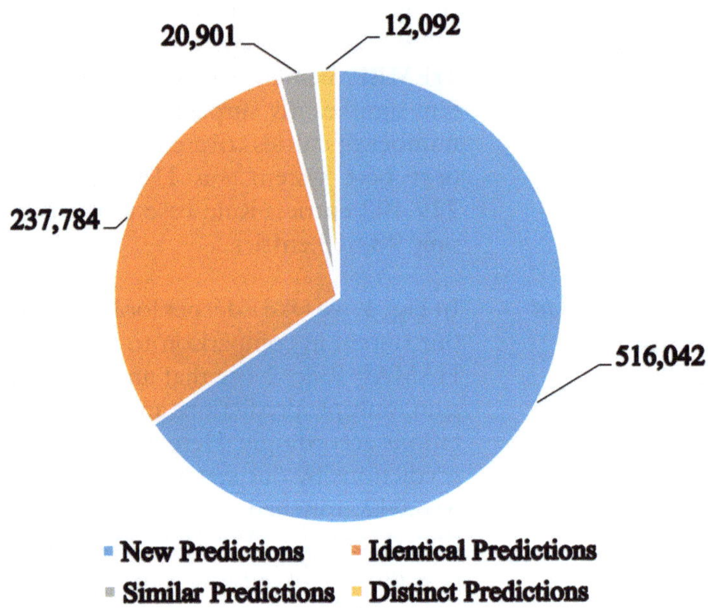

Fig. 4 Comparison of predictions applied on UniProtKB/TrEMBL reference proteome prokaryotic entries relative to three main automatic annotation systems present in UniProtKB/TrEMBL which are HAMAP-Rule, SAAS and Rule-base

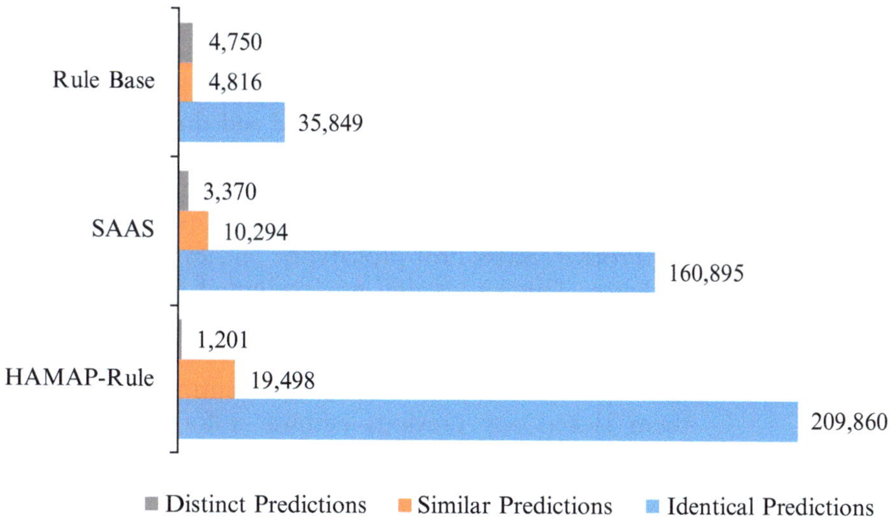

Fig. 5 Comparison of predictions corresponding to UniProtKB/TrEMBL reference proteome prokaryotic entries touched by our system, HAMAP-Rule, SAAS, and Rule Base

SAAS, we compare the predictions that correspond to entries touched by our system and the three other systems. Fig. 5 compares the distribution of annotations produced by our system and those provided by Rule-base, HAMAP-Rule, and SAAS systems. For instance, there were 35,849 annotations in Rule-base identical to those predicted by our system, while there were 209,860 and 160,895 predictions identical to those made by HAMAP-Rule and SAAS respectively. On the other hand, we found 4816 Rule-base, 19,498 HAMAP-Rule and 10,294 SAAS annotations that were similar to those annotated by our system. The similarity occurs due to the hierarchical property of pathway annotations that renders some annotations to be either more general or more specific. Moreover, we observed 4,750 Rule-base annotations, 1,201 HAMAP-Rule annotations, and 3,370 SAAS annotations that were completely different to those annotations provided by our system. These results indicate that for those entries touched by both our system and the other two systems, the majority of predictions were identical and similar. This provides an insight into the behavior of our system as an automatic annotation tool. This shared similarity supports the validity of our prediction models and their relevance on UniProtKB/TrEMBL entries.

4 Conclusions

In this chapter, we introduce a new approach to computationally assign pathway membership for protein sequences. The approach is based on the generation of association rules by the well-known Apriori algorithm and subsequent selection of significant (undominated and incomparable) rules by recently developed Sky-Rule software. It was demonstrated that specific combinations of protein domains (recorded in our rules) strongly determine pathways in which proteins are involved and thus provide information that let us very accurately assign pathway membership to proteins of a given prokaryotic taxon. Our system implemented in the ARBA4-Path software gains its knowledge on pathway identification from UniProtKB/Swiss-Prot prokaryotic entries with manual assertion evidence. This knowledge was presented in the form of human readable prediction models to annotate UniProtKB/TrEMBL prokaryotic reference proteomes entries. Using ARBA4Path we annotated 551,418 UniProtKB/TrEMBl entries, where 371,265 of them lacked any previous pathway information. Furthermore, cross-validation testing demonstrated a very high accuracy of ARBA4Path pathway identification with an F_1-measure of 0.987 and an AUC of 0.99. Future development of this system includes studying the obtained pathway models to unveil pathways presence patterns across prokaryotic taxa and possible extension of the system to the annotation of eukaryotic proteins.

5 Notes

1. Since many entries in UniProtKB/Swiss-Prot are cross-referenced with KEGG or MetaCyc databases of pathways, the user may choose to apply the proposed method using KEGG or MetaCyc identifiers as pathway target instead of UniProt textual representation of pathways as described in this chapter.
2. There are different implementations proposed for Apriori. The user may wish to select another implementation for the target task.
3. The choice of threshold values for support and confidence of association rules depends on several factors such as the initial size of the dataset and how frequent the target type is present in the dataset. The user may wish to experiment with these parameters in order to obtain an optimal threshold that satisfies the requirements of high coverage and high quality annotations.
4. The user may choose another set of additional statistical evaluation measures based on his/her preference.
5. The full list of prediction models obtained is available at: http://www.ebi.ac.uk/~rsaidi/arba/prokaryotapathway/learningdetails in JSON format and can be viewed using any JSON viewer.
6. A Java Archive (JAR) package for our system ARBA4Path to apply the prediction models on various UniProtKB/TrEMBL

prokaryotic entries is provided at: http://www.ebi.ac.uk/~rsaidi/arba/software. This module was built based on a case study describing an application of our system on UniProtKB prokaryotic data.

7. Predictions applied on some prokaryotic organisms present in UniProtKB/TrEMBL along with graphical reports illustrating ARBA4Path's prediction compared to those made by other systems present in UniProtKB/TrEMBL are available at: http://www.ebi.ac.uk/~rsaidi/arba/prokaryotapathway/organisms/comparison.
8. In case the user is not satisfied with the size of the input data that has manual assertion evidence, the user may choose to neglect filtering the input dataset by evidence.

Acknowledgments

The second author conducted this work as part of a research internship at the European Bioinformatics Institute, UniProt team. The funding for this internship was provided by King Abdullah University of Science and Technology. The authors would also like to thank UniProt Consortium for their valuable support and feedback on the development of this work.

References

1. Kretschmann E, Fleischmann W, Apweiler R (2001) Automatic rule generation for protein annotation with the c4.5 data mining algorithm applied on swiss-prot. Bioinformatics 17(10):920–926. doi:10.1093/bioinformatics/17.10.920
2. Quinlan JR (1993) C4.5: programs for machine learning. Morgan Kaufmann Publishers Inc., San Francisco, CA
3. The UniProt Consortium (2015) Uniprot: a hub for protein information. Nucleic Acids Res 43(D1):D204–D212. doi:10.1093/nar/gku989
4. Biswas M, O'Rourke JF, Camon E, Fraser G, Kanapin A, Karavidopoulou Y, Kersey P, Kriventseva E, Mittard V, Mulder N, Phan I, Servant F, Apweiler R (2002) Applications of interpro in protein annotation and genome analysis. Brief Bioinform 3(3):285–295. doi:10.1093/bib/3.3.285
5. Pedruzzi I, Rivoire C, Auchincloss AH, Coudert E, Keller G, de Castro E, Baratin D, Cuche BA, Bougueleret L, Poux S, Redaschi N, Xenarios I, Bridge A, The UniProt Consortium (2013) Hamap in 2013, new developments in the protein family classification and annotation system. Nucleic Acids Res 41(D1):D584–D589. doi:10.1093/nar/gks1157
6. Muller S, Leser U, Fleischmann W, Apweiler R (1999) Edittotrembl: a distributed approach to high-quality automated protein sequence annotation. Bioinformatics 15(3):219–227. doi:10.1093/bioinformatics/15.3.219
7. Wu CH, Huang H, Arminski L, Castro-Alvear J, Chen Y, Hu ZZ, Ledley RS, Lewis KC, Mewes HW, Orcutt BC, Suzek BE, Tsugita A, Vinayaka CR, Yeh LSL, Zhang J, Barker WC (2002) The protein information resource: an integrated public resource of functional annotation of proteins. Nucleic Acids Res 30(1):35–37. doi:10.1093/nar/30.1.35
8. Campbell N, Reece J (2002) Biology. In: Addison-Wesley world student series, vol 1. Benjamin Cummings, San Francisco, CA, USA
9. Chen X, Xu J, Huang B, Li J, Wu X, Ma L, Jia X, Bian X, Tan F, Liu L, Chen S, Li X (2011) A sub-pathway-based approach for identifying drug response principal network. Bioinformatics 27(5):649–654. doi:10.1093/bioinformatics/btq714

10. Chen Y, Hu Y, Zhou T, Zhou KK, Mott R, Wu M, Boulton M, Lyons TJ, Gao G, Ma JX (2009) Activation of the wnt pathway plays a pathogenic role in diabetic retinopathy in humans and animal models. Am J Pathol 175 (6):2676–2685. doi:10.2353/ajpath.2009.080945
11. Silberberg Y, Gottlieb A, Kupiec M, Ruppin E, Sharan R (2012) Large-scale elucidation of drug response pathways in humans. J Comput Biol 19(2):163–174. doi:10.1089/cmb.2011.0264
12. Parkes M, Cortes A, van Heel DA, Brown MA (2013) Genetic insights into common pathways and complex relationships among immune-mediated diseases. Nat Rev Genet 14 (9):661–673. doi:10.1038/nrg3502
13. Bebek G, Yang J (2007) Pathfinder: mining signal transduction pathway segments from protein-protein interaction networks. BMC Bioinformatics 8(1):335. doi:10.1186/1471-2105-8-335
14. Klopman G, Tu M, Talafous J (1997) Meta. 3. A genetic algorithm for metabolic transform priorities optimization. J Chem Inf Comput Sci 37(2):329–334. doi:10.1021/ci9601123
15. Jaworska J, Dimitrov S, Nikolova N, Mekenyan O (2002) Probabilistic assessment of biodegradability based on metabolic pathways: catabol system. SAR QSAR Environ Res 13 (2):307–323. doi:10.1080/10629360290002794
16. Hou B, Ellis L, Wackett L (2004) Encoding microbial metabolic logic: predicting biodegradation. J Ind Microbiol Biotechnol 31 (6):261–272. doi:10.1007/s10295-004-0144-7
17. Button WG, Judson PN, Long A, Vessey JD (2003) Using absolute and relative reasoning in the prediction of the potential metabolism of xenobiotics. J Chem Inf Comput Sci 43 (5):1371–1377. doi:10.1021/ci0202739
18. Karp P, Latendresse M, Caspi R (2011) The pathway tools pathway prediction algorithm. Stand Genomic Sci 5(3):424–429
19. Karp PD, Riley M, Saier M, Paulsen IT, Paley SM, Pellegrini-Toole A (2000) The ecocyc and metacyc databases. Nucleic Acids Res 28 (1):56–59. doi:10.1093/nar/28.1
20. Dale J, Popescu L, Karp P (2010) Machine learning methods for metabolic pathway prediction. BMC Bioinformatics 11(1):15. doi:10.1186/1471-2105-11-15
21. Creighton C, Hanash S (2003) Mining gene expression databases for association rules. Bioinformatics 19(1):79–86. doi:10.1093/bioinformatics/19.1.79
22. Georgii E, Richter L, Rckert U, Kramer S (2005) Analyzing microarray data using quantitative association rules. Bioinformatics 21 (suppl 2):ii123–ii129. doi:10.1093/bioinformatics/bti1121
23. Bodenreider O, Aubry M, Burgun A (2005) Non-lexical approaches to identifying associative relations in the gene ontology. In: Altman RB, Jung TA, Klein TE, Dunker AK, Hunter L (eds) Pacific symposium on biocomputing, World Scientific, pp 104–115
24. Artamonova II, Frishman G, Gelfand MS, Frishman D (2005) Mining sequence annotation databanks for association patterns. Bioinformatics 21(Suppl 3):iii49–iii57. doi:10.1093/bioinformatics/bti1206
25. Boudellioua I, Saidi R, Hoehndorf R, Martin MJ, Solovyev V (2016) Prediction of Metabolic Pathway Involvement in Prokaryotic UniProtKB Data by Association Rule Mining. PLOS ONE 11(7)
26. The InterPro Consortium, Mulder NJ, Apweiler R, Attwood TK, Bairoch A, Bateman A, Binns D, Biswas M, Bradley P, Bork P, Bucher P, Copley R, Courcelle E, Durbin R, Falquet L, Fleischmann W, Gouzy J, Griffith-Jones S, Haft D, Hermjakob H, Hulo N, Kahn D, Kanapin A, Krestyaninova M, Lopez R, Letunic I, Orchard S, Pagni M, Peyruc D, Ponting CP, Servant F, Sigrist CJA (2002) Interpro: an integrated documentation resource for protein families, domains and functional sites. Brief Bioinform 3(3):225–235. doi:10.1093/bib/3.3.225
27. Agrawal R, Srikant R (1994) Fast algorithms for mining association rules in large databases. In: Bocca JB, Jarke M, Zaniolo C (eds) VLDB 94, proceedings of 20th international conference on very large data bases, September 12–15, 1994, Morgan Kaufmann, Santiago de Chile, Chile, pp 487–499
28. Bouker S, Saidi R, Yahia SB, Nguifo EM (2012) Ranking and selecting association rules based on dominance relationship. In: IEEE 24th international conference on tools with artificial intelligence, ICTAI 2012, Athens, Greece, November 7–9, 2012, pp 658–665. doi:10.1109/ICTAI.2012.94
29. Bouker S, Saidi R, Yahia SB, Nguifo EM (2014) Mining undominated association rules through interestingness measures. Int J Artif Intell Tools 23(4). doi:10.1142/S0218213014600112
30. Borgelt C, Kruse R (2002) Induction of association rules: apriori implementation. In: Proceedings of the 15th conference on computational statistics (COMPSTAT), Physica Verlag, pp 395–400

31. Agrawal R, Srikant R (1994) Fast algorithms for mining association rules in large databases. In: Proceedings of the 20th international conference on very large data bases, VLDB 94, Morgan Kaufmann Publishers Inc., San Francisco, CA, pp 487–499
32. Borgelt C (2003) Efficient implementations of apriori and eclat. In: Proceedings of the 1st IEEE ICDM workshop on frequent item set mining implementations (FIMI 2003, Melbourne, FL). CEUR workshop proceedings 90, p 90
33. Borgelt C (2004) Recursion pruning for the apriori algorithm. In: Bayardo RJ Jr., Goethals B, Zaki MJ (eds) FIMI, CEUR workshop proceedings, vol. 126. CEUR-WS.org
34. Brin S, Motwani R, Silverstein C (1997) Beyond market baskets: generalizing association rules to correlations. In: Proceedings of the 1997 ACM SIGMOD international conference on management of data, SIGMOD 97, ACM, New York, NY, pp 265–276. doi:10.1145/253260.253327
35. Kirsch A, Mitzenmacher M, Pietracaprina A, Pucci G, Upfal E, Vandin F (2009) An efficient rigorous approach for identifying statistically significant frequent itemsets. In: Proceedings of the twenty-eighth ACM SIGMOD-SIGACT-SIGART symposium on principles of database systems, PODS 09, ACM, New York, NY, pp 117–126. doi:10.1145/1559795.1559814
36. Huntley RP, White O, Blake JA, Lewis SE, Giglio M (2014) Standardized description of scientific evidence using the evidence ontology (eco). Database 2014. doi:10.1093/database/bau075
37. Pesquita C, Faria D, Falco AO, Lord P, Couto FM (2009) Semantic similarity in biomedical ontologies. PLoS Comput Biol 5(7): e1000443. doi:10.1371/journal.pcbi.1000443
38. The Gene Ontology Consortium (2015) Gene ontology consortium: going forward. Nucleic Acids Res 43(D1):D1049–D1056. doi:10.1093/nar/gku1179
39. Harispe S, Ranwez S, Janaqi S, Montmain J (2014) The semantic measures library and toolkit: fast computation of semantic similarity and relatedness using biomedical ontologies. Bioinformatics 30(5):740–742. doi:10.1093/bioinformatics/btt581
40. Resnik P (1995) Using information content to evaluate semantic similarity in a taxonomy. In: Proceedings of the 14th international joint conference on artificial intelligence, IJCAI'95, vol 1, Morgan Kaufmann Publishers Inc., San Francisco, CA, pp. 448–453

Chapter 13

ArrayTrack: An FDA and Public Genomic Tool

Hong Fang, Stephen C. Harris, Zhenjiang Su, Minjun Chen, Feng Qian, Leming Shi, Roger Perkins, and Weida Tong

Abstract

A robust bioinformatics capability is widely acknowledged as central to realizing the promises of toxicogenomics. Successful application of toxicogenomic approaches, such as DNA microarrays, inextricably relies on appropriate data management, the ability to extract knowledge from massive amounts of data and the availability of functional information for data interpretation. At the FDA's National Center for Toxicological Research (NCTR), we are developing a public microarray data management and analysis software, called ArrayTrack that is also used in the routine review of genomic data submitted to the FDA. ArrayTrack stores a full range of information related to DNA microarrays and clinical and nonclinical studies as well as the digested data derived from proteomics and metabonomics experiments. In addition, ArrayTrack provides a rich collection of functional information about genes, proteins, and pathways drawn from various public biological databases for facilitating data interpretation. Many data analysis and visualization tools are available with ArrayTrack for individual platform data analysis, multiple omics data integration and integrated analysis of omics data with study data. Importantly, gene expression data, functional information, and analysis methods are fully integrated so that the data analysis and interpretation process is simplified and enhanced. Using ArrayTrack, users can select an analysis method from the ArrayTrack tool box, apply the method to selected microarray data and the analysis results can be directly linked to individual gene, pathway, and Gene Ontology analysis. ArrayTrack is publicly available online (http://www.fda.gov/nctr/science/centers/toxicoinformatics/ArrayTrack/index.htm), and the prospective user can also request a local installation version by contacting the authors.

Key words ArrayTrack, Bioinformatics, MAQC, Pharmacogenomics, VXDS, VGDS, Microarray, Toxicogenomics, Systems Toxicology, Database, Genomics

Abbreviation

CDISC	Clinical Data Interchange Standard Consortium
DEG	Differentially Expressed Gene
FDA	Food and Drug Administration
GO	Gene Ontology
GOFFA	Gene Ontology For Functional Analysis
HCA	Hierarchical Cluster Analysis
IPA	Ingenuity Pathway Analysis

Tatiana V. Tatarinova and Yuri Nikolsky (eds.), *Biological Networks and Pathway Analysis*, Methods in Molecular Biology, vol. 1613, DOI 10.1007/978-1-4939-7027-8_13, © Springer Science+Business Media LLC 2017

KEGG	Kyoto Encyclopedia of Genes and Genomes
MAQC	MicroArray Quality Control
MIAME	Minimum Information About a Microarray Experiment
NCTR	National Center for Toxicological Research
PCA	Principal Component Analysis
PGx	Pharmacogenomics
SDTM	Study Data Tabulation Model
TGx	Toxicogenomics
VGDS	Voluntary Genomic Data Submission

1 Introduction

Genomics, proteomics, and metabonomics (collectively called omics), along with other emerging methodologies, e.g., high-density genotyping for Genome Wide Association Study, contribute to our understanding of disease and health. The use of "omics" technologies to assess the gene/protein expression changes in chemical- and/or environment-induced toxicity, with emphasis on determination of corresponding gene/protein functions, pathways, and regulatory networks, are driving the emergence of the new research field of toxicogenomics [1]. DNA microarray is one of the main technological advances that has revolutionized both the theory and practice of addressing toxicological questions at the molecular level [2–4].

A DNA microarray experiment proceeds through hypothesis, experimental design and gene expression measurement in a manner similar to a conventional toxicology study. The amount and nature of data associated with a microarray experiment, however, impose far more substantial bioinformatics support requirements. There are three major bioinformatics requirements for the microarray experiment:

- Data management—This step acquires, organizes, and enables access to description of data from a microarray experiment. A microarray experiment involves multiple steps and the data in each step needs to be appropriately managed, annotated, and, most importantly, stored in an appropriate data structure for ready access. This enables efficient and reliable access for subsequent data analysis normally done by a multidisciplinary group of scientists. This is the same for periodic reexamination of the data in light of continual evolution of gene annotation information in the public domain. Furthermore, reanalysis is likely to be needed as new or more accepted analytic methods evolve, a process much more easily carried out with a well-managed and annotated dataset can be easily reanalyzed.

- Data analysis—A single experiment can produce a large amount of data and a formidable analysis undertaking. Normally, the immensity of data analysis scales directly with the complexity of the experiment, such as the number of technical and biological replicates, and temporal and dose response parameters. The ability to search, filter, and apply mathematical and statistical operations and graphically visualize data quickly with an intuitive user interface facilitates the laborious process.
- Data interpretation—Experiment interpretation is a highly contextual process incorporating known and unknown functions of genes, proteins, and pathways. The inherent noise in microarray data and a plethora of potential sources of variability inevitably complicate and possibly confound interpretation. Efficient and effective interpretation demands that relevant knowledge residing in public sources for gene annotations, protein functions, and pathways are readily available and integrated with the data analysis process.

The National Center for Toxicological Research (NCTR) of the US Food Drug Administration (FDA) has developed an integrated software system meeting the aforementioned bioinformatics requirements related to recently advanced high throughput and/or high content genomic assays, with emphasis on DNA microarrays [5]. ArrayTrack was originally conceived and developed to provide a one-stop bioinformatics solution for DNA microarray experiments, a capability now extended to integrated analysis of multiple "omics" expression profiles, such as proteomics and metabonomics.

2 A Brief History of ArrayTrack—Its Role in FDA and Public Use

2.1 Early Mission

NCTR has the mission of conducting peer-reviewed research to support the FDA regulatory mission. NCTR earned its reputation in the toxicological research community by conducting diverse toxicology studies, to which toxicogenomics (TGx) was added in 2000. Like many other institutes in the nation that invested early in TGx, NCTR began by printing its own two-color arrays and inexpensive filter arrays; ArrayTrack was initially developed as a research tool to support in-house DNA microarray experiments done with these platforms.

The following criteria were considered at ArrayTrack's inception and remain salient during continuing development: (1) A rich collection of gene, protein, and pathway functional information to provide context in data interpretation; (2) A software environment that automatically integrates gene expression data with functional information and visual and analytic tools for efficient and effective data analysis and interpretation; (3) Ability to cross-link gene expression and conventional toxicological data for phenotypic-driven exploration of underlying mechanisms of toxicity; and (4) modularization for easy extensibility to other types of "omics" data (e.g., proteomic and metabonomic data) to enable systems toxicology research.

The early ArrayTrack has progressively evolved to serve more roles inside and outside FDA, to accommodate additional data types, to provide ever richer analytic tools and functionality, and improved ease of use.

2.2 Roles in FDA

Over 7 years in development at this writing, ArrayTrack has had increasing and demonstrable impacts in FDA programs, of which the Voluntary Genomics Data Submission (VGDS) program [6] and the MicroArray Quality Control (MAQC) project [7] are notable examples. The program roles and demands have, in turn, led to identification and implementation of new capabilities and functionalities.

The VGDS is a novel data submission mechanism within FDA. Through VGDS, the sponsor can interact with FDA by submitting the genomic data on the voluntary basis. ArrayTrack became the FDA genomic tool to support VGDS in early 2004. All VGDS DNA microarray data received from 2004 on has exclusively been from Affymetrix GeneChip technology. Accordingly, significant ArrayTrack development has been oriented to improve GeneChip data handling and analysis. New functionality includes (1) direct loading of CEL files into ArrayTrack; (2) choice of converting probe level data to any or all of the probe-set level data types including MAS5, RMA, DChip, and Plier; (3) data filtering based on the presence/absence call; (4) mapping the affy ID to other types of gene IDs (e.g., Entrez Gene ID), protein ID (e.g., SwissProt Accession number), and different array platform ID (e.g., Agilent ID); and (5) providing annotations (e.g., pathways, functions) for all Affymetrix chips.

A primary goal in VGDS is better understanding of how sponsors reach biological conclusions from genomics data, a process requiring reproducing the sponsors' analysis methods. Reanalyses together with reviewing PGx/TGx studies in the literature enabled delineation of many issues, including (1) Array quality—what degree of experiment quality and individual array platform technical performance should be deemed achievable and adequate? (2) Data analysis issues—what results can be anticipated from different algorithms and approaches, and its corollary: can consensus be reached for a baseline approach to microarray data analysis? and (3) cross-platform issue—what consistency can be expected among different microarray experimental platforms?

Addressing the above issues were major motivators for initiating the MAQC program in 2005 [7]. MAQC is FDA led, but has a huge collaborative community spanning public, private, and academic communities. MAQC Phase I used six different commercial and one institutionally developed microarray platforms, a scope requiring significant expansion of ArrayTrack functionalities to manage data. As a result, a generalized data management scheme was implemented that can handle data from most if not all commercial array platforms. Since most commercial array types are preloaded in ArrayTrack (available from ChipLib in ArrayTrack), a cross-chip comparison can be carried out to assess commonality and difference between chips provided by the same company (e.g., Affymetrix) as well as the chips provided by different companies (e.g., Affymetrix versus Agilent).

Importantly, VGDS and MAQC emphasize interaction and collaboration among FDA, private industry and elements of the entire research community with the stated objective of moving toward consensus on best practices for microarray data management, analysis, and interpretation. The programs are similarly geared toward advancing the science and consensus. The lessons learned from both VGDS and MAQC are paving the way for development of a Best Practice Guidance Document for future voluntary as well as regular submissions of PGx data to the FDA. Recently, such a best practice document draft, a companion document to "Guidance for Industry—Pharmacogenomic Data Submission" was released for comments [8]. ArrayTrack both supports VGDS and MAQC, and benefits from the programs, contributing to an ever more powerful and versatile FDA integrated bioinformatics infrastructure to support data management, analysis, and interpretation. Synchronizing ArrayTrack development with VGDS and MAQC will assure the platform meets agency needs to routinely employ PGx/TGx data in regulatory review and decision making (Fig. 1), when that time arrives.

2.3 Beyond DNA Microarrays

ArrayTrack development initially focused on management, analysis, and interpretation for DNA microarray data. By the end of 2006, however, the VGDS program has seen proteomics and metabonomics data appearing as voluntary submissions. ArrayTrack was subsequently modified to accommodate significant lists of proteins and metabolites, and a new systems biology function called CommonPathway was added that enabled examination of common pathways and functional categories (e.g., Gene Ontology terms) shared by different data types (*see* Subheading 5 below).

VGDS submissions normally came with a large amount of both clinical and non-clinical information. To manage these traditional data types, a general mechanism for handling study data was implemented in ArrayTrack using the Study Data Tabulation Model (SDTM) for nonclinical data and clinical data standard suggested by the Clinical Data Interchange Standard Consortium (CDISC) [9]. Additionally, functions were developed to facilitate interpretation of multiple data types (nonclinical, clinical, and "omics") in the

Fig. 1 A schematic presentation about the integrated nature of an array of pharmacogenomic effort at FDA: (*1*) the FDA genomic software, ArrayTrack; (*2*) the FDA Voluntary eXploratory Data Submission (VXDS); (*3*) the MicroArray Quality Control (MAQC) project; and (*4*) the best practice presented in the draft companion document to "FDA Guidance for Industry: Pharmacogenomic Data Submission." VXDS and MAQC are program mechanisms allowing FDA interaction in a collaborative environment with the private sector and research community, respectively. Both programs are aimed at gaining consensus on analysis methods for and valid applications of recently advanced molecular technologies in drug development and regulation. The collective lessons learned from both programs formed the basis to develop the companion document. *ArrayTrack* provides primary support to VXDS and MAQC, thereby continuing its evolution to be the software vehicle that translates best practices into routine application for regulatory review and decision making in the FDA

context of phenotypic anchoring, which, in turn, enabled identification of possible molecular level mechanisms related to phenotype (*see* Subheading 5 below).

2.4 Public Use

ArrayTrack has been a key genomic tool for the VGDS program and genomic submission in FDA. By now, over 100 FDA reviewers and scientists have attended the ArrayTrack training. However, the need of making the tool publicly available to the research community was identified early on and has been a continuing priority throughout the planning and development phases of ArrayTrack. As with VGDS, the feedback from the wide user community has reciprocally benefited ArrayTrack through linking its development to emerging common practices, and providing validations of functions and usefulness. ArrayTrack was made openly available to the public in 2003, where users can gain access either through the FDA website [5] or by requesting media for local installation, which would then normally entail local provision of backend database support with ORACLE.

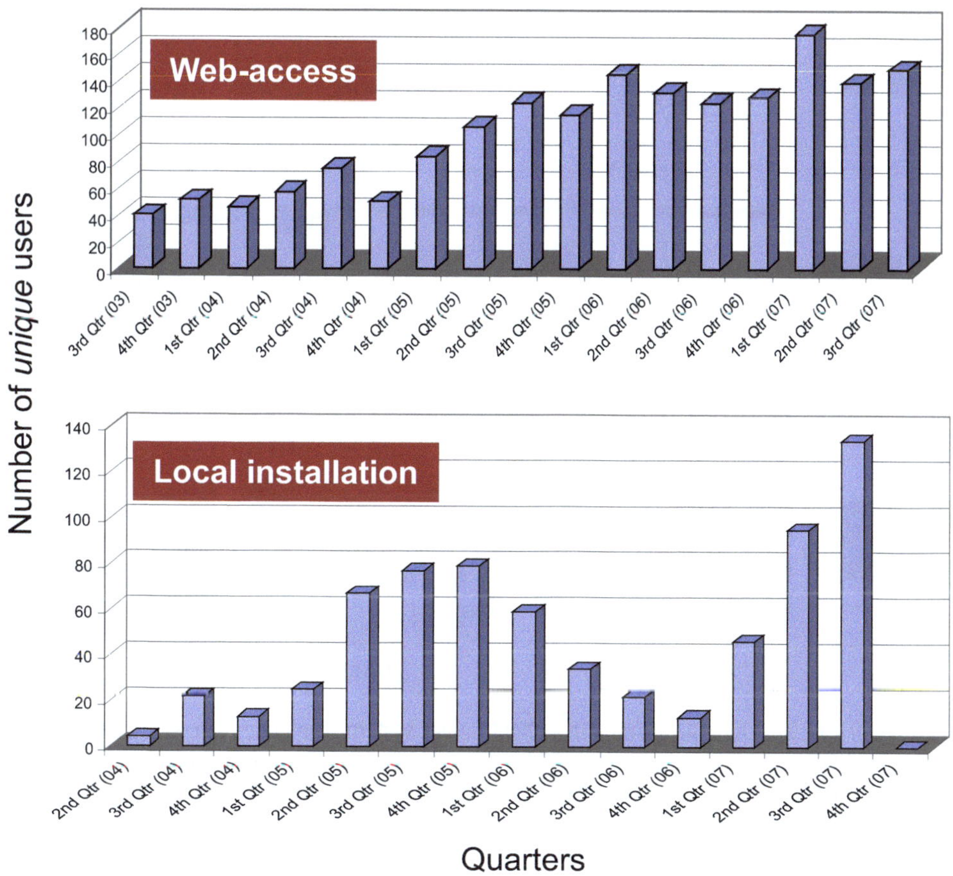

Fig. 2 A summary of unique users accessing ArrayTrack in the quarterly basis. There are two types of users: (*1*) Using the FDA ArrayTrack and (*2*) Accessing ArrayTrack installed in their respective institutes or companies

In addition to its broad use within FDA in various regulatory-driven programs, ArrayTrack is also freely available to the entire scientific community. ArrayTrack user base has steadily grown (Fig. 2), and has been adopted by several government agencies (e.g., EPA, CDC, and NIH), academia, and private sector. At this writing, ArrayTrack version 3.4 can be accessed through http://cdkb.fda.gov/webstart/arraytrack (http://weblaunch.nctr.fda.gov/jnlp/arraytrack for FDA users). The full user manual, quick-start manual, and tutorial are available from the ArrayTrack website http://www.fda.gov/nctr/science/centers/toxicoinformatics/ArrayTrack/

3 ArrayTrack Architecture

As depicted in Fig. 3, ArrayTrack is a client-server system. The ORACLE server stores and integrates in-house omics data, study data and data from public resources about genes, proteins, and

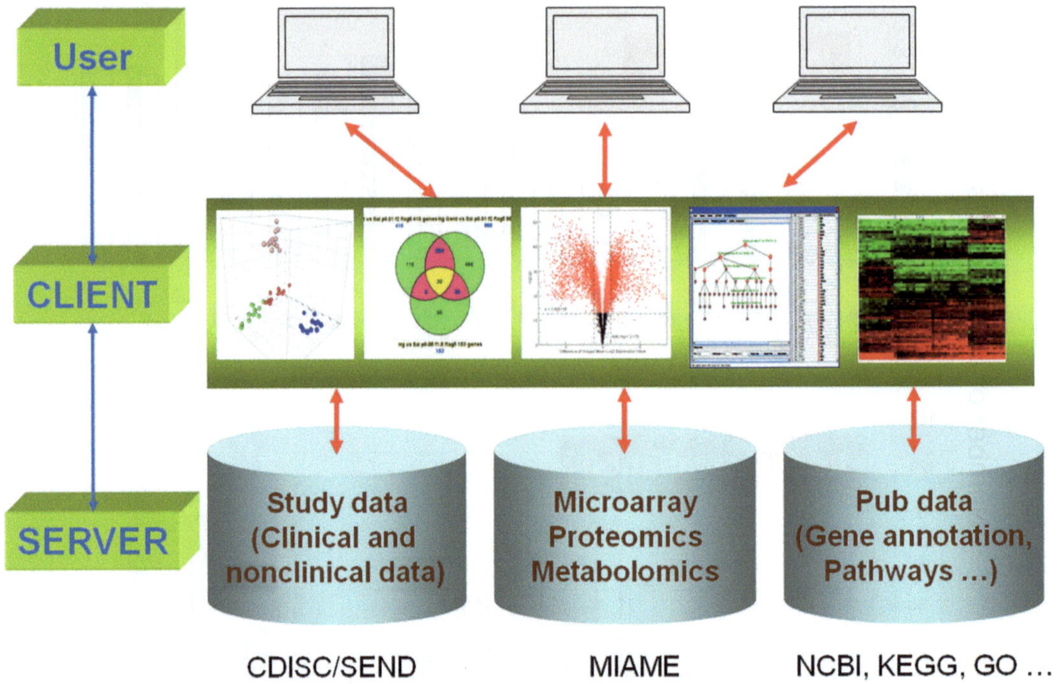

Fig. 3 ArrayTrack Architecture. ArrayTrack is a client-server system. The omic data, study data, and the data from the public domain are managed by the ORACLE database while the visualization and analysis tools are available from the client side mainly using Java. The tools in the client side can also be directly applied to the data outside of ArrayTrack such as these stored in the local hard drive

pathways. The Java language was used to construct the entire user interface, query mechanism, and data visualization and analysis tools. ArrayTrack was implemented using Java Webstart technology that allows installation through a single web link, with updates of the software performed automatically whenever the application is run.

ArrayTrack has a modular architecture. Each module for each application has been constructed independently, such that existing or new capabilities can be enhanced, changed or added in accordance with priorities and evolving experimental progress. In this manner ArrayTrack has remained in continuous development and updating.

The client-server connection in ArrayTrack is accomplished through JDBC (Java Database Connectivity). The use of JDBC makes it easy for ArrayTrack to use other relational databases for backend storage. Currently, database support is in the process of being extended to open source PostgreSQL. Because ArrayTrack's client-server implementation uses the fat client, performance of ArrayTrack is largely dependent on the client computer. A benefit of this architecture is the option to apply the analysis functions in ArrayTrack to data stored in the local machine instead of the server.

Use of Java ensures portability of ArrayTrack to all major computer operating systems. Integration with non-Java applications can readily be made through socket-based communication on a local computer. In this manner, ArrayTrack has been interfaced with a number of other open and commercial applications, including R program, JMP Genomics, GeneGo MetaCore, Ingenuity Pathway Analysis (IPA), and others.

4 ArrayTrack Core Components

ArrayTrack comprises three major integrated components (Fig. 4): (1) MicroarrayDB that stores essential data associated with a microarray experiment, including information on slides, samples, treatments, and experimental results; (2) TOOL that provides analysis capabilities for data visualization, normalization, significance analysis, clustering, and classification; and (3) LIB that contains information (e.g., gene annotation, protein function and

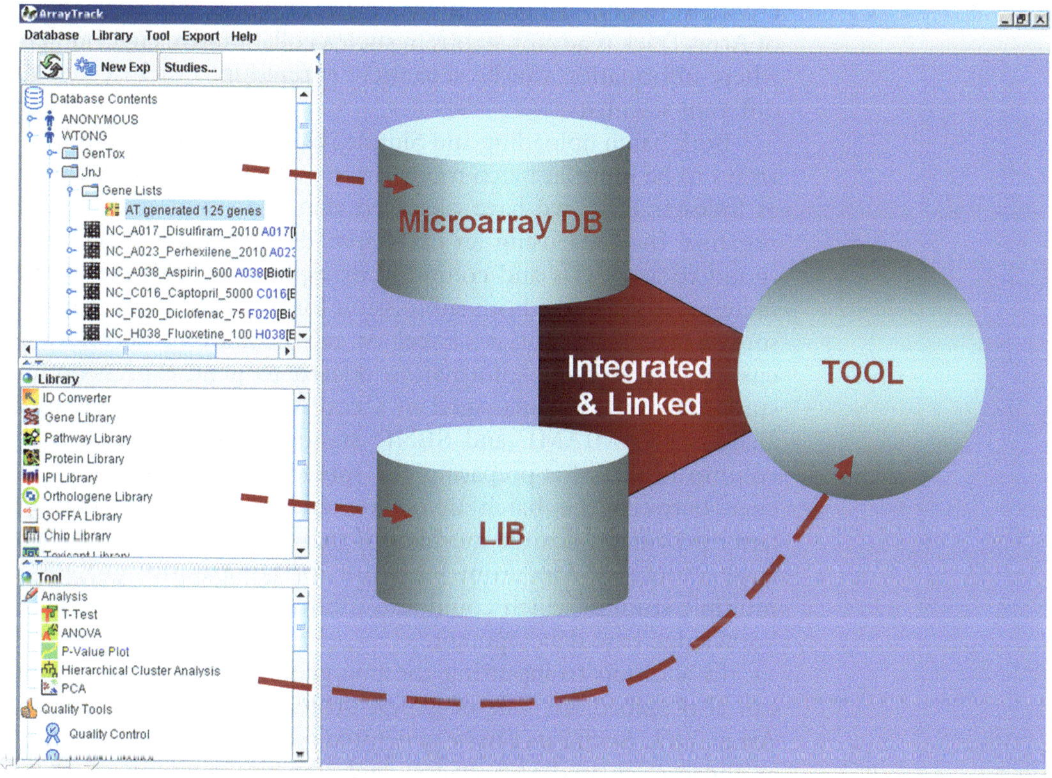

Fig. 4 ArrayTrack core components. The software consists of three integrated components that are organized as three panels in the left side of interface: (*1*) MicroarrayDB captures toxicogenomic data associated with a microarray experiment; (*2*) TOOL provides data visualization and analysis capabilities; and (*3*) LIB contains annotated information on genes, proteins, and pathways

pathways) from public repositories. Through a user-friendly interface, the user can select an analysis method from the TOOL, apply the method to selected microarray data stored in the MicroarrayDB, and the analysis results can be directly linked to associated functional annotations in the LIB.

The key functionalities associated with these three components are discussed below and the full list of functions is available from the ArrayTrack Website [10].

4.1 MicroarrayDB

ArrayTrack supports the MIAME (Minimal Information About a Microarray Experiment) guideline. MIAME defines essential information for a microarray experiment that enables the results to be interpretable and the experiment to be reproducible [11]. Microarray information along with a study data can be input through three submission formats, manual submission, batch uploading and SimpleTox format.

The manual data submission is through a comprehensive data submission form in ArrayTrack [10]). It is common that hypothesis generation, hybridization experiment, and sample preparation is done by different groups of people within an organization, especially in one that has a microarray core facility. The form design of ArrayTrack is advantageous in such a collaborative environment, where information can be separately entered into each section by different scientists.

Both batch uploading and SimpleTox allows a larger number of arrays to be input in batch mode. Input schemas and rationales are as follows. First, we have observed that most biologists tend to organize the data using an excel spreadsheet, where rows correspond to array IDs and columns correspond to experiment parameters. Accordingly, both submission formats directly accept such spreadsheet formats (i.e., Excel or tab delimited). Secondly, to ensure that essential information related to gene expression and study data is being managed in a consistent way for cross-study analysis, the MIAME and SEND standards are enforced as the column headers for preparing the spreadsheet. The major difference between the batch uploading and SimpleTox is that the latter provides a flexible mechanism that can be used to manage a large variety of data from literature for comparative analysis of multiple studies, which could also ultimately serve as a means for knowledge base development.

In addition to inputting the raw gene expression data, a user can also upload any lists of genes, proteins, and metabolites into ArrayTrack. Such lists can be generated outside of ArrayTrack, such as those calculated in a customized statistical method or simply assembled from literature or other knowledge sources. This function is useful in many ways. First, any statistical analysis tool implemented in ArrayTrack has the option to be applied only to a specified gene list such that, for example, the grouping of the treated samples across different time points and doses can be examined using a cluster analysis based on a preloaded gene list. Secondly, the preloaded gene list can be directly compared with the gene list generated using the ArrayTrack tool for comparative analysis. In VGDS, for example, significant genes chosen by the ArrayTrack tool are often then compared with the list provided by the sponsor to assess the commonalities and differences in biological interpretation. Thirdly, if the lists of genes, proteins, and metabolites from a multi-omic experiment are input independently into ArrayTrack, the common pathways and/or functional categories shared by three lists can be examined (*see* Subheading 5 below).

4.2 LIB

The ArrayTrack LIB comprises of a number of libraries. Each library contains the content-specific information that is organized in such a way that they are not only convenient for interpretation of omics data but also useful for other genomic research. Each library has a common look-and-feel. Specifically, the main part of a library is an Excel-like spreadsheet, where each row is associated with an entity of interest that can be gene, protein, chemical, pathway, etc., depending on the content of a library. Each column presents particular information for each entity in the row, such as functional annotation, chromosomal location, pathways, etc. The query function is on left side of the spreadsheet, where the user can quickly identify the functional information for a set of significant genes derived from the analysis by searching the library. In addition, a set of functions available on the top of the spreadsheet allows the information in a library to be mapped to other libraries in ArrayTrack or to external resources such as GeneGo, MetaCore, IPA, etc.

ArrayTrack contains libraries that partially mirror the contents of GenBank, SWISS-PROT, LocusLink, KEGG (Kyoto Encyclopedia of Genes and Genomes), GO, and others. We extract the functional information from these databases to construct several enriched libraries, such as GeneLib, ProteinLib, and PathwayLib that, as the names suggest, concentrate functional information on genes, proteins, and pathways, respectively [5]. ChipLib contains all functional information for the probes on a chip provided by the array manufacturers. Since understanding the function and biological characteristics of the probes (genes) present on a microarray could be essential for interpretation of microarray results, genes present on the array are also directly linked with other libraries for facilitating biological interpretation of experiment results.

4.3 TOOL

Microarray data analysis normally starts with data normalization and quality control, followed by class comparison, class discovery, and/or class prediction. At this time, ArrayTrack provides all the functionalities associated with data analysis except class prediction (which will be available soon).

4.3.1 Normalization

ArrayTrack provides several normalization methods to convert the probe level data to the probe-set level data for the Affymetrix GeneChip, including MAS5, RMA, DChip, and Plier. The raw gene expression data from other array platforms can be processed using several global normalization approaches, such as total intensity normalization [12], log ratio mean scale normalization [13], and LOWESS normalization.

4.3.2 Quality Control

A QA/QC tool was developed to assist quality control of two-color array results. The tool summarizes most relevant information into one interface to facilitate the process of quality control. The user can determine the quality of individual microarray results by visualizing data, applying statistical measures and viewing experimental annotation. Statistical measures are provided to assess the quality of a hybridization result based on the raw expression data, including signal-to-noise ratio and the percentage of non-hybridized spots. The experimental annotations associated with the processes of hybridization, RNA extraction and labeling are also available to the end-user. Additionally, a scatter plot of Cy3 vs. Cy5 together with the original image is available for visual inspection for quality control purposes [10].

4.3.3 Class Comparison

One of the most common data analyses in DNA microarrays is determining a list of genes that are differentially expressed by comparing, for example, the treated group with the control group, and then using this subset of differentially expressed genes (DEGs) for biological interpretation. Over the years, a number of methods have been proposed to identify DEGs. ArrayTrack offers many such methods, ranging from the simple t-test, to ANOVA, the Volcano plot, and more advanced statistical approaches such as False Discovery Rate (FDR) and Significance Analysis of Microarrays (SAM) [14].

4.3.4 Class Discovery

Two commonly employed tools for class discovery and pattern identification, Principal Component Analysis (PCA) and Cluster Analysis are available. PCA generates the linear combination of the genes, named principal components, using a mathematical transformation. The algorithm ensures that the first principal component explains the maximal amount of variance of the data. The second principal component explains the maximal remaining variance in the data subject to being orthogonal to the first principal component, and so on, such that all principal components taken together explain all the variance of the original data. The PCA plot of the first three principal components, which usually explains the majority of variance in the data, is used to inspect the inter-sample and inter-gene relationships. ArrayTrack offers both 2D and 3D views of the PCA results, along with the loading tables.

ArrayTrack also provides two cluster analysis methods, a two-way Hierarchical Cluster Analysis (HCA) and k-mean clustering, to investigate the grouping of samples in terms of their similarities in gene expression profiles, as well as the grouping of genes in terms of their similarity of samples. The primary purpose of two-way HCA analysis is to present data in such a manner that genes with similar expression level across the samples are clustered together along one axis while the samples with similar gene expression patterns are grouped together along another axis. Since the genes in the same cluster are likely to share similar functions, this analysis could reveal the relationships of molecular functions and phenotypes. In contrast, k-mean clustering is mainly used to assess the gene expression profiles across different experiment conditions defined in the experiment design.

5 ArrayTrack Use Cases

Four examples are provided below to illustrate the utility of *ArrayTrack* in addressing the bioinformatics challenges in the FDA VGDS program and research.

5.1 A Common Workflow

Drug X was being evaluated for treatment of cancer in a Phase II clinical trial with 100 cancer patients. Before treatment, samples of peripheral blood mononuclear cells were obtained from individual patients and gene expression in peripheral blood mononuclear cells measured with Affymetrix microarrays. Treatment benefit was observed for 80 patients, but not for the rest. The purpose of this study was to identify a testable hypothesis to explain the treatment outcome. Thus, the analysis required identification of DEGs by comparing patients responsive to treatment with Drug X with those who were not, followed by an interpretation of the biological significance of the comparison.

Figure 5 depicts a prototypical workflow in *ArrayTrack* to carry out the required bioinformatics (i.e., data management, analysis, and interpretation), all of which can be done in the single *ArrayTrack* software platform, precluding the need for cumbersome import and export of data between software. *ArrayTrack* was designed a priori to provide such a one-stop solution. Using *ArrayTrack*, the user can select an analysis method from the TOOL and apply the method to selected omics data stored in DB; the analysis results can then be linked directly to pathways, Gene Ontology database and other functional information stored in LIB. To further facilitate the data interpretation, *ArrayTrack* also provides a direct link of analysis results to the external public data repositories, such as OMIM, UniGene, Chromosomal Map, and GeneCard. Finally, the power and flexibility of *ArrayTrack* is furthered by its

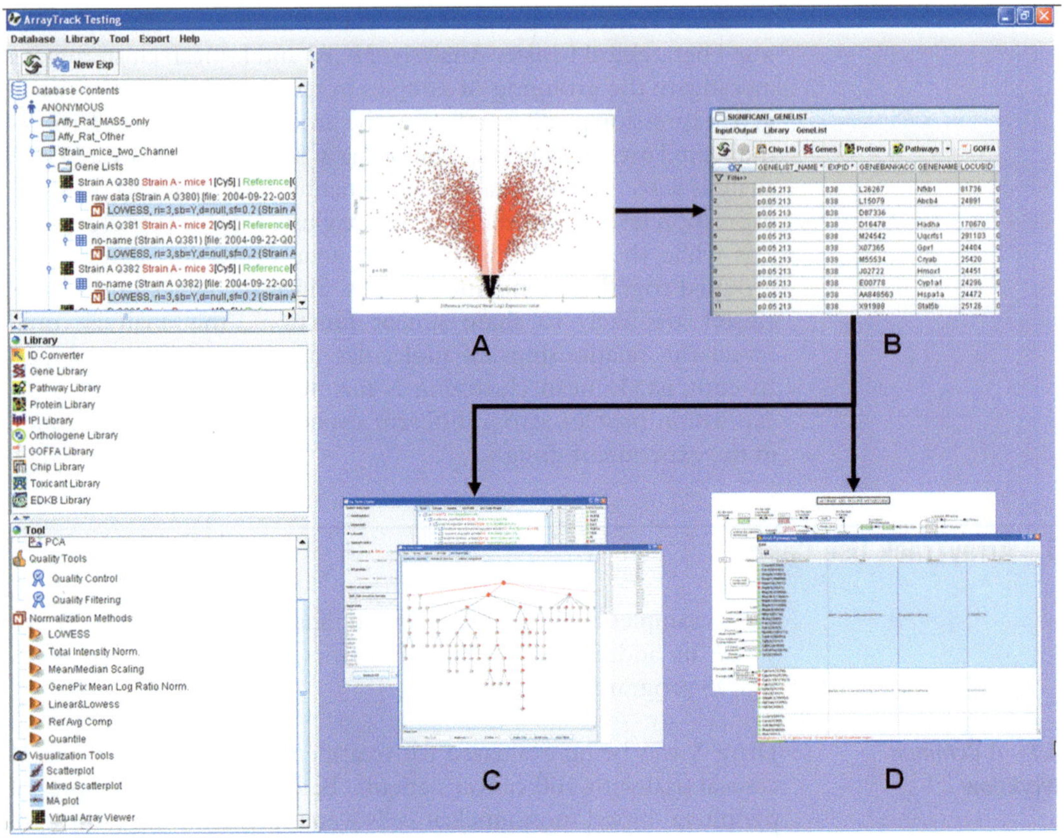

Fig. 5 A typical workflow using *ArrayTrack* to identify differentially expressed genes (DEGs) distinguishing treatment and control groups, followed by pathway and Gene Ontology (GO) analyses. (**a**) DEGs are identified using the Volcano plot or other means in *ArrayTrack*. DEGs can also be identified using other commercial or public tools and uploaded into ArrayTrack; (**b**) DEGs are summarized in a table format and can be readily linked to *ArrayTrack* library functions for biological interpretation; (**c**) Significant altered KEGG pathways are identified based on DEGs; (**d**) DEGs are submitted to Gene Ontology For Functional Analysis (GOFFA) tool in ArrayTrack to identify GO terms associated with significantly altered gene expression

interface to, or integration with, many commercial and public software systems, including IPA, GeneGO MetaCore, PathArt, JMP Genomics, and R package.

5.2 Gene Ontology Analysis Using GOFFA

Gene Ontology (GO) which characterizes and categorizes the functions of genes and their products according to biological processes, molecular functions and cellular components has played an increasingly important role in interpretation of data from high-throughput genomics and proteomics technologies. A FDA GO tool named as Gene Ontology for Functional Analysis (GOFFA) was implemented in ArrayTrack. With GOFFA, the user can dynamically incorporate ArrayTrack analysis functions with the GO data in the context of biological interpretation of gene expression data.

GOFFA first ranks GO terms in the order of prevalence for a list of selected genes or proteins, and then it allows the user to interactively select GO terms according to their significance and specific biological complexity within the hierarchical structure. GOFFA provides five interactive functions (Tree view, Terms View, Genes View, GO Path, and GO TreePrune) to analyze the GO data. Among the five functions, GO Path and GO TreePrune are unique. The GO Path ranks the GOFFA Tree Paths based on statistical analysis. The GO TreePrune provides a visualization of a reduced GO term set based on user's statistical cutoffs. Therefore, the GOFFA can provide an intuitive depiction of the most likely relevant biological functions.

A dataset from a toxicogenomics study was used to demonstrate the utility of GOFFA. In this study, the renal toxicity and carcinogenicity associated with the treatment of aristolochic acid (AA) in rats was studied using DNA microarray [15]. The DEG list was determined in ArrayTrack and then directly passed to GOFFA for functional analysis. Of 1176 identified genes, 417 genes had GO information for analysis [16]. The GOFFA results are summarized in Fig. 6.

Fig. 6 In GOFFA, lists of genes or proteins from an experiment are analyzed by five functional modules, Tree View, Terms View, Genes View, GO Path, and GO TreePrune. (**a**) GO Path identified the significant GO term based on its path. The most significant ten paths are graphically displays and a color key for the top ten paths is located beneath the plot. Clicking either a circle in a path in the plot or its corresponding color key launches a Tree View (**b**) with the selected path highlighted in *blue*. (**c**) GO TreePrune display allows the user to filter out nodes and thus reduce the complexity of a tree by specifying the *p*- and *E*-value as well as the user-defined number of genes in the end node

The statistics based on a combination of Fisher's Exact Test ($p < 0.05$) and Relevant Enrichment Factor ($E > 2$) identified 52 enriched GO terms in the GO biological process. The majority of the terms are related to four functional categories, induction of apoptosis, defense response, response to stress, and amino acid metabolism. These four functional categories reflect the known biological and pharmacological responses of kidney to the AA treatment [17]. Out of these four functional categories, GO Path ranked "defense response" as an important mechanism associated with the AA treatment (Fig. 6a), and similar results were obtained from GO TreePrune as well (Fig. 6c). This finding is consistent with the general understanding that defense response, which includes immune response, is a complex network response of a tissue to toxins and carcinogens (such as AA) for defending the body. Figure 6b gives the GO Path results in the Tree window, where the majority of genes involved in the defense response are upregulated to oppose damage by AA. For example, the *inhba* gene (first gene in the right panel) is a growth factor with 4.1-fold increase in expression in kidney. This is a tumor-suppressor gene and it produces protein that increases arrest in the G1 phase of tumor cells [18]. Therefore, its induction inhibits tumorigenesis in kidney treated with AA.

5.3 Analysis of Microarray Gene Expression Data with Conventional Toxicological Endpoints

A number of drugs were recently removed in post market due to liver toxicity. In fact, hepatotoxicity is recognized as such a significant problem that its study is prevalent in both public and private research communities. The VGDS program has observed considerable effort by sponsors to identify relevant preclinical biomarkers for drug-induced liver toxicity.

This example used DNA microarrays to identify a set of genes with differential expression correlating with clinical pathology parameters associated with, and thus possibly biomarkers for, hepatotoxicity. Specifically, rats were treated with a single high dose of Drug Y and sacrificed at days 2, 4, 8, 16, and 24. Each time point contained five treated animals along with five matched controls. The liver samples were collected for both treated rats and controls at each time point and analyzed by using Affymetrix microarrays, and clinical pathology.

This example required integrating conventional toxicological endpoints with gene expression data in such a way that phenotype-anchored toxicogenomic analysis could be performed. *ArrayTrack* enables such analyses because a "study domain" is definable based on SDTM developed by CDISC [9]. Using SDTM, *ArrayTrack* is able to concurrently manage disparate clinical and non-clinical data types together with PGx and other biomarker data. Moreover, various statistical analyses at the toxicological data level, gene expression level, or in combination can be conducted.

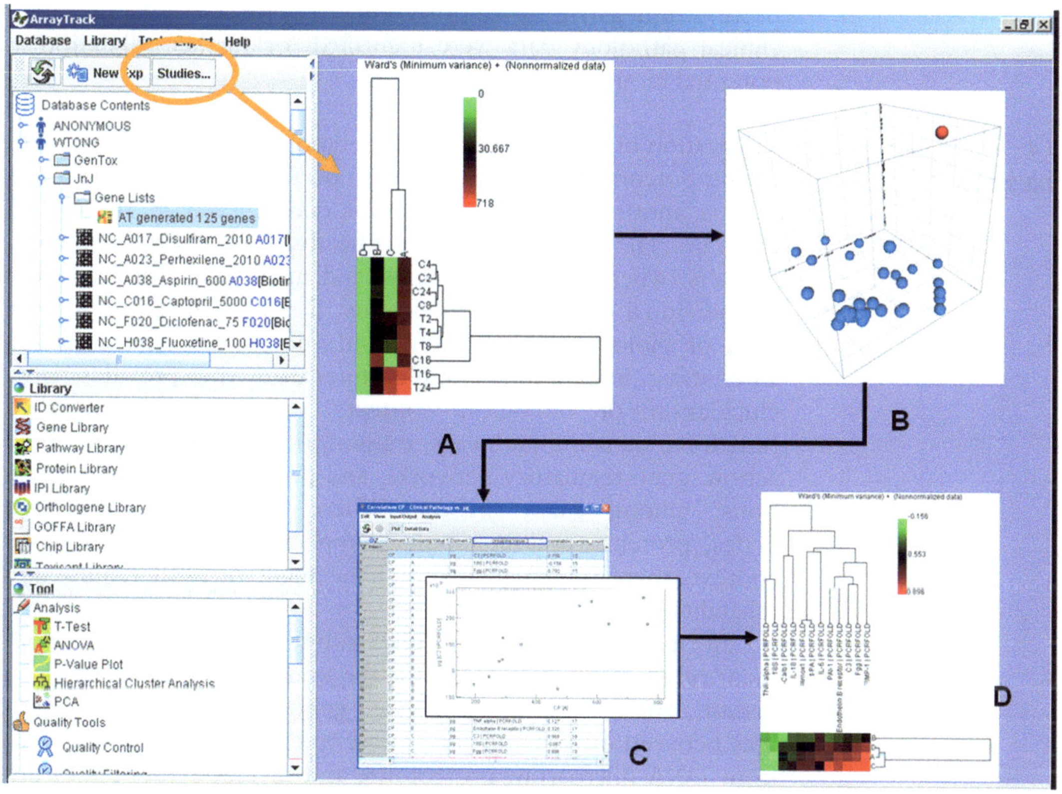

Fig. 7 A typical data analysis procedure and results for Example Study 3 correlating gene expressions at multiple time points with conventional toxicological endpoints. (**a**) Hierarchical Cluster Analysis is used to assess the ability of clinical pathology to distinguish treatment and control groups. (**b**) Principal Component Analysis of the clinical pathology data enable an anomalous outlier in the control group to be identified. (**c**) The DEGs at each time point are correlated with each corresponding set of clinical pathology data. The correlation coefficients are summarized in a table format and each correlation can also be displayed in a pairwise plot. (**d**) The correlation results between the clinical pathology data and gene expression data is summarized in a heat map, where each cell represents a specific pair (a clinical pathology observation and a gene) in the correlation analysis with magnitude of correlation represented with color (*red* for the positive correlation and *green* for the negative correlation)

In this example study, the first step to identify relevant biomarker genes was determining whether the clinical pathology data contained sufficient biological information to distinguish time points, as well as to separate the control and treatment groups. As illustrated in Fig. 7a, HCA based on four clinical pathology parameters clearly separated all treatment groups, but not the control group sacrificed on day 16. Further analysis using PCA indicated that one of the five control animals had anomalous clinical pathology (Fig. 7b) and should be considered for removal before differential expression analysis. Next, the DEGs at each time point were identified, and these genes were correlated with each type of the clinical pathology data (Fig. 7c). Genes that showed the highest positive or negative correlations (Fig. 7d) with any of the measured clinical pathology data were identified for further validation as potential biomarkers.

5.4 Omics Data Integration

Integration of gene, protein, and metabolite information for identifying potential biomarkers through perturbed pathways or function is another type of application encountered in the VGDS program. The rationale is that, in the absence of data integration, markers (whether genes, proteins, and metabolites) derived from an individual omics platform are just lists providing but a single level of biological information, and subject to Type 1 errors. In contrast, integrating multiple omics data types provides richer elucidation of biological contexts such as the perturbed functions, signaling pathways, transcription-factor mechanisms of action, gene regulatory networks, and posttranslational modifications, among many others. Where differentially expressed genes, proteins, and metabolites implicate the same biological context, there is a qualitative enhancement of both validity and reliability [19].

In this example study, a VGDS submission proposed development of a testable hypothesis for the underlying mechanisms of a disease. The differentially expressed genes, proteins, and metabolites between disease and the disease-free patients were generated from DNA microarray, proteomics, and metabolomics platforms, respectively. The hypothesis was that pathways common to significant gene, protein, and metabolite lists are more likely to be disease-relevant pathways than pathways identified by a single significance list.

The CommonPathway function in *ArrayTrack* was used to identify the common pathways or functions shared by a combination of genes/proteins/metabolites differentially expressed between disease and disease free groups. Figure 8 depicts a typical *ArrayTrack* workflow for required analyses. Once differentially expressed genes, proteins, and metabolites were independently identified from corresponding data, each profile was independently mapped to the pathways to determine which pathways were significantly altered for each data type. The separate pathway lists from the gene, protein, and metabolite profiles were then compared in a Venn diagram to determine the commonly altered pathways. The statistical significance of each pathway was estimated using Fisher's Exact Test. Each significant pathway's details were also displayed with its differentially expressed genes, proteins, and metabolites highlighted in different colors. The same process can be equally applied to GO data to identify commonly altered GO terms (i.e., gene functions).

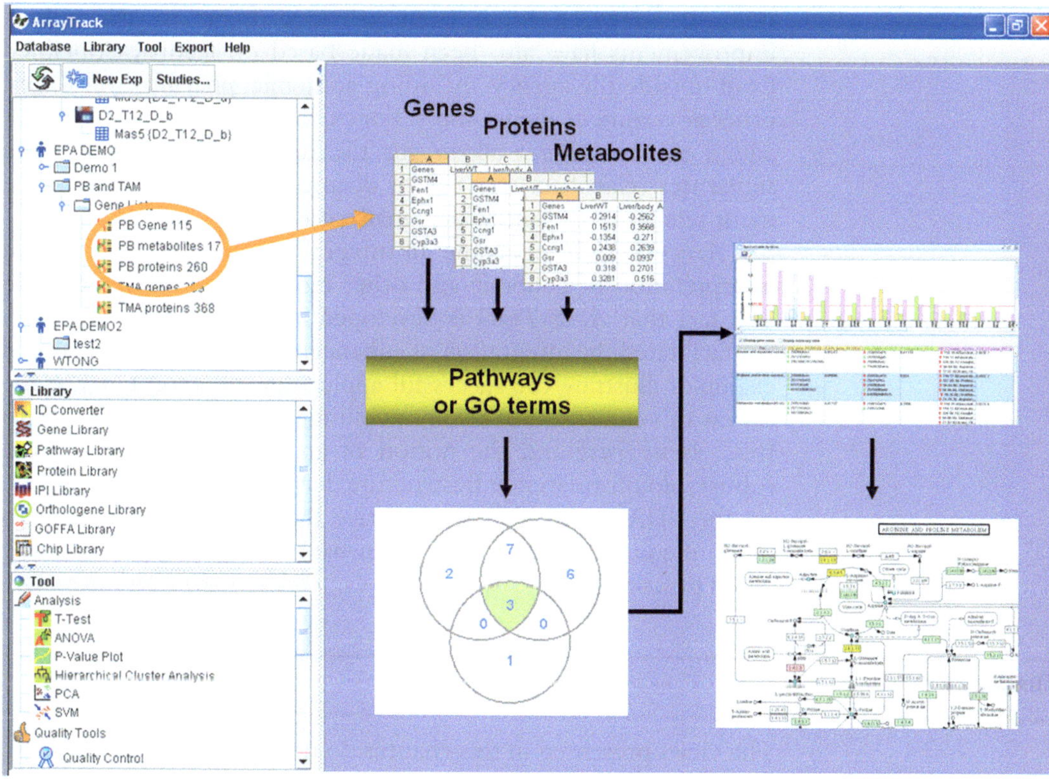

Fig. 8 An illustration of omics data integration logic in *ArrayTrack*. First, differentially expressed genes, proteins, and metabolites are generated or uploaded/stored in *ArrayTrack*. Then genes, proteins, and metabolites are each independently mapped to pathways or GO terms which are considered to also be significantly altered. Altered pathways or GO terms common between data types are next identified using a Venn diagram. The statistical significance of each common pathways or GO terms is estimated and displayed in a bar chart or spreadsheet. For each common pathway, the detailed pathway map can be viewed where the differentially expressed genes, proteins, and metabolites are highlighted in different colors

6 Conclusions

New high-throughput molecular technologies play an increasingly important role in both basic research and in drug discovery and development, and widespread anticipation exists that this trend will continue. The FDA has gained experience in analyzing new omics data through the VGDS program. The management, analysis, and interpretation of these data constitute a formidable effort for regulatory review. An efficient and integrated bioinformatics infrastructure within the agency is therefore essential to review and understand how sponsors reach their biological conclusions, to enable effective interactions with sponsors, and to ensure the incorporation of PGx data into regulatory processes.

ArrayTrack continues to undergo constant refinement and enhancement based on the feedback and needs of reviewers.

Because *ArrayTrack* has been provided freely to the public, improvements have also been made based on feedback obtained from outside the agency, including academic, pharmaceutical, and other government agency users. For example, one function recently added to *ArrayTrack* allows for the development of predictive signatures (classifiers) for use of diagnosis, prognosis, and treatment selection relevant to personalized medicine.

ArrayTrack has become an integral tool for the analysis and interpretation of genomic and other biomarker data at the FDA. The fact that *ArrayTrack* is developed internally within the FDA has facilitated the integration of enhancements and updates. Several examples illustrate the successful application of *ArrayTrack* in the review of voluntary, but also nonvoluntary data submissions. With this, *ArrayTrack* and the notion of an integrated, flexible, and robust bioinformatics infrastructure have become a cornerstone on the FDA's Critical Path Initiative that is aimed at helping to move medicine from a population-based to a more individually based practice.

Disclaimer

The views presented in this chapter do not necessarily reflect those of the US Food and Drug Administration.

References

1. Schmidt CW (2002) Toxicogenomics: an emerging discipline. Environ Health Perspect 110:A750–A755
2. Afshari CA, Nuwaysir EF, Barrett JC (1999) Application of complementary DNA microarray technology to carcinogen identification, toxicology, and drug safety evaluation. Cancer Res 59:4759–4760
3. Nuwaysir EF, Bittner M, Trent J, Barrett JC, Afshari CA (1999) Microarrays and toxicology: the advent of toxicogenomics. Mol Carcinog 24:153–159
4. Hamadeh HK, Amin RP, Paules RS, Afshari CA (2002) An overview of toxicogenomics. Curr Issues Mol Biol 4:45–56
5. Tong W, Cao X, Harris S, Sun H, Fang H, Fuscoe J, Harris A, Hong H, Xie Q, Perkins R, Shi L, Casciano D (2003) ArrayTrack—supporting toxicogenomic research at the U.S. Food and Drug Administration National Center for Toxicological Research. Environ Health Perspect 111:1819–1826
6. Frueh FW (2006) Impact of microarray data quality on genomic data submissions to the FDA. Nat Biotechnol 24:1105–1107
7. Shi L, Reid LH, Jones WD, Shippy R, Warrington JA, Baker SC, Collins PJ, de Longueville F, Kawasaki ES, Lee KY, Luo Y, Sun YA, Willey JC, Setterquist RA, Fischer GM, Tong W, Dragan YP, Dix DJ, Frueh FW, Goodsaid FM, Herman D, Jensen RV, Johnson CD, Lobenhofer EK, Puri RK, Schrf U, Thierry-Mieg J, Wang C, Wilson M, Wolber PK, Zhang L, Amur S, Bao W, Barbacioru CC, Lucas AB, Bertholet V, Boysen C, Bromley B, Brown D, Brunner A, Canales R, Cao XM, Cebula TA, Chen JJ, Cheng J, Chu TM, Chudin E, Corson J, Corton JC, Croner LJ, Davies C, Davison TS, Delenstarr G, Deng X, Dorris D, Eklund AC, Fan XH, Fang H, Fulmer-Smentek S, Fuscoe JC, Gallagher K, Ge W, Guo L, Guo X, Hager J, Haje PK, Han J, Han T, Harbottle HC, Harris SC, Hatchwell E, Hauser CA, Hester S, Hong H, Hurban P, Jackson SA, Ji H, Knight CR, Kuo WP, LeClerc JE, Levy S, Li QZ, Liu C, Liu Y, Lombardi MJ, Ma Y, Magnuson SR, Maqsodi B, McDaniel T, Mei N, Myklebost O, Ning B, Novoradovskaya N, Orr MS, Osborn TW, Papallo A, Patterson TA, Perkins RG, Peters EH, Peterson R, Philips KL, Pine PS, Pusztai L, Qian F, Ren H,

Rosen M, Rosenzweig BA, Samaha RR, Schena M, Schroth GP, Shchegrova S, Smith DD, Staedtler F, Su Z, Sun H, Szallasi Z, Tezak Z, Thierry-Mieg D, Thompson KL, Tikhonova I, Turpaz Y, Vallanat B, Van C, Walker SJ, Wang SJ, Wang Y, Wolfinger R, Wong A, Wu J, Xiao C, Xie Q, Xu J, Yang W, Zhong S, Zong Y, Slikker W Jr (2006) The MicroArray Quality Control (MAQC) project shows inter- and intraplatform reproducibility of gene expression measurements. Nat Biotechnol 24:1151–1161

8. Guidance for Industry: Pharmacogenomic data submissions–Companion Guidance: Department of Health and Human Services (HHS), Food and Drug Administration (FDA) (August 2007) http://www.fda.gov/cder/guidance/7735dft.pdf
9. Clinical Data Interchange Standard Consortium (CDISC): CDISC Inc., 15907 Two Rivers Cove, Austin, Texas 78717 (2007) http://www.cdisc.org/index.html
10. Tong W, Harris S, Cao X, Fang H, Shi L, Sun H, Fuscoe J, Harris A, Hong H, Xie Q, Perkins R, Casciano D (2004) Development of public toxicogenomics software for microarray data management and analysis. Mutat Res 549:241–253
11. Brazma A, Hingamp P, Quackenbush J, Sherlock G, Spellman P, Stoeckert C, Aach J, Ansorge W, Ball CA, Causton HC, Gaasterland T, Glenisson P, Holstege FC, Kim IF, Markowitz V, Matese JC, Parkinson H, Robinson A, Sarkans U, Schulze-Kremer S, Stewart J, Taylor R, Vilo J, Vingron M (2001) Minimum Information About a Microarray Experiment (MIAME)-toward standards for microarray data. Nat Genet 29:365–371
12. Quackenbush J (2002) Microarray data normalization and transformation. Nat Genet 32 (Suppl):496–501
13. Fielden MR, Halgren RG, Dere E, Zacharewski TR (2002) GP3: GenePix post-processing program for automated analysis of raw microarray data. Bioinformatics 18:771–773
14. Tusher VG, Tibshirani R, Chu G (2001) Significance analysis of microarrays applied to the ionizing radiation response. Proc Natl Acad Sci U S A 98:5116–5121
15. Chen T, Guo L, Zhang L, Shi LM, Fang H, Sun YM, Fuscoe JC, Mei N (2006) Gene expression profiles distinguish the carcinogenic effects of aristolochic acid in target (kidney) and non-target (liver) tissues in rats. BMC Bioinformatics 7(Suppl 2):S20
16. Sun H, Fang H, Chen T, Perkins R, Tong W (2006) GOFFA: Gene Ontology for Functional Analysis—a FDA Gene ontology tool for analysis of genomic and proteomic data. BMC Bioinformatics 7(Suppl 2):S23
17. Arlt VM, Ferluga D, Stiborova M, Pfohl-Leszkowicz A, Vukelic M, Ceovic S, Schmeiser HH, Cosyns JP (2002) Is aristolochic acid a risk factor for Balkan endemic nephropathy-associated urothelial cancer? Int J Cancer 101:500–502
18. Shav-Tal Y, Zipori D (2002) The role of activin a in regulation of hemopoiesis. Stem Cells 20:493–500
19. Fang H, Perkins R, Tong W (2007) Omics integrating systems using ArrayTrack and other bioinformatics tools. Am Drug Discov 2:49–52

Chapter 14

Identification of Transcriptional Regulators of Psoriasis from RNA-Seq Experiments

Alena Zolotarenko, Evgeny Chekalin, Rohini Mehta, Ancha Baranova, Tatiana V. Tatarinova, and Sergey Bruskin

Abstract

Psoriasis is a common inflammatory skin disease with complex etiology and chronic progression. To provide novel insights into the molecular mechanisms of regulation of the disease we performed RNA sequencing (RNA-Seq) analysis of 14 pairs of skin samples collected from psoriatic patients. Subsequent pathway analysis and an extraction of transcriptional regulators governing psoriasis-associated pathways was executed using a combination of MetaCore Interactome enrichment tool and cisExpress algorithm, and followed by comparison to a set of previously described psoriasis response elements. A comparative approach has allowed us to identify 42 core transcriptional regulators of the disease associated with inflammation (NFkB, IRF9, JUN, FOS, SRF), activity of T-cells in the psoriatic lesions (STAT6, FOXP3, NFATC2, GATA3, TCF7, RUNX1, etc.), hyperproliferation and migration of keratinocytes (JUN, FOS, NFIB, TFAP2A, TFAP2C), and lipid metabolism (TFAP2, RARA, VDR). After merging the ChIP-seq and RNA-seq data, we conclude that the atypical expression of FOXA1 transcriptional factor is an important player in psoriasis, as it inhibits maturation of naive T cells into this Treg subpopulation (CD4+FOXA1+CD47+CD69+PD-L1(hi)FOXP3−), therefore contributing to the development of psoriatic skin lesions.

Key words Psoriasis, RNA-Seq, FOXA1, Transcriptional regulation, Inflammation, Signaling pathways

1 Introduction

Psoriasis is a common chronic immune-mediated inflammatory condition characterized by complex alterations of cell signaling leading to the progression of the disease. The observed synergy between the aberrant activation of immune cells and an abnormal proliferation and differentiation of keratinocytes leads to the development of typical psoriatic symptom—red scaly thickened plaques on the skin surface. Another feature of psoriasis is a "cytokine storm" that begins locally within the skin, and then spreads throughout the body in form of a systemic inflammation that

Tatiana V. Tatarinova and Yuri Nikolsky (eds.), *Biological Networks and Pathway Analysis*, Methods in Molecular Biology, vol. 1613, DOI 10.1007/978-1-4939-7027-8_14, © Springer Science+Business Media LLC 2017

contributes to a development of comorbidities such as heart disease, stroke, diabetes, and psoriatic arthritis.

In order to identify the key signaling cascades and gene expression alternations causing the disease development and progression, we have performed RNA-Seq analysis of skin transcriptome in 14 patients with psoriasis [1]. As compared to other methods of gene expression analysis, RNA-Seq provides a more precise measurement of transcription levels, wider dynamic range of detection, and higher reproducibility of results. It was pointed out in Quigley [2], while for the most abundant transcripts both microarray and RNA-Seq produce similar results, RNA-Seq is capable of identification of a large number of transcripts expressed at low levels that could not be confidently called as differentially expressed when using microarrays to analyze the same number of samples.

In this study, we present the results of RNA-Seq analysis that allowed us to identify important signaling cascades enriched with differentially expressed genes (DEGs) that highlight potential transcription regulators contributing to the development of the disease. To identify transcriptional regulators of psoriatic pathology, we utilized two knowledge-based tools, MetaCore [3] and cisExpress [4, 5]. Modulation of the identified signaling pathways may be a promising approach for development of novel management strategies of psoriasis and other diseases commonly associated with this condition [6–8].

In psoriasis, the observed changes in gene expression levels may be due to two different disease-associated phenomena: the change in transcription and degradation rates of mRNA, and alterations in cell composition within the lesion that is usually characterized by the epidermal thickening, the accumulation of immune cells, and the thinning of subcutaneous fat layer. Therefore, whether the differential expression observed in comparison of lesional and non-lesional psoriatic samples truly reflects alterations of intracellular signaling remains unclear [9].

We have identified 1564 genes differentially expressed in psoriatic lesions (psoriatic DEGs), 938 of them were upregulated and 626 downregulated [1]. Analysis of the Top 20 upregulated DEGs highlighted the importance of the unspecific immune defense mechanisms, inflammatory response, taxis and chemotaxis of immune cells, and epidermal differentiation. It seems that impairment of these pathways contributes to the pathogenesis of psoriasis. This analysis suggests that the genes with the largest magnitude of expression changes are the "response" genes that contribute to pathophysiological manifestations of psoriasis rather than an initiation of the disease. In particular, a majority of the top 20 overexpressed genes were linked to lipid biosynthesis and lipid metabolism. Interestingly, among the top 20 downregulated genes, we detected a number of poorly characterized expression units, including possible pseudogenes, and noncoding RNAs. It remains an open question if noncoding RNAs enrichment is of any functional significance or just an indication of technology bias.

In order to identify the molecular underpinning of the psoriatic pathology, the gene ontology (GO) analysis and the MetaCore-guided pathway analysis were performed [3]. The results of the Gene ontology (GO) analysis generally supported the findings by MetaCore pathway enrichment (Fig. 1). The most DEG-enriched GO processes pointed at signaling alternations either relevant to psoriasis or to the cell populations contributing to the development of the disease. Top ten signaling pathways highlighted the importance of activation and chemotaxis of immune cells mediated by local enhanced production of pro-inflammatory cytokines and chemokines and the escalating reinforcement of inflammation. Below we discuss a number of psoriasis-associated pathways in details.

Skin serves as a first line of defense against the pathogen invasion. The stimulation of different pathogen-sensing receptors (such as PRRs) leads to activation of antimicrobial defense that is orchestrated by a number of key transcription factors, including nuclear factor kappaB (NF-κB), activator protein 1 (AP-1), cAMP response element-binding protein (CREB), and interferon-regulatory factor (IRF). Although the putative antigen leading to

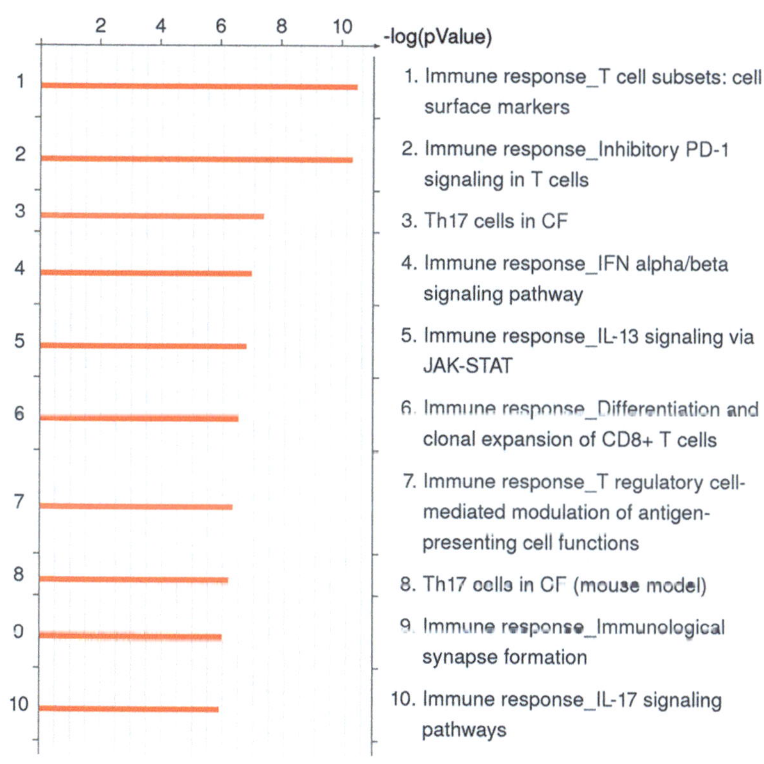

Fig. 1 Top ten DEG-enriched signaling pathways, as sorted by statistical significance of the findings

the activation of pro-inflammatory signaling in psoriasis has not yet been identified, it is widely accepted that the signaling cascades activated in course of psoriatic inflammation are mainly the same as these stimulated during the pathogen invasion. Activation of inflammatory and anti-apoptotic proteins ultimately alerts the immune system of invasion and induces the recruitment of leukocytes to the site of infection [10].

In order to ascertain the key regulatory "hub points" of the psoriatic networks, two independent approaches were used. The first of them utilizes the MetaCore Interactome enrichment tool [3] that evaluates levels of connectivity between the nodes (that can be either proteins or genes), identifies overconnected nodes and, according to the node function, suggests possible transcriptional regulators that drive the observed pattern of the gene expression in entire dataset. The second approach relies on the cisExpress algorithm [4, 5] that allows one to perform de novo discovery of the motif within the putative promoter regions of DEGs by means of comparison of these regions with the content of HOCOMOCO v9 [11], JASPAR 2014 [12], HumanTF 1.0 [13], and footprintDB [14] databases of known transcription factor binding sites (TFBSs). Next, the identified lists of transcriptional regulators were compared to the data compiled by Swindell et al. [15, 16] using meta-analysis of transcriptomes of 237 psoriatic patients and a dictionary of 2935 putative TFBSs and the sites for unconventional DNA-binding proteins (uDBPs). Swindell et al. [16] identified psoriasis response elements (PREs) overrepresented upstream of psoriasis DEGs in putative promoters that were defined as sequences starting at 5 kb upstream and ending at 500 bp downstream from the major transcription start site (TSS).

2 Materials

2.1 Patients and Samples

The patients in the RNA-Seq study were unrelated Caucasian individuals with plaque form of psoriasis from the Bryansk regional STD and Dermatology Center. Two 4 mm punch biopsy specimens were taken from skin of the patients with psoriasis, one from the lesional (LS sample) and another from nonlesional (NL sample) skin 3–4 cm apart from the lesion, so as the area does not have any visual signs of psoriasis. Patients did not obtain any systemic or PUVA/UV treatment 1 month before the biopsy taking. All biopsy samples were immediately transferred to the liquid nitrogen until RNA extraction.

2.2 RNA Sequencing

TissueLyser LT homogenizer (Qiagen, USA) was used to homogenize biopsy specimens. Total RNA was extracted with ExtractRNA reagent (Evrogen, Russia) according to the manufacturer's protocol. Isolated RNA was dissolved in RNase free water, rRNA was depleted using RiboMinus™ Eukaryote Kit for RNA-Seq (Life Technologies, USA), and the samples were stored at −80 °C. The quality of total RNA was evaluated with RNA 6000 Pico Chip Kit and Agilent 2100 Bioanalyzer (Agilent Technologies Inc., USA) and with Quant-iT™ RNA Assay Kit and Qubit Fluorometer (Life-Technologies, USA). The average RNA integrity number (RIN) of samples was ≥7. Library preparation and sequencing were performed using SOLiD 4 System platform and sequencing chemistry according to the manufacturer's instructions (Life Technologies, USA).

2.3 Processing and Mapping of RNA-Seq Reads and Differential Expression Analysis

Raw pair-end reads (50 + 25 bp) were obtained from SOLID4 System (Applied Biosystems) in color space format were filtered for quality, the adaptor sequences were trimmed and the reads were aligned to the UCSC human reference genome (hg19) using the Applied Biosystems's Bioscope software to obtain reads in the BAM format. Mapping to multiple locations was permitted. The aligned read BAM files were assembled into transcripts, their abundance was estimated and tests for differential expression were processed by Bioconductor DESeq package [17]. FDR correction for multiple testing was performed according to Benjamini et al. [18–20].

2.4 Pathway Analysis and Identification of Transcriptional Regulators

List of differentially expressed genes (FC > 1.5, FDR < 0.05) was used for gene ontology analysis with DAVID tool (Database for Annotation, Visualization and Integrated Discovery ver. 6.7) [21] and pathway analysis as well as Interactome analysis with MetaCore database from Thomson Reuters (ver. 6.11, build 41105, GeneGo, Thomson Reuters, USA) [3]. Using the cutoffs of fold change (FC) > 1.5 and false discovery rate of (FDR) <0.05, only genes that had read counts at all samples were listed, we have identified 1564 DEG, 938 of them were upregulated and 626 downregulated.

MetaCore Pathway analysis tool was used to perform gene network enrichment analysis, MetaCore Interactome tool was used for identification of transcriptional regulators of DEG-enriched pathways [3]. cisExpress algorithm [4, 5] was used for identification promoter motifs and discovery of cis-elements in promoter sequences that are statistically associated with expression patterns of DEG.

FOXA1 target list was obtained by merging ChIP-seq data from GSE39241 [22] and GSM1099031 [23] using edgeR [24] and DESeq [17] packages as recommended by authors, and then compared to DEG list.

3 Methods

3.1 MetaCore Guided Identification of the Transcriptional Regulators

MetaCore Interactome tool determines density of interactions between each protein from a dataset of interest, and all other proteins, evaluates statistically significant interactions within the set, and analyzes the functions of the selected interacting proteins [3]. Since proteins usually work in groups (such as protein complexes and pathways), which are defined by protein interactions, it is assumed that relative connectivity of each hub reflects its relevance, or importance, and may be used for identification of transcriptional regulators of DEG-enriched signaling cascades. Even if the expression levels of mRNA that encodes transcriptional factor itself is not altered, for example, when the TF in question is predominantly regulated posttranscriptionally, the number of targets it interacts depends on the state of its activation or suppression. Hence, the enrichment or the depletion of interacting protein networks indicates activation or suppression of the TF that orchestrates this network.

We identified possible transcriptional regulators of DEGs, computed their ranking according to the enrichment of interactions between the analyzed datasets, calculated from the normalized difference between the obtained number of targets and the expected number of expressed proteins, allowed us to identify the "Top" transcriptional regulators of the differentially expressed genes (Fig. 2) that lead to the development of the main distinctive features of psoriasis. In the "Top" TF list there are both cell-type specific transcriptional factors (e.g., PU.1 that is a master regulator of myeloid cells [25]) and ubiquitously expressed transcriptional factors associated with inflammatory pathways (e.g., NFkB and IRF [26]) that reflects alternations in cell populations in a plaque compared to unaffected skin. In addition to the transcription factors commonly associated with psoriasis, we have identified transcriptional factors that have not been previously associated with this disease. Further analysis is needed to determine their roles in the disease progression.

3.2 De Novo Analysis of Transcriptional Regulation of DEGs Using cisExpress Algorithm

Another approach that we have utilized for identification of the transcriptional regulators of DEGs was based on the cisExpress algorithm [4, 5]. cisExpress finds putative regulatory elements using a combination of sequence and expression information. Promoter sequences were obtained from the EPDnew database [27], which is a collection of experimentally validated promoters in human, mouse, fruit fly and zebrafish genomes. Evidence comes from TSS-mapping from high-throughput experimental techniques such as CAGE [28] and Oligocapping [29]. Positions of promoters were validated using the NPEST algorithm [30]. We identified 16,542 *H. sapiens* promoters that have corresponding RNA-Seq

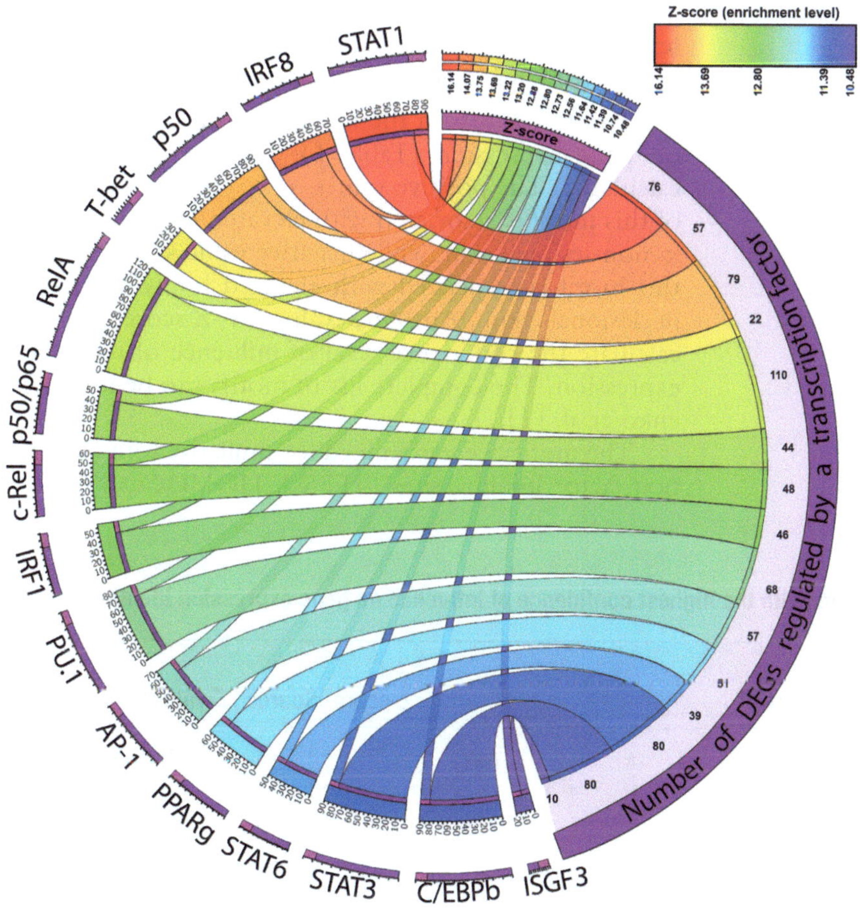

Fig. 2 Top 15 transcriptional regulators of DEG genes. Transcriptional factors ranked according to their Z-score (the level of connectivity of the TF to the DEG list). The *colors* from *green* to *pink* indicate the number of target genes for this transcriptional factor in the DEG list

gene expression measurements in lesional and non-lesional skin. Relative expression values for every gene were calculated from gene expression data for lesional ($n = 14$) and non-lesional ($n = 14$) skin according to the formula:

$$\text{Expression} = \ln \frac{\text{Avg NL expression value}}{\text{Avg LS expression value}}.$$

The length of a promoter region varies from gene to gene, and the identification of a "promoter window" containing the most important regulatory sequences for each gene is a separate challenge. Hence, for our analysis we used the "core promoter-5′ UTR" region of +500 –500 bp around the TSS of each gene. The analyzed set of the promoters was examined for the presence of motifs (putative transcription factor binding sites), and the corresponding gene expression values were compared for genes whose promoters did and did not contain the motifs using the *t*-test. We compiled a ranked list of over 100 position-specific motifs in the promoter regions of DEGs.

Top ten motifs in the ranked list have the highest influence on gene expression (Table 1). For every motif, *Z*-score was calculated. The positive values of *Z*-score suggests that presence of this motif is associated with elevated levels of gene expression in non-lesional skin, while negative value shows that presence of this motif is associated with elevated levels of gene expression in lesional skin. Absolute value of *Z*-score can be used to calculate the confidence level of influence of the motif on gene expression. The complete list of motifs can be found in Zolotarenko et al. [1].

The motifs were examined for similarity with known transcription factor binding sites (TFBSs). HOMOCO v9 [11], JASPAR

Table 1
Top ten motifs with the highest confidence of influence on gene expression identified with cisExpress

From...to, bp	Motif		*Z*-score	Source	*E*-value	Associated proteins
10...30	AAGATG	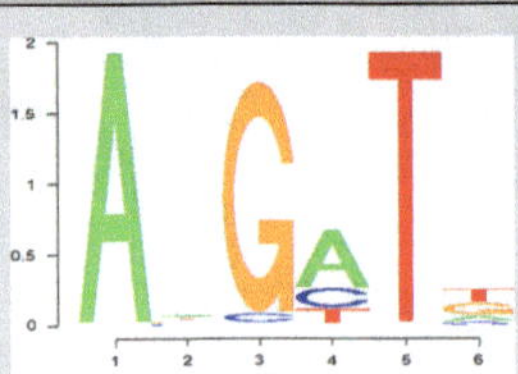	7.08	1 2	1.3e−05 1.4e−05	ETS1, p54 AP-1, p39, AP1
−20...0	CCGGAA		5.87	3	3.1e−08	ELK4
−10...10	CAC[CT]C	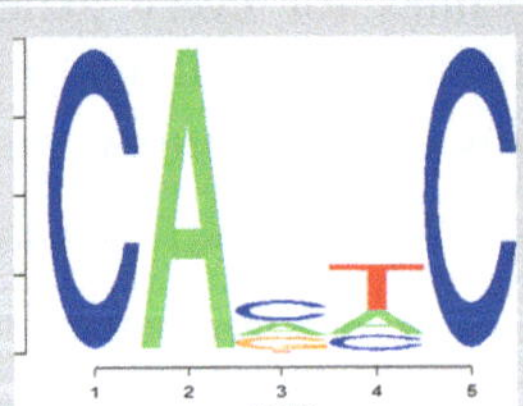	−5.71	4 4 2 4 4	2.3e−06 6.1e−06 1.3e−05 3.8e−04 0.002	ZIC1, ZIC2, ZIC3 GLI3, KLF1 SREBF1 NKX25 AP-1, p39
−60...−40	CGGAA		5.64	1 4 3	2.7e−08 5.8e−06 5.9e−06	NFATC2, ETS2, ELK4 NRF-2/GABP1 ELK1

(continued)

Table 1
(continued)

From. . .to, bp	Motif		*Z*-score	Source	*E*-value	Associated proteins
10. . .30	TGGCGG	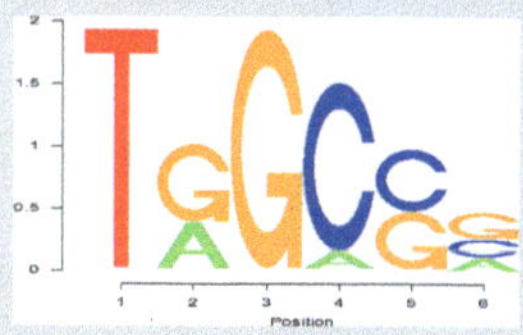	5.64	3 4	8.6e−08 1.7e−06	E2F8, NF-E1 TFDP1
0. . .20	AAGAT		5.56	3 4 2	1.9e−05 0.001 0.001	TCF7L1 GATA2, GATA6 JUN
−50. . .−30	CGGAA		5.44	1 4 3	2.7e−08 5.8e−06 5.9e−06	NFATC2, ETS2, ELK4 NRF-2/GABP1 ELK1
300. . .320	CCGGT	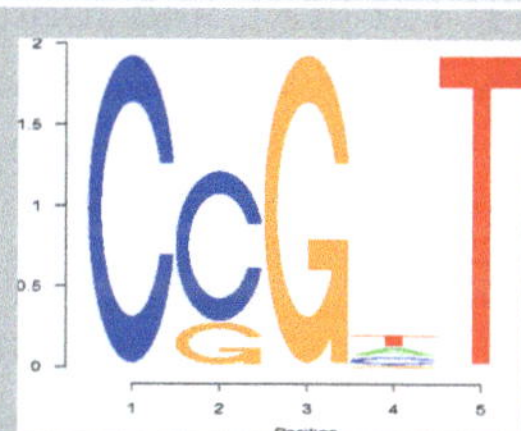	5.43	4 3	6.0e−06 6.0e−06	ELK4 GRHL1
−10. . .10	GCCAT		5.37	4 3	6.9e−05 8.1e−05	RFX3 E2F2
0. . .20	GATGGC	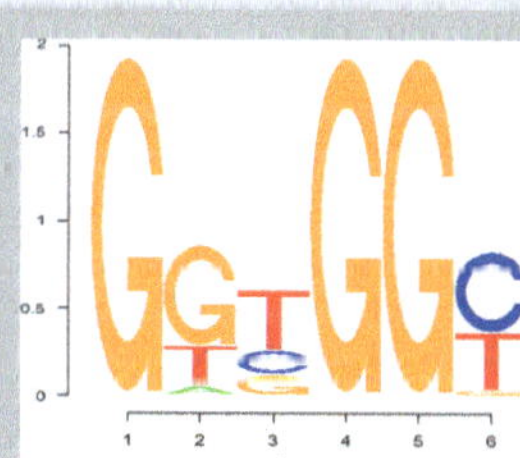	5.31	4 4 4	8.3e−07 4.9e−05 1.5e−04	ZBTB4 HXA1, HXB1 TAL1

"From. . .to" is the position of the "window" where the motif has been discovered, relative to the gene transcription start. *Source*: *1* footprintDB, *2* JASPAR, *3* HumanTF 1.0, *4* HOCOMOCO. The positive values of *Z*-score suggest that presence of the motif promotes the expression of a gene in non-lesional skin while negative values of *Z*-score suggest that motif acts in lesional skin

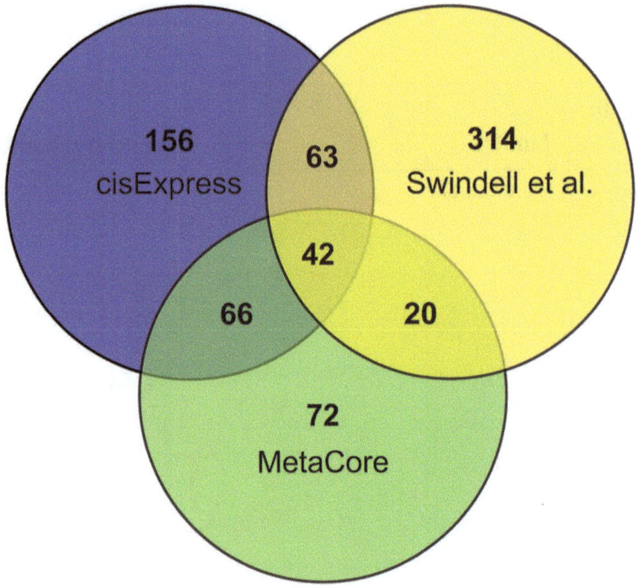

Fig. 3 *Venn diagram* showing overlap between lists of transcriptional regulators of DEG. *Violet*—TFs, identified by cisExpress tool; *yellow*—identified in Swindell et al. [16], *green*—identified by the MetaCore software

2014 [12], HumanTF 1.0 [13], and footprintDB [14] databases were used for this analysis. For example, one of the identified motifs (AAGATG) is related to the *ETS1* transcription factor, which was associated with psoriasis because *ETS1* is a negative regulator of *Th17* cells [31] and *GATA-3*, differentially expressed in Th1/Th2 cells during psoriasis [32]. CCGGAA motif is associated with ELK4, which is highly expressed in lesional psoriasis skin [15].

3.3 Identification of the Key Transcriptional Regulators of Psoriatic Transcriptome

In order to find the key transcriptional regulators of DEG we compared lists identified by the two computational approaches (MetaCore and cisExpress), as well as the results of Swindell et al. [16]. Comparison of the three groups of transcription regulators (327 cisExpress-identified, 200 MetaCore-identified, and 439 identified by Swindell et al.) (Fig. 3) found 42 common transcriptional factors representing the "core" TF regulators of psoriatic transcriptome (Table 2).

The majority of elements of the "core TF" list are transcription factors associated with inflammation (NFkB, IRF9, JUN, FOS, SRF), activity of T-cells in the psoriatic lesions (STAT6, FOXP3, NFATC2, GATA3, TCF7, RUNX1 etc.), hyperproliferation and migration of keratinocytes (JUN, FOS, NFIB, TFAP2A, TFAP2C), and with lipid metabolism (TFAP2, RARA, VDR). There were several FOX (Forkhead box) family proteins in the list, containing the evolutionary conserved "fork-head" or "winged-helix" DNA-binding domain (DBD). These proteins

Table 2
Core TF regulators if psoriatic transcriptome

Ensembl gene ID	Gene symbol	Transcription factor name
ENSG00000067955	CBFB	Core-binding factor, beta subunit
ENSG00000105516	DBP	D site of albumin promoter (albumin D-box) binding protein
ENSG00000101412	E2F1	E2F transcription factor 1
ENSG00000164330	EBF1	Early B-cell factor 1
ENSG00000120738	EGR1	Early growth response 1
ENSG00000134954	ETS1	v-ets erythroblastosis virus E26 oncogene homolog 1 (avian)
ENSG00000170345	FOS	v-fos FBJ murine osteosarcoma viral oncogene homolog
ENSG00000129514	FOXA1	Forkhead box A1
ENSG00000125798	FOXA2	Forkhead box A2
ENSG00000111206	FOXM1	Forkhead box M1
ENSG00000150907	FOXO1	Forkhead box O1
ENSG00000128573	FOXP2	Forkhead box P2
ENSG00000049768	FOXP3	Forkhead box P3
ENSG00000107485	GATA3	GATA binding protein 3
ENSG00000162676	GFI1	Growth factor independent 1 transcription repressor
ENSG00000135100	HNF1A	HNF1 homeobox A
ENSG00000101076	HNF4A	Hepatocyte nuclear factor 4, alpha
ENSG00000213928	IRF9	Interferon regulatory factor 9
ENSG00000177606	JUN	Jun oncogene
ENSG00000169926	KLF13	Kruppel-like factor 13
ENSG00000106689	LHX2	LIM homeobox 2
ENSG00000099326	MZF1	Myeloid zinc finger 1
ENSG00000101096	NFATC2	Nuclear factor of activated T cells, cytoplasmic, calcineurin-dependent 2
ENSG00000147862	NFIB	Nuclear factor I/B
ENSG00000165030	NFIL3	Nuclear factor, interleukin 3 regulated
ENSG00000109320	NFKB1	Nuclear factor of kappa light polypeptide gene enhancer in B-cells 1
ENSG00000143190	POU2F1	POU class 2 homeobox 1
ENSG00000131759	RARA	Retinoic acid receptor, alpha
ENSG00000159216	RUNX1	Runt-related transcription factor 1

(continued)

Table 2 (continued)

Ensembl gene ID	Gene symbol	Transcription factor name
ENSG00000186350	RXRA	Retinoid X receptor, alpha
ENSG00000175387	SMAD2	SMAD family member 2
ENSG00000143842	SOX13	SRY (sex determining region Y)-box 13
ENSG00000125398	SOX9	SRY (sex determining region Y)-box 9
ENSG00000112658	SRF	Serum response factor (c-fos serum response element-binding transcription factor)
ENSG00000166888	STAT6	Signal transducer and activator of transcription 6, interleukin-4 induced
ENSG00000162367	TAL1	T-Cell acute lymphocytic leukemia 1
ENSG00000071564	TCF3	Transcription factor 3 (E2A immunoglobulin enhancer binding factors E12/E47)
ENSG00000081059	TCF7	Transcription factor 7 (T-cell specific, HMG-box)
ENSG00000148737	TCF7L2	Transcription factor 7-like 2 (T-cell specific, HMG-box)
ENSG00000137203	TFAP2A	Transcription factor AP-2 alpha (activating enhancer binding protein 2 alpha)
ENSG00000087510	TFAP2C	Transcription factor AP-2 gamma (activating enhancer binding protein 2 gamma)
ENSG00000111424	VDR	Vitamin D (1,25-dihydroxy vitamin D3) receptor

could work as active regulators of cell proliferation and metabolism and also serve as pioneer factors that decondense chromatin, therefore facilitating binding of other sequence-specific transcription factors to target enhancers, repressors and promoters, wiring global gene networks essential for cell fate decisions [33]. We also found 294 DEG-associated transcription factors not identified by Swindell et al. [16]. Sixty-six of them were identified by cisExpress as well as by MetaCore (Fig. 3).

3.4 Pathway Analysis

In order to investigate relationships between the DEGs, we performed gene network enrichment analysis with the MetaCore software. The top ten signaling networks enriched with DEGs were mainly associated with different alternations in immune signaling present in the psoriatic lesions (Fig. 1), e.g., the map "Immune response_IL-17 signaling pathway" (Fig. 4) This agrees with the hypothesis that the main feature of psoriasis is the cytokine storm and alternated balance of cytokines, chemokines, and growth factors regulating various immune and inflammatory responses (*see* Zolotarenko et al. [1] for a detailed discussion). The main

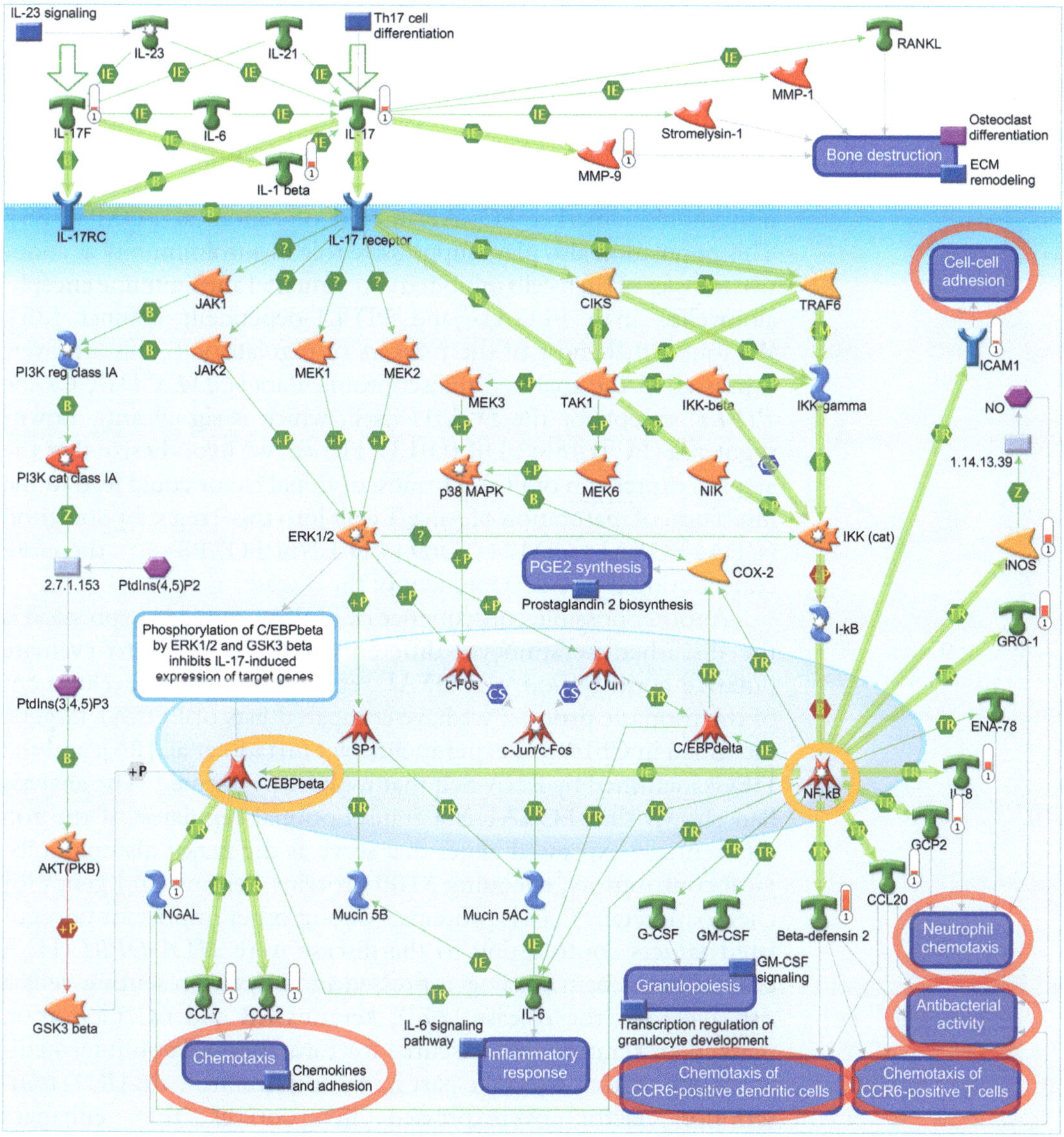

Fig. 4 Immune response_IL-17 signaling pathway. Illustration generated with MetaCore pathway analysis tool (GeneGO/Thomson Reuters)

transcriptional regulators of this map are NFkB, C/EBPb and different proteins from the AP1 superfamily. All these transcription factors were identified in the Top10 list of MetaCore analysis (Fig. 2) as well as in the cisExpress analysis.

The importance of different populations of T-cells in the pathogenesis of the disease is illustrated by the DEG-enriched map “Immune response: T-cell subsets secreted signals.” It shows the shift of T-cell populations to the IL-17-producing types, hence, being a sign of activation and enhanced migration of psoriasis-specific populations of T-cells to the lesional skin.

One of the main regulators of T cell polarization and population maintenance is regulatory T-cell population (Treg). One of the hypotheses points out the alternations in presence and activity of regulatory T cells in skin of psoriasis patients as a possible reason of the development of the disease [34]. Recent studies uncovered a new population of regulatory T FOXA1+ Treg cells that carry noncanonical marker FOXA1 instead of canonical FOXP3 [35]. This population also plays suppressive role in autoimmunity as adoptive transfer of such cells inhibited experimental autoimmune encephalomyelitis in a FOXA1- and PD-L1-dependent manner [35]. We found that most of the markers of regulatory T cells are overexpressed in the analyzed transcriptomic data (*CD4*, *CD47*, *CD69*, *PD-L1*) except for the *FOXA1* itself, which is significantly downregulated (FC 0.485; FDR 0.013). Hence, we hypothesize that the atypical expression of FOXA1 transcriptional factor could lead to the inhibition of maturation of naive T cells into this Treg subpopulation (CD4+FOXA1+CD47+CD69+PD-L1(hi)FOXP3−), therefore contributing to the development of the disease.

Another possible consequence of FOXA1 reduced expression is the disturbed keratinocyte differentiation. In order to evaluate putative contribution of FOXA1 regulation to the development of the psoriatic process, we have compared lists of FOXA1 targets, identified in ChIP-Seq experiments by Hurtado et al. [36], and the DEGs identified by RNA-Seq that we have performed. The analysis had shown that FOXA1 is a transcriptional regulator of the top differentially expressed genes that serve as the major histopathological contributors, encoding S100 proteins, serpins, and genes for chemoattractant CXCL proteins. Among other important upregulated targets contributing to the disease were *HLA-DPB1* (HLA class II beta chain paralog expressed in antigen presenting cells a risk allele for the disease [37]); keratins 6B and 6C (activation markers of keratinocytes essential for formation keratin intermediate filaments that also take part in wound healing); *PPARD* (transcription factor overexpressed in psoriasis that enhances proliferation of keratinocytes and is induced by JUNB in keratinocytes) [38]. Among the downregulated targets, there were genes associated with lipid disturbances observed in psoriatic lesions, including known components of fatty acid metabolism acyl-CoA wax alcohol acyltransferase gene *AWAT2*, fatty acid elongase gene *ELOVL3*, fatty acid binding protein FABP4, and many others. Ontology analysis of FOXA1 target DEGs had shown a similar list of ontologies as the whole DEG list discussed above. Comparison of the lists of ontologies has shown that FOXA1 is a part of regulatory complex accounting for the most important psoriatic alternations and signaling cascades important for the pathology, so this transcription factor is a promising candidate for future investigation in the context of psoriasis.

4 Conclusions

In conclusion, the performed analysis has highlighted the importance of immune system alternations for the development of the disease. We present a list of identified core transcriptional regulators of psoriatic transcriptome that should be further investigated as a source of insights into the mechanisms of pathology-specific gene regulation. We have also identified novel transcriptional regulators of psoriasis-associated pathways previously not suspected to play a role in the pathology. The comparison of our data with public ChIP-seq data has allowed us to formulate a hypothesis explaining the role of FOXA1 transcription factor in psoriasis.

References

1. Zolotarenko A et al (2016) Integrated computational approach to the analysis of RNA-seq data reveals new transcriptional regulators of psoriasis. Exp Mol Med 48(11):268
2. Quigley D (2014) RNA-seq permits a closer look at normal skin and psoriasis gene networks. J Invest Dermatol 134(7):1789–1791
3. Bessarabova M et al (2012) Knowledge-based analysis of proteomics data. BMC Bioinformatics 13(Suppl 16):13
4. Triska M et al (2013) cisExpress: motif detection in DNA sequences. Bioinformatics 29 (17):2203–2205
5. Troukhan M et al (2009) Genome-wide discovery of cis-elements in promoter sequences using gene expression. OMICS 13 (2):139–151
6. Liu B et al (2015) A cytoplasmic NF-kappaB interacting long noncoding RNA blocks IkappaB phosphorylation and suppresses breast cancer metastasis. Cancer Cell 27(3):370–381
7. Mayer TZ et al (2013) The p38-MSK1 signaling cascade influences cytokine production through CREB and C/EBP factors in human neutrophils. J Immunol 191(8):4299–4307
8. Pallandre JR et al (2007) Role of STAT3 in CD4+CD25+FOXP3+ regulatory lymphocyte generation: implications in graft-versus-host disease and antitumor immunity. J Immunol 179(11):7593–7604
9. Li B et al (2014) Transcriptome analysis of psoriasis in a large case-control sample: RNA-seq provides insights into disease mechanisms. J Invest Dermatol 134(7):1828–1838
10. Hodgson A, Wan F (2015) Interference with nuclear factor kappaB signaling pathway by pathogen-encoded proteases: global and selective inhibition. Mol Microbiol 99(3):439–452
11. Kulakovskiy IV et al (2013) HOCOMOCO: a comprehensive collection of human transcription factor binding sites models. Nucleic Acids Res 41(Database issue):D195–D202
12. Mathelier A et al (2014) JASPAR 2014: an extensively expanded and updated open-access database of transcription factor binding profiles. Nucleic Acids Res 42(Database issue): D142–D147
13. Jolma A et al (2013) DNA-binding specificities of human transcription factors. Cell 152 (1–2):327–339
14. Kirsanov DD et al (2013) NPIDB: nucleic acid-protein interaction DataBase. Nucleic Acids Res 41(Database issue):D517–D523
15. Swindell WR et al (2011) Genome-wide expression profiling of five mouse models identifies similarities and differences with human psoriasis. PLoS One 6(4):e18266
16. Swindell WR et al (2015) Psoriasis drug development and GWAS interpretation through in silico analysis of transcription factor binding sites. Clin Transl Med 4:13
17. Anders S, Huber W (2010) Differential expression analysis for sequence count data. Genome Biol 11(10):R106
18. Benjamini Y et al (2001) Controlling the false discovery rate in behavior genetics research. Behav Brain Res 125(1–2):279–284
19. Benjamini Y, Heller R (2008) Screening for partial conjunction hypotheses. Biometrics 64 (4):1215–1222
20. Benjamini Y, Hochberg Y (1995) Controlling the false discovery rate: a practical and powerful approach to multiple testing. J R Stat Soc B 57 (1):289–300
21. Huang da W, Sherman BT, Lempicki RA (2009) Systematic and integrative analysis of

large gene lists using DAVID bioinformatics resources. Nat Protoc 4(1):44–57
22. Caravaca JM et al (2013) Bookmarking by specific and nonspecific binding of FoxA1 pioneer factor to mitotic chromosomes. Genes Dev 27 (3):251–260
23. Ni M et al (2013) Amplitude modulation of androgen signaling by c-MYC. Genes Dev 27 (7):734–748
24. Robinson MD, McCarthy DJ, Smyth GK (2010) edgeR: a Bioconductor package for differential expression analysis of digital gene expression data. Bioinformatics 26 (1):139–140
25. Vangala RK et al (2003) The myeloid master regulator transcription factor PU.1 is inactivated by AML1-ETO in t(8;21) myeloid leukemia. Blood 101(1):270–277
26. Iwanaszko M, Kimmel M (2015) NF-kappaB and IRF pathways: cross-regulation on target genes promoter level. BMC Genomics 16:307
27. Dreos R et al (2013) EPD and EPDnew, high-quality promoter resources in the next-generation sequencing era. Nucleic Acids Res 41(Database issue):D157–D164
28. Takahashi H et al (2012) CAGE (cap analysis of gene expression): a protocol for the detection of promoter and transcriptional networks. Methods Mol Biol 786:181–200
29. Maruyama K, Sugano S (1994) Oligo-capping: a simple method to replace the cap structure of eukaryotic mRNAs with oligoribonucleotides. Gene 138(1–2):171–174
30. Tatarinova T et al (2013) NPEST: a nonparametric method and a database for transcription start site prediction. Quant Biol 1(4):261–271
31. Lee PH et al (2014) The transcription factor E74-like factor 4 suppresses differentiation of proliferating CD4+ T cells to the Th17 lineage. J Immunol 192(1):178–188
32. Zhang P et al (2014) Analysis of Th1/Th2 response pattern for erythrodermic psoriasis. J Huazhong Univ Sci Technolog Med Sci 34 (4):596–601
33. Lam EW et al (2013) Forkhead box proteins: tuning forks for transcriptional harmony. Nat Rev Cancer 13(7):482–495
34. Keijsers RR et al (2013) Balance of Treg vs. T-helper cells in the transition from symptomless to lesional psoriatic skin. Br J Dermatol 168 (6):1294–1302
35. Liu Y et al (2014) FoxA1 directs the lineage and immunosuppressive properties of a novel regulatory T cell population in EAE and MS. Nat Med 20(3):272–282
36. Hurtado CW, Furuta GT, Kramer RE (2011) Etiology of esophageal food impactions in children. J Pediatr Gastroenterol Nutr 52(1):43–46
37. Kim TG et al (2000) The association of psoriasis with human leukocyte antigens in Korean population and the influence of age of onset and sex. J Invest Dermatol 114(2):309–313
38. Romanowska M et al (2008) PPARdelta enhances keratinocyte proliferation in psoriasis and induces heparin-binding EGF-like growth factor. J Invest Dermatol 128(1):110–124

Chapter 15

Comprehensive Analyses of Tissue-Specific Networks with Implications to Psychiatric Diseases

Guan Ning Lin, Roser Corominas, Hyun-Jun Nam, Jorge Urresti, and Lilia M. Iakoucheva

Abstract

Recent advances in genome sequencing and "omics" technologies are opening new opportunities for improving diagnosis and treatment of human diseases. The precision medicine initiative in particular aims at developing individualized treatment options that take into account individual variability in genes and environment of each person. Systems biology approaches that group genes, transcripts and proteins into functionally meaningful networks will play crucial role in the future of personalized medicine. They will allow comparison of healthy and disease-affected tissues and organs from the same individual, as well as between healthy and disease-afflicted individuals. However, the field faces a multitude of challenges ranging from data integration to statistical and combinatorial issues in data analyses. This chapter describes computational approaches developed by us and the others to tackle challenges in tissue-specific network analyses, with the main focus on psychiatric diseases.

Key words Psychiatric diseases, Autism, Genetics, Gene expression, Protein–protein interactions, Alternatively spliced isoforms, Copy number variants, De novo mutations, Network analyses, Systems biology

1 Introduction to Psychiatric Disease Networks

Recent large-scale genetic studies of patients and family cohorts have begun to unravel the genetic architecture of psychiatric disorders, including autism spectrum disorders (ASD), schizophrenia (SCZ), and intellectual disability (ID) [1]. Hundreds to thousands of genetic *loci* have been identified as putative risk factors for these diseases, with only a handful of them being strongly implicated as causative. To understand how this overwhelming number of identified genetic risk factors contributes to abnormal functioning of the brain and ultimately leads to disease phenotypes, it is necessary to adopt rigorous data-driven framework that operates at the system or network levels [2]. Over the past decade, rapid progress has been made in our understanding that biological

Tatiana V. Tatarinova and Yuri Nikolsky (eds.), *Biological Networks and Pathway Analysis*, Methods in Molecular Biology, vol. 1613, DOI 10.1007/978-1-4939-7027-8_15, © Springer Science+Business Media LLC 2017

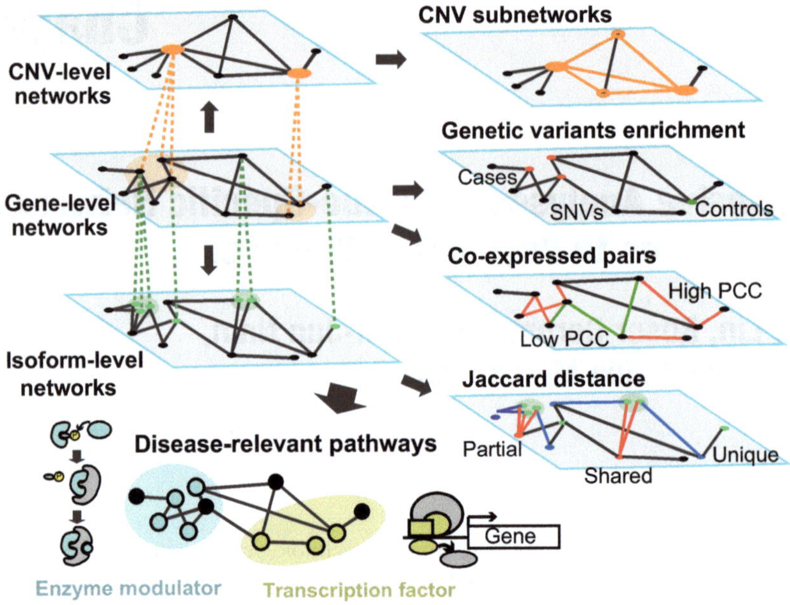

Fig. 1 Schematic representation of the multilayer analyses of disease networks leading to the identification of the disease-relevant pathways. Three layers of network complexity are considered (*left panels*): *top*, the CNV-level network, where proteins encoded by genes from the same copy number variant (CNV) are grouped into one network node and the interactions of these proteins are merged; *middle*, the gene-level network, where each network node represents one gene/protein; *bottom*, the isoform-level network, where a new layer of complexity is added by splitting gene nodes into multiple splicing isoform nodes. (*Right panels*) Various types of analyses carried out on the networks. Examples of disease-relevant pathways are shown at the *bottom* and represent potential new disease biomarkers or drug targets. Abbreviations: CNV - Copy Number Variant, SNV - Single Nucleotide Variant, PCC - Pearson Correlation Coefficient

networks formed by complex sets of interactions between numerous genes, transcripts and proteins play important role in deciphering disease phenotypes [3, 4]. In particular, gene expression networks have been increasingly used to obtain systematic views about an immensely complex molecular landscape across brain development [5–10]. However, a dramatic increase in high-throughput experimental and computational data created a need for further improvement of effective network analytical techniques in order to unravel the molecular basis of brain disorders. This chapter describes computational approaches developed by us and the others for analyzing brain-specific biological networks related to psychiatric disorders (Fig. 1). The approaches described below are generally applicable to other human diseases for which genetic, transcriptomic, and protein interaction data are readily available.

2 Gene-Level Networks for Psychiatric Disorders

2.1 Construction of Protein–Protein Interaction Networks

In order to build a protein–protein interaction (PPI) network relevant to a specific disease, it is necessary to first select a set of the disease risk factors, and then obtain a set of PPIs connecting these factors. The list of disease candidate genes could be obtained either by literature curation of multiple studies with diverse sources of experimental evidence, or by extracting relevant genes from the high-throughput genetic studies, such as whole exome sequencing (WES) [11–13] or whole genome sequencing (WGS) of patients or family cohorts [14, 15], or from the genome-wide association studies (GWAS) [16]. The set of PPIs for these genes can be obtained experimentally [17, 18], predicted computationally [19], or alternatively downloaded from public databases such as BioGRID [20], HPRD [21], IntAct [22], and similar. The literature-curated protein interaction databases aim to aggregate all known interactions between proteins from multiple experimental sources. However, individual experiments generally focus on a selected subset of target proteins, and typically use a specific method for data collection, such as yeast two-hybrid (Y2H) system, tandem affinity purification, or co-immunoprecipitation followed by mass spectrometry proteomics. In addition, the majority of PPIs are not collected in a tissue-specific manner. This complicates interpretation of the results relevant to the disease networks that likely operate in a tissue-specific manner. Therefore, selection of appropriate control (or background) networks for the analyses are crucial for obtaining meaningful insights into specific disease mechanisms.

2.2 Selection of Control Networks for the Analyses

The PPIs from the public databases are intrinsically biased toward highly studied proteins, for example those implicated in cancer. These well-studied proteins accumulate more interactions than less studied ones, and consequently tend to become hubs in the PPI networks. Although hub proteins may be highly relevant to some disease networks, they may not have similar strong relevance to other disease networks. In order to minimize the biases introduced by well-studied proteins, one needs to carefully select subsets of compatible background data for network analyses in order to draw meaningful conclusions about a specific disease.

Other important sources of network biases are the intrinsic properties of the genes/proteins within the network. It has been noted that the number of interactions of a protein is correlated with the length of a protein, with longer proteins having a greater number of interactions [23]. Thus, when using genetic data to construct and analyze networks, it is important to take into account the properties of the genes, particularly gene length and GC content [24]. It is especially important to ensure that the control networks chosen for the comparison have similar properties with the disease network under consideration. One possible way to control for biases is to limit the control or background networks to proteins that are most similar to the subset of studied proteins from the disease network, as detailed below.

2.2.1 Experimentally Compatible Control Protein Interaction Networks

In order to control for experimental biases of the PPIs networks, one could limit the PPIs to only those interactions that are obtained through the high-throughput systematic screens [17, 25, 26]. These and similar studies identify thousands of PPIs by systematically testing large number of human proteins against the entire human ORFeome [27] using consistent experimental procedures with follow-up pair-wise verification by independent methods [28]. Although such selection strategy may decrease the number of PPIs that could be used for network construction, compared to those available from large public databases such as STRING [29] or InWeb [30], selection of unbiased PPIs would likely increase confidence and reliability of the resulting networks.

In cases, when the PPIs for building disease network were obtained through a new experimental screen, it is essential to use PPIs generated in a similar manner as a control for network analyses. In our recent work [31], we constructed an Autism Spliceform Interaction Network (ASIN) by screening over 400 isoforms of autism risk factors (baits) against ~15,000 human ORFs (preys) using a systematic yeast-two-hybrid (Y2H) screen. We used a similar, but much larger independent control dataset, the Human Interactome (HI) [17] for comparison with ASIN. HI is a set of ~14,000 high-quality binary human PPIs obtained in an unbiased reciprocal screen of ~15,000 human ORFs [17]. Both, the ASIN and the HI, shared the same prey search space and were generated in the same laboratory using similar experimental pipelines. Therefore, HI served as a perfect control network for enrichment analysis in ASIN [31].

Alternatively, in the absence of an experimental control dataset, a set of PPIs with comparable quality curated from the published literature or public databases can be used. To assemble such a dataset, one should restrict the curated data to only human high-quality interactions and preferably only those obtained using the same experimental method (e.g., binary physical interactions) [32, 33]. For example, in case of ASIN, in addition to HI control network, we have also used a human binary literature-curated interaction (LCI) dataset consisting of ~40,000 PPIs. The LCI was assembled by updating a previously used PPI dataset [32] to include all newly reported PPIs from major databases with the identical filtering criteria. To verify that the control networks are comparable to ASIN, we tested for gene length and GC content biases and observed no statistically significant differences between the ASIN and both control datasets [31].

2.2.2 Randomized Control Networks

Another common strategy for analyzing disease networks when biologically comparable control networks are unavailable is to create randomized networks and then estimate the null distribution of the test statistic (i.e., number of interacting partners of certain nodes, network connectivity and others). The statistics of the null distribution can be estimated from a set of randomized networks (e.g., 10,000 random networks with comparable properties). Such randomized networks could be generated by randomly selecting one protein of an interacting pair and replacing it with another protein from a randomly selected interaction pair, and repeating the replacement X number of times (where X is at least four times the total number of nodes in the network) to achieve complete randomization. Each randomized network preserves the degree distribution of the corresponding real network. As in all statistical hypothesis tests, the significance of a permutation test is represented by its P-value, which is estimated by the probability of obtaining an observed value at least as extreme as the test statistic given that the null hypothesis is true. In our work [31], we applied this strategy to create random control networks for assessing ASIN quality using different functional annotations (i.e., co-expression and co-regulation).

2.3 Comparing Properties of Disease and Control Networks

After the disease and control networks are constructed, it is sometimes desirable to compare the properties of these two networks. This could be achieved through calculating various parameters: (1) *network degree* is the number of interacting partners (i.e., the number of edges) of a network protein (i.e., a node); (2) *network shortest path* is the minimal number of protein–protein interactions (i.e., edges) needed to connect two nodes (i.e., the minimal number of edges traversed); (3) *network clustering coefficient* (C_x) calculates the degree of interconnectivity of the partners of one protein. The clustering coefficient is defined as:

$$C_x = \frac{2p_x}{(N_x(N_x - 1))},$$

where N_x is the number of neighbors of x and p_x is the number of the connected pairs between all neighbors of x, which ranges from zero (the partners of the node are not connected) to one (the partners are fully connected); and (4) *network betweenness centrality* is a measure that could be used for ranking a protein node within a network considering the number of shortest paths that pass through that node. The betweenness centrality is calculated as follows:

$$B(v) = \sum_{s-1}^{n} \sum_{p-1}^{s-1} \frac{n_{sd}(v)}{n_{sd}}$$

where n_{sd} is the total number of shortest paths from a protein s to a protein d, and $n_{\text{sd}}(v)$ is the number of the paths that traverse the node v. This value is then normalized by dividing the number of all possible edges between all proteins in the network (excluding s): $((n - 1)(n - 2))/2$. The parameters that are statistically different between the disease and carefully chosen control networks could represent biologically important findings.

2.4 Obtaining Biological Insights from Disease Networks

The disease networks could be useful in many aspects, from predicting the pathways disrupted by the disease mutations, to discovering new disease risk factors. The latter task could be accomplished by ranking binding partners (i.e., preys) of the confident disease-implicated proteins. The prey ranking method can be applied successfully when an unbiased control network with similar properties to the disease network is available. This method involves estimation of preys' biological connectivity. First, the number of confident disease proteins, to which a prey binds in the disease network is calculated. Then, the disease proteins are mapped to the control network, and binding enrichment is subsequently calculated using one-tailed Fisher's exact test with 5% FDR correction (Fig. 2). This type of ranking is sensitive to network degree variations; therefore, the ranking should be repeated by decreasing/increasing the degree of each node tested in the disease network and for each partner in the control networks. The subset of preys consistently ranked high after applying these corrections could be potentially relevant new disease risk factors (Fig. 2).

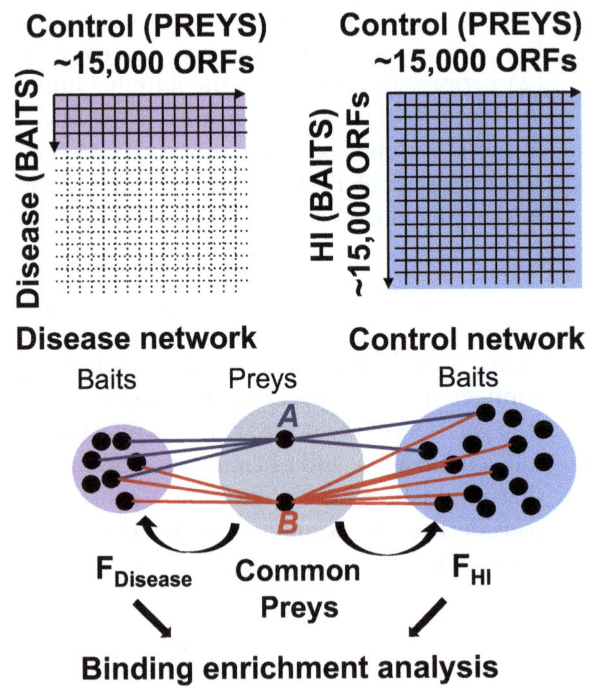

Fig. 2 Schematic representation of the binding enrichment analysis of disease network. The *top grids* represent disease-related network (*left*) and the human interactome (HI) as a control network (*right*). The Y2H screens share the same prey search space (~15,000 ORFs). The *lower figure* shows how the binding enrichment is calculated for two preys (A and B) present in both, disease and control networks. Prey A, but not prey B, interacts with a greater number of disease baits than expected from the control HI network. The counts should be normalized by the bait search space. For each prey, $F_{Disease}$ is the fraction of all disease network baits binding to the common preys, and F_{HI} is the fraction of all control HI baits binding to the common preys

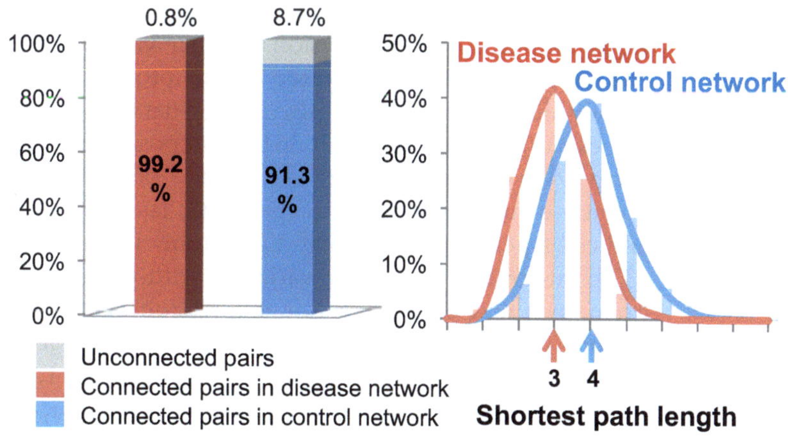

Fig. 3 Disease network connectivity analyses. Interacting partners of the disease genes are tightly connected at the protein level. (*Left panel*) A significantly higher number of interactors of the disease proteins are connected when mapped to the control background network compared to the random control. (*Right panel*) Shorter paths among the interacting partners of the disease network can be detected by mapping them to the control network and comparing the path length to the empirical null distribution of connectivity of 10,000 sets of preys randomly selected from the control network

Another approach for analyzing disease networks is to compare their properties to those of the compatible control networks. Connectivity among binding partners of the disease risk factors (i.e., prey connectivity) could be compared using the control networks, thereby identifying preys that serve as biological linkers of the disease genes. First, the binding partners of the disease proteins from the disease network are mapped to the prey space of the control network (Fig. 3). Then, the statistical significance is calculated by comparing connectivity of the binding partners in the control network with the empirical null distribution of connectivity of 10,000 or more sets of partners randomly selected from the control network. For example, we observed that in the background unbiased control network (HI), significantly more ASIN preys were connected to each other with the shorter path lengths [31] (Fig. 3). This suggests that ASD risk factors form a highly connected group and may converge on similar functions or processes.

2.4.1 Cell-Type and Tissue-Specific Networks

When performing network analyses of a disease affecting a particular tissue or organ, it is important to restrict the networks to only those genes/proteins that are expressed in the tissue of interest. Given the lack of tissue-specific PPI data, other methods could be utilized to construct tissue-specific networks. The results of such restrictive types of analyses are more biologically meaningful, as the PPIs identified experimentally by the in vitro assays in principle could occur only if both interacting genes/proteins are expressed in the same tissue and/or developmental point.

Collaborative and consortium-level efforts, including Illumina Body Map 2.0 (GEO accession number GSE30611), ENCODE [34], and GTEx [35], collected transcriptomes of various human tissues and cell lines. Recently, a large dataset of brain transcriptomes has been generated [5, 36] and deposited into the BrainSpan (http://www.brainspan.org/). Other brain-related resources include quantification of the epigenetic landscape in CNS tissues and cell types by the Roadmap Epigenomics Mapping Consortium [37], integration of genetic variation with gene expression in the brain by the Genotype-Tissue Expression (GTEx) project [38] (http://www.gtexportal.org/), as well as datasets from the BrainCloud [39] and CommonMind Consortiums (http://www.commonmind.org/). All these datasets are applicable for the construction and analyses of the brain-specific disease networks, as described below.

2.4.2 Co-expression Network Analyses

Gene expression has long been used to elucidate biological and functional relationships between human genes. Gene co-expression analysis in particular was designed to identify shared patterns of expression across different experiments, tissues, or species [40–43]. Co-expression network analysis uses gene expression as a proxy for the biological and functional state of the system under investigation.

One popular approach for analyzing gene co-expression networks is to identify Topological Overlap (TO) between functional modules or subnetworks that are relevant to the disease. The module discovery using the *W*eighted *G*ene *C*o-expression *N*etwork *A*nalysis (WGCNA) package [44] is a widely used method for this purpose. WGCNA aims to identify modules of genes that are highly correlated based on their expression patterns, with the goal to facilitate the identification of potential therapeutic targets or biomarkers. The usefulness of this method has been widely demonstrated as evidenced by multiple publications including diseases such as cancer, Alzheimer, and ASD among many others [10, 45, 46].

In order to perform WGCNA analyses, the pair-wise correlation matrix is computed for each gene pair, and an adjacency matrix is calculated by raising the correlation matrix to a power of 10 using the scale-free topology criterion. Modules are defined as branches of the clustering tree and are characterized based on the expression of the module eigengene (ME), or the first principle component of the module. To obtain moderately large and distinct modules, the minimum module size could be set at five genes and the minimum height for merging modules at 0.25. Genes are then assigned to a module if they have a high module membership (kME $>$ 0.7). In ASIN, we have applied the WGCNA method to our PPI network genes and successfully detected five modules, with two of them enriched in brain-relevant functions [31] (Fig. 4).

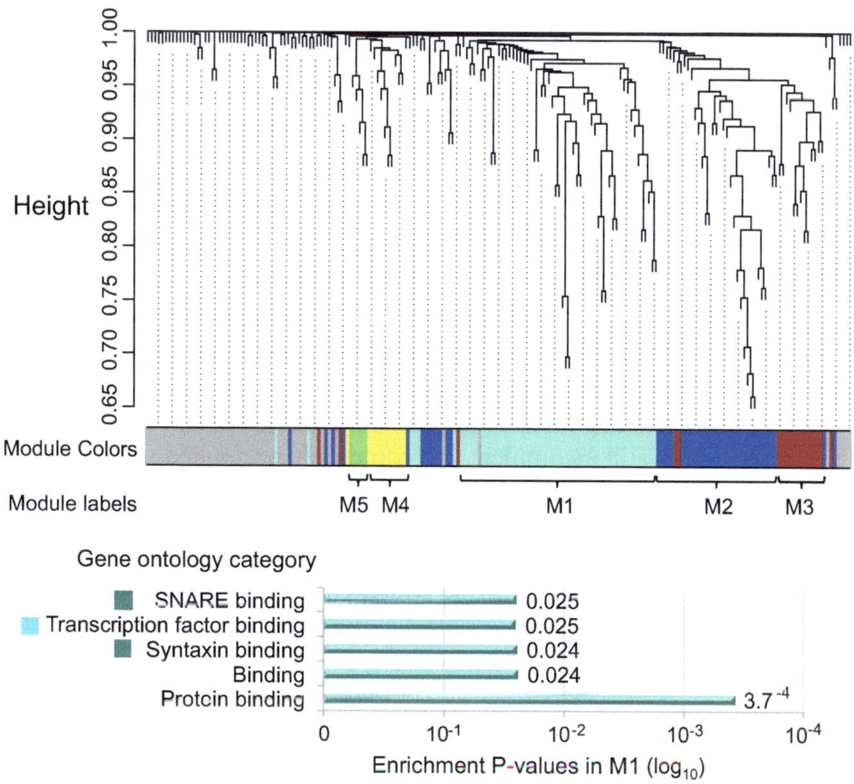

Fig. 4 Co-expression network analysis using WGCNA. Network analysis dendrogram shows the modules based on the co-expression topological overlap of genes within the network. *Color bars* below the *dendrogram* give information on module membership; the enriched GO terms for largest module (M1) are shown below the *dendrogram* with the turquoise *horizontal bar* indicating the significance of the enrichments.

2.4.3 Integration of Co-expression and Protein Interaction Networks

Despite several recent efforts to generate genome-wide complete interactomes [17, 18], the PPI datasets are still largely incomplete and lack tissue specificity, especially with regard to the brain tissue. In the absence of brain-specific PPIs, one approach for building brain-relevant networks is to integrate brain-specific RNA expression (or co-expression) data with the PPIs [9]. The aim for integrating transcriptional (RNA-seq) and translational (PPIs) information is to render the networks more biologically relevant and dynamic, by restricting PPIs to only those that have a higher probability to occur within the brain tissue. Since two types of independent lines of evidence are used for network construction, the Pearson Correlation Coefficient could be used to identify co-expressed gene pairs. Then, the edges between two genes/proteins are drawn only when these two nodes are co-expressed (PCC ≥ 0.5), and there is also a reported physical PPI between these nodes/proteins. More stringent PCC cutoffs (0.7–0.9) could also be used if desired.

Using this strategy, we have recently generated dynamic spatio-temporal networks by integrating gene expression data from the developing human brain (obtained from BrainSpan) with the PPIs for genes from different Copy Number Variants (CNVs) implicated in various psychiatric diseases [9]. Based on the constructed networks, we observed that different CNVs showed distinct spatio-temporal signatures. This suggested that the dynamic CNV networks can reveal changes during the brain development.

2.4.4 Integration of GO Annotations with Protein Interaction Networks

Another way to render networks more biologically relevant is by adding functional information extracted from public databases such as Gene Ontology (GO), KEGG Pathways, and others. Although the information about biological processes, functions or pathways is still incomplete and could only be applied to a limited number of genes, using currently available annotations together with PPIs has a potential to identify novel gene functions, modules and connections.

One interesting strategy that we used in our work is to utilize GO Biological Process annotations for identification of functional modules within networks, and then integrate PPI information to construct co-GO networks [31]. GO database can be downloaded from the GO website (http://geneontology.org/page/lead-database-downloads), and we suggest to exclude GO annotations inferred from Electronic Annotation (IEA) entries as less confident since they were not annotated by a human curator. Three GO branches, Molecular Function (MF), Biological Process (BP) and Cellular Component (CC), could be used for the analyses. GO annotations should first be filtered based on information content (IC). The IC of a GO term t is defined as:

$$\mathrm{IC}(t) = -\ln(p(t))$$

where $p(t)$ is the fraction of genes annotated with term t or its descendants. GO terms with $\mathrm{IC} < 0.95$ (i.e., those shared by more than 5% of all the annotated genes in one GO branch) should be discarded to avoid the "shallow annotation problem." After filtering, G-SESAME method [47] could be implemented to calculate the similarity score of gene pairs in each GO branch. Once the functional modules are defined based on GO, the PPI information should be combined to either detect novel connections or to support existing functional links.

We applied this strategy to the ASIN network [31] with the aim to test the functionality of the autism risk factors that we have initially selected for the analyses. We investigated how known autism risk factors are functionally related, and whether adding newly discovered PPIs would provide new functional insights. We constructed the co-GO networks as described earlier, and the majority of the candidate genes were grouped into three functional

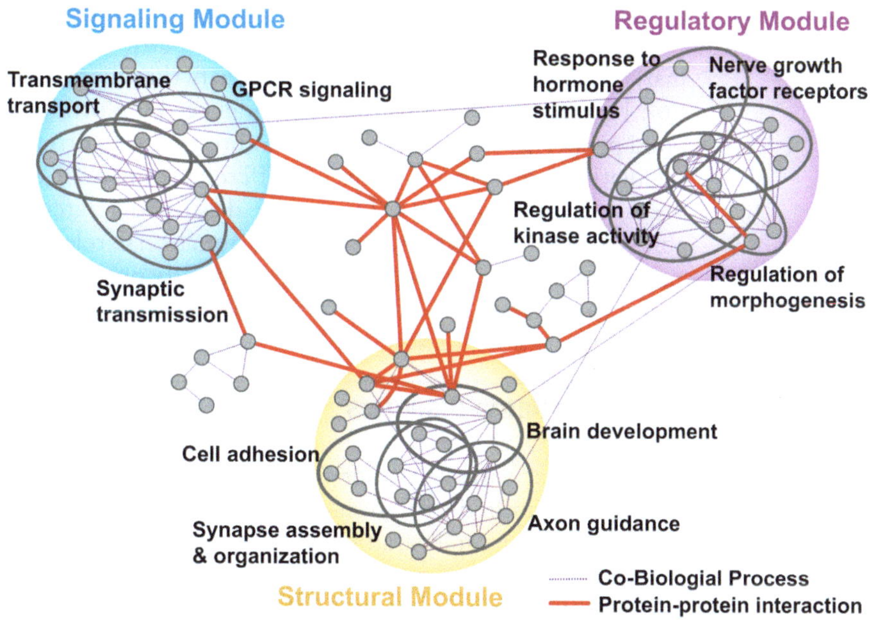

Fig. 5 Protein–protein interactions create a more comprehensive connectivity map among disease candidate genes. The disease network is clustered into three distinct subnetworks (*colored spheres*) consisting of functionally related groups of genes (*grey ovals*) based on the shared co-GO annotations (*dotted edges*). The newly identified PPIs (*red edges*) link these subnetworks into a single connected component

modules (Fig. 5). The new PPI data provided novel, previously unknown links for connecting these modules (Fig. 5).

2.4.5 Functional Network Enrichment Using Independent Sources of Evidence

In addition to general information extracted from public databases such as gene expression, PPIs, GO and others, the independent sources of information that are specific to a particular disease and are curated by expert investigators could also be incorporated into the networks. The expertise of the investigators in a specific disease is a crucial part of this process.

With the advent of whole exome (WES) and whole genome (WGS) sequencing technologies, genetic data for many psychiatric and other human disorders are becoming available. Integrating genetic data into the networks and testing for genetic enrichment of the disease networks in genes mutated in the patients is a feasible way to identify key functional modules or disease pathways and to prioritize disease genes.

For example, when analyzing autism disease networks [31], we used the following lines of evidence to perform the enrichment analyses: (1) genes present in the de novo autism CNVs; (2) genes impacted by the de novo mutations from autism patients; (3) genes preferentially expressed in the brain; (4) post-synaptic density (PSD) genes [48]; (5) genes annotated with psychiatric phenotypes in the Online Mendelian Inheritance of Men (OMIM) database;

(6) Fragile X Mental Retardation Protein (FMRP) targets [49]; (7) genes that are sensitive to mutations [50]; and (7) neuronal marker genes [51]. It is important to verify that there is no correlation between multiple sources of evidence, by plotting the annotated properties against each other and testing for correlation (e.g., calculating correlations between different sources of evidence as previously shown [31]). Using the described strategy we were able to not only better understand functional relevance of the autism network that we have built, but also predict new protein partners that may be relevant to this disease.

3 CNV-Level Networks for Psychiatric Disorders

Copy number variants comprising large deletions and duplications that span hundreds of thousands of DNA bases are the most widespread structural variation in the human genome [52]. Multiple studies have consistently demonstrated that CNVs play a major role in psychiatric disorders [53]. De novo CNV discovery and CNV burden studies have been successful in identifying precise genomic regions and even individual CNV genes conferring high risk for multiple neuropsychiatric disorders [54–60]. The CNV-level networks integrated with protein interaction networks have been previously used by us and others to investigate the mechanisms of psychiatric diseases [9, 31, 59]. The first step is to assemble the list of the disease-relevant CNVs defined by the genomic coordinates, and then to extract the list of genes spanning these CNVs' boundaries. The next step is to identify functional links among the CNV genes, which could be represented by co-expression, protein interactions, or other functional measures.

3.1 Co-expression Networks of CNV Genes

Gene co-expression can be used to investigate how different CNVs conferring high risk for psychiatric disorders are related to each other in the context of brain development. The knowledge of when and where CNV genes are co-expressed in the brain will help to find the potential convergence points of different CNVs and to identify convergent molecular pathways that are disrupted by different CNVs. We used this approach to investigate 11 CNVs, each strongly implicated in more than one neuropsychiatric disorder (ASD, SCZ, ID or bipolar disorder). Using 169 genes from 11 CNVs, we constructed CNV–CNV co-expression networks using BrainSpan gene expression data (5) (Fig. 6). Co-expressed gene pairs were defined as those with the pairwise PCC of ≥ 0.7. The weight of the edges connecting different CNV nodes was calculated as a fraction of co-expressed gene pairs between CNVs normalized by the total fraction of co-expressed gene pairs between all CNVs within the network. Specifically, the edge weight ($W_{a,b}$) is defined as:

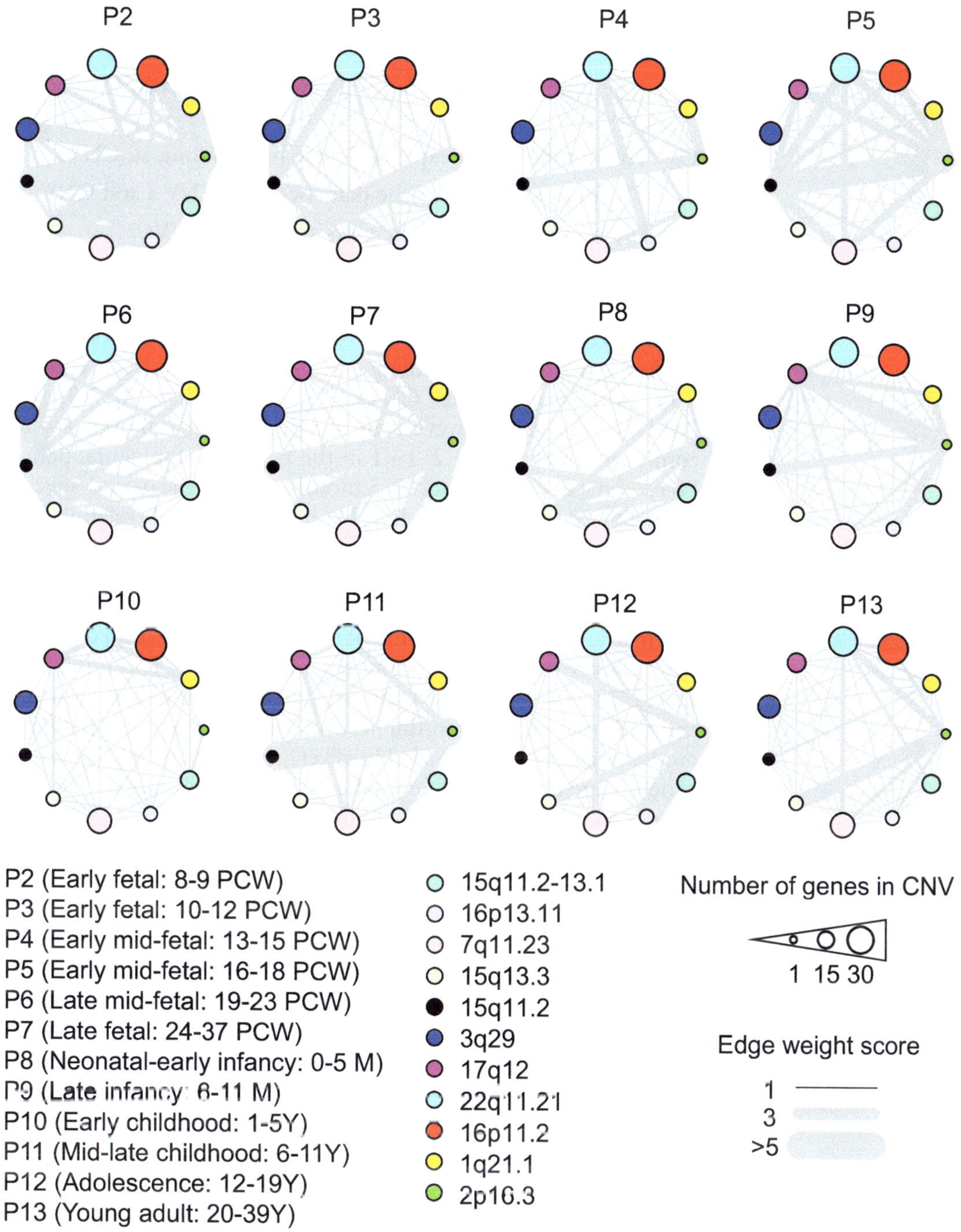

Fig. 6 The CNV–CNV co-expression networks. Each CNV is shown as a *circle* with different *color* and *size*. The size of the *circle* reflects the number of genes in each CNV. The *edge* represents the normalized fraction of co-expressed gene pairs between different CNVs. *PCW* post conception weeks, *M* months, *Y* years.

$$W_{a,b} = \left(\frac{C_{a,b}}{T_{a,b}}\right) / \left(\frac{C_{all}}{T_{all}}\right)$$

$C_{a,b}$: The number of co-expressed gene pairs between CNV a and CNV b.

C_{all}: The number of co-expressed gene pairs among all CNVs.

$T_{a,b}$: The number of total gene pairs between CNV a and CNV b.

T_{all}: The number of total gene pairs among all CNVs.

The CNV–CNV co-expression networks demonstrate period-specific connectivity among different CNVs. For example, in the brain developmental period P2 (early fetal: 8–9 post conception weeks), 2p16.3 CNV (spanning *NRXN1* gene) has strong connections with many other CNVs, e.g., 15q12.2, 15q13.3, and 7q11.23 (Fig. 6). However, the same CNV has only one strong connection with 15q11.2–13.1 in the brain developmental period P8 (Neonatal–early infancy: 0–5 months). Thus, the period-specific connectivity can provide insight into functional relationships between CNV genes implicated in multiple psychiatric disorders.

3.1.1 CNV Genes Prioritization Using Co-expression

The CNVs have been implicated as causative mutations for multiple psychiatric disorders, and some of them have demonstrated phenotypic effects in humans and model organisms. For example, the 16p11.2 CNV is associated with macrocephaly in deletion carriers and with microcephaly in duplication carriers in humans [61] and model organisms [62, 63]. However, the mechanisms by which gene dosage changes within CNVs cause these diseases and specific phenotypes are still unknown in majority of cases.

One previous study investigated transcriptomic effect of CNVs in human lymphoblastoid cell lines (LCLs) from different human populations [64]. The authors demonstrated that the changes of gene copy number within CNVs explained approximately 20% of transcriptional variation [64]. Other studies [65, 66] investigated the transcriptional changes caused by the deletion and duplication of 16p11.2 and 22q11.2 CNVs, respectively. They show that CNV genes affected by the deletions demonstrate ~30% expression level decrease, whereas those affected by the duplications demonstrate ~30% increase. Therefore, one could expect *cis*-effect of CNVs on expression to be approximately at the level of ~20–30%, usually in the same direction as CNV gene dosage changes. This approximation could be used to simulate the effect of CNVs on gene expression in order to prioritize the genes within CNVs that are most sensitive to dosage changes. Below, we simulated expression levels of 169 genes within 11 high-risk CNVs implicated in psychiatric disorders (Fig. 7a).

To estimate relative expression level changes of CNV genes, we first calculated the percentile rank of the expression levels of each CNV gene by comparing its expression with expression levels of all human genes during 12 periods of brain development (Fig. 7b).

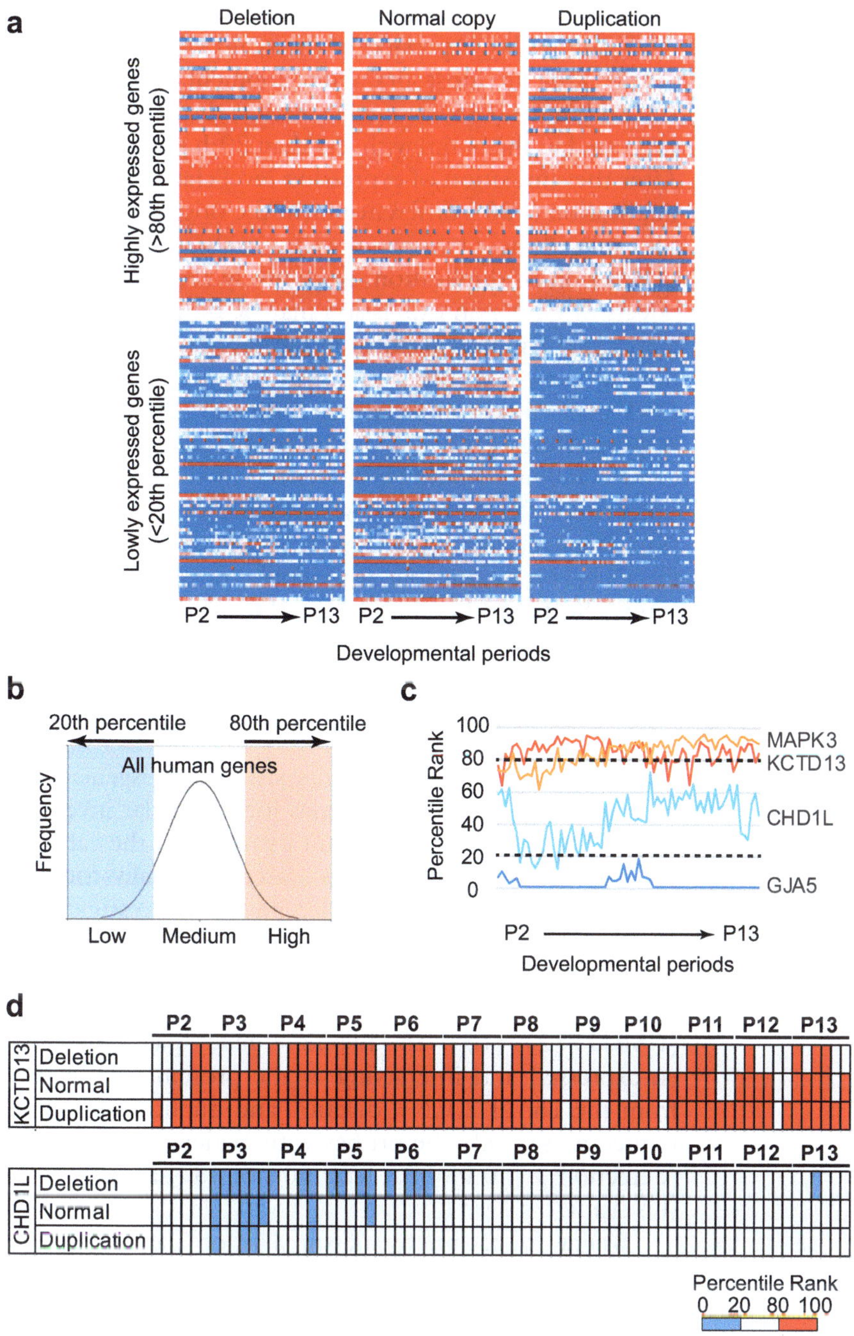

Fig. 7 Simulation of the expression level changes of CNV genes. (**a**) Simulated expression level changes of highly and lowly expressed CNV genes as a result of duplications and deletions. (**b**) Percentile ranking of the expression levels of all human genes. The CNV genes were defined as highly or lowly expressed based on the 80th and 20th percentiles cutoff relative to all human genes. (**c**) *MAPK3* and *KCTD13* are highly expressed CNV genes (>80th percentile rank), whereas *CHD1L* and *GJA5* are lowly expressed CNV genes (<20th percentile rank). (**d**) Examples of simulated expression changes of *KCTD13* and *CHD1L* across developmental periods as a result of the deletion and duplication.

For example, *MAPK3* and *KCTD13*, two genes within the 16p11.2 CNV, are highly expressed in the brain with more than 80th percentile rank in most brain developmental periods. In contrast, *CHD1L* and *GJA5*, two genes within the 1q21.1 CNV, are lowly expressed in the brain, with less than 20th percentile rank (Fig. 7c). After the simulation of expression level changes due to deletion or duplication (~30% difference from the original values), the percentile ranks of these two genes were largely reduced or increased due to deletion or duplication, respectively (Fig. 7d). These opposing effects of CNVs on expression may be different for different CNV genes, depending on their initial level of expression under the normal copy number conditions. Although the ranking of the CNV genes based on a specified 80th and 20th percentile threshold is quite arbitrary, it is reasonable to speculate that highly and lowly expressed genes would be influenced by the deletions and duplications to a different degree. When more experimental gene expression data for either CNV carriers or CRISPR cell lines with specific CNVs becomes available, it would be possible to model transcriptomic effects of different CNVs and to prioritize genes that have the largest *cis-* and even *trans-*effects. The availability of such models will improve our understanding of the functional mechanisms behind high-risk CNVs implicated in psychiatric disorders.

3.1.2 CNV Genes Prioritization Using Expression Data from Patients and Controls

One caveat that greatly impairs studies of psychiatric diseases is inaccessibility of the brain tissues for molecular investigation. The availability of postmortem brain tissues from the carriers with specific genetic mutations is also scarce, especially for diseases with early onset such as autism. Therefore, the investigators frequently rely on the peripheral tissues (i.e., blood) for gene expression studies. We have used gene expression dataset derived from lymphoblast cell lines (LCLs) of autism patients and controls (GSE37772) [65] to identify gene pairs that are highly co-expressed and interacting at the protein level with the genes from the 16p11.2 CNV deletion and duplication carriers [9]. To reduce noise, only the probes with evidence of robust expression (detection $p \leq 0.05$ in at least 50% of 439 samples) were used for this study.

Briefly, for each network pair, a list of "partner-alike genes" was assembled by selecting genes that are highly co-expressed ($SCC \geq 0.7$) with each partner of the 16p11.2 genes in the healthy control siblings (Fig. 8). Then, the expression profiles of the "partner-alike genes" from the control siblings were compared with those for the same "partner-alike genes" from the 16p11.2 deletion and duplication carriers. The network pair was considered to have a significantly reduced expression in the deletion carriers, or a significantly increased expression in the duplication carriers, if its co-expression correlation coefficient was lower than 5th percentile

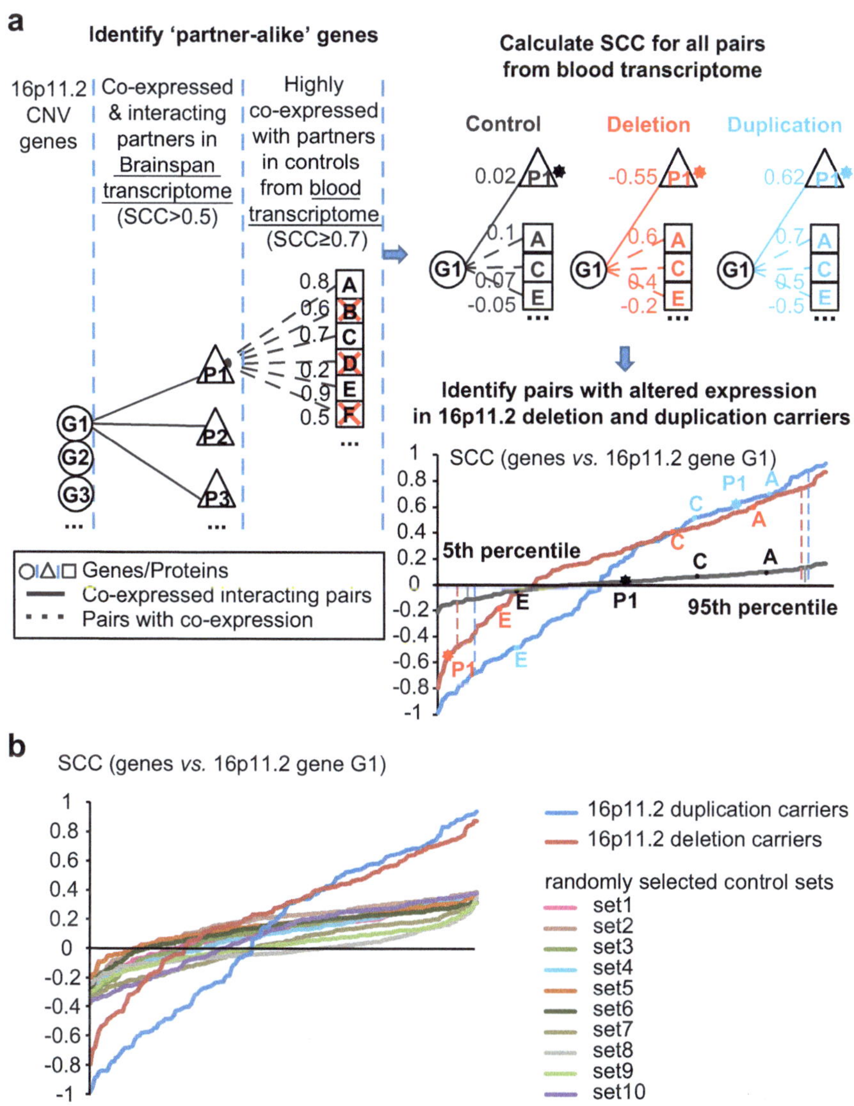

Fig. 8 Overview of the method for identifying co-expressed and interacting pairs with altered expression in the 16p11.2 deletion and duplication carriers. (**a**) *Left panel*: The procedure for identification of the "partner-alike genes" to build the expected distributions of correlation coefficients for control, 16p11.2 deletion and duplication carriers. *Upper right panel*: Pairwise Spearman correlation coefficients (SCCs) were calculated between CNV genes (G1, G2, G3) and their partners (P1, P2, P3), as well as CNV genes with partner-alike genes (A, C, E). *Lower right panel*: The distributions of SCCs between CNV genes and "partner-alike genes" in the control (*gray line*), deletion (*red line*), and duplication (*blue line*) carriers as well as SCCs between G1 and P1 in the same datasets. (**b**) The distribution of SCCs for KCTD13-CUL3 pair in ten randomly selected sets (six samples each) of control subjects (*multicolored lines*), six deletion carriers (*red line*), and six duplication carriers (*blue line*). Reprinted from [9]

or higher than 95th percentile of the background distribution of the correlation coefficients for the "partner-alike genes" (Fig. 8). We observed that some highly co-expressed interacting pairs, including *KCTD13-Cul3*, have a significantly reduced expression in the lymphoblasts of deletion carriers, whereas other pairs have a significantly increased expression in the duplication carriers. This allowed identification of gene pairs that are dysregulated by the 16p11.2 CNV deletions and duplications in the ASD patients [9].

3.2 CNV-Level Protein Interaction Networks

The CNV-level networks could provide additional insights into functional relationships between different mutations involved in human diseases. In addition to the gene-level networks, individual genes from the same CNV could be combined into the same node and the CNV networks could be constructed (Fig. 1). Various types of functional data could be used to draw the edges between different CNVs. Below, we describe the construction of the CNV networks using protein interaction data derived from the Y2H studies.

To generate CNV-level protein interaction network, each protein-encoding gene of interest is first mapped to either a specific CNV (for example, the CNV implicated in a disease), or to a non-CNV region. The genes that overlap CNVs are denoted as "baits," and their non-CNV interacting partners as "preys." Two types of CNV networks can be built: (1) CNV-prey networks, with CNV nodes as baits and interacting partners as preys; and (2) CNV–CNV networks, with CNV nodes as both, baits and preys.

To build a CNV–prey network, all baits from the same CNVs are merged to create a new network CNV node, and all edges become links between the CNV node and interacting prey proteins. This type of network allows identification of new disease risk factors that link more than expected number of disease CNVs. In order to generate compatible background network for enrichment analysis, random genomic regions need to be generated to mimic real CNVs. Two alternative randomization procedures of human genomic regions can be used: (1) either preserving the genomic size of CNVs and the total number of interactions for each CNV; or (2) by randomly selecting genomic regions (i.e., random CNVs) with the same number of genes (controlling for gene length and GC content) and similar interaction degree distribution as observed in the real CNV network. Likewise, the CNV–CNV network is constructed by merging all network genes into the new CNV nodes and preserving only edges between different CNV nodes. Next, the null distribution of the test statistic can be estimated by generating a large set of simulated CNV networks (~10,000).

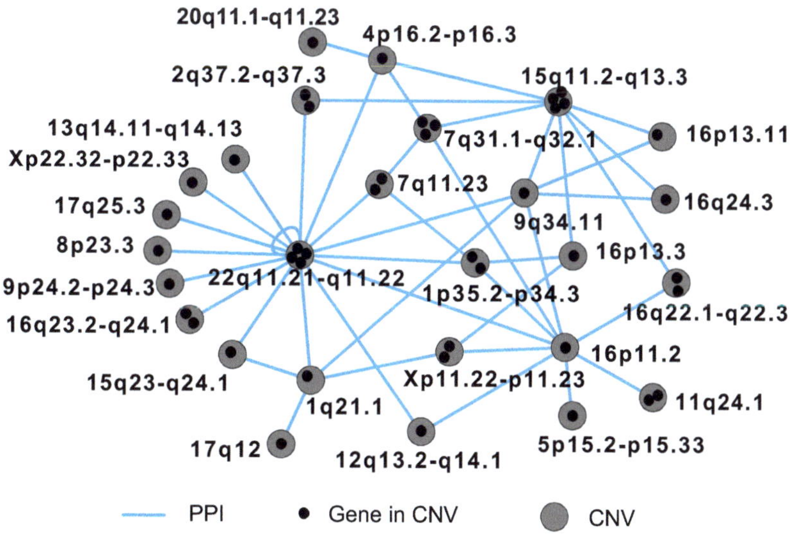

Fig. 9 The CNV–CNV network links CNV nodes into a single connected component. Genes (*small black circles*) within the same CNV node are grouped into nodes (*larger grey circles*), edges (*blue lines*) correspond to PPIs. The disease CNV–CNV network connects all 27 CNVs into a single connected component

3.3 Biological Applications for the CNV-Level Networks

In our hands, the CNV-prey network has been a useful tool for detecting new autism risk factors [31]. By generating the CNV-prey network, we were able to identify 25 network preys that bind to at least two different de novo CNVs, with some of them linking five or even six different CNVs. The null distribution of 10,000 random CNV-prey networks with simulated CNVs could find no random networks with any prey binding to more than 4 CNVs. The preys that connected significantly more CNVs than expected by chance represent potentially new autism risk factors [31]. In the same study we also generated CNV–CNV network that directly connected 27 de novo autism CNVs into a single connected component. The largest connected component of the control network generated from randomly simulated CNVs with the same number of genes and PPIs as the real CNV network comprised only eight CNVs, a significantly smaller number than in the real network (Fig. 9). The autism CNV–CNV network connected several individually rare autism CNVs with each other at the protein level, thereby pointing toward common molecular networks shared among different ASD patients.

4 Isoform-Level Networks for Psychiatric Disorders

4.1 Introduction to Isoform–Isoform Networks

The majority of the multi-exon human genes undergo alternative splicing or use alternative promoters to increase proteomic diversity [67, 68]. The brain, in particular, is one of the most complex tissues

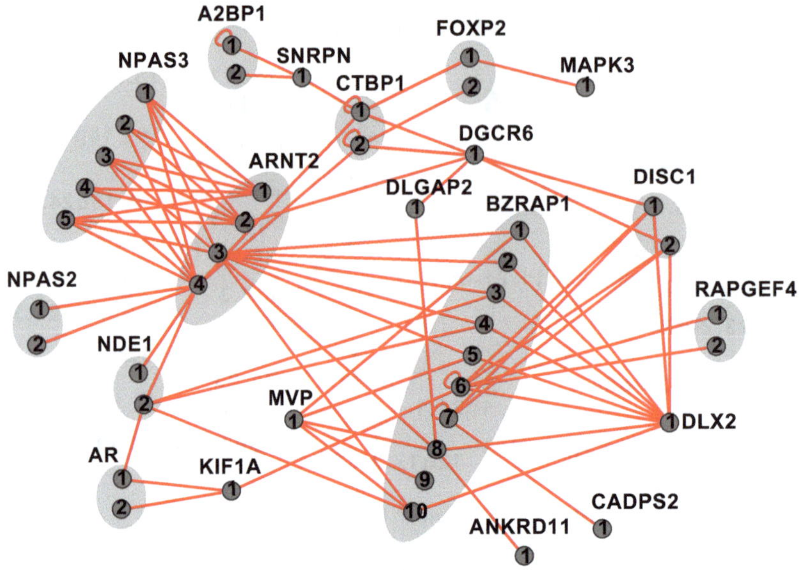

Fig. 10 Isoform-level protein–protein interactions network. Bidirectional isoform-level network can only be constructed when isoform-level information for both, baits and preys, is available. Each isoform is shown as a *small dark circle* with the number inside, and isoforms from the same gene are grouped into *grey ellipses*. The *red edges* represent isoform-level protein–protein interactions. Different isoforms may have different interaction patterns.

with finely regulated mechanisms of alternative splicing [35, 69, 70]. As a result, it is important to consider the diversity of interaction patterns of protein variants encoded by different isoforms of the same gene. Until now, only a limited number of studies, including one by us, have considered this variability in interaction networks, and these studies were also based on various subsets of human genes [31, 71, 72]. Recently, we have performed a global systematic screen of over a thousand of protein isoforms for interactions [73]. We demonstrated that the majority of isoform pairs, encoded by the same gene, share less than 50% of their interactions. In the global context of interactome network maps, alternative isoforms tend to behave like distinct proteins rather than minor variants of each other. We also showed that interaction partners specific to alternative isoforms tend to be expressed in a highly tissue-specific manner and belong to distinct functional modules [73].

To build an isoform-level PPI network, it is necessary to obtain isoform-level experimental data. Public databases are still scarce in any type of information at the isoform level, with the most comprehensive database recently designed by us (see http://isoform.dfci.harvard.edu/). Alternatively, the data could be generated experimentally for a selected subset of disease gene candidates by cloning their splicing isoforms from a relevant tissue and then testing them for PPIs either with the Y2H system [31], or by using cloned isoforms for pull-down experiments combined with mass-spectrometry [74]. It should be noted, however, that the latter approach does not provide binary interaction information that is critical for building accurate isoform-level networks.

While building isoform-level networks using Y2H, it is crucial to plan the experiments in a way that minimizes false positive and false negative interactions. In our recent study [73], each cloned isoform of a gene was tested for interactions against the entire human ORFeome (first-pass, ~15,000 ORFs), and subsequently each isoform was retested in triplicate against a union of all interaction partners for all isoforms of the same gene from the first-pass. This ensured that the observed lack of interaction for a specific isoform is a true negative event.

Using the described methods, it is possible to construct two types of isoform-level networks. First, if isoforms of only disease candidates (i.e., baits) are available for testing, then the isoform-level network would be unidirectional. However, if the isoform clones are also available for the interaction partners (i.e., preys), then bidirectional isoform networks could be constructed (Fig. 10).

We observed that having isoform-level information for both interacting partners increases the number of interactions and the level of depth and complexity of the networks. For example, when we included the isoform-level PPI information in our autism network, it was expanded by 30% compared to the network constructed from only a single reference isoform of each gene [31].

4.2 Evaluating Interaction Profile Similarity of Different Isoforms Encoded by the Same Gene

More than 90% of multi-exon human genes encode multiple isoforms. These isoforms may share and/or have unique interacting partners at the protein level. The differences in interaction profiles of isoforms can be calculated using the Jaccard distance. This value measures the dissimilarity between sample sets, i.e., the differences in interaction profiles of multiple isoforms of a protein (Fig. 11). The Jaccard distance $D_J(\mathrm{A, B})$ is calculated as follows:

$$D_J(\mathrm{A, B}) = 1 - \mathrm{J}(\mathrm{A, B}) = 1 - \frac{|\mathrm{A} \cup \mathrm{B}| - |\mathrm{A} \cap \mathrm{B}|}{|\mathrm{A} \cup \mathrm{B}|}$$

where J(A, B) is the Jaccard Score. When all isoforms interact with exactly the same partners, the Jaccard distance is 0; alternatively, when isoforms share no interacting partners the Jaccard distance is 1. In our recent study we observed that ~16% of the isoform pairs shared no interacting partners, whereas ~63% shared less than half of their interacting partners [73]. This suggests that different protein isoforms encoded by the same gene have strikingly different functional properties, and that they can behave like completely different proteins in interaction networks. We also observed that different isoforms can participate in different cellular pathways, thereby further emphasizing the importance of isoform networks for improving our understanding of human disease pathways.

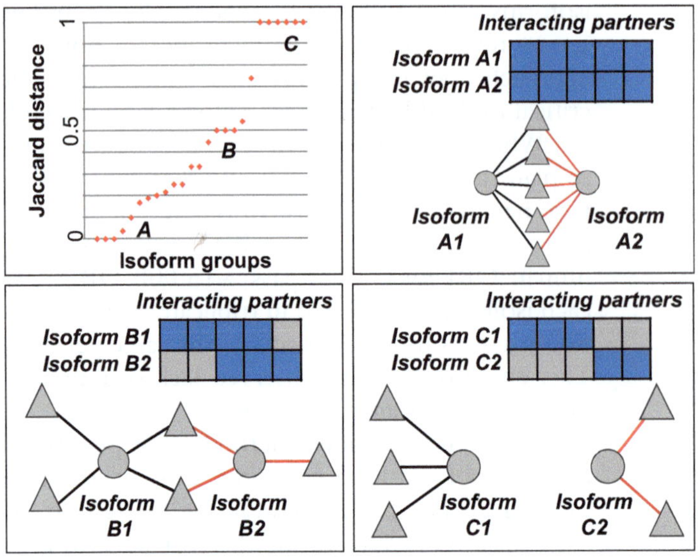

Fig. 11 The Jaccard distance measures interaction profile dissimilarity of protein isoforms encoded by the same gene. The network genes with two or more isoforms with PPIs can be arranged in order of increasing fraction of differences in interacting partners (Jaccard distance). The PPI matrices are shown for three proteins, A, B, and C. The *blue squares* within a matrix represent positive interactions, the *grey squares* represent negative interactions. In network representations, nodes corresponding to multiple isoforms of the same gene are shown as *grey circles*, and their interacting partners are shown as *grey triangles*. The *dark grey edges* are interactions of one isoform, and the *red edges* are interactions of the other isoform of the same gene

4.3 Isoform-Level Networks of Co-expressed and Interacting Proteins

Many of the isoform variants are differentially regulated, and their expression is often restricted to certain organs, tissues or cells [68]. The differences in isoform expression have also been observed across tissues or organs development [75]. A given isoform may even exhibit dominant negative effects over other isoforms encoded by the same gene, be up- or down-regulated instead of being constitutively active, or even have opposing cellular functions [76]. Traditionally, and similar to protein interaction networks, the isoform-level expression data are less abundant in the public databases than the gene-level data. The majority of previous publications that use isoform-level expression data are typically only focused on a selected subset of genes of interested [77, 78].

Although both, isoform-level PPI and isoform-level expression data, started to accumulate during last couple of years, the direct integration of these two types of datasets to create the isoform-level networks of co-expressed and physically interacting proteins still pose a great challenge. This is mainly due to the fact that these datasets are usually generated separately in different laboratories, and often different sets of isoforms are selected for the experiments.

4.3.1 Integration of Isoform-Level Co-expression and Protein Interaction Data

As a feasible alternative to collecting both types of data, one can computationally estimate the expression levels of isoforms by quantifying it from the available tissue-specific RNA-Seq data, extracted from the Human Body Map (GEO accession number GSE30611), BrainSpan [5] (http://www.brainspan.org/) or GTEx [35] (http://www.gtexportal.org/). The dataset of known isoforms for each gene could be assembled from public databases, such as RefSeq, GenCode, UCSC known genes, and others.

When experimental isoform-level PPI networks are available for the genes of interest [31, 73], one can first computationally quantify the expression levels of the network isoforms using tissue-specific RNA-seq data and the tools, such as TopHat [79], RSEM [80], or eXpress [81]. Next, co-expression between all interacting isoforms can be calculated by a correlation measurement (i.e., Pearson's correlation coefficient, Spearman's rank correlation coefficient, Mutual information, or Euclidean distance). Alternatively, in the absence of the isoform-level PPI network for the genes of interest, one can start with a literature-curated gene-level PPI network. Then, all gene pairs in the network could be expanded to include all possible isoform pairs. Since isoform expression levels could be quantified using RNA-seq data, the isoform-level co-expression network can be generated by calculating co-expression coefficient for all isoform pairs. Typically, we connect a pair of isoforms with an edge in the network when co-expression coefficient between them is ≥ 0.5 (or ≥ 0.7 if a more stringent network is desired), and if they are also shown to physically interact (i.e., the PPI is reported). Figure 12 shows the example of two different isoforms of a *CTBP1* gene that are co-expressed and interacting with different isoforms of four other genes during brain development. Note that the isoform-level interaction networks change depending on the brain developmental period (P2–P13). Other examples of spatio-temporal networks are shown in in our recent publication [9]. It is likely that the detailed isoform-level co-expressed PPI networks could provide more detailed information about healthy or disease states compared to the gene-level networks.

5 Summary

This chapter describes approaches to data integration that have been developed over the years in our laboratory for building and analyzing disease-relevant networks. Heterogeneous data sources that include genetic, gene expression, GO, pathway annotations, disease variants (DNMs, CNV) could be used to build disease networks. Here, we used three levels of data abstraction (gene, CNV, and isoform) (Fig. 1) to demonstrate how to gain insights into molecular mechanisms of psychiatric diseases using tissue-specific dynamic

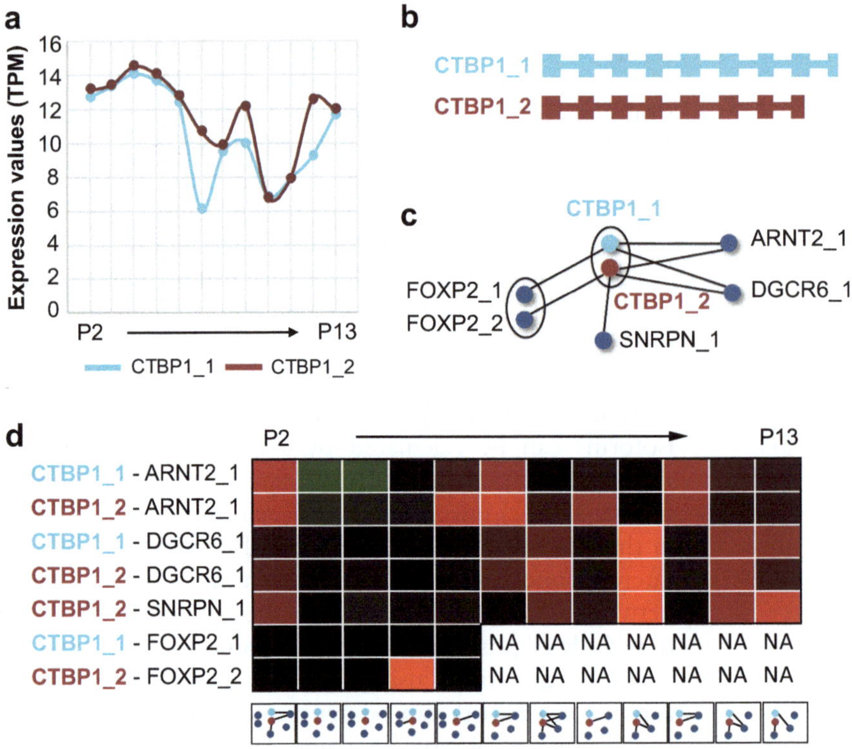

Fig. 12 The example of different isoforms of the *CTBP1* gene interacting with different isoform partners during human brain development. (**a**) Expression values of two different isoforms of *CTBP1* gene across different brain developmental periods (P2–P13, fetal to adult). (**b**) Exon-intron structure of *CTBP1* isoforms. (**c**) Static PPIs between *CTBP1* isoforms and the isoforms of their interacting partners. (**d**) The heatmap shows co-expression between interacting isoform pairs across brain developmental periods (P2–P13). *Green color* represents negative correlation and *red color* represents positive correlation. The dynamic networks below the heatmap are constructed by integrating co-expression with PPIs; note network changes in different brain developmental periods.

networks. In addition, we also described how to use appropriate controls to eliminate potential biases, as well as summarized the advantages and limitations of the developed approaches.

Overall, integrating co-expression data with protein–protein interactions networks proved particularly useful in our hands for inferring disease pathways and functional modules, and leading to improved understanding of autism biology [9, 31]. Other studies also used networks to discover disease mechanisms by integrating multiple data types (e.g., RNA expression, genetic variation, epigenetics, or PPIs) and data sets (e.g., human postmortem brain, mouse data, in vitro assays) [82]. Data integration is becoming the method of choice for many human disease studies, because data from diverse datasets can reveal unexpected and distinct relationships between genes, RNAs, and proteins that are invisible when only a single dataset is used for the analyses. Developing reliable approaches to data integration is especially important in light of the precision medicine initiative. Large patient-derived datasets such as WGS, transcriptomic, metabolomic, and proteomic, will soon be produced with the aim of predicting disease susceptibility, or the course of disease development. Integrating these datasets is currently presenting enormous challenges that likely will be solved during the next decade of research. We hope that approaches described in this chapter will be contributing to the bright future of precision medicine.

References

1. Gratten J, Wray NR, Keller MC, Visscher PM (2014) Large-scale genomics unveils the genetic architecture of psychiatric disorders. Nat Neurosci 17(6):782–790. doi:10.1038/nn.3708
2. Parikshak NN, Gandal MJ, Geschwind DH (2015) Systems biology and gene networks in neurodevelopmental and neurodegenerative disorders. Nat Rev Genet 16(8):441–458. doi:10.1038/nrg3934. http://www.nature.com/nrg/journal/v16/n8/abs/nrg3934.html#supplementary-information
3. Goh KI, Cusick ME, Valle D, Childs B, Vidal M, Barabasi AL (2007) The human disease network. Proc Natl Acad Sci U S A 104(21):8685–8690
4. Vidal M, Cusick ME, Barabasi AL (2011) Interactome networks and human disease. Cell 144(6):986–998. doi:10.1016/j.cell.2011.02.016. S0092-8674(11)00130-9 (pii)
5. Kang HJ, Kawasawa YI, Cheng F, Zhu Y, Xu X, Li M, Sousa AM, Pletikos M, Meyer KA, Sedmak G, Guennel T, Shin Y, Johnson MB, Krsnik Z, Mayer S, Fertuzinhos S, Umlauf S, Lisgo SN, Vortmeyer A, Weinberger DR, Mane S, Hyde TM, Huttner A, Reimers M, Kleinman JE, Sestan N (2011) Spatio-temporal transcriptome of the human brain. Nature 478(7370):483–489. doi:10.1038/nature10523. nature10523 (pii)
6. Willsey AJ, Sanders SJ, Li M, Dong S, Tebbenkamp AT, Muhle RA, Reilly SK, Lin L, Fertuzinhos S, Miller JA, Murtha MT, Bichsel C, Niu W, Cotney J, Ercan-Sencicek AG, Gockley J, Gupta AR, Han W, He X, Hoffman EJ, Klei L, Lei J, Liu W, Liu L, Lu C, Xu X, Zhu Y, Mane SM, Lein ES, Wei L, Noonan JP, Roeder K, Devlin B, Sestan N, State MW (2013) Coexpression networks implicate human midfetal deep cortical projection neurons in the pathogenesis of autism. Cell 155(5):997–1007. doi:10.1016/j.cell.2013.10.020. S0092-8674(13)01296-8 (pii)
7. Parikshak NN, Luo R, Zhang A, Won H, Lowe JK, Chandran V, Horvath S, Geschwind DH (2013) Integrative functional genomic analyses implicate specific molecular pathways and circuits in autism. Cell 155(5):1008–1021. doi:10.1016/j.cell.2013.10.031. S0092-8674(13)01349-4 (pii)
8. Gulsuner S, Walsh T, Watts AC, Lee MK, Thornton AM, Casadei S, Rippey C, Shahin H, Nimgaonkar VL, Go RC, Savage RM, Swerdlow NR, Gur RE, Braff DL, King MC, McClellan JM (2013) Spatial and temporal mapping of de novo mutations in schizophrenia to a fetal prefrontal cortical network. Cell 154(3):518–529. doi:10.1016/j.cell.2013.06.049. S0092-8674(13)00831-3 (pii)
9. Lin GN, Corominas R, Lemmens I, Yang X, Tavernier J, Hill DE, Vidal M, Sebat J, Iakoucheva LM (2015) Spatiotemporal 16p11.2 protein network implicates cortical late midfetal brain development and KCTD13-Cul3-RhoA pathway in psychiatric diseases. Neuron 85(4):742–754. doi:10.1016/j.neuron.2015.01.010. S0896-6273(15)00036-7 (pii)
10. Voineagu I, Wang X, Johnston P, Lowe JK, Tian Y, Horvath S, Mill J, Cantor RM, Blencowe BJ, Geschwind DH (2011) Transcriptomic analysis of autistic brain reveals convergent molecular pathology. Nature 474(7351):380–384. doi:10.1038/nature10110. nature10110 (pii)
11. Iossifov I, O'Roak BJ, Sanders SJ, Ronemus M, Krumm N, Levy D, Stessman HA, Witherspoon KT, Vives L, Patterson KE, Smith JD, Paeper B, Nickerson DA, Dea J, Dong S, Gonzalez LE, Mandell JD, Mane SM, Murtha MT, Sullivan CA, Walker MF, Waqar Z, Wei L, Willsey AJ, Yamrom B, Lee YH, Grabowska E, Dalkic E, Wang Z, Marks S, Andrews P, Leotta A, Kendall J, Hakker I, Rosenbaum J, Ma B, Rodgers L, Troge J, Narzisi G, Yoon S, Schatz MC, Ye K, McCombie WR, Shendure J, Eichler EE, State MW, Wigler M (2014) The

contribution of de novo coding mutations to autism spectrum disorder. Nature 515 (7526):216–221. doi:10.1038/nature13908. nature13908 (pii)

12. Buxbaum JD, Daly MJ, Devlin B, Lehner T, Roeder K, State MW, Autism Sequencing Consortium (2012) The autism sequencing consortium: large-scale, high-throughput sequencing in autism spectrum disorders. Neuron 76 (6):1052–1056. doi:10.1016/j.neuron.2012.12.008
13. De Rubeis S, He X, Goldberg AP, Poultney CS, Samocha K, Cicek AE, Kou Y, Liu L, Fromer M, Walker S, Singh T, Klei L, Kosmicki J, Shih-Chen F, Aleksic B, Biscaldi M, Bolton PF, Brownfeld JM, Cai J, Campbell NG, Carracedo A, Chahrour MH, Chiocchetti AG, Coon H, Crawford EL, Curran SR, Dawson G, Duketis E, Fernandez BA, Gallagher L, Geller E, Guter SJ, Hill RS, Ionita-Laza J, Jimenz Gonzalez P, Kilpinen H, Klauck SM, Kolevzon A, Lee I, Lei I, Lei J, Lehtimaki T, Lin CF, Ma'ayan A, Marshall CR, McInnes AL, Neale B, Owen MJ, Ozaki N, Parellada M, Parr JR, Purcell S, Puura K, Rajagopalan D, Rehnstrom K, Reichenberg A, Sabo A, Sachse M, Sanders SJ, Schafer C, Schulte-Ruther M, Skuse D, Stevens C, Szatmari P, Tammimies K, Valladares O, Voran A, Li-San W, Weiss LA, Willsey AJ, Yu TW, Yuen RK, Cook EH, Freitag CM, Gill M, Hultman CM, Lehner T, Palotie A, Schellenberg GD, Sklar P, State MW, Sutcliffe JS, Walsh CA, Scherer SW, Zwick ME, Barett JC, Cutler DJ, Roeder K, Devlin B, Daly MJ, Buxbaum JD (2014) Synaptic, transcriptional and chromatin genes disrupted in autism. Nature 515 (7526):209–215. doi:10.1038/nature13772. nature13772 (pii)
14. Michaelson JJ, Shi Y, Gujral M, Zheng H, Malhotra D, Jin X, Jian M, Liu G, Greer D, Bhandari A, Wu W, Corominas R, Peoples A, Koren A, Gore A, Kang S, Lin GN, Estabillo J, Gadomski T, Singh B, Zhang K, Akshoomoff N, Corsello C, McCarroll S, Iakoucheva LM, Li Y, Wang J, Sebat J (2012) Whole-genome sequencing in autism identifies hot spots for de novo germline mutation. Cell 151 (7):1431–1442. doi:10.1016/j.cell.2012.11.019. S0092-8674(12)01404-3 (pii)
15. Yuen RK, Thiruvahindrapuram B, Merico D, Walker S, Tammimies K, Hoang N, Chrysler C, Nalpathamkalam T, Pellecchia G, Liu Y, Gazzellone MJ, D'Abate L, Deneault E, Howe JL, Liu RS, Thompson A, Zarrei M, Uddin M, Marshall CR, Ring RH, Zwaigenbaum L, Ray PN, Weksberg R, Carter MT, Fernandez BA, Roberts W, Szatmari P, Scherer SW (2015) Whole-genome sequencing of quartet families with autism spectrum disorder. Nat Med 21 (2):185–191. doi:10.1038/nm.3792. nm.3792 (pii)
16. Lee SH, Ripke S, Neale BM, Faraone SV, Purcell SM, Perlis RH, Mowry BJ, Thapar A, Goddard ME, Witte JS, Absher D, Agartz I, Akil H, Amin F, Andreassen OA, Anjorin A, Anney R, Anttila V, Arking DE, Asherson P, Azevedo MH, Backlund L, Badner JA, Bailey AJ, Banaschewski T, Barchas JD, Barnes MR, Barrett TB, Bass N, Battaglia A, Bauer M, Bayes M, Bellivier F, Bergen SE, Berrettini W, Betancur C, Bettecken T, Biederman J, Binder EB, Black DW, Blackwood DH, Bloss CS, Boehnke M, Boomsma DI, Breen G, Breuer R, Bruggeman R, Cormican P, Buccola NG, Buitelaar JK, Bunney WE, Buxbaum JD, Byerley WF, Byrne EM, Caesar S, Cahn W, Cantor RM, Casas M, Chakravarti A, Chambert K, Choudhury K, Cichon S, Cloninger CR, Collier DA, Cook EH, Coon H, Cormand B, Corvin A, Coryell WH, Craig DW, Craig IW, Crosbie J, Cuccaro ML, Curtis D, Czamara D, Datta S, Dawson G, Day R, De Geus EJ, Degenhardt F, Djurovic S, Donohoe GJ, Doyle AE, Duan J, Dudbridge F, Duketis E, Ebstein RP, Edenberg HJ, Elia J, Ennis S, Etain B, Fanous A, Farmer AE, Ferrier IN, Flickinger M, Fombonne E, Foroud T, Frank J, Franke B, Fraser C, Freedman R, Freimer NB, Freitag CM, Friedl M, Frisen L, Gallagher L, Gejman PV, Georgieva L, Gershon ES, Geschwind DH, Giegling I, Gill M, Gordon SD, Gordon-Smith K, Green EK, Greenwood TA, Grice DE, Gross M, Grozeva D, Guan W, Gurling H, De Haan L, Haines JL, Hakonarson H, Hallmayer J, Hamilton SP, Hamshere ML, Hansen TF, Hartmann AM, Hautzinger M, Heath AC, Henders AK, Herms S, Hickie IB, Hipolito M, Hoefels S, Holmans PA, Holsboer F, Hoogendijk WJ, Hottenga JJ, Hultman CM, Hus V, Ingason A, Ising M, Jamain S, Jones EG, Jones I, Jones L, Tzeng JY, Kahler AK, Kahn RS, Kandaswamy R, Keller MC, Kennedy JL, Kenny E, Kent L, Kim Y, Kirov GK, Klauck SM, Klei L, Knowles JA, Kohli MA, Koller DL, Konte B, Korszun A, Krabbendam L, Krasucki R, Kuntsi J, Kwan P, Landen M, Langstrom N, Lathrop M, Lawrence J, Lawson WB, Leboyer M, Ledbetter DH, Lee PH, Lencz T, Lesch KP, Levinson DF, Lewis CM, Li J, Lichtenstein P, Lieberman JA, Lin DY, Linszen DH, Liu C, Lohoff FW, Loo SK, Lord C, Lowe JK, Lucae S, Macintyre DJ, Madden PA, Maestrini E, Magnusson PK, Mahon PB, Maier W, Malhotra AK, Mane SM, Martin CL, Martin NG, Mattheisen M, Matthews K, Mattingsdal M, McCarroll SA, McGhee KA, McGough JJ, McGrath PJ, McGuffin P, McInnis MG, McIntosh A, McKinney R, McLean AW, McMahon

FJ, McMahon WM, McQuillin A, Medeiros H, Medland SE, Meier S, Melle I, Meng F, Meyer J, Middeldorp CM, Middleton L, Milanova V, Miranda A, Monaco AP, Montgomery GW, Moran JL, Moreno-De-Luca D, Morken G, Morris DW, Morrow EM, Moskvina V, Muglia P, Muhleisen TW, Muir WJ, Muller-Myhsok B, Murtha M, Myers RM, Myin-Germeys I, Neale MC, Nelson SF, Nievergelt CM, Nikolov I, Nimgaonkar V, Nolen WA, Nothen MM, Nurnberger JI, Nwulia EA, Nyholt DR, O'Dushlaine C, Oades RD, Olincy A, Oliveira G, Olsen L, Ophoff RA, Osby U, Owen MJ, Palotie A, Parr JR, Paterson AD, Pato CN, Pato MT, Penninx BW, Pergadia ML, Pericak-Vance MA, Pickard BS, Pimm J, Piven J, Posthuma D, Potash JB, Poustka F, Propping P, Puri V, Quested DJ, Quinn EM, Ramos-Quiroga JA, Rasmussen HB, Raychaudhuri S, Rehnstrom K, Reif A, Ribases M, Rice JP, Rietschel M, Roeder K, Roeyers H, Rossin L, Rothenberger A, Rouleau G, Ruderfer D, Rujescu D, Sanders AR, Sanders SJ, Santangelo SL, Sergeant JA, Schachar R, Schalling M, Schatzberg AF, Scheftner WA, Schellenberg GD, Scherer SW, Schork NJ, Schulze TG, Schumacher J, Schwarz M, Scolnick E, Scott LJ, Shi J, Shilling PD, Shyn SI, Silverman JM, Slager SL, Smalley SL, Smit JH, Smith EN, Sonuga-Barke EJ, St Clair D, State M, Steffens M, Steinhausen HC, Strauss JS, Strohmaier J, Stroup TS, Sutcliffe JS, Szatmari P, Szelinger S, Thirumalai S, Thompson RC, Todorov AA, Tozzi F, Treutlein J, Uhr M, van den Oord EJ, Van Grootheest G, Van Os J, Vicente AM, Vieland VJ, Vincent JB, Visscher PM, Walsh CA, Wassink TH, Watson SJ, Weissman MM, Werge T, Wienker TF, Wijsman EM, Willemsen G, Williams N, Willsey AJ, Witt SH, Xu W, Young AH, Yu TW, Zammit S, Zandi PP, Zhang P, Zitman FG, Zollner S, Devlin B, Kelsoe JR, Sklar P, Daly MJ, O'Donovan MC, Craddock N, Sullivan PF, Smoller JW, Kendler KS, Wray NR (2013) Genetic relationship between five psychiatric disorders estimated from genome-wide SNPs. Nat Genet 45 (9):984–994. doi:10.1038/ng.2711. ng.2711 (pii)

17. Rolland T, Tasan M, Charloteaux B, Pevzner SJ, Zhong Q, Sahni N, Yi S, Lemmens I, Fontanillo C, Mosca R, Kamburov A, Ghiassian SD, Yang X, Ghamsari L, Balcha D, Begg BE, Braun P, Brehme M, Broly MP, Carvunis AR, Convery-Zupan D, Corominas R, Coulombe-Huntington J, Dann E, Dreze M, Dricot A, Fan C, Franzosa E, Gebreab F, Gutierrez BJ, Hardy MF, Jin M, Kang S, Kiros R, Lin GN, Luck K, MacWilliams A, Menche J, Murray RR, Palagi A, Poulin MM, Rambout X, Rasla J, Reichert P, Romero V, Ruyssinck E, Sahalie JM, Scholz A, Shah AA, Sharma A, Shen Y, Spirohn K, Tam S, Tejeda AO, Trigg SA, Twizere JC, Vega K, Walsh J, Cusick ME, Xia Y, Barabasi AL, Iakoucheva LM, Aloy P, De Las RJ, Tavernier J, Calderwood MA, Hill DE, Hao T, Roth FP, Vidal M (2014) A proteome-scale map of the human interactome network. Cell 159(5):1212–1226. doi:10.1016/j.cell.2014.10.050. S0092-8674(14) 01422-6 (pii)

18. Hein MY, Hubner NC, Poser I, Cox J, Nagaraj N, Toyoda Y, Gak IA, Weisswange I, Mansfeld J, Buchholz F, Hyman AA, Mann M (2015) A human interactome in three quantitative dimensions organized by stoichiometries and abundances. Cell 163(3):712–723. doi:10.1016/j.cell.2015.09.053

19. Zhang QC, Petrey D, Deng L, Qiang L, Shi Y, Thu CA, Bisikirska B, Lefebvre C, Accili D, Hunter T, Maniatis T, Califano A, Honig B (2012) Structure-based prediction of protein-protein interactions on a genome-wide scale. Nature 490(7421):556–560. doi:10.1038/nature11503. nature11503 (pii)

20. Chatr-Aryamontri A, Breitkreutz BJ, Oughtred R, Boucher L, Heinicke S, Chen D, Stark C, Breitkreutz A, Kolas N, O'Donnell L, Reguly T, Nixon J, Ramage L, Winter A, Sellam A, Chang C, Hirschman J, Theesfeld C, Rust J, Livstone MS, Dolinski K, Tyers M (2015) The BioGRID interaction database: 2015 update. Nucleic Acids Res 43(Database issue):D470–D478. doi:10.1093/nar/gku1204

21. Keshava Prasad TS, Goel R, Kandasamy K, Keerthikumar S, Kumar S, Mathivanan S, Telikicherla D, Raju R, Shafreen B, Venugopal A, Balakrishnan L, Marimuthu A, Banerjee S, Somanathan DS, Sebastian A, Rani S, Ray S, Harrys Kishore CJ, Kanth S, Ahmed M, Kashyap MK, Mohmood R, Ramachandra YL, Krishna V, Rahiman BA, Mohan S, Ranganathan P, Ramabadran S, Chaerkady R, Pandey A (2009) Human protein reference database—2009 update. Nucleic Acids Res 37(Database issue):D767–D772. doi:10.1093/nar/gkn892

22. Orchard S, Ammari M, Aranda B, Breuza L, Briganti L, Broackes-Carter F, Campbell NH, Chavali G, Chen C, Del-Toro N, Duesbury M, Dumousseau M, Galeota E, Hinz U, Iannuccelli M, Jagannathan S, Jimenez R, Khadake J, Lagreid A, Licata L, Lovering RC, Meldal B, Melidoni AN, Milagros M, Peluso D, Perfetto L, Porras P, Raghunath A, Ricard-Blum S, Roechert B, Stutz A, Tognolli M, van Roey K, Cesareni G, Hermjakob H (2014) The

MIntAct project—IntAct as a common curation platform for 11 molecular interaction databases. Nucleic Acids Res 42(Database issue):D358–D363. doi:10.1093/nar/gkt1115

23. Zhang W, Landback P, Gschwend AR, Shen B, Long M (2015) New genes drive the evolution of gene interaction networks in the human and mouse genomes. Genome Biol 16:202. doi:10.1186/s13059-015-0772-4
24. Raychaudhuri S, Korn JM, McCarroll SA, Altshuler D, Sklar P, Purcell S, Daly MJ (2010) Accurately assessing the risk of schizophrenia conferred by rare copy-number variation affecting genes with brain function. PLoS Genet 6(9):e1001097. doi:10.1371/journal.pgen.1001097
25. Rual JF, Venkatesan K, Hao T, Hirozane-Kishikawa T, Dricot A, Li N, Berriz GF, Gibbons FD, Dreze M, Ayivi-Guedehoussou N, Klitgord N, Simon C, Boxem M, Milstein S, Rosenberg J, Goldberg DS, Zhang LV, Wong SL, Franklin G, Li S, Albala JS, Lim J, Fraughton C, Llamosas E, Cevik S, Bex C, Lamesch P, Sikorski RS, Vandenhaute J, Zoghbi HY, Smolyar A, Bosak S, Sequerra R, Doucette-Stamm L, Cusick ME, Hill DE, Roth FP, Vidal M (2005) Towards a proteome-scale map of the human protein-protein interaction network. Nature 437(7062):1173–1178
26. Stelzl U, Worm U, Lalowski M, Haenig C, Brembeck FH, Goehler H, Stroedicke M, Zenkner M, Schoenherr A, Koeppen S, Timm J, Mintzlaff S, Abraham C, Bock N, Kietzmann S, Goedde A, Toksoz E, Droege A, Krobitsch S, Korn B, Birchmeier W, Lehrach H, Wanker EE (2005) A human protein-protein interaction network: a resource for annotating the proteome. Cell 122(6):957–968
27. ORFeome Collaboration (2016) The ORFeome Collaboration: a genome-scale human ORF-clone resource. Nat Methods 13 (3):191–192. doi:10.1038/nmeth.3776
28. Venkatesan K, Rual JF, Vazquez A, Stelzl U, Lemmens I, Hirozane-Kishikawa T, Hao T, Zenkner M, Xin X, Goh KI, Yildirim MA, Simonis N, Heinzmann K, Gebreab F, Sahalie JM, Cevik S, Simon C, de Smet AS, Dann E, Smolyar A, Vinayagam A, Yu H, Szeto D, Borick H, Dricot A, Klitgord N, Murray RR, Lin C, Lalowski M, Timm J, Rau K, Boone C, Braun P, Cusick ME, Roth FP, Hill DE, Tavernier J, Wanker EE, Barabasi AL, Vidal M (2009) An empirical framework for binary interactome mapping. Nat Methods 6 (1):83–90
29. Szklarczyk D, Franceschini A, Wyder S, Forslund K, Heller D, Huerta-Cepas J, Simonovic M, Roth A, Santos A, Tsafou KP, Kuhn M, Bork P, Jensen LJ, von Mering C (2015) STRING v10: protein-protein interaction networks, integrated over the tree of life. Nucleic Acids Res 43(Database issue):D447–D452. doi:10.1093/nar/gku1003
30. Lage K, Mollgard K, Greenway S, Wakimoto H, Gorham JM, Workman CT, Bendsen E, Hansen NT, Rigina O, Roque FS, Wiese C, Christoffels VM, Roberts AE, Smoot LB, Pu WT, Donahoe PK, Tommerup N, Brunak S, Seidman CE, Seidman JG, Larsen LA (2010) Dissecting spatio-temporal protein networks driving human heart development and related disorders. Mol Syst Biol 6:381. doi:10.1038/msb.2010.36
31. Corominas R, Yang X, Lin GN, Kang S, Shen Y, Ghamsari L, Broly M, Rodriguez M, Tam S, Trigg SA, Fan C, Yi S, Tasan M, Lemmens I, Kuang X, Zhao N, Malhotra D, Michaelson JJ, Vacic V, Calderwood MA, Roth FP, Tavernier J, Horvath S, Salehi-Ashtiani K, Korkin D, Sebat J, Hill DE, Hao T, Vidal M, Iakoucheva LM (2014) Protein interaction network of alternatively spliced isoforms from brain links genetic risk factors for autism. Nat Commun 5:3650. doi:10.1038/ncomms4650. ncomms4650 (pii)
32. Cusick ME, Yu H, Smolyar A, Venkatesan K, Carvunis AR, Simonis N, Rual JF, Borick H, Braun P, Dreze M, Vandenhaute J, Galli M, Yazaki J, Hill DE, Ecker JR, Roth FP, Vidal M (2009) Literature-curated protein interaction datasets. Nat Methods 6(1):39–46
33. Wang X, Wei X, Thijssen B, Das J, Lipkin SM, Yu H (2012) Three-dimensional reconstruction of protein networks provides insight into human genetic disease. Nat Biotechnol 30 (2):159–164. doi:10.1038/nbt.2106. nbt.2106 (pii)
34. Consortium EP (2012) An integrated encyclopedia of DNA elements in the human genome. Nature 489(7414):57–74. doi:10.1038/nature11247
35. Consortium GT (2015) Human genomics. The genotype-tissue expression (GTEx) pilot analysis: multitissue gene regulation in humans. Science 348(6235):648–660. doi:10.1126/science.1262110
36. Miller JA, Ding SL, Sunkin SM, Smith KA, Ng L, Szafer A, Ebbert A, Riley ZL, Royall JJ, Aiona K, Arnold JM, Bennet C, Bertagnolli D, Brouner K, Butler S, Caldejon S, Carey A, Cuhaciyan C, Dalley RA, Dee N, Dolbeare TA, Facer BA, Feng D, Fliss TP, Gee G, Goldy J, Gourley L, Gregor BW, Gu G, Howard RE, Jochim JM, Kuan CL, Lau C, Lee CK, Lee F, Lemon TA, Lesnar P, McMurray B, Mastan N,

Mosqueda N, Naluai-Cecchini T, Ngo NK, Nyhus J, Oldre A, Olson E, Parente J, Parker PD, Parry SE, Stevens A, Pletikos M, Reding M, Roll K, Sandman D, Sarreal M, Shapouri S, Shapovalova NV, Shen EH, Sjoquist N, Slaughterbeck CR, Smith M, Sodt AJ, Williams D, Zollei L, Fischl B, Gerstein MB, Geschwind DH, Glass IA, Hawrylycz MJ, Hevner RF, Huang H, Jones AR, Knowles JA, Levitt P, Phillips JW, Sestan N, Wohnoutka P, Dang C, Bernard A, Hohmann JG, Lein ES (2014) Transcriptional landscape of the prenatal human brain. Nature 508(7495):199–206. doi:10.1038/nature13185. nature13185 (pii)

37. Kundaje A, Meuleman W, Ernst J, Bilenky M, Yen A, Heravi-Moussavi A, Kheradpour P, Zhang Z, Wang J, Ziller MJ, Amin V, Whitaker JW, Schultz MD, Ward LD, Sarkar A, Quon G, Sandstrom RS, Eaton ML, Wu YC, Pfenning AR, Wang X, Claussnitzer M, Liu Y, Coarfa C, Harris RA, Shoresh N, Epstein CB, Gjoneska E, Leung D, Xie W, Hawkins RD, Lister R, Hong C, Gascard P, Mungall AJ, Moore R, Chuah E, Tam A, Canfield TK, Hansen RS, Kaul R, Sabo PJ, Bansal MS, Carles A, Dixon JR, Farh KH, Feizi S, Karlic R, Kim AR, Kulkarni A, Li D, Lowdon R, Elliott G, Mercer TR, Neph SJ, Onuchic V, Polak P, Rajagopal N, Ray P, Sallari RC, Siebenthall KT, Sinnott-Armstrong NA, Stevens M, Thurman RE, Wu J, Zhang B, Zhou X, Beaudet AE, Boyer LA, De Jager PL, Farnham PJ, Fisher SJ, Haussler D, Jones SJ, Li W, Marra MA, McManus MT, Sunyaev S, Thomson JA, Tlsty TD, Tsai LH, Wang W, Waterland RA, Zhang MQ, Chadwick LH, Bernstein BE, Costello JF, Ecker JR, Hirst M, Meissner A, Milosavljevic A, Ren B, Stamatoyannopoulos JA, Wang T, Kellis M (2015) Integrative analysis of 111 reference human epigenomes. Nature 518 (7539):317–330. doi:10.1038/nature14248. nature14248 (pii)

38. Consortium GT (2013) The genotype-tissue expression (GTEx) project. Nat Genet 45 (6):580–585. doi:10.1038/ng.2653

39. Colantuoni C, Lipska BK, Ye T, Hyde TM, Tao R, Leek JT, Colantuoni EA, Elkahloun AG, Herman MM, Weinberger DR, Kleinman JE (2011) Temporal dynamics and genetic control of transcription in the human prefrontal cortex. Nature 478(7370):519–523. doi:10.1038/nature10524

40. Stuart JM, Segal E, Koller D, Kim SK (2003) A gene-coexpression network for global discovery of conserved genetic modules. Science 302 (5643):249–255. doi:10.1126/science.1087447. 1087447 (pii)

41. Zhang B, Horvath S (2005) A general framework for weighted gene co-expression network analysis. Stat Appl Genet Mol Biol 4:Article 17

42. Henegar C, Tordjman J, Achard V, Lacasa D, Cremer I, Guerre-Millo M, Poitou C, Basdevant A, Stich V, Viguerie N, Langin D, Bedossa P, Zucker JD, Clement K (2008) Adipose tissue transcriptomic signature highlights the pathological relevance of extracellular matrix in human obesity. Genome Biol 9(1):R14. doi:10.1186/gb-2008-9-1-r14

43. Prifti E, Zucker JD, Clement K, Henegar C (2010) Interactional and functional centrality in transcriptional co-expression networks. Bioinformatics 26(24):3083–3089. doi:10.1093/bioinformatics/btq591

44. Horvath S, Zhang B, Carlson M, Lu KV, Zhu S, Felciano RM, Laurance MF, Zhao W, Qi S, Chen Z, Lee Y, Scheck AC, Liau LM, Wu H, Geschwind DH, Febbo PG, Kornblum HI, Cloughesy TF, Nelson SF, Mischel PS (2006) Analysis of oncogenic signaling networks in glioblastoma identifies ASPM as a molecular target. Proc Natl Acad Sci U S A 103 (46):17402–17407. doi:10.1073/pnas.0608396103. 0608396103 (pii)

45. Presson AP, Yoon NK, Bagryanova L, Mah V, Alavi M, Maresh EL, Rajasekaran AK, Goodglick L, Chia D, Horvath S (2011) Protein expression based multimarker analysis of breast cancer samples. BMC Cancer 11:230. doi:10.1186/1471-2407-11-230

46. Levine AJ, Miller JA, Shapshak P, Gelman B, Singer EJ, Hinkin CH, Commins D, Morgello S, Grant I, Horvath S (2013) Systems analysis of human brain gene expression: mechanisms for HIV-associated neurocognitive impairment and common pathways with Alzheimer's disease. BMC Med Genet 6:4. doi:10.1186/1755-8794-6-4

47. Wang JZ, Du Z, Payattakool R, Yu PS, Chen CF (2007) A new method to measure the semantic similarity of GO terms. Bioinformatics 23(10):1274–1281. doi:10.1093/bioinformatics/btm087. btm087 (pii).

48. Bayes A, van de Lagemaat LN, Collins MO, Croning MD, Whittle IR, Choudhary JS, Grant SG (2011) Characterization of the proteome, diseases and evolution of the human postsynaptic density. Nat Neurosci 14 (1):19–21. doi:10.1038/nn.2719. nn.2719 (pii)

49. Darnell JC, Van Driesche SJ, Zhang C, Hung KY, Mele A, Fraser CE, Stone EF, Chen C, Fak JJ, Chi SW, Licatalosi DD, Richter JD, Darnell RB (2011) FMRP stalls ribosomal translocation on mRNAs linked to synaptic function and autism. Cell 146(2):247–261. doi:10.

1016/j.cell.2011.06.013. S0092-8674(11)00655-6 (pii)

50. Samocha KE, Robinson EB, Sanders SJ, Stevens C, Sabo A, McGrath LM, Kosmicki JA, Rehnstrom K, Mallick S, Kirby A, Wall DP, MacArthur DG, Gabriel SB, DePristo M, Purcell SM, Palotie A, Boerwinkle E, Buxbaum JD, Cook EH Jr, Gibbs RA, Schellenberg GD, Sutcliffe JS, Devlin B, Roeder K, Neale BM, Daly MJ (2014) A framework for the interpretation of de novo mutation in human disease. Nat Genet 46(9):944–950. doi:10.1038/ng.3050
51. Pasca SP, Portmann T, Voineagu I, Yazawa M, Shcheglovitov A, Pasca AM, Cord B, Palmer TD, Chikahisa S, Nishino S, Bernstein JA, Hallmayer J, Geschwind DH, Dolmetsch RE (2011) Using iPSC-derived neurons to uncover cellular phenotypes associated with Timothy syndrome. Nat Med 17 (12):1657–1662. doi:10.1038/nm.2576. nm.2576 (pii)
52. Sebat J, Lakshmi B, Troge J, Alexander J, Young J, Lundin P, Maner S, Massa H, Walker M, Chi M, Navin N, Lucito R, Healy J, Hicks J, Ye K, Reiner A, Gilliam TC, Trask B, Patterson N, Zetterberg A, Wigler M (2004) Large-scale copy number polymorphism in the human genome. Science 305(5683):525–528
53. Malhotra D, Sebat J (2012) CNVs: harbingers of a rare variant revolution in psychiatric genetics. Cell 148(6):1223–1241. doi:10.1016/j.cell.2012.02.039. S0092-8674(12)00277-2 (pii)
54. Pinto D, Delaby E, Merico D, Barbosa M, Merikangas A, Klei L, Thiruvahindrapuram B, Xu X, Ziman R, Wang Z, Vorstman JA, Thompson A, Regan R, Pilorge M, Pellecchia G, Pagnamenta AT, Oliveira B, Marshall CR, Magalhaes TR, Lowe JK, Howe JL, Griswold AJ, Gilbert J, Duketis E, Dombroski BA, De Jonge MV, Cuccaro M, Crawford EL, Correia CT, Conroy J, Conceicao IC, Chiocchetti AG, Casey JP, Cai G, Cabrol C, Bolshakova N, Bacchelli E, Anney R, Gallinger S, Cotterchio M, Casey G, Zwaigenbaum L, Wittemeyer K, Wing K, Wallace S, van Engeland H, Tryfon A, Thomson S, Soorya L, Roge B, Roberts W, Poustka F, Mouga S, Minshew N, McInnes LA, McGrew SG, Lord C, Leboyer M, Le Couteur AS, Kolevzon A, Jimenez Gonzalez P, Jacob S, Holt R, Guter S, Green J, Green A, Gillberg C, Fernandez BA, Duque F, Delorme R, Dawson G, Chaste P, Cafe C, Brennan S, Bourgeron T, Bolton PF, Bolte S, Bernier R, Baird G, Bailey AJ, Anagnostou E, Almeida J, Wijsman EM, Vieland VJ, Vicente AM, Schellenberg GD, Pericak-Vance M, Paterson AD, Parr JR, Oliveira G, Nurnberger JI, Monaco AP, Maestrini E, Klauck SM, Hakonarson H, Haines JL, Geschwind DH, Freitag CM, Folstein SE, Ennis S, Coon H, Battaglia A, Szatmari P, Sutcliffe JS, Hallmayer J, Gill M, Cook EH, Buxbaum JD, Devlin B, Gallagher L, Betancur C, Scherer SW (2014) Convergence of genes and cellular pathways dysregulated in autism spectrum disorders. Am J Hum Genet 94 (5):677–694. doi:10.1016/j.ajhg.2014.03.018
55. Walsh T, McClellan JM, McCarthy SE, Addington AM, Pierce SB, Cooper GM, Nord AS, Kusenda M, Malhotra D, Bhandari A, Stray SM, Rippey CF, Roccanova P, Makarov V, Lakshmi B, Findling RL, Sikich L, Stromberg T, Merriman B, Gogtay N, Butler P, Eckstrand K, Noory L, Gochman P, Long R, Chen Z, Davis S, Baker C, Eichler EE, Meltzer PS, Nelson SF, Singleton AB, Lee MK, Rapoport JL, King MC, Sebat J (2008) Rare structural variants disrupt multiple genes in neurodevelopmental pathways in schizophrenia. Science 320(5875):539–543
56. Kumar RA, KaraMohamed S, Sudi J, Conrad DF, Brune C, Badner JA, Gilliam TC, Nowak NJ, Cook EH Jr, Dobyns WB, Christian SL (2008) Recurrent 16p11.2 microdeletions in autism. Hum Mol Genet 17(4):628–638
57. Itsara A, Wu H, Smith JD, Nickerson DA, Romieu I, London SJ, Eichler EE (2010) De novo rates and selection of large copy number variation. Genome Res 20(11):1469–1481. doi:10.1101/gr.107680.110
58. Levy D, Ronemus M, Yamrom B, Lee YH, Leotta A, Kendall J, Marks S, Lakshmi B, Pai D, Ye K, Buja A, Krieger A, Yoon S, Troge J, Rodgers L, Iossifov I, Wigler M (2011) Rare de novo and transmitted copy-number variation in autistic spectrum disorders. Neuron 70 (5):886–897. doi:10.1016/j.neuron.2011.05.015. S0896-6273(11)00396-5 (pii)
59. Gilman SR, Iossifov I, Levy D, Ronemus M, Wigler M, Vitkup D (2011) Rare de novo variants associated with autism implicate a large functional network of genes involved in formation and function of synapses. Neuron 70 (5):898–907. doi:10.1016/j.neuron.2011.05.021. S0896-6273(11)00439-9 (pii)
60. Sanders SJ, He X, Willsey AJ, Ercan-Sencicek AG, Samocha KE, Cicek AE, Murtha MT, Bal VH, Bishop SL, Dong S, Goldberg AP, Jinlu C, Keaney JF III, Klei L, Mandell JD, Moreno-De-Luca D, Poultney CS, Robinson EB, Smith L, Solli-Nowlan T, Su MY, Teran NA, Walker MF, Werling DM, Beaudet AL, Cantor RM, Fombonne E, Geschwind DH, Grice DE, Lord C, Lowe JK, Mane SM, Martin DM,

Morrow EM, Talkowski ME, Sutcliffe JS, Walsh CA, Yu TW, Autism Sequencing C, Ledbetter DH, Martin CL, Cook EH, Buxbaum JD, Daly MJ, Devlin B, Roeder K, State MW (2015) Insights into autism spectrum disorder genomic architecture and biology from 71 risk loci. Neuron 87(6):1215–1233. doi:10.1016/j.neuron.2015.09.016

61. McCarthy SE, Makarov V, Kirov G, Addington AM, McClellan J, Yoon S, Perkins DO, Dickel DE, Kusenda M, Krastoshevsky O, Krause V, Kumar RA, Grozeva D, Malhotra D, Walsh T, Zackai EH, Kaplan P, Ganesh J, Krantz ID, Spinner NB, Roccanova P, Bhandari A, Pavon K, Lakshmi B, Leotta A, Kendall J, Lee YH, Vacic V, Gary S, Iakoucheva LM, Crow TJ, Christian SL, Lieberman JA, Stroup TS, Lehtimaki T, Puura K, Haldeman-Englert C, Pearl J, Goodell M, Willour VL, Derosse P, Steele J, Kassem L, Wolff J, Chitkara N, McMahon FJ, Malhotra AK, Potash JB, Schulze TG, Nothen MM, Cichon S, Rietschel M, Leibenluft E, Kustanovich V, Lajonchere CM, Sutcliffe JS, Skuse D, Gill M, Gallagher L, Mendell NR, Craddock N, Owen MJ, O'Donovan MC, Shaikh TH, Susser E, Delisi LE, Sullivan PF, Deutsch CK, Rapoport J, Levy DL, King MC, Sebat J (2009) Microduplications of 16p11.2 are associated with schizophrenia. Nat Genet 41(11):1223–1227
62. Golzio C, Willer J, Talkowski ME, Oh EC, Taniguchi Y, Jacquemont S, Reymond A, Sun M, Sawa A, Gusella JF, Kamiya A, Beckmann JS, Katsanis N (2012) KCTD13 is a major driver of mirrored neuroanatomical phenotypes of the 16p11.2 copy number variant. Nature 485(7398):363–367. doi:10.1038/nature11091. nature11091 (pii)
63. Horev G, Ellegood J, Lerch JP, Son YE, Muthuswamy L, Vogel H, Krieger AM, Buja A, Henkelman RM, Wigler M, Mills AA (2011) Dosage-dependent phenotypes in models of 16p11.2 lesions found in autism. Proc Natl Acad Sci U S A 108(41):17076–17081. doi:10.1073/pnas.1114042108. 1114042108 (pii)
64. Stranger BE, Forrest MS, Dunning M, Ingle CE, Beazley C, Thorne N, Redon R, Bird CP, de Grassi A, Lee C, Tyler-Smith C, Carter N, Scherer SW, Tavare S, Deloukas P, Hurles ME, Dermitzakis ET (2007) Relative impact of nucleotide and copy number variation on gene expression phenotypes. Science 315 (5813):848–853. doi:10.1126/science.1136678
65. Luo R, Sanders SJ, Tian Y, Voineagu I, Huang N, Chu SH, Klei L, Cai C, Ou J, Lowe JK, Hurles ME, Devlin B, State MW, Geschwind DH (2012) Genome-wide transcriptome profiling reveals the functional impact of rare de novo and recurrent CNVs in autism spectrum disorders. Am J Hum Genet 91 (1):38–55. doi:10.1016/j.ajhg.2012.05.011. S0002-9297(12)00267-4 (pii)
66. Rees E, Kirov G, Sanders A, Walters JT, Chambert KD, Shi J, Szatkiewicz J, O'Dushlaine C, Richards AL, Green EK, Jones I, Davies G, Legge SE, Moran JL, Pato C, Pato M, Genovese G, Levinson D, Duan J, Moy W, Goring HH, Morris D, Cormican P, Kendler KS, O'Neill FA, Riley B, Gill M, Corvin A, Wellcome Trust Case Control C, Craddock N, Sklar P, Hultman C, Sullivan PF, Gejman PV, McCarroll SA, O'Donovan MC, Owen MJ (2014) Evidence that duplications of 22q11.2 protect against schizophrenia. Mol Psychiatry 19(1):37–40. doi:10.1038/mp.2013.156
67. Pan Q, Shai O, Lee LJ, Frey BJ, Blencowe BJ (2008) Deep surveying of alternative splicing complexity in the human transcriptome by high-throughput sequencing. Nat Genet 40 (12):1413–1415. doi:10.1038/ng.259. ng.259 (pii)
68. Wang ET, Sandberg R, Luo S, Khrebtukova I, Zhang L, Mayr C, Kingsmore SF, Schroth GP, Burge CB (2008) Alternative isoform regulation in human tissue transcriptomes. Nature 456(7221):470–476
69. Yeo G, Holste D, Kreiman G, Burge CB (2004) Variation in alternative splicing across human tissues. Genome Biol 5(10):R74. doi:10.1186/gb-2004-5-10-r74
70. Mele M, Ferreira PG, Reverter F, DeLuca DS, Monlong J, Sammeth M, Young TR, Goldmann JM, Pervouchine DD, Sullivan TJ, Johnson R, Segre AV, Djebali S, Niarchou A, Wright FA, Lappalainen T, Calvo M, Getz G, Dermitzakis ET, Ardlie KG, Guigo R (2015) Human genomics. The human transcriptome across tissues and individuals. Science 348 (6235):660–665. doi:10.1126/science.aaa0355. 348/6235/660 (pii)
71. Buljan M, Chalancon G, Eustermann S, Wagner GP, Fuxreiter M, Bateman A, Babu MM (2012) Tissue-specific splicing of disordered segments that embed binding motifs rewires protein interaction networks. Mol Cell 46(6):871–883. doi:10.1016/j.molcel.2012.05.039. S1097-2765(12)00484-4 (pii)
72. Ellis JD, Barrios Rodiles M, Colak R, Irimia M, Kim T, Calarco JA, Wang X, Pan Q, O'Hanlon D, Kim PM, Wrana JL, Blencowe BJ (2012) Tissue-specific alternative splicing remodels protein-protein interaction networks. Mol Cell 46(6):884–892. doi:10.1016/j.molcel.2012.05.037. S1097-2765(12)00482-0 (pii)

73. Yang X, Coulombe-Huntington J, Kang S, Sheynkman GM, Hao T, Richardson A, Sun S, Yang F, Shen YA, Murray RR, Spirohn K, Begg BE, Duran-Frigola M, MacWilliams A, Pevzner SJ, Zhong Q, Trigg SA, Tam S, Ghamsari L, Sahni N, Yi S, Rodriguez MD, Balcha D, Tan G, Costanzo M, Andrews B, Boone C, Zhou XJ, Salehi-Ashtiani K, Charloteaux B, Chen AA, Calderwood MA, Aloy P, Roth FP, Hill DE, Iakoucheva LM, Xia Y, Vidal M (2016) Widespread expansion of protein interaction capabilities by alternative splicing. Cell 164(4):805–817. doi:10.1016/j.cell.2016.01.029

74. Liu C, Song X, Nisbet R, Gotz J (2016) Co-immunoprecipitation with tau isoform-specific antibodies reveals distinct protein interactions, and highlights a putative role for 2N tau in disease. J Biol Chem. doi:10.1074/jbc.M115.641902

75. Kalsotra A, Cooper TA (2011) Functional consequences of developmentally regulated alternative splicing. Nat Rev Genet 12 (10):715–729. doi:10.1038/nrg3052

76. Schwerk C, Schulze-Osthoff K (2005) Regulation of apoptosis by alternative pre-mRNA splicing. Mol Cell 19(1):1–13. doi:10.1016/j.molcel.2005.05.026

77. Liu J, McCleland M, Stawiski EW, Gnad F, Mayba O, Haverty PM, Durinck S, Chen YJ, Klijn C, Jhunjhunwala S, Lawrence M, Liu H, Wan Y, Chopra V, Yaylaoglu MB, Yuan W, Ha C, Gilbert HN, Reeder J, Pau G, Stinson J, Stern HM, Manning G, Wu TD, Neve RM, de Sauvage FJ, Modrusan Z, Seshagiri S, Firestein R, Zhang Z (2014) Integrated exome and transcriptome sequencing reveals ZAK isoform usage in gastric cancer. Nat Commun 5:3830. doi:10.1038/ncomms4830

78. Barrett CL, DeBoever C, Jepsen K, Saenz CC, Carson DA, Frazer KA (2015) Systematic transcriptome analysis reveals tumor-specific isoforms for ovarian cancer diagnosis and therapy. Proc Natl Acad Sci U S A 112(23): E3050–E3057. doi:10.1073/pnas.1508057112

79. Trapnell C, Pachter L, Salzberg SL (2009) TopHat: discovering splice junctions with RNA-Seq. Bioinformatics 25(9):1105–1111. doi:10.1093/bioinformatics/btp120

80. Li B, Dewey CN (2011) RSEM: accurate transcript quantification from RNA-Seq data with or without a reference genome. BMC Bioinformatics 12:323. doi:10.1186/1471-2105-12-323

81. Roberts A, Pachter L (2013) Streaming fragment assignment for real-time analysis of sequencing experiments. Nat Methods 10 (1):71–73. doi:10.1038/nmeth.2251

82. Pathway Analysis working group of the International Cancer Genome Consortium (2015) Pathway and network analysis of cancer genomes. Nat Methods 12(7):615–621. doi:10.1038/nmeth.3440

Chapter 16

Semantic Data Integration and Knowledge Management to Represent Biological Network Associations

Sascha Losko and Klaus Heumann

Abstract

The vast quantities of information generated by academic and industrial research groups are reflected in a rapidly growing body of scientific literature and exponentially expanding resources of formalized data, including experimental data, originating from a multitude of "-omics" platforms, phenotype information, and clinical data. For bioinformatics, the challenge remains to structure this information so that scientists can identify relevant information, to integrate this information as specific "knowledge bases," and to formalize this knowledge across multiple scientific domains to facilitate hypothesis generation and validation. Here we report on progress made in building a generic knowledge management environment capable of representing and mining both explicit and implicit knowledge and, thus, generating new knowledge. Risk management in drug discovery and clinical research is used as a typical example to illustrate this approach. In this chapter we introduce techniques and concepts (such as ontologies, semantic objects, typed relationships, contexts, graphs, and information layers) that are used to represent complex biomedical networks. The BioXM™ Knowledge Management Environment is used as an example to demonstrate how a domain such as oncology is represented and how this representation is utilized for research.

Key words Knowledge management, Bioinformatics, Biomarkers, Biological networks, Semantic technologies, Data integration, Ontologies, Oncology

1 Introduction

Today, the life sciences generate an ever-increasing amount of information. This information explosion is mainly driven by two factors. First, the life sciences are highly complex fields of research. There are millions of enzymes, genes, chemical compounds, diseases, species, cell types, and organs that interact and are related in many different ways. Second, new experimental methods are continuously being developed and as their throughput increases, the amount of raw data generated increases with overwhelming speed.

Any system aiming to support a scientist in "understanding" large amounts of data should "speak" the language of the scientist's research domain. Information technology (IT) solutions are

Tatiana V. Tatarinova and Yuri Nikolsky (eds.), *Biological Networks and Pathway Analysis*, Methods in Molecular Biology, vol. 1613, DOI 10.1007/978-1-4939-7027-8_16, © Springer Science+Business Media LLC 2017

needed to support the knowledge generation cycle to ultimately gain an adequate understanding of whole biological systems [1]. Modern semantic technologies provide a conceptual foundation that helps to meet these demanding requirements. Promising advances like the "Semantic Web" and current progress in ontology development [2, 3] are expected to contribute to the next generation of software for the life sciences by enabling the scientist to actually voice specific questions instead of having to "construct" technical database queries.

Biomax Informatics AG has developed the BioXM™ Knowledge Management Environment, an enterprise platform for semantic data integration focusing on the life science industry [4]. In the BioXM system, knowledge is conceptualized as *typed relationships* between semantic objects representing "elements of a scientific domain" (such as genes or drugs). Those relations are supplemented by the annotation of evidence to provide validation. For the related objects, further validated relations to other "elements of a scientific domain" (such as cell types or diseases) may exist and, thus, expand the knowledge network. Specific parts of the knowledge model may be organized in subnetwork contexts (such as a particular signal transduction pathway in an organism of interest), which allow hierarchical structuring of knowledge.

The conceptualization of entire areas of interest in ontologies allows the use of inherent inference relationships for the exploration of knowledge networks. Entities from external public or proprietary databases, accessible through either the embedded BioRS™ Integration and Retrieval System or external relational database management systems (RDBMS), can serve as "virtual semantic objects" in the knowledge network. They can also be used as "read-only" annotation of the "real" semantic objects. All semantic objects (such as elements, relations, contexts, ontology instances, or external database entries) can be annotated with additional information. Annotations are form based and support hierarchical organization of information.

The BioXM system provides graphical browsing through the network. An advanced query builder allows flexible exploration of the knowledge with complex queries that use a natural-language-like syntax. Flexible reporting allows specified sets of information relevant to the particular semantic objects to be displayed in one view. A versatile data management system allows the information networks to be modified and expanded without the need for additional programming. In this way, research projects can be modeled and extended dynamically.

While the main client application of the BioXM system provides a graphical user interface for all purposes—administration, querying and data mining, graphical exploration and reporting—research scientists often require a task-centric and project-specific graphical user interface. The BioXM system, thus, allows configuration of specialized and easy-to-use web apps that can be deployed quickly to large user groups and communities via Intranet or Internet. These web apps can be used with any modern web browser on all devices including desktop and tablet computers as well as smart phones.

Applications for platforms, such as the BioXM Knowledge Management Environment, that use semantic technologies, are manifold. The BioXM system has been used in many different domains in the last 10 years, including oncology research and drug development, toxicology, food and nutrition, clinical research and healthcare in both Europe and the USA, and has been constantly extended to address new scientific areas.

2 Materials

The key to semantic network definitions is to be able to unite two requirements: (a) to formulate a descriptive model of the world and (b) to relate data resources to that model. Formulating a descriptive model in a systematic way requires a set of well-defined building blocks. The model should be extendable, like a model made of LEGO® building blocks; by combining the pieces, the model evolves (*see* Subheading 3.2.1 and **Note 1**). Definition of the building blocks is essential for the design (*see* Subheading 3.2.2 and **Note 2**).

The objective is to come up with a universal tool kit, i.e., with a set of building-block concepts which constitute the foundation of a generic semantic network building system. In the BioXM system, the set of semantic objects provides this foundation.

2.1 Semantic Objects

The set of semantic objects formulates the principles of what can be expressed in the system. Table 1 shows the semantic objects defined in the BioXM system. Each semantic object implements a concept of expression.

2.1.1 Elements

Elements represent the basic units in a knowledge model. Once an element type has been specified, elements can be defined and imported or created. For example, the "gene" and "disease" elements could be created to represent genes and disease information in a project for studying genetic diseases. Elements are the generic nodes in the network. Note that each instance of an element should reflect exactly one unique real-world object: one gene, one protein, etc.

2.1.2 Relations

Relations are semantic objects that describe a relationship between two semantic objects. For example, the "gene-disease" relation could be created to represent the participation of a gene in a known disease by associating elements of type "gene" and

Table 1
Fundamental semantic objects

Semantic object	Description	Example
Element	Represents a basic unit of a knowledge model	"Gene" element type can be used to create the "STAT3" gene element
		"Disease" element type can be used to create the "pancreatic tumor" disease element
Relation	Describes a relationship between semantic objects	"Gene-disease" relation class can be used to create the "STAT3 is associated with disease pancreatic tumor" relation
Annotation	Extends the properties of a semantic object by a set of attributes	Gene report Patient record Protein entry Literature abstract Experimental data (evidence)
Ontology	Classifies semantics objects according to a defined hierarchical nomenclature of concepts	Gene Ontology to classify biological function NCI Thesaurus of disease terms taxonomy
Context	Represents sets of semantic objects	Metabolic pathways Protein complexes A disease process or pattern

"disease." Relations are generic edges in the network. Relations are directed. Note that relations are typed in terms of which objects they are allowed to connect. This does not mean that exactly one element type is connected with another distinct element type. There can be more than one type at each side of the link; however, the set of related objects is defined and establishes constraints on what instances of semantic objects may be connected.

2.1.3 Annotations

In addition to the nodes and edges in the network tool kit, annotations allow supplementary information to be assigned to a semantic object and managed by the BioXM system. This "data about data" (metadata) is used to describe the annotated object with specific information from various sources, such as analyses and experiments as well as proprietary and common knowledge. Annotation is assigned to objects with user-defined annotation forms. For example, the "Patient information" annotation form could be created to assign annotation to elements of type "Patient." The assigned annotation might contain information such as "Name" and "Date of birth," Annotations are generic content containers in the network. They add substance to the semantic network.

Annotations do not necessarily need to be assigned to only one semantic object; in fact, an annotation can be shared by multiple semantic objects. An annotation can consist of multiple, hierarchically organized annotations. In this way, an annotation constitutes a data structure in itself.

2.1.4 Ontologies

Ontologies, a central concept in knowledge management, relate the conceptualization of a domain to the knowledge model. Ontologies are the link between the semantic network and knowledge management. An ontology may be linked to any semantic network including another ontology. In contrast to the relations used to create the BioXM semantic network, relationships within an ontology are typically defined using a formal semantic (e.g., meronymy: "A *is_part_of* B", hypernomy: "A *is_a* B", or synonymy: "A *is_the_same_as* B") that allows rule-based inference. The BioXM system allows any relation type defined in an ontology, supporting both transitive and reflexive relation types. Transitive relation types need to adhere to the constraints of directed acyclic graphs (DAG), while non-transitive relation types are allowed to form cycles within an ontology. The BioXM system allows single-type and multi-type inference on transient relation types.

Ontologies are often developed by domain experts as a set of "scientific nomenclature" and are widely used in the life sciences. Linking ontology entries to specific instances of semantic objects is an art in itself. For example, linguistic analysis using highly sophisticated recognizers/taggers [5, 6] is often applied to provide this link. The challenge is often described as the "mapping problem."

2.1.5 Contexts

Knowledge networks can become quite extensive. Different levels of abstraction are often represented within the network. Contexts are a means to define a set of semantic objects and to treat that set of objects as a single object. Contexts are subnetworks, though a context may be related as an entity to other semantic objects, including another context. In that sense, contexts provide a link between different levels of abstraction.

2.2 Additional Concepts

Besides classical user management and work organization in projects, a set of additional concepts is available within the BioXM system to complete the required functionality (Table 2).

2.2.1 Queries

Queries are formulated on the basis of the knowledge model. Consequently, everything that is described in the model can be an argument in a query expression. Because the knowledge model is the basis of the query building process, any change to the knowledge model has immediate effect on the expressive power of the system with respect to the queries that can be formulated. The advantage of a domain-specific knowledge model is that

Table 2
Additional concepts

Concept	Description	Example
Alias	Alternative identifier composed of "alias source" and "alias id"; aliases can be assigned to all semantic objects and are not type-specific	UMLS:1234 is assigned to OntologyEntry: DOID:1234 and UMLS:2345 is assigned to OntologyEntry:NCI Thesaurus:25432
Query	Allows exploring the knowledge network using a natural-language-style mechanism	A query to "Find Genes which are in relationship to a disease with a name like Cancer"
Information layer	Organizes different levels of complexity as a semantic context	Layers of metabolic pathways, expression data, or signaling pathways
Graph	Renders the knowledge network as an interactive whiteboard	*See* Fig. 6
Experimental data	Provides a numerical data matrices of experimental results	A gene expression chip result or a protein analysis assay
External database entry	Integrates entries from external relational database tables or views as external semantic objects	An entry from the "Physical Entities" table of the Reactome Simplified Database is integrated as a semantic object
BioRS databank entry	Integrates entries from BioRS databanks as external semantic objects or, alternatively, as metadata associated with native semantic objects	A DrugBank entry is integrated as a semantic object; The PubChem entry with CID 2244 is assigned as metadata to the BioXM Element:Compound:Aspirin
Report	Provides tables or documents of compiled information	A table to compare the gene function of two organisms or a clinical record of a patient
Import/ Export	Enables two-way data interchange	An Excel® spreadsheet can be loaded to the system to map the semantics of the columns and the rows to the knowledge network or a report of all information about a gene can be exported in Portable Document Format (PDF)

queries based on such a model are relatively easy to read. For example, a query may read as follows: "*give me all genes which are related to a disease which has the name 'lung cancer.'*"

Queries can be accessed in three ways: by a query builder exporting the model, by templates in which only specific query variables need to be inserted by the user, and through the so-called smart folders. Smart folders are canned (i.e., predefined) queries which behave like a folder, but render a dynamic query result. The three levels of query formulation reflect the levels of user skill necessary to interact with the system. Smart folders are the easiest method, since no understanding of the knowledge model is required. Using the templates and the query builder require more experience.

BioXM queries may also extend to external data resources to make any external resource searchable. This expands the explicit knowledge model, de facto, to a transient model representing external resources.

2.2.2 Information Layers

Information layers are similar to contexts in that they also allow the management of complexity. In an information layer, certain semantic objects can be grouped on the same level of complexity, whereas a context organizes semantic objects by meaning. One can imagine information layers to be a stack of transparencies, which can be placed on top of each other. For example, a metabolic pathway may constitute a context. The proteins and metabolites can be defined as a layer that establishes the rough picture. Further, the side reactions can be defined as a second layer, flux in the network as a third, and expression activity as a fourth.

Information layers allow information to be overlaid depending on a particular point of focus. This becomes relevant when dealing with complex graphs (*see* Subheading 2.2.3) and helps to maintain an overview and manage complexity.

2.2.3 Graphs

A graph can be used to visualize a semantic object with any associated objects. This tool provides functions that are central to understanding and using semantic objects such as elements and relations as well as associated objects of other types. A graph is primarily a visualization tool for the network, but it is also used to explore and navigate within the network. It provides paths, which can be followed virtually. Furthermore, the graph can be used to formulate questions, such as the following: Are there connections, either direct or indirect, between any given node in the graph? For example, given a compound and a disease: Can a connection between them be found? In the graph, the items can be selected and, if a connection exists, the system will render the edges that represent the connecting paths. Taking the example further, the graph may also show additional relevant information, for example, that the compound regulates a gene which is known (from literature) to be associated with the disease and that there is a clinical study which used the compound in the disease context.

Any type of classical biological network, such as metabolic and signaling pathways, can also be rendered as a graph. Information layers are typically used in the context of graph exploration to manage complex graphs and make them comprehensible (*see* Subheading 2.2.2).

2.2.4 Experimental Data

Experimental data are, in fact, a special type of annotation—typically of samples taken from a patient, a plant or an animal. Experimental data are defined as a distinct semantic object type "Experiment" in the system because they may be large and require specific mathematical operations or interaction with external analytical tools, such as the "R" package. Experiment objects represent a design pattern optimized for high-throughput experiments. One experiment contains all experimental measurements for every single element (e.g., probes or proteins) being measured. Aggregate functions can be used with experimental data, enabling queries like "give me all probes which are at least twofold over-expressed in all experiments owned by my project of interest."

2.2.5 External Database Entries

External database entries allow integration of external relational database tables and views. External database entries have all properties of a native semantic object in the BioXM system; they can be annotated, become the source or target for relationships, and more.

The capability to integrate existing Relational Database Management Systems (RDBMS) is especially helpful in existing research infrastructures that already maintain, for example, LIMS databases, corporate gene or compound indices which are continuously updated and are an integral part of existing business processes.

2.2.6 BioRS Databank Entries

The BioRS Integration and Retrieval System [7] is a middleware developed by Biomax that allows the integration of typical life science databases. In particular, flat-file databases, sequence databases and databases that require efficient full-text indexing are efficiently managed using the BioRS system. The BioRS system automatically maintains all cross-references between the integrated databases, enabling efficient traversal of "chains" of multiple public databases.

The BioXM system can directly incorporate any BioRS databank, such as Genbank, UniProtKB, or PubChem. In contrast to external database entries, BioRS databank entries can be used in two different ways. They can be used as semantic objects, similar to native semantic objects such as elements, or they can be used as metadata that is assigned to native semantic objects, similar to annotation (*see* Subheading 2.1.3).

2.2.7 Reports

Reports have two principal forms: tables or documents. Tables report on sets of semantic objects and documents report on a specific semantic object. Both types are rendered through configured views; multiple table and document report styles can be defined. For both types of reports anything that can be reached within the network can be compiled. This feature facilitates rendering knowledge in condensed form, exporting the information to external applications and so forth. Table reports can also be used with the reimport mechanism of the system, which allows semantic objects to be repopulated with different content and results from external applications to be integrated permanently. The reporting mechanism is also used to define object-type-specific report labels for visualization in graphs (*see* Subheading 2.2.3).

2.2.8 Import/Export

The import functionality provides a way to connect the BioXM system to external resources. The most frequently used external format is Excel® spreadsheet. The importer allows mapping the meaning of the rows, columns, and cells in the spreadsheet to the semantic network. This can be used to populate the knowledge network with information and puts the pieces of information into place (*see* Subheading 3.2.2).

The export functionality is based on the reports concept (*see* Subheading 2.2.7); any view formulated can be exported to the file system or an external application.

2.3 Representing Explicit and Implicit Knowledge

Though the BioXM system is based on the concepts of semantic networks, it significantly extends these concepts to handle current technologies and requirements in the life science, biomedical, and clinical domains.

Assertions retrieved from the knowledge network maintained in the system are represented mostly as edges in the network. These edges may have been created explicitly, but they may also be implied logically or generated on-the-fly as virtual edges representing the result of arbitrary algorithms. The BioXM system offers the following main technical concepts to model both explicit and implicit knowledge:

- *Relations* (*see* Subheading 2.1.2) represent typed associations that are *explicitly* instantiated. Relations are the main contributor to most usage scenarios that manage explicit knowledge.
- *Ontology relations* (*see* Subheading 2.1.4) define the semantic rules for inference within an ontology. Instantiated assertions allow *implied* assertions to be inferred with a level of confidence that reflects the quality of design and content of the ontology used.
- *Virtual relations* defined by an associated query (*see* Subheading 2.2.1) are used to *imply* associations by special business logic. The associated query implying an association can be composed using both the built-in search criteria derived from the knowledge model as well as external search criteria that contribute any type of external logic (analysis results, tool integration, etc.).

While relation classes are exclusively used to manage knowledge explicitly, the other two concepts—inference using ontologies and implicit associations resulting from for example statistical analyses or other types of calculations or algorithms—are the main mechanisms offered by the BioXM system to generate new knowledge.

Compared to other semantic integration platforms, the BioXM system is unique because it generates implications using its main query mechanisms which go beyond explicit paths or paths that are "only" logically implied by the semantic inherent to the ontology used.

Because the BioXM query language supports extension with "external" search criteria that can virtually implement any algorithm, analysis or tool, the user can also leverage implicit knowledge generated analytically.

2.4 Searching on a Semantic Network Meta Index

While the knowledge network can be searched very effectively by constructing an exact query tree using knowledge-model-derived search criteria for querying and mining both explicit and implicit knowledge, building a query requires some understanding of the underlying knowledge model.

To allow for efficient information retrieval without prior knowledge about the knowledge model, the system offers a global full-text index on top of the knowledge graph to allow for unstructured search expressions that essentially return a subgraph of the global knowledge network maintained by the system. The default views configured are used to build this index. Alternative views can be configured for specific semantic object types, allowing users to optimize the index.

The user interaction model is similar to that of familiar Internet search engines. Simple keyword entries, supporting an optional Boolean search syntax, result in retrieval of semantic assertions which can be reviewed both as textual reports and graphically as knowledge networks.

3 Methods

Components of a knowledge system are modeled in the BioXM environment using semantic objects (*see* Subheading 2.1). The flexibility of the system allows semantic objects to be defined and constrained to the knowledge model used. This is summarized in Fig. 1. The process of building a biomedical knowledge network utilizes the components and concepts described in Subheading 2 and maps them to the requirements of a specific application. This is a defined three-step process, which is continuously reiterated.

- *Step 1—Modeling*: Define the domain-specific knowledge model.
- *Step 2—Implementation*: Populate the knowledge model and, thus, instantiate the semantic network with data and information from external resources and user interaction.
- *Step 3—Use*: Use the semantic network by querying, exploring the graph, and reporting.

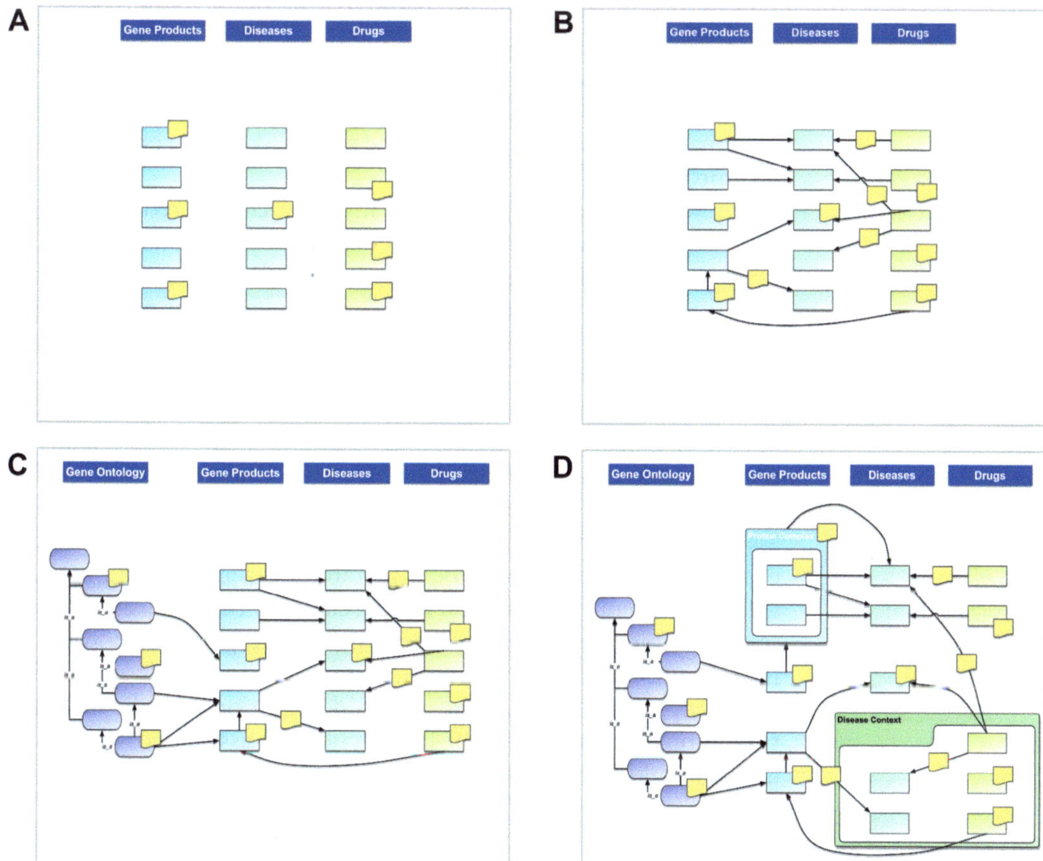

Fig. 1 The BioXM system supports user-defined semantic objects representing elements of a scientific domain. Elements, such as gene products, diseases, or drugs, can be annotated with additional information using configurable forms (**a**). In the BioXM system, knowledge is conceptualized as relationships between elements. Those relations are supplemented by the annotation of evidence, which provides validation (**b**). For the related objects, further validated relations with other elements (such as cell types or diseases) may exist, thus expanding the knowledge network. The conceptualization of entire areas of interest in ontologies like the Gene Ontology or other ontologies allows the use of inherent inference relations for the exploration of knowledge networks (**c**). Specific parts of the knowledge model may be organized in subnetwork contexts (e.g., a particular signal transduction pathway in an organism of interest) allowing for hierarchical structuring of knowledge (**d**). Note that all semantic objects (not only elements but also relations, contexts, ontology concepts, or external database entries) can be annotated with additional information using user-defined annotation forms

The three steps are interdependent and should be conducted in a closed feedback loop (*see* **Note 3**). In the following section, the steps are described using an example.

3.1 Example Application

3.1.1 Scenario

A typical usage scenario for the BioXM Knowledge Management Environment is the integration of clinical research data with information about the molecular background of the disease of interest and the actual results of experiments, e.g., gene expression analyses.

The following example demonstrates how the BioXM system can be configured to manage translational research in a clinical setting, integrating study data with molecular data from various sources. This configured example is then deployed as a "systems medicine" information portal to be used by the intended end users directly in a web browser.

A wide range of information must be considered. The study provides patient data, including detailed information about patient demographics, diagnosis, and treatment. Biopsy material taken from various organ sources requires management of tissue sample information. Furthermore, gene expression analysis experiments have been performed for all tissue samples and must be incorporated. Though analysis of primary experimental data has been done with statistical software packages, the actual results of those analyses should be evaluated within the context of existing knowledge about the molecular processes of involved genes and associated cancers. Existing knowledge about tissue-specific gene expression patterns should be integrated as well, allowing patient-specific genomic information to be related with established tissue-specific biological interaction networks. The genomic context of those genes needs to be accessible to allow for single nucleotide polymorphisms (SNPs) to be analyzed. Additional public information about drugs that might be functionally associated with those genes should be taken into account.

3.1.2 Public Data Sets

The example that is presented here depends on a number of publicly available data sets:

- Human reference genome (RefSeq Homo sapiens GRCh38.p7).
- Physical and genetic protein–protein interactions [8] (BioGRID version 3.4.140).
- DisGeNet Gene–Disease Interactions [9].
- GTEx Gene–Tissue Expression data [10] (GTEx Analysis V6p).

Additional public databases such as EntrezGene, ENSEMBL, UniProtKB are integrated to allow a richer environment of biological entities with associated identifiers.

3.2 Modeling

3.2.1 Configuration of a Knowledge Model

The established BioXM knowledge model represent a "set of rules" describing a particular scientific domain as seen by the scientists. It represents a *hypothesis* of how things interact and work together. This hypothesis will change as the way things are viewed/understood evolves over time. The BioXM system allows the domain model to be changed at any time and provides supporting mechanisms to update existing knowledge according to the changed model.

In implementing a BioXM knowledge model suitable to represent the above scenario, the following entities are defined based on the semantic object type "Element": "Patient," "Disease," "Gene," "Drug," and "Tissue Sample" (*see* Subheading 2.1.1). Classification of biological function and diseases is enabled by importing the Gene Ontology Plus [11], IUPAC Enzyme Nomenclature, the NCI Thesaurus [12], and the Human Disease Ontology [13] as BioXM Ontologies: "GO," "EC," "NCI Thesaurus," and "DOID" (*see* Subheading 2.1.4). The public database "ClincialTrials.gov" is integrated virtually using an embedded version of the BioRS Integration and Retrieval System (*see* Subheading 2.2.6). Gene expression analysis experiments are managed in the user-defined object "Expression Experiment" based on the static object type "Experiment" (*see* Subheading 2.2.4). For more information about building the knowledge model, *see* **Note 1**.

Figure 2 shows an excerpt of the actual knowledge model as configured in the BioXM graphical knowledge model editor. Based on the initial object configuration, the user defines relationships to specify rules about how the objects interact (*see* Subheading 2.1.2). As shown in Fig. 2, BioXM "Relations" have different types which represent different semantics. For example, the relation "Interaction" is defined to connect Genes and is used to represent the BioGRID protein–protein interactions (*see* Subheading 3.1.2).

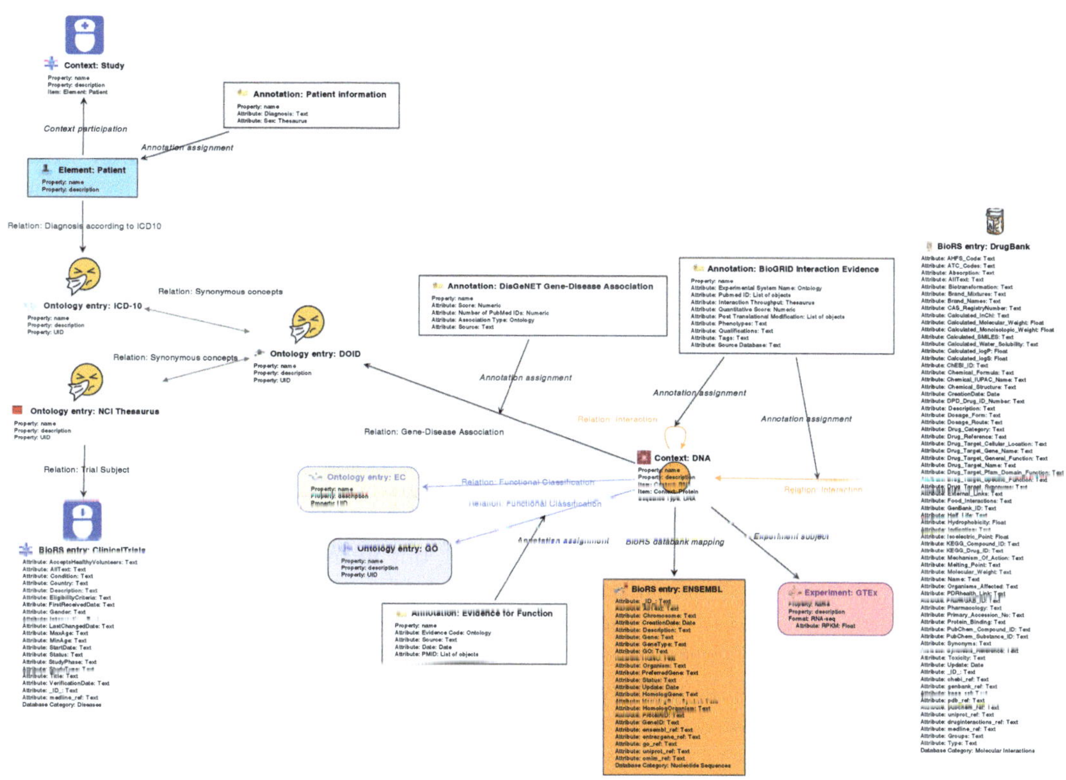

Fig. 2 Excerpt of a knowledge model displaying the relevant semantic objects and relationships for a biomedical research setting

The relation "Gene-disease association" connects drugs with genes based on the content of the DisGeNet data set (*see* Subheading 3.1.2), pointing out potential disease markers. For scenarios like this, there is a lot of metadata associated with each "object of interest" (patients, genes, diseases, gene–disease associations, etc.), which need to be integrated accordingly. For more information about integrating metadata, *see* **Note 2**.

As described in Subheading 2.1.3, the BioXM system offers form-based annotation that allows configuration of any type of property typically needed to describe a scientific entity. Because such scientific entities can also be modeled as relations in the BioXM system, annotation can provide the evidence required to further assess the validity of such relations. This allows, for example, patient demographics to be collected in one annotation form, and experimental information about protein–protein interactions to be assembled in another annotation form.

Annotation forms can be used to supplement all static BioXM object types (such as elements, database entries, relations, ontologies, and contexts) with user-defined properties. Many different attribute types are supported, e.g., simple attributes like "numeric" or complex attributes like "ontologies." Attribute types such as "numerical value with physical unit" allow physical properties to be stored efficiently and accessed with on-the-fly interconversion of units. An attribute type "file" supports the import of images and text files, for example, and is used to annotate the "Tissue sample" element type with microscopy images of histological sections. Imported PDF files are indexed and the user can search within the files' content, which can be used, for example, to provide quick full-text access to electronic pathology reports.

3.2.2 Population of the Knowledge Model

Once a knowledge model is established, populating it is straightforward: one resource at a time is attached to the semantic objects of the model. After an initial model that captures the specifics of an ongoing project or research environment is configured, users can import their own data.

The BioXM system supports direct import of various XML-based (Extensible Markup Language) files such as RDF (Resource Description Framework) or OWL (Web Ontology Language) and other structured file formats, but in many cases tabular data, e.g., Excel spreadsheets, need to be imported. For this example, the study data, the BioGRID data [8] and the RNA-seq data from GTEx [10] were imported as Excel spreadsheets or plain text tables. The BioXM system implements a versatile importer for tabular data, enabling the user to define the semantic of the table columns and graphically build instruction sets ("scripts") guiding the data transformation process. During the import, all information contained in the input data sheet is transformed according to the semantics of the knowledge model. This mapping process between

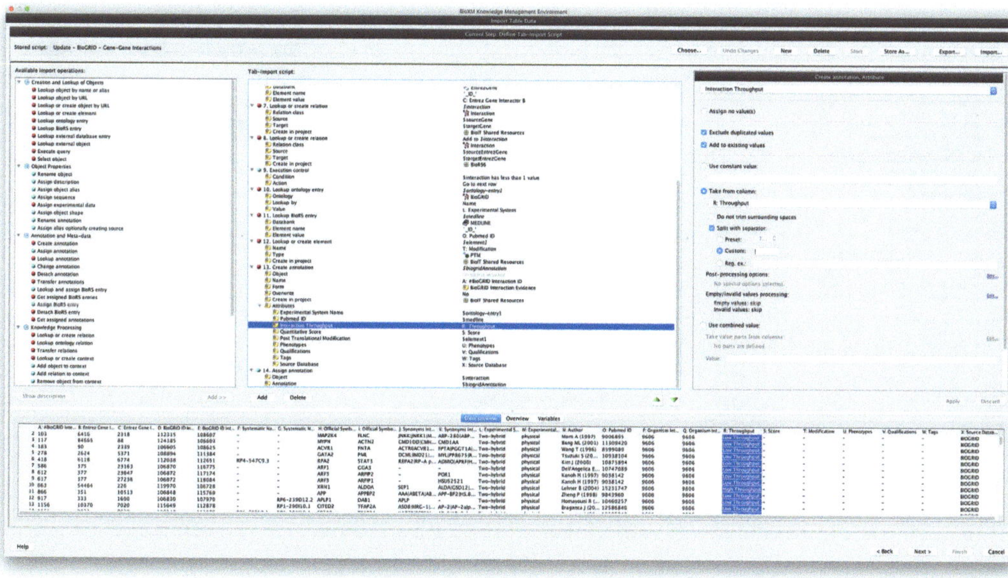

Fig. 3 This screenshot of the BioXM tab importer shows a typical example of an import script to transform the "flat" semantic of tabular data into the network representation of the BioXM system. The script is built by dragging an import operation from the left list of available operations into the growing script located in the *middle panel*. Parameters of single operations can be specified in the *right panel*. A preview of the table to be imported is available at the *bottom* of the *window*. Import scripts can be saved as templates and reused in a simplified import wizard. In addition, import scripts can be used from the system's APIs and published to the web portal framework to be used by end users

the defined knowledge model and the data records ensures consistency. Figure 3 gives an example of how the link between the spreadsheet data and the model is established.

With respect to the clinical study used here, three main tables have been imported. The first table contained extensive clinical information (e.g., demographic data, information about diagnosis, result of lung function tests, and treatment data). The second table contained information about the tissue sample preparation process (e.g., sample quality) with reference to standard operating procedures (SOPs) for Affymetrix gene expression array experiments. The third table contained all primary results from the expression analysis (e.g., expression levels, *p*-values).

Other resources, e.g., disease and treatment information from the ClinicalTrials.gov database are tied to the specific elements and relations through the BioRS system, which makes external data resources accessible (*see* Subheading 2.2.6).

3.2.3 Using the Knowledge Network

Using the knowledge network includes exploring the network through the graph, querying, and reporting (*see* Subheading 2.2). These actions are based on the knowledge model.

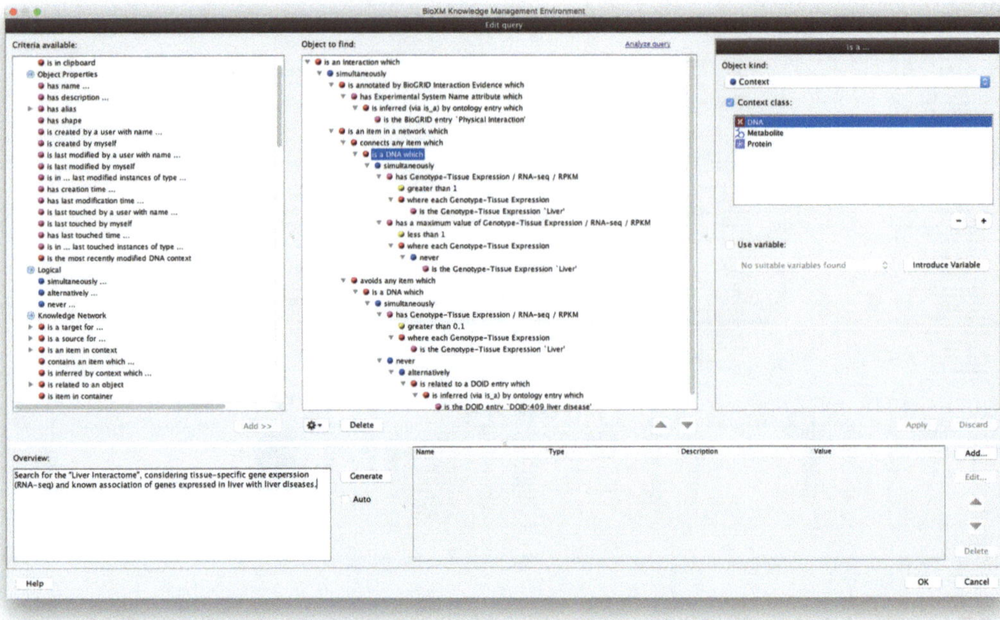

Fig. 4 This screenshot shows the BioXM advanced query builder. Similar to the tab importer, this GUI allows a query to be built graphically (*middle panel*) by using available search criteria found in the *left panel*. The query builder automatically offers only search criteria that are valid in the context of the selected criterion in the *middle panel*. This example generates a "Liver Interactome" network from physical protein–protein interactions (BioGRID [8]) in just one query expression, considering tissue-specific gene expression (GTEx [10] RNA-seq data) and known associations of genes expressed in liver with any liver disease (DisGeNet [9])

An example of the query builder (*see* Subheading 2.2.1), which allows users to take full advantage of the knowledge model with a natural-language representation, is shown in Fig. 4. In it, a biological network is generated representing the "Liver Interactome." The knowledge model defines the query space, and thus any information that is maintained in the system can be found and returned to the user. The query in Fig. 4 spans a substantial portion of the global knowledge network maintained by the BioXM system and puts conditions on what attributes of semantic objects need to be satisfied to qualify as a result.

The result of a query is usually a set of semantic objects that become nodes and edges in a graphical representation of the search result. For each instance of a semantic object, a report with multiple views can be configured (*see* Subheading 2.2.7). Figure 5 gives an example of such a report, which is configured using the populated annotation forms of an element found and related elements. A report is a specific aggregation of the knowledge network from the perspective of the semantic object being reported. The example given in Fig. 5 is configured using the web portal framework (*see* **Note 4**).

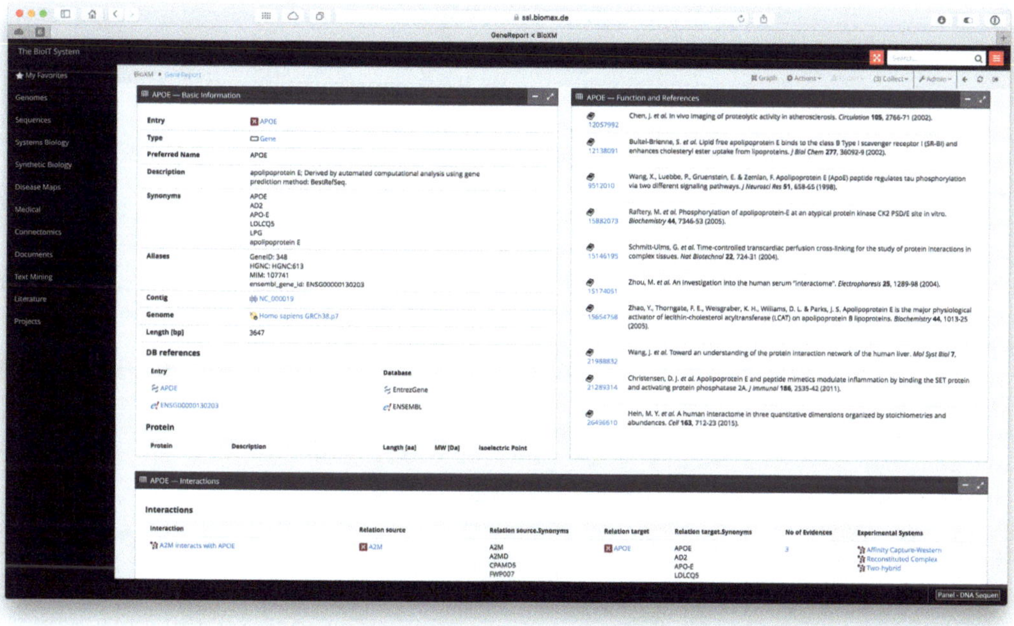

Fig. 5 This report presents information corresponding to the apoE gene, which is part of the "Liver Interactome" as retrieved by the query shown in Fig. 4. The report shown is presented in the web portal framework that was used to configure a dedicated "systems medicine" portal

Knowledge networks are made of relationships between semantic objects. The graph (*see* Subheading 2.2.3) is the interactive visualization of the network. Figure 6 is an example of a network instance based on the designed and populated network, which is expandable at any node of the graph.

Graphs, reports, and queries are the ultimate point of feedback for the user, and the test for a successful design (*see* **Note 5**); however, the intellectual work is the design of the model.

4 Notes

1. On designing the model

   When you design the model, reflect on the fact that you are dealing with a LEGO building-block type of system in which you are allowed to formulate the shape and properties of your pieces. It is important to understand the problem you want to solve and how that reflects on the basic concepts detailed in Subheading 2. This process is similar to an agile software development process, with a focus on the modeling phase. The following questions will help you understand and model your domain: What are the main semantic objects and how do they

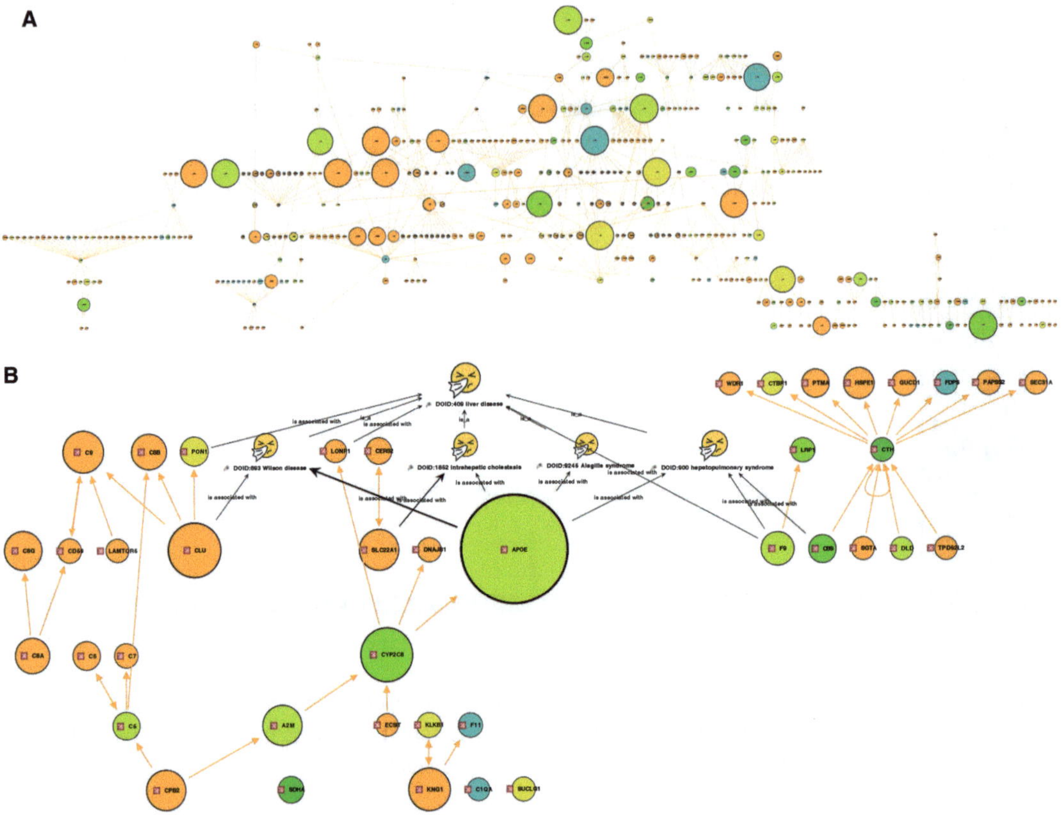

Fig. 6 (**a**) The "Liver Interactome" network retrieved by the query shown in Fig. 4. The network nodes represent genes expressed in liver. The *size* of the node represents the measured median RPKM (reads per kilobase transcript per million reads) for the corresponding gene. The *color* represents the number of drugs known to be associated with the corresponding genes, based on DrugBank information. (**b**) An excerpt of the Liver Interactome expanded to show known disease interactions. The focus is on the apoE genes, for which the corresponding report is shown in Fig. 5

relate? What attributes are relevant and what questions would you like to answer with the system? Stay close to your scientific domain and your scientific objects of interest, do not compromise on clarity of your design for technical reasons in your first iterations. Once your design is validated, i.e., the implementation starts to provide relevant answers to your scientific questions, optimize your knowledge model in additional iterations.

2. On data definition and populating the model

   Start simple: for example, "a gene expresses a protein." In the BioXM system, this means you need to model elements and a relationship. What attributes constitute a gene: the unique name, the species, and perhaps the chromosomal location? For the protein this is similar; What attributes constitute a protein: a name and functional properties? A gene is usually represented by an identifier in a specific database, but what do you do if a gene does not exist in that database? Use databases to populate your knowledge network, not to determine the semantics of the network. Make sure the names are always readable and meaningful to the user, not necessarily to the modeler of the system. Be aware that names reflect identity both from a design point of view (e.g., uniqueness) and the scientist's point of view (e.g., common use).

It has been said that when you have two scientists, you will have three opinions on how a gene should be named. If this happens, do not try to follow the user blindly, follow the anticipated usage and try to reflect the diversity of opinions. If nomenclature is disrupted, you could say there is one gene with one name in each species and take (or make) an ontology or defined vocabulary, which reflects the standard. That ontology can be used to assign the name and all other variants can be indicated as synonyms. The synonyms will be treated as equivalent to the given name within the system. When representing the gene, information from different resources and synonyms may be combined in multiple representations in different report views.

Note, if your elements are not well defined within the BioXM system there is the risk of ambiguity. Ambiguities may have consequential effects, because elements have many information resources available and when you extend the model and populate it, "ambiguous" elements become your anchors for new data or elements.

3. An interactive process

   Once you have an element or a relation or a question, start to populate the knowledge network early. This makes the knowledge network more concrete with respect to the intended purpose and allows for feedback. Try to make the full round-trip cycle of the three-step development process as short as possible. Embrace the Manifesto of Agile Software Development [14]: interactivity is important. The faster the iteration cycle moves the better. See yourself as the translator, who describes the world of science in the BioXM system and mirrors it back to science. The theme is to listen, think, and act. Use the knowledge network to explore ideas and hypotheses. Imagine yourself in an ultra-extreme programming environment, only that you do not write code, but configure your knowledge system.

4. On creating web apps for the intended users

   You are likely implementing applications for a larger group of intended users (sometimes called "end users"). Keep in mind that their usage scenarios differ significantly from those of any user implementing and maintaining a solution. Their main focus is not to model and build the knowledge network, but to address their main scientific questions. They need to benefit from your efforts—ideally effortlessly without the need to dive deep into the details of the implementation.

The BioXM system provides the easily configurable portal system for deployment of web apps targeting specific use cases. When you configure such a browser-based web app, you should deliberately change to the end-user perspective. Your intended users need to solve very specific problems, try to address them as specific as possible. If necessary, configure "wizards" that guide the users through a sequence of interactive steps leading to the intended result. While focusing on the narrow needs of the end-user-specific application, you can leverage the broad scope of the knowledge network maintained by the BioXM system. Hide its complexity, but use it to enable end users to get easy answers to difficult questions.

5. A good design metric

   Maintain a close link to the questions you want to answer. Reflect these questions in queries and easy-to-use smart folders. The best guidance is to continuously validate your design: Do I get meaningful answers to my scientific questions?

Acknowledgments

The ideas and concepts outlined in this chapter have evolved over an extended period of time and have benefited from discussions with numerous friends and colleagues. The authors would especially like to thank Wenzel Kalus and Martin Wolff. Without their work the BioXM system would not have become a reality in its current form. The authors would also like to thank Sheridon Sauer for her very helpful assistance during the work on the manuscript.

References

1. Searls DB (2005) Data integration: challenges for drug discovery. Nat Rev Drug Discov 4:45–58
2. Mukherjea S (2005) Information retrieval and knowledge discovery utilising a biomedical Semantic Web. Brief Bioinform 6:252–262
3. Kashyap V (2003) The UMLS Semantic Network and the Semantic Web. AMIA Annual Symposium proceedings/AMIA Symposium AMIA Symposium, pp 351–355
4. Losko S et al (2006) Knowledge networks of biological and medical data: an exhaustive and flexible solution to model life science domains. In: Data integration in the life sciences, Lecture notes in computer science, vol 4075. Springer, New York, NY, pp 232–239
5. Settles B (2005) ABNER: an open source tool for automatically tagging genes, proteins and other entity names in text. Bioinformatics 21:3191–3192
6. Rocktäschel T, Weidlich M, Leser U (2012) ChemSpot: a hybrid system for chemical named entity recognition. Bioinformatics 28:1633–1640
7. Kaps A et al (2006) The BioRS™ Integration and retrieval system: an open system for distributed data integration. J Integr Bioinform 3
8. Stark C et al (2006) BioGRID: a general repository for interaction datasets. Nucleic Acids Res 34:D535–D539
9. Piñero J et al (2015) DisGeNET: a discovery platform for the dynamical exploration of human diseases and their genes. Database (Oxford) 2015:bav028–bav028

10. GTEx Consortium (2015) Human genomics. The Genotype-Tissue Expression (GTEx) pilot analysis: multitissue gene regulation in humans. Science 348:648–660
11. Gene Ontology Consortium (2015) Gene Ontology Consortium: going forward. Nucleic Acids Res 43:D1049–D1056
12. Sioutos N et al (2007) NCI Thesaurus: a semantic model integrating cancer-related clinical and molecular information. J Biomed Inform 40:30–43
13. Kibbe WA et al (2015) Disease Ontology 2015 update: an expanded and updated database of human diseases for linking biomedical knowledge through disease data. Nucleic Acids Res 43:D1071–D1078
14. Fowler M, Highsmith J (2001) The agile manifesto. Software Dev 9(8):28–32

Chapter 17

Knowledge-Based Compact Disease Models: A Rapid Path from High-Throughput Data to Understanding Causative Mechanisms for a Complex Disease

Anatoly Mayburd and Ancha Baranova

Abstract

High-throughput profiling of human tissues typically yields the gene lists composed of a variety of more or less relevant molecular entities. These lists are riddle by false positive observations that often obstruct generation of mechanistic hypothesis that may explain complex phenotype. From general probabilistic considerations, the gene lists enriched by the mechanistically relevant targets can be far more useful for subsequent experimental design or data interpretation. Using Alzheimer's disease as example, the candidate gene lists were processed into different tiers of evidence consistency established by enrichment analysis across subdatasets collected within the same experiment and across different experiments and platforms. The cutoffs were established empirically through ontological and semantic enrichment; resultant shortened gene list was reexpanded by Ingenuity Pathway Assistant tool. The resulting subnetworks provided the basis for generating mechanistic hypotheses that were partially validated by mined experimental evidence. This approach differs from previous consistency-based studies in that the cutoff on the Receiver Operating Characteristic of the true–false separation process is optimized by flexible selection of the consistency building procedure. The resultant Compact Disease Models (CDM) composed of the gene list distilled by this analytic technique and its network-based representation allowed us to highlight possible role of the protein traffic vesicles in the pathogenesis of Alzheimer's. Considering the distances and complexity of protein trafficking in neurons, it is plausible to hypothesize that spontaneous protein misfolding along with a shortage of growth stimulation may provide a shortcut to neurodegeneration. Several potentially overlapping scenarios of early-stage Alzheimer pathogenesis are discussed, with an emphasis on the protective effects of Angiotensin receptor 1 (AT-1) mediated antihypertensive response on cytoskeleton remodeling, along with neuronal activation of oncogenes, luteinizing hormone signaling and insulin-related growth regulation, forming a pleiotropic model of its early stages. Compact Disease Model generation is a flexible approach for high-throughput data analysis that allows extraction of meaningful, mechanism-centered gene sets compatible with instant translation of the results into testable hypotheses.

Key words Signature, Network, Knowledge-based algorithms, Alzheimer's, Protein traffic vesicles, Affymetrix, Illumina, Antihypertensive drugs

---

Electronic supplementary material: The online version of this chapter (doi:10.1007/978-1-4939-7027-8_17) contains supplementary material, which is available to authorized users.

Tatiana V. Tatarinova and Yuri Nikolsky (eds.), *Biological Networks and Pathway Analysis*, Methods in Molecular Biology, vol. 1613, DOI 10.1007/978-1-4939-7027-8_17, © Springer Science+Business Media LLC 2017

1 Introduction

In developed economies, the costs of medical services are constantly rising, stifling the economic growth and projecting to become unsustainable if the trend remains unchanged [1, 2]. Some solutions propose the shift of the focus to early diagnostics of the diseases with the highest societal impact, to designing the strategies for reliable risk assessment and to tailoring prophylaxis efforts to the highest risk groups [3]. Another approach seeks to streamline the process of drug development by focusing the effort on the most promising targets and preclinical drug candidates. Both solutions may be assisted by the methods of bioinformatics and chemoinformatics that operate within the realm of systems biology [4–6].

The most common type of the data analyzed by bioinformaticians is a set of differentially expressed genes obtained by microarray or RNAseq. Typical candidate list derived from these kinds of studies contains hundreds to thousands differentially expressed genes. However valuable, these sets are riddled with false positives that changed their expression levels due to compensation for an overall increase in cellular stress or as a secondary effect of certain regulatory events, for example, the suppression of transcription factor activity or the shift in histone modification landscape. In other words, the differential expression of given gene often is a passive consequence of stress rather than a critical event directly contributing to disease pathogenesis.

Obviously, the focus of the research efforts should be on genes most essential in pathogenesis of given disease. However, this focusing is not trivial, as every chronic disease is studied by multiple research groups that customarily formulate multiple competing hypotheses [7], thus populating the lists of potential candidates with thousands of entries. For Alzheimer's disease alone, the Gene Cards compiled by Weizmann Institute of Science list 1890 molecules of relevance [8]. With <25,000 genes known to comprise a genome, and no more than a third of them being expressed in a single tissue [9], this number is indicative that the long gene lists of today reflect rather poor target prioritization. Thus, there is a need for highly prioritized shortlists of potential targets directly linked to major pathogenic processes. Such lists, contracted by ontological enrichment, reexpanded by interaction network, and validated by network clustering and alignment with literature, were termed here Compact Disease Models (CDMs).

The reproducibility of a result in an independent experiment with at least slightly varied technical settings is the typical verification criterion for any scientific derivation [10]. In accordance to that, the gene-specific probes differentially expressed in the same direction in independently analyzed multiple subsets of the same experimental dataset and also in different experiments are less likely to report noise. Filtering of biological signals by consistency of gene-expression changes already demonstrated its value for enrichment analyses of genes mechanistically important for tumorigenesis [11, 12] and metastasis [13]. As compared to gene lists generated using *t*-test, the lists generated using consistency of differential expression are target-enriched [4, 11–14] and, thus, are more mechanistically relevant. For example, a reliable cancer mortality signature was produced by meta-analysis for the consensus changes observed in a variety of experiments across a number of model organisms [15]. An enrichment of gene expression signatures with mechanistically relevant targets was also attempted for neurodegeneration studies, such as Alzheimer's and Parkinson's diseases [16].

While any enrichment technique is capable of demonstrating the target enrichment, the utility of this enrichment is determined by the Receiver Operating Characteristic (ROC) of the process and the point of ROC cutoff. Importantly, nonoverlapping components of individual datasets may be disease-specific while remaining related to pathogenic mechanism. Therefore, the requirement of consistency has to be imposed with minimally stringent cutoffs [17]. Here we present an approach that provides an improvement over previously described techniques. In that, we implemented sorting out the gene lists into the Consistency Tiers, thus gaining control of the extent of information loss in the nonoverlapping subsets. Each Tier can be assessed further by functional enrichment and alignment with independent literature data. The optimal size of the consensus signature could be selected depending on the nature of the disease [17]. Altogether, our process includes a three-step noise filter composed of (1) prioritization of candidates by consistency of reported directional gene expression changes, (2) functional enrichment and (3) co-clustering of candidates in a network [18].

The resultant Compact Disease Model (CDM) provides a significant saving of research effort. Assuming N independent platforms being included in given analysis and m intersections needed to provide a robust mechanism-related gene list, the number of potential contributions becomes:

$$\mathrm{REC} = {C_N}^m P(m) \tag{1}$$

where REC is the recall number (the number of totally available true positives), ${C_N}^m$ is the number of contributing platform combinations, and $P(m)$ is the number of strong mechanistic associates extracted per a single platform combination. In this case, the multiplatform nature of analyzed datasets would compensate for a low ROC curve area observed due to low recall (yield) component, while enrichment is high. Additional platforms to consider are comparative proteomics and quantitative PCR studies, massive parallel genome sequencing, promoter methylation arrays or others. To further improve the recall rate and polish the mechanistic details, the aggregated gene lists produced by all platforms combined are subjected to an interaction network building algorithm, subnetwork identification and detailed assessment of the most relevant subnetworks by literature review.

Compacting mechanistically relevant genes into distilled short-lists may have a major impact on the routine verification of individual mechanistic hypotheses. Assume that a hypothesis H assigns correct connectivity between the functions X, Y, Z, the X being a receptor, Y being a G-protein and Z being a kinase. Relevance of the gene list is measured by factor of q, where q is the decimal fraction of bona-fide mechanistically relevant genes in the total list. The assignment will receive experimental verification only if all members X, Y, Z are bona-fide mechanistically related. For a 3-member sequence, the relationship is PEC $= q^3$ which can be generalized into:

$$\mathrm{PEC} = q^n \tag{2}$$

where PEC (probability of experimental correctness) is the probability that the mechanistic hypothesis is correct for a n-member sequence; q is the distillation factor of the list, n is the number of steps in a sequential mechanistic hypothesis. Under all other factors and techniques being equal, the exclusion of false positives from the gene lists is especially important for the mechanisms studied to a lesser degree (low g) and for complex hypotheses (high n).

To test CDM approach, we selected an example of Alzheimer disease, the most common form of adult-onset dementia. We were especially interested in addressing the earliest stages of this disease, when the pathological changes are still reversible and/or preventable. The particular focus of our analysis was at previously demonstrated anti-Alzheimer effects of antihypertensive drugs [19–22]. In our study, an application of CDM resulted in the distilled, tiered list of Alzheimer's disease-related genes integrated into a biological network model. As potential players in early disease, a group of genes that encode proteins associated with traffic vesicles, oncogenes, G-protein regulators, gonadotropin hormones and insulin-related signaling molecules was identified. Insights gained by an analysis of this CDM may aid in shifting the therapeutic efforts to the reversible stages of neurodegenerative disease, when the neuronal damage is mild and self-perpetuating misfolded protein oligomers do not yet form.

2 Materials

2.1 Selection of Datasets

Datasets for the study included (A) GSE5281 on GPL570 HG-U133 Plus 2 Affymetrix Human Genome U133 Plus 2.0 Array including 71 normal controls and 91 disease related samples (N = 162); (B) GSE15222 on GPL2700 Illumina Sentrix Human Ref-8 Expression Bead Chip, including 187 normal controls and 176 disease samples (N = 363); and (C) GSE26927 on GPL6255 Illumina human Ref-8 v2.0 expression bead chip platform, including 58 normal controls and 60 disease samples (N = 118). The latter dataset comprises differential expression data covering several neuropathies: Alzheimer's disease; Amyotrophic lateral sclerosis (ALS); Huntington Disease (HD); Multiple Sclerosis (MS); Parkinson Disease (PD); and Schizophrenia (SHIZ) of approximately equal size. The patient histories and disease severities were extracted from the information that accompanies the public domain datasets at GEO, NCBI at http://www.ncbi.nlm.nih.gov/geo/. Other datasets covering Alzheimer's disease and dementia on GPL96 and GPL90 Affymetrix platforms were explored but not included due to failure to pass the quality controls, namely, large number of missing genes, evidence of data imputation or evidence of weak hybridization/weak signal. The primary data describing datasets A, B, and C are presented in Additional file 1.

2.2 Forming of a Distilled Gene List (CDM): Consistency Profiling Step

The dataset GSE5281 comprises several distinct tissue subsets: EC—entorhinal cortex; HIP—hippocampus; MTG—Medial Temporal Gyrus; PC—Posterior Singulate; SFG—Superior Frontal Gyrus; VCX—Primary Visual Cortex, each being composed of control and Alzheimer's disease samples. The expression values were averaged for each anatomical locus for norm and disease. The averaged signal intensities were sorted by their magnitude and the upper 40% of the entries were included in the analysis on assumption that the expression levels for the remaining low-intensity signals is unlikely to exceed experimental noise. The ratios of the averages produce either upregulated or downregulated fold change values. The primary data were subjected to two-tail, different variance hypothesis T-tests between normal control and disease subsets for each brain tissue type. The p-values of these T tests were converted into negative logarithms and the logarithms were averaged across all *tissue types*. For GSE5281, these averages formed the Primary Consistency Scores (PCSs). In addition to differential expression, for each gene, absolute expression levels were also tracked.

The dataset GSE15222 comprises normal controls separated into two subsets, numbers 1–85 and 86–178. The disease samples were also separated into two subsets, numbers 1–85 and 86–176.

Absolute averaged expression levels were computed for each normal and disease subset, separately. Similarly to analysis of previous dataset, the upper 40% of entries by their expression level intensities were considered significant and included in the analysis. The difference in expression between the normal and disease subsets was assessed by *T*-tests as described above to compare each disease subset to each normal subset, four separate values were produced, and the negative decimal logarithms of *T*-test *p*-values were computed. The average of four negative logarithms formed Primary Consistency scores for GSE15222.

All gene-specific labels in GSE5281 and GSE15222 were ranked according to their Primary Consistency scores (PCSs) and the top 10% were selected. The highest ranking probes in GSE5281 and GSE15222 were assigned to a Consistency Tier 3, if the functionally related molecules (members of the same pathway) were also displaying high PCS. Assignment to Consistency Tier 2 was made in either of two situations: (a) two Affymetrix probes representing the same gene were displaying high rank PCS, and the direction of differential change was the same for both probes (all downregulated or all upregulated) in the group; (b) Affymetrix and Illumina probes representing the same gene were displaying high rank PCS and the direction of differential change was identical for both platforms. Consistency Tier 1 was assigned if either of three situations: (a) to the genes that displayed high PCS on both Affymetrix and Illumina platforms as well as multiple probes on Affymetrix platform, when the direction of differential expression changes was the same for all gene-specific probes; (b) to the probes that simultaneously qualify for Tier 3 and Tier 2; (c) to the three or more probes on Affymetrix platform that simultaneously showed high PCS ranking and the direction of expression changes was identical for all probes. Tiered consistency scores for all scored genes are available as the dataset D of the Additional file 1. The Tier 0 was produced by overlapping the Tier 1 and Tier 2 genes with the highest PCS ranks of GSE26927, thus identifying a group of genes commonly participating in a number of neuropathies in addition to Alzheimer's disease.

3 Methods

3.1 Forming of a Distilled Gene List (CDM): Ontological Enrichment Analysis Step

Quantitative ontological analysis was performed using GO-MINER tool (http://discover.nci.nih.gov.gominer) using high-throughput online computing option at http://discover.nci.nih.gov/gominer/GoCommandWebInterface.jsp"GoCommandWebInterface.jsp. This technique measures preferential enrichment of the differentially expressed gene lists in one or more of approx. 9300 functional categories, organized in a tiled partially overlapping manner, with smaller specific categories merging into greater ones. To compute enrichment in a given category, the genes that are known to be related are tracked in the differential expressed gene list and in the total list. The enrichment coefficient can be estimated as:

$$\mathrm{ENR} = [\mathrm{CG}/L]/[\mathrm{TG}/T] \qquad (3)$$

Where: ENR—enrichment coefficient, CG—the number of genes with detected expression changes in a given functional category in the experimental gene list L, L—the number of genes in the experimental gene list, TG—total number of genes in a given functional category, T—the total number of genes assessed. Robustness of the enrichment coefficients is established by permuting the composition of L and expressed as p-value and False Discovery Rate (FDR). The current implementation of GoMiner uses a one-sided p-value calculated from a Fisher's exact test. To get a low p-value, good enrichment and a fairly large size of category are required. The FDR approach addresses the multiple comparison problem, and protects against over-interpreting p-values that do not have a biological meaning.

The combination of Affymetrix and Illumina probe populations was used as the "Total file" or T. Since highly expressed genes are more likely to produce consistent differential expression signatures, the total file (T) was normalized to ensure equal average expression level as compared to the gene lists (L) under study, compensating this bias. Specifically, the total list in each case was ranked by expression levels and the upper rank populations of T producing equal averages to the given L were retained as expression-adjusted total files, and the rest were discarded from the analysis, effectively decreasing the number of genes in T. Tier 0, Tier 1, Tier 1 + Tier 2, and Tier 1 + Tier 2 + Tier 3 gene lists were used as "Changed file." An option "All" was elected for "Data source." Evidence Code was elected as "All," accepting either experimental, curator inferred or computed data of functional involvement. Lookup setting for gene searching in the GO Consortium database was accepted as achieved by both cross-referencing and by use of synonyms. Both p-value of a functional enrichment category and false discovery rate (FDR) were elected as statistical criteria for including the qualifying genes in the summary report. The prospective functional enrichment categories were validated by 100 randomization cycles according to GO-MINER protocol. The smallest size for a functional enrichment category was accepted as 5. The GO-MINER output was sorted by FDR with the cutoff FDR <0.2. The functional categories with the lowest false discovery rates were re-sorted by enrichment coefficients in the descending order. The relative functional enrichment coefficients reflect the extent of association of the differentially expressed genes with the pathological mechanism that caused the differential expression event in first place. The outputs of GO-MINER ontological analysis to the genes within Consistency Tiers 0–3 are available in the Additional files 2, 3, 4, and 5 datasets G–J.

The extent of ontological enrichment provides a cut-off for selection of the Consistency Tier levels to be submitted to network association step. The Tier 1 provided a conditionally optimal ROC cut-off due to high ontological enrichment and preserved pathway diversity. The Tier 1 + 2 + 3 was accepted, but considered less preferential due to a substantially lower proportion of mechanistically relevant genes based on ontological enrichment step.

3.2 Forming of a Distilled Gene List (CDM): Network Analysis Step

To organize sets of genes into biological networks, Ingenuity Pathway Assistant (IPA) tool was utilized (http://ingenuity.com/). Briefly, the tool places a gene list in the context of experimental and computed interactions systematized in its database. The functional links between the members of a gene list under study form a network with high clustering coefficients for members of the same biological pathway while clustering coefficients for random associates are low. Indeed, the members of the same pathway must be functionally associated with multiple other members of the same pathway, directly or via intermediates, thus producing nonrandom clustering. The extent of observed clustering is compared with a random model and the extent of observed clustering is expressed as a p-value of a network. The p-value matches a probability that the associations in the network have emerged randomly. The network is partitioned into subnetworks based on global optimization of clustering when a gene under consideration is shifted between the subnetworks as a test. The formed subnetworks are ranked based on the score, the latter being the negative decimal logarithm of subnetwork nonrandomness p-values.

The sets of subnetworks were built using gene lists comprising Consistency Tier 1 and a joint list composed of all three numbered Tiers (Tier 1 + Tier 2 + Tier 3), the latter as a benchmark control to illustrate the loss of the priority rank by the subnetwork comprising the genes relevant to neurological diseases. In each analysis, the highest ranking subnetworks were selected, merged and plotted as connectivity graphs. The genes displaying experimentally observed differential expression were co-plotted with sets of known network interaction partners. The possible network hubs were expanded, producing additional connections to more distant members. The Additional files 6 and 7 comprise the subnetwork compositions for the Tier 1 and (Tier 1 + Tier 2 + Tier 3), including both experimental and inferred members.

3.3 Validation of CDM by Semantic Tag Enrichment Analysis

The quantitative evaluation of enrichment of the microarray-derived dataset with literature-derived associations was applied as a validation criterion. Each gene lists was converted into Boolean (OR) statement, for example: (gene name 1) OR (gene name 2) OR ...etc. and used as a search query in Pubmed. The hits produced by PubMed were defined as Total. Additional delimiting search queries were imposed: A. ((disease or pathology or disorder)); B. (cancer); C ((disease or pathology or disorder) and stress; D. ((disease or pathology or disorder) and (Alzheimer's or Alzheimer or neuropathy or neuropathic or neuro-degeneration or neurodegeneration or neurodegenerative or dementia)). The numbers of hits for each delimited strategy were enumerated and related to the number for the total list based on gene names only. Variation in the datasets was taken into account by dividing each consistency tier list into subsets and repeating the procedure independently for each subset, pooling the variation and distributing it equally per each subset (within each consistency tier).

In this technique, both (cancer) and ((disease or pathology or disorder)) strings were used as controls accounting for nonspecific organism or tissue-level stress that generally accompanies any severe pathology, while the string ((disease or pathology or disorder) and stress) was controlling for explicit gene association with stress in pathological conditions and the string ((disease or pathology or disorder) and ((Alzheimer's) or (Alzheimer or neuropathy or neuropathic or neuro-degeneration or neurodegeneration or neurodegenerative or dementia)) was controlling the expected specific association of the gene lists and the disease of interest.

3.4 Validation of CDM by Global Differential Expression Pattern Consistent with Broad Mechanistic Picture of the Disease

Affymetrix GPL570 platform comprises approximately 54,000 probes, while Illumina platform comprises ~22,000 probesets. Of the ~24,000 independent expressed genes measured by both platforms, 78 sets were satisfying criteria of the Tier 0, 105 sets were satisfying the criteria of the Tier 1, 85 sets were satisfying the criteria of the Tier 2, 450 sets matched the Tier 3 and 1298 sets were demonstrating high PCS without being validated by other consistency criteria. On both platforms, the genes within the top 40% range by their absolute expression served as random control. Of the 190 probe-sets in Tier 1 + Tier 2, the Tier 0 comprised 78, indicating that >40% of genes robustly reported as being differentially expressed in Alzheimer's disease also produced robust detection in other neuropathies in agreement with [16]. In all Consistency Tiers, the fold differences of differential expression effects were relatively small, rarely exceeding threefold. In Tier 1, 20 out of 105 probe-sets were upregulated and the remaining 85 being downregulated. In Tier 2, 12 out of 97 probe-sets were upregulated, the remaining 85 being downregulated. In Tier 3, the downregulated pattern was shown by 316 genes and 176 genes were upregulated. In the high PCS/unconfirmed group, 715 genes were downregulated and 570 were upregulated. In random control, the ratio of upregulation and downregulation signals was close to 1. The extent of relative downregulation was strongly correlating with the extent of differential expression detection consistency. These numbers show a greater tendency for downregulation in

Alzheimer's disease-related genes and support functional significance of the consistency profiled gene lists, in agreement with degenerative character of the Alzheimer's process [23].

3.5 Exclusion of Transcript Copy-Number Bias in Producing a Condensed Gene List for CDM Building

The absolute expression levels were also positively correlating with consistency of differential expression detection. Thus, Tier 0 average signal was ~4300 arbitrary units, the remaining (Tier 1 + Tier 2) signals were, after exclusion of Tier 0, at ~2330 arbitrary units, while the random control genes were at ~1725 arbitrary units for the top 40% of ranked intensities and ~730 arbitrary units for the entire array (*see* the Dataset C in Additional file 1). The three- to fivefold increase in average absolute expression in the Consistent gene lists vs. Random Control may be an artifact: the genes with higher expression levels may also display higher signal-to-noise ratio at hybridization. Also, at a higher concentration of transcript the thermodynamic quotient and Gibbs energy of binding increases. For genes with higher expression levels the relative proportion of binding at nonspecific sites is lower. However, it is also known that mechanistically important targets tend to be the hubs of the biological network that also tend to be expressed at higher levels than non-hub entities [14, 24]; this feature of biological networks provides additional robustness [25, 26]. However, to account for possible gene intensity bias, further functional enrichment analysis was conducted after respective normalization. Specifically, the Consistent gene list (Subheading 3.2) was compared with a sample of Random gene list with the average copy number equal to the copy number of the Consistent gene list. Both sets were subjected to Ontological Enrichment step (Subheading 3.3). If the Consistent gene list emerged due to above listed artifacts, it would have demonstrated ontological enrichments comparable to the magnitude and *p*-values of the enrichments in the copy-number adjusted random control, emerging due to random drawing of the gene population. In fact, the resulting ontological enrichments for the copy-number adjusted control did not differ substantially from non-adjusted random controls, but differed dramatically from the Consistent gee list. Thus, a confident conclusion can be made that the consistency profile does not emerge due to higher copy-number bias and corresponding artifacts. On the contrary, the differential expressed genes are the network hubs or stand in proximity to the hubs and thus are relatively overexpressed to ensure greater network robustness.

3.6 Exclusion of Low Variation Bias in the Consistency Selected Gene Lists for CDM Building

Another concern was a possibility of a bias due to decreased inherent variation within consistently reporting gene sets, as could be expected for tightly regulated pathways. If this is true, the consistency in differential expression of these genes would reflect not a prevalence of their biological relevancy, but lower levels of respective backgrounds. To rule this scenario out, variations were

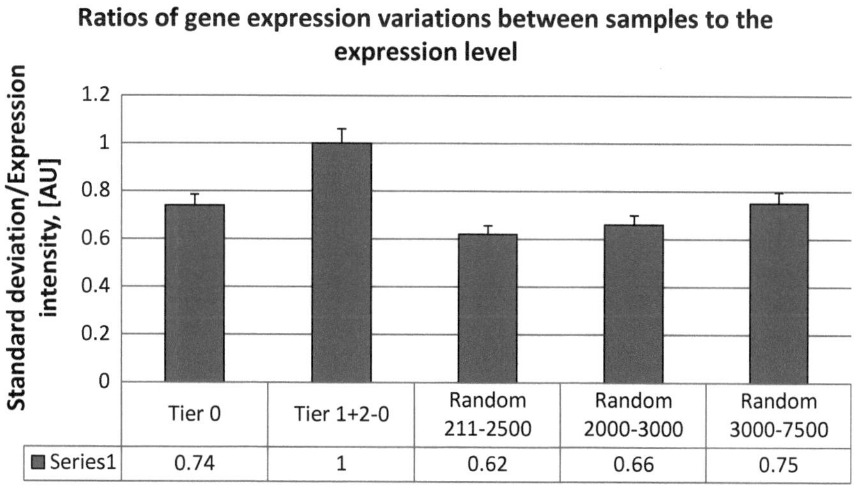

Fig. 1 Tier 0 and Tier (1 + 2) genes differentially expressed in Alzheimer's disease and other neuropathies are compared with significantly expressed random genes on Illumina platform, dataset C. Tier 0 is produced by an overlap of Tier (1 + 2) in Alzheimer's disease panel (Datasets A and B) with the multiple neurodegeneration disease panel (Dataset C). Tier (1 + 2–0) is produced by the balance of Tier (1 + 2) genes with the subtraction of Tier 0. Random genes ranked by intensity were sampled based on the position in the rank. Expression intensities in the groups of genes formed as described above were measured and plotted

measured among all Consistency Tiers and were compared to variations observed in Random Control genes both globally and in the subsets selected by matching of their expression intensities. The results are presented in Fig. 1. This analysis points at higher variation in consistent differentially expressing datasets. Thus, the biased scenario was ruled out and the consistency of the gene expression changes, indeed, was found to reflect the difference between the disease and the norm.

3.7 Exclusion of Stress Response Bias in the Consistency Selected Gene Lists for CDM Building

Still, there might be a concern that the differentially expressed data represent stress responses at both organism and tissue-specific levels, in other words, the responses expected to be pertinent to any severe pathology rather than to reflect a disease-specific mechanism. Figure 2 shows the extracted Consistency Tiers as analyzed by the methodology described above. The method comprises the Boolean presentation of the gene list crossed with the delimiting statement reflecting either association with a nonspecific stress or with a specific disease. The statements like ((disease or disorder or pathology)) crossed with the corresponding Boolean representations of the consistency tiers would locate the literature publications associating the genes of interest with any disease, nonspecific to the study. The query statements like (cancer) crossed with the corresponding Boolean representations of the consistency tiers would locate the literature publications associating the genes of interest with cancer as another proxy for a nonspecific multiple

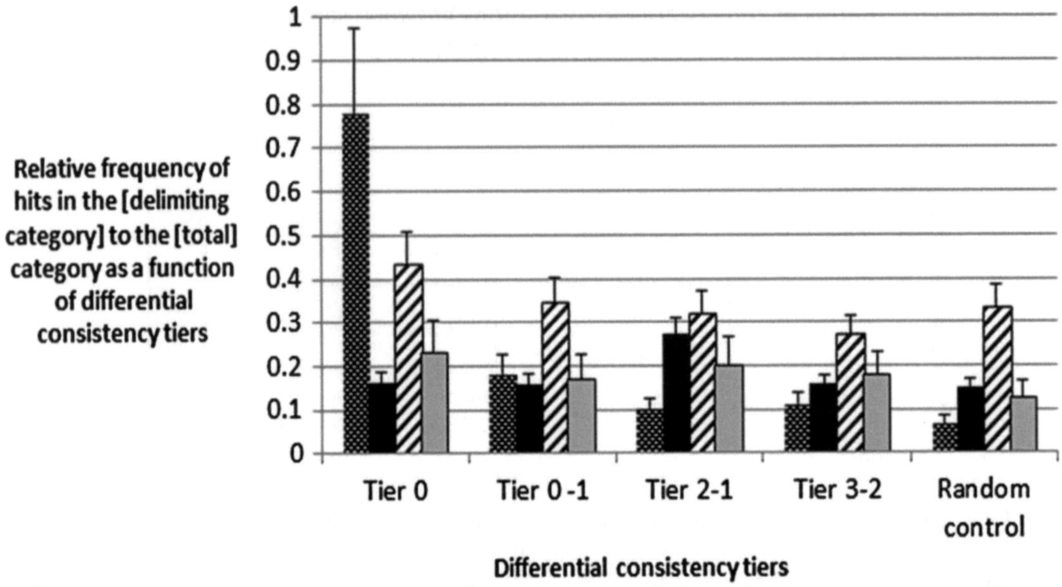

Fig. 2 Dependence of the relative enrichment in literature-inferred gene roles as a function of detection consistency. For each gene list (Tier 0, Tier 0–1, Tier 2–1, Tier 3–2, Random control), the gene symbols were converted into a Boolean representation (simply connected by the operator (OR)). Each Boolean-converted list was used as a query in PubMed and the number of hits was detected. The primary query for each gene list was modified by four subqueries, from *left* to *right*: *Checkered bars*: (gene list) + ((disease or pathology or disorder) and (Alzheimer's or Alzheimer or neuropathy or neurodegeneration)); *black bars*: (gene list) + (cancer); *striped bars*: (gene list) + ((disease or disorder or pathology)); *grey bars*: (gene list) + ((disease or pathology or disorder) and stress). The modified queries produced the numbers of hits smaller than the number of hits produced by undelimited gene list in Boolean form. The ratios of the database responses for the modified vs. unmodified query were plotted for each group of four bars representing a consistency tier. The relative frequencies (ratios) for the queries ((disease or pathology or disorder) and stress) and ((disease or pathology or disorder) and (Alzheimer's or Alzheimer or neuropathy or neurodegeneration)) were multiplied by 10 for convenience of representation and analysis. The Tiers 0–3 represent the lists of genes obtained as disclosed in the Methods; the Tier 1–0 is the result of subtracting the Tier 0 list from the Tier 1 list; the Tier 2–1 is the result of subtracting the Tier 1 from the Tier 2 list; the Tier 3–2 is the result of subtracting the Tier 2 from the Tier 3 list; the Random control set was obtained by randomly selecting the genes among Affymetrix and Illumina total lists and the list is not expression intensity normalized

roles played by many signaling molecules. The statement relating the gene list of interest to stress-response comprises the negative control. Thus, relative enrichment of the disease-specific vs. disease nonspecific PubMed hits for certain levels of consistency would represent a measure of ensuring the mechanistic involvement of the genes in the disease-specific pathogenesis. Figure 2 represents Venn-transformed enrichment diagrams for all Consistency Tiers.

Based on the results presented in Fig. 2, it is apparent that only the Tier 0 produces a highly enriched disease associated gene list, the Venn Tier 1–0 is still significantly more enriched than the Random Control gene list, while an enrichments in Venn Tiers 2–1 and 3–2 were marginal. In fact, the enrichment for the delimiter query ((disease or pathology or disorder) and (Alzheimer's or Alzheimer or neuropathy or neurodegeneration)) in the Tier 0 was approximately tenfold as compared to the Random Control. The degrees of enrichment against all nonspecific disease-related controls were similar in each Consistency Tier and remained within a margin of experiment error. Altogether, against the non-delimited gene list' background and against the panel of negative controls, the disease-specific genes demonstrate the relative enrichment of ~10 in the Tier 0, ~3 in the Tier 1–0, ~1.5–2 in the Tiers 1–2 and 3–2.

3.8 Applicability of Subheadings 3.1–3.9 to Other Sources of Data Beyond Transcriptomics

Differential expression and integration of individual experiments as well as multiple platforms in a data fused consistency profile is available not only for microarrays. Any protocol relying on differential signals between the disease state and healthy control can be subjected to the same processing, also including ontological enrichment and network building. The primary methods of differential data collection include SAGE and EST tag libraries, quantitative PCR, differential proteomics on protein arrays, differential immunocytochemistry, differential immunohistochemistry, parallel sequencing projects comparing healthy control groups and disease, differential polymorphism detection studies, differential phosphorylation arrays, differential data by miRNA arrays, metabolomics data traceable to the genes in the active pathways, differential promoter methylation and demethylation studies, differential G-protein assays, differential intron arrays, differential alternative splicing events. In all these categories the genes important in the disease would display some distinctions—for example prevalent polymorphisms, prevalent splice forms, metabolic products traceable to a gene product. The statistical significance of these differences can be further validated by comparing the direction of change across multiple research groups, emphasizing consistency and penalizing discrepancies in the total score. Every individual signature can be subjected to ontological enrichment and the most consistent tiers integrated in the common network, producing multi-source CDMs, possibly superior to the single source transcription based CDM in this chapter.

3.9 The Link Between Higher Consistency Score and Causation

In this report, Alzheimer's Compact Disease Model (CDM) was built using both Illumina and Affymetrix platforms through extraction of differential expression data followed by tiering the gene expression evidence by its consistency. Both microarray platforms rely on oligonucleotide multi-probe approach; however, the experimental workflow, probe length, probe choice and signal processing statistics between the two platforms substantially differ [27, 28]. In our approach, this inter-platform discrepancy is expected to serve as a filter that eliminates the signals that display poor consistency due to low reproducibility of expression level changes or due to elevated person-to-person expression variability, thus cutting out the probability to detect meaningless (i.e., false positive) signals.

Inferring clinically relevant insights from the complex picture of the quantitative changes in gene expression/polymorphism/ transcript processing/function levels remains a major challenge of systems biology. An interpretation of the disease signature remains the least standardized part of analytic procedures. In most cases, this analysis is riddled with subjective inference about whether given change in expression levels should be classified as causal, passively associated with observed phenotype or simply incidental to study design. Recent introduction of knowledge-based algorithms is expected to aid in producing reasonable hypotheses linking altered pathways to phenotypic changes. We assume that molecular targets pertinent to pathogenesis of certain chronic disease may be recognized by their consistent visibility (differential expression, association of SNPs, functional evidence, etc.) across most of independently designed experiments. In other words, a molecular target highlighted in a majority of studies (high-prevalence target) is more likely to be mechanistically important than the target detected in a minority of studies (low-prevalence target), although this relationship may be not so straightforward [17]. For example, the comparative prevalence of particular SNPs in a disease set vs. norm relatively clearly points to the pathways determining predisposition. However, the prevalence of the comparative signal at differential expression, differential methylation, transcript processing, phosphorylation and metabolic levels may—at least in theory—reflect a convergence of analogous secondary changes caused by diverse etiologies within the same broad mechanism (*see* Sect. 4.5 for the proposed broad mechanism). To rule this alternative out, an alignment with literature data is required pointing to causative nature of the most prevalent comparative signals. Such alignment was conducted and in general confirms the hypothesis (*see* Subheading 4). In addition, the genes prioritized by relatively simple approach described in our study are later validated by a highly clustered network, amplifying causative evidence for each member of the cluster through "guilty by association" principle. In other consistency profiling reports (mostly in cancer studies [11, 12]), the prevalent signature was therapeutic target-rich and the ability of the targets to influence the outcome of the disease points to proximity to causation and at the very least to practical utility.

3.10 How Effective Is the Benchmark T-Test in the Analysis of Differential Data?

Producing the cutoff at Tier 1 in the consistency profiling and validating it by network building allowed us to obtain a putatively true result and thus enabled to reassess alternative methodological approaches. One of such approaches is selection based on high log value of differential expression and low p-value of significance for each gene, enough to pass the stringent Bonferroni correction. The T-test filtered data were later validated by network aggregation. As shown in Fig. 2, the Tier 3–2 data corresponding to this approach are better than random control set in terms of relevant functional enrichment in literature-based tags (compare the checkered leftmost bars, see the legend). However, consistency distilled Tier 0 and Tier 1–0 demonstrate much greater level of functional enrichment. Furthermore, the current networking algorithm cannot overcome the noise in *T*-test only set based on analyzing the composition of the network reconstructed from the inferior consistency tiers. Such models are dominated by stress response and inflammation pathways—important as a consequence and for self-perpetuation of neurodegeneration—but not as a cause according to the more distilled network construct defined above. Apparently, to equal the potency of consistency based "compartmentalized" approach, the benchmark *T*-test must compare much greater groups of samples which directly effects resource economy of the research. An informal (qualitative) statistical explanation of the efficiency of consistency profiling may stem from treating the subsets with the greatest noise. Instead of full contribution in the noise as is the case in the benchmark method, in consistency approach such noise-rich subsets contribute just one vote (compartmentalization of noise), improving resolution. Considering the reliance on Big Data and the substantial investment in the field, the improved downstream extraction of translational information from high-throughput datasets would amplify already significant potential of the new methodologies. With the increase in the number of distilled gene lists by diverse methods, the quality of the final CDM should improve through improvement of recall rate of bona fide causative players.

4 Practical Example: Applications of CDM Methodology to Understanding of Early Alzheimer's Disease

4.1 Overview of Differential Expression Consistency in Alzheimer's Disease: Biases Are Ruled Out

The data of the report show a pattern of downregulation for the majority of the genes comprising the consistency tiers (*see* Subheading 3.6), with the proportion of downregulated genes increasing with the stringency of consistency. This observation is in agreement with the degenerative nature of the disease. The bias potentially associated with high copy-number artifacts was eliminated by the controls described in Subheading 3.7. The bias associated with the possibility of the initial strong *T*-test (in differential expression data) arising due to a potential decreased expression variation across data points for network hubs was tested and the results pointed to actually increased expression variation across consistent gene list. Thus the consistency of gene detection in mechanism related signatures occurs despite higher variation, explicable by a more complex regulation of potent hubs and a greater error propagation in these control loops as compared to the regulation of random genes (Fig. 1). To ensure relevance to the true mechanism of pathology and not to nonspecific chronic inflammation and stress response, the tiers of the Consistent gene list were screened against known literature and the enrichment in neurodegeneration related tags was evaluated (Fig. 2). The result shows that the Tiers 0 and 1 are strongly enriched in disease-specific tags suggesting a rational cutoff in the initial distilled gene list intended for subsequent network analysis.

4.2 Functional Enrichment Analysis

Specific pathological mechanisms manifest by differential expression of genes that belong to just a few selected pathways. The effected functional categories may develop high and statistically significant enrichment coefficients in the changed gene list. The higher the extent of enrichment, the stronger is the link between the disease mechanism and the functional category of interest, pointing to greater specificity of the signal. Table 1 below combines expression-normalized functional enrichment coefficients computed for Top 20 most enriched ontological categories for gene lists in the Tiers 0–3.

In general, the enrichment coefficients positively correlate with consistency scores, decreasing in the direction from Tier 1 (highest consistency) to Tier 3 (lowest consistency). The trend reversal from Tier 0 to Tier 1 can be explained by significant reduction (by 90%) of the total gene number in the Tier 0 as a result of expression normalization. In the Top 20 categories, the enrichment coefficients were in the range of 4.5–19, with a tendency to an upper side of the range. In non-normalized datasets, the stringent normalization by absolute expression masks the extent of functional enrichment as it may reach the values of ~100 for Tier 0 and ~40 for Tier 1, being far above the typical values observed in traditional microarray experiments [29]. Thus, much higher distillation coefficient q of the model (1)–(2) may be reached. Based on comparison of the functional enrichments in CDM approach and the benchmark exemplified by [29], the increase in network-assisted capability to infer relevant mechanisms may be quite dramatic and certainly merits further study.

Based on the Table 1, the function of traffic vesicle formation dominates in the Tiers 0 and 1 and Venn Tier 1 + 2. This function includes subfunctions of kinesin binding (synuclein-α; actin β; actin γ1; kinesin-associated protein 3), clathrine vesicle formation (synaptotagmin 1; synaptotagmin 13; synaptic vesicle glycoprotein 2B), calcium release (calmodulin 2; synuclein-α; thymus cell antigen 1, θ; cholecystokinin B receptor; guanine nucleotide binding protein (G protein), γ 3). Synaptic vesicle development categories were prominent in the Tier 0, while an axon development category was prominent in the Venn Tier 1 + 2 (synaptotagmin 1; synaptotagmin 13; synaptic vesicle glycoprotein 2B). Vesicle formation-related functionalities displayed the highest

Table 1
Enrichment coefficients for top significance functional categories in different consistency tiers

Tier	Go category	TG	CG	ENR	LOG10(*p*)	FDR
0	GO:0019894_kinesin_binding	8	4	15	−4.1	0.011
0	GO:0007269_neurotransmitter_secretion	38	7	5.7	−3.8	0.018
0	GO:0030426_growth_cone	30	6	6	−3.5	0.024
0	GO:0008021_synaptic_vesicle	42	7	5	−3.5	0.030
0	GO:0007204_elevation_of_cytosolic_calcium_ion_concentration	20	5	8	−3.5	0.028
0	GO:0051480_cytosolic_calcium_ion_homeostasis	22	5	7	−3.3	0.03
0	GO:0051279_regulation_of_release_of_sequestered_calcium_ion_into_cytosol	6	3	15	−3.2	0.05
0	GO:0030672_synaptic_vesicle_membrane	24	5	6.4	−3.1	0.06
0	GO:0010522_regulation_of_calcium_ion_transport_into_cytosol	7	3	13.	−3.0	0.06
0	GO:0048854_brain_morphogenesis	7	3	13	−3.0	0.0
0	GO:0050852_T_cell_receptor_signaling_pathway	15	4	8	−3.0	0.06
0	GO:0051648_vesicle_localization	16	4	8	−2.9	0.06
0	GO:0051209_release_of_sequestered_calcium_ion_into_cytosol	8	3	12	−2.8	0.06
0	GO:0051282_regulation_of_sequestering_of_calcium_ion	8	3	12	−2.8	0.06
0	GO:0051283_negative_regulation_of_sequestering_of_calcium_ion	8	3	12	−2.8	0.06
0	GO:0002429_immune_response-activating_cell_surface_receptor_signaling_pathway	17	4	7	−2.7	0.06
0	GO:0002768_immune_response-regulating_cell_surface_receptor_signaling_pathway	17	4	7	−2.7	0.07
0	GO:0050851_antigen_receptor-mediated_signaling_pathway	17	4	7	−2.7	0.07
0	GO:0007281_germ_cell_development	29	5	5	−2.7	0.07
0	GO:0030594_neurotransmitter_receptor_activity	9	3	10	−2.6	0.08
1	GO:0060198_clathrin_sculpted_vesicle	5	3	19	−3.6	0.03

(continued)

Table 1
(continued)

Tier	Go category	TG	CG	ENR	LOG10(*p*)	FDR
1	GO:0019894_kinesin_binding	9	5	18	−5.5	0.004
1	GO:0042288_MHC_class_I_protein_binding	6	3	16	−3.3	0.05
1	GO:0042287_MHC_protein_binding	7	3	14	−3.0	0.06
1	GO:0051279_regulation_of_release_of_sequestered_calcium_ion_into_cytosol	7	3	14	−3.0	0.06
1	GO:0005834_heterotrimeric_G-protein_complex	12	5	13	−4.8	0.0049
1	GO:0010524_positive_regulation_of_calcium_ion_transport_into_cytosol	5	2	13	−2.0	0.12
1	GO:0035267_NuA4_histone_acetyltransferase_complex	5	2	13	−2.0	0.12
1	GO:0043113_receptor_clustering	5	2	13	−2.0	0.12
1	GO:0045576_mast_cell_activation	5	2	13	−2.0	0.12
1	GO:0046173_polyol_biosynthetic_process	5	2	13	−2.0	0.12
1	GO:0051322_anaphase	5	2	13	−2.0	0.12
1	GO:0051668_localization_within_membrane	5	2	13	−2.0	0.12
1	GO:0010522_regulation_of_calcium_ion_transport_into_cytosol	8	3	12	−2.8	0.06
1	GO:0031177_phosphopantetheine_binding	8	3	12	−2.8	0.06
1	GO:0008277_regulation_of_G-protein_coupled_receptor_protein_signaling_pathway	12	4	11	−3.5	0.03
1	GO:0008298_intracellular_mRNA_localization	9	3	11	−2.7	0.07
1	GO:0032410_negative_regulation_of_transporter_activity	9	3	11	−2.7	0.07
1	GO:0010676_positive_regulation_of_cellular_carbohydrate_metabolic_process	6	2	11	−1.9	0.14
1	GO:0045298_tubulin_complex	6	2	11	−1.9	0.14
1 + 2	GO:0019894_kinesin_binding	9	6	12	−5.7	0.007

1 + 2	GO:0060198_clathrin_sculpted_vesicle	5	3	11	−2.8	0.06
1 + 2	GO:0008298_intracellular_mRNA_localization	9	5	10	−4.4	0.009
1 + 2	GO:0005871_kinesin_complex	8	4	9	−3.3	0.03
1 + 2	GO:0042288_MHC_class_I_protein_binding	6	3	9	−2.5	0.08
1 + 2	GO:0045298_tubulin_complex	6	3	9	−2.5	0.08
1 + 2	GO:0005834_heterotrimeric_G-protein_complex	12	5	7.5	−3.5	0.02
1 + 2	GO:0005881_cytoplasmic_microtubule	15	5	6	−3.0	0.04
1 + 2	GO:0008088_axon_cargo_transport	15	5	6	−3.0	0.04
1 + 2	GO:0008277_regulation_of_G-protein_coupled_receptor_protein_signaling_pathway	12	4	6	−2.5	0.08
1 + 2	GO:0014047_glutamate_secretion	12	4	6	−2.5	0.08
1 + 2	GO:0032182_small_conjugating_protein_binding	16	5	5.6	−2.9	0.045
1 + 2	GO:0043130_ubiquitin_binding	16	5	5.6	−2.9	0.045
1 + 2	GO:0006458_'de_novo'_protein_folding	26	8	5.5	−4.3	0.009
1 + 2	GO:0006941_striated_muscle_contraction	13	4	5.5	−2.3	0.09
1 + 2	GO:0072384_organelle_transport_along_microtubule	13	4	5.5	−2.3	0.09
1 + 2	GO:0051084_'de_novo'_posttranslational_protein_folding	24	7	5	−3.6	0.02
1 + 2	GO:0005876_spindle_microtubule	18	5	5	−2.6	0.07
1 + 2	GO:0005200_structural_constituent_of_cytoskeleton	33	9	5	−4.3	0.01
1 + 2	GO:0010970_microtubule-based_transport	26	7	5	−3.4	0.02
1 + 2 + 3	GO:0033180_proton-transporting_V-type_ATPase__V1_domain	7	6	6	−4.2	0.0018
1 + 2 + 3	GO:0004708_MAP_kinase_kinase_activity	6	5	6	−3.4	0.006
1 + 2 + 3	GO:0042288_MHC_class_I_protein_binding	6	5	6	−3.4	0.006
1 + 2 + 3	GO:0042777_plasma_membrane_ATP_synthesis_coupled_proton_transport	6	5	6	−3.4	0.006

(continued)

Table 1
(continued)

Tier	Go category	TG	CG	ENR	LOG10(*p*)	FDR
1 + 2 + 3	GO:0030897_HOPS_complex	5	4	5	−2.7	0.025
1 + 2 + 3	GO:0031338_regulation_of_vesicle_fusion	5	4	5	−2.7	0.025
1 + 2 + 3	GO:0035542_regulation_of_SNARE_complex_assembly	5	4	5	−2.7	0.025
1 + 2 + 3	GO:0060198_clathrin_sculpted_vesicle	5	4	5	−2.7	0.025
1 + 2 + 3	GO:0046933_hydrogen_ion_transporting_ATP_synthase_activity__rotational_mechanism	15	11	5	−6.3	0
1 + 2 + 3	GO:0046961_proton-transporting_ATPase_activity__rotational_mechanism	18	13	5	−7.2	0
1 + 2 + 3	GO:0004712_protein_serine_threonine_tyrosine_kinase_activity	7	5	5	−2.9	0.014
1 + 2 + 3	GO:0042287_MHC_protein_binding	7	5	5	−2.9	0.014
1 + 2 + 3	GO:0033178_proton-transporting_two-sector_ATPase_complex__catalytic_domain	17	12	5	−6.5	0
1 + 2 + 3	GO:0019894_kinesin_binding	9	6	4.5	−3.2	0.008
1 + 2 + 3	GO:0010676_positive_regulation_of_cellular_carbohydrate_metabolic_process	6	4	4.5	−2.2	0.059
1 + 2 + 3	GO:0035493_SNARE_complex_assembly	6	4	4.5	−2.3	0.059
1 + 2 + 3	GO:0045261_proton-transporting_ATP_synthase_complex__catalytic_core_F(1)	6	4	4.5	−2.3	0.059
1 + 2 + 3	GO:0045739_positive_regulation_of_DNA_repair	6	4	4.5	−2.3	0.059
1 + 2 + 3	GO:0045913_positive_regulation_of_carbohydrate_metabolic_process	6	4	4.5	−2.3	0.059
1 + 2 + 3	GO:0009135_purine_nucleoside_diphosphate_metabolic_process	8	5	4	−2.6	0.032

TG total genes in a given functional category within the total list *T*, *CG* changed genes within the list *L*, *ENR* enrichment coefficient, *LOG10*(*p*) logarithm of *p*-value of one-sided Fisher's test of significance of a given ENR at a given group size, *FDR* false discovery rate

Venn Tier 1 + 2 is the combination of the Tier 1 and Tier 2, Venn Tier 1 + 2 + 3 is the combination of the Tiers 1, 2, and 3.

enrichment coefficients among all consistency tiers and were accompanied by the lowest *p*-values and FDRs. Microtubule and cytoskeleton development were prominent in the Tier 1 and Venn Tiers 1 + 2 and 1 + 2 + 3 (tubulin, γ complex associated protein 3; tubulin, γ complex associated protein 2; tubulin, β 3 class III; tubulin, β 2C; tubulin β, class I; tubulin, α1c; tubulin, α1b). A related function of cell motility was predominately populated by the molecules that relate to mast cell activation (tyrosine 3-monooxygenase/tryptophan 5-monooxygenase activation protein, ζ polypeptide; thymus cell antigen 1, θ; synuclein-α). Another prominent functional category, common to the Tiers 0–3, was MHC binding, receptor binding, ubiquitin targeting and other forms of protein binding mediated by proteasome subunits, cytoskeleton and chaperones (ubiquitin-conjugating enzyme E2N; p21 protein (Cdc42/Rac)-activated kinase 1; thymus cell antigen 1, θ; actin β; actin γ1). Regulation of G-protein signaling was highly enriched category in the most conserved Venn Tiers 0–2 (regulator of G-protein signaling 4; regulator of G-protein signaling 6; regulator of G-protein signaling 7; synuclein-α, calmodulin 2; cholecystokinin B receptor; γ-aminobutyric acid (GABA) B receptor, 2; guanine nucleotide binding protein (G protein), γ3). Less surprisingly, an importance of GABA neurotransmission was detected (γ-aminobutyric acid (GABA) B receptor, 2; γ-aminobutyric acid (GABA) A receptor, gamma 2), as well as related glutamate secretion (glutaminase; synaptotagmin 1; synuclein-α), neurotransmitter binding (cholinergic receptor, muscarinic 1; cholecystokinin B receptor; γ-aminobutyric acid (GABA) A receptor, γ2) and brain morphogenesis (platelet-activating factor acetylhydrolase 1b, regulatory subunit 1; McKusick–Kaufman syndrome; presenilin 2) functionalities. Remarkably, consistent datasets lacked amyloid β (A4) precursor protein APP. One possible explanation is that differential expression of APP monomer is negligible, while its pathological role unfolds at the level of toxic oligomers [30–32].

4.3 Biological Network Modeling

The molecules populating Consistency Tier 1 that was optimal in terms of balance between functional enrichment and recall rate of the most relevant mechanistic participant entries were analyzed using Ingenuity Pathway Assistant (IPA) tool. The first order interactions were imputed automatically, the partners being the hubs of the cell signaling pathways in network proximity to the changed genes. The addition provided by the IPA network is valuable, since this feature partially compensates for low recall rate observed when the consistency criteria are applied. The distinctions between the subnetworks are algorithm-generated and therefore somewhat artificial. We retained for further analysis all significant hubs regardless of the subnetwork they were assigned to by IPA. A subnetwork 1 of the interaction network is shown in Fig. 3 and the composition of the entire network is provided in Table 2.

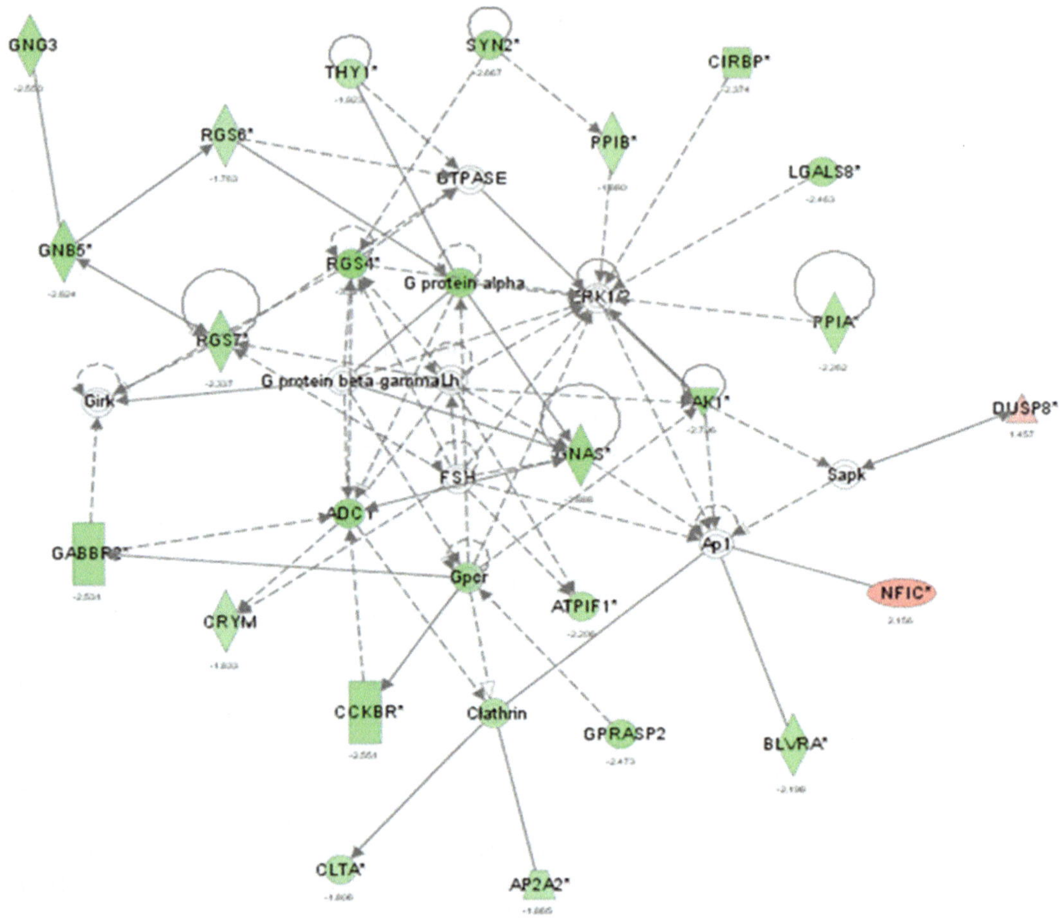

Fig. 3 Biological network-based model of interactions between the most essential Alzheimer's disease-related genes representing the subnetwork 1 of Table 2. *Green* figures indicate downregulation, *red* figures indicate upregulation, *grey* figures mean unchanged expression levels. *Rectangular* figures indicate receptors, *rombi*—peptidases, *triangles*—kinases/phosphatases, *circles*—other; *solid connecting line*—binding only, *solid connecting arrow*—acts upon, *dotted lines* indicate indirect functional relationships (such as co-regulation of expression of both genes in cell lines)

Molecules in network comprise both experimentally discovered molecules and known/predicted close interaction partners. The Score is a measure of clustering coefficient between the subnetwork components. Focus Molecules are the differentially expressed gene products with experimental evidence of the linkage to disease pathogenesis and are denoted by capital letters, while small letters designate inferred interactions.

According to the Table 2, the subnetworks 1 and 2 demonstrate significant scores associated with *p*-value of 10^{-49} and 10^{-47}, respectively, where the score being the probability that the genes associated at this extent of clustering were drawn randomly. Per IPA functional assignment, the highest score subnetwork 1 corresponds

Table 2
Composition and scoring of Alzheimer's disease molecular subnetworks generated by IPA

ID	Molecules in network	Score	Focus molecules	Top functions
1	ADCY, Ap1, AP2A2, ATPIF1, BLVRA, CCKBR, CIRBP, Clathrin, CLTA, CRYM, DUSP8, ERK1/2, FSH, GABBR2, Girk, GNAS, GNB5, GNG3, GPRASP2, GTPASE, LGALS8, Lh, PAK1, PPIA, PPIB, RGS4, RGS5, RGS7, Sapk, SYN2, THY1, TNPO1	49	23	Neurological Disease, Reproductive System Development and Function, Cell Death
2	14-3-3, ACTB, ACTG1, Actin, ACTR1A, aldo, Alpha tubulin, ATP5C1, ATP5G1, ATP6V1D (includes EG:299159), ATP6V1E1, ATP6V1E2, Beta Tubulin, Cofilin, Dynein, EIF3K, ELAVL4, F Actin, GAPDH, H+-transporting two-sector ATPase, Hsp90, NFkB (complex), PFDN5, SMARCC1, SNCA, SORBS1, STMN2, TUBA1B, TUBA1C, TUBB3, TUBB, TUBB2C, Tubulin, Vacuolar H+ATPase, ZBTB20	47	22	Cellular Assembly and Organization, Cellular Function and Maintenance, Immunological Disease
3	26sProteasome, Akt, ATP8A2, BAG6, CD3, COPS4, COX5B (includes EG:100002334), DNAJB12, DNAJC6, DNAJC8, ERK, GABRG2, Hsp70, HSP, HSPA8, HSPB3, Ikb, Insulin, Jnk, Laminin, LSM14B, Mapk, NDFIP2, P38MAPK, Pka, PPME1, PSMD4, PTP4A2, SNRPN, Sos, SYT1 (includes EG:20979), TCR, UBE2N, Ubiquitin, YWHAZ	31	16	Cellular Compromise, Cell-To-Cell Signaling and Interaction, Cellular Growth and Proliferation
4	ADORA2A, ATP, ATP6V1H, CACNA1E, CACNA2D3, CACNB2, DYNC1I1, EIF1, EIF5B, FOS, GPR1 (includes EG:100004124), IDS, KDM5B, KIFAP3, KLF15, L-glutamic acid, LANCL1, MAP2K7, MAPK3, MYC, MZT1, PNO1, PPP1R7, RAD51C, REEP1, RPL15, SLC17A7, SLC17A, SRSF2, TGFB1 (includes EG:21803), TIMM23, TMEM97, TUBGCP2, TUBGCP3, XRCC3	29	17	Hematological System Development and Function, Hematopoiesis, Tissue Development
5	AGRP, APC, ARHGEF9, CHCHD3, CHCHD6, CYB5D1, EIF3C/EIF3CL, IDH3G, LAMA3, MC4R, MRPS7, MRPS22, NAV1, NAV2, NFIA, POLH, PRDM5, PSMA1, PSMB2, PSMB3, PSMB6, RPL5, RPL6, RPL17, RPL19, RPL30, RPL31, RPL10A, RPLP0, RPLP2, RUNX2, SLC35E1, UBC, UBL7, USP3	21	12	Connective Tissue Development and Function, Tissue Morphology, Genetic Disorder

(continued)

Table 2 (continued)

ID	Molecules in network	Score	Focus molecules	Top functions
6	ACACB, ATP5A1, ATP5C1, ATP5D, BBX, BCL2L1, CDC16, CUEDC1, CUL3 (includes EG:26554), DHX30, E2F4, FAM162A, GSK3A, HINT1, IL4 (includes EG:16189), KCNAB2, LAMP2, mir-451, NACC1, NFIC, OTUD7B, PEBP1, PRKCQ, PSEN1, RELA, RTN1 (includes EG:104001), SLC11A2, SMAD3, SYT13, TMEM85, TNFSF10, TPT1 (includes EG:100043703), TTK, TUBA3C/TUBA3D, ZNF83	21	12	Nucleic Acid Metabolism, Small Molecule Biochemistry, Cellular Compromise
7	FAM63A, NAA38	2	1	
8	GNB2L1, SLC9A6	2	1	Cell Cycle, Connective Tissue Development and Function, Developmental isorder
9	DDX19B, GADD45G, RWDD2B	2	1	Tissue Development, Cell Cycle, DNA Replication, Recombination, and Repair

to neurological diseases and comprises vesicle-forming components in agreement with GO-MINER analysis, validating the CDM approach from the point of internal consistency. The subnetwork 2 comprises mostly cytoskeleton and mitochondrial components. The subnetwork 3 displays a score of 31 and comprises other components such as chaperones, ubiquitin pathway members and proteasome subunits. Subnetworks 4–9 were significant but displayed lower scores.

The abundance of oncogenes in the associated hub subset was remarkable. To quantify the extent of association with oncogenes, the symbol *T* was defined as Pubmed response to the gene symbol, assumed to be proportional to the total number of biological interactions mediated by the gene and its products. High *T* numbers correspond to the hubs of biological network.

To put it in a larger genomic context, a random sample of 118 gene names was extracted and the *T*-values were determined, producing two hits with $1000 < T < 5000$, two hits with $5000 < T < 10{,}000$ and one hit with $T > 10{,}000$. Based on this sampling and the total number of genes ~20,000, an estimate of ~500 hubs with $T > 5000$ was shown for the human interaction network. In the network sample associated with the Tier 1 consistently expressed gene list, 36 hubs of the comparable connectivity was present per 96 network associates. This is not a remarkable finding, considering that Ingenuity databank is likely biased in favor of hub enrichment. However, the finding that the ratio of oncogenes to tumor suppressors is skewed toward oncogenes is counterintuitive for a degenerative disease.

To assess the background ratio of oncogenes vs. suppressors, several databases were enquired. Search of OMIM (www.ncbi.nlm.nih.gov/omim) leads to 647 hits responding to the query ((“oncogene or oncogenes”)), while 882 hits responded to the queries ((“tumor suppressor” or “tumor suppressors”)). These numbers correspond to ~7:5 ratio of tumor suppressors to oncogenes in the global network. Similar search with the databases “Genes” and “Proteins” at NCBI produced ~1:2 ratios. An analysis of the database GeneCards at www.genecards.org leads to the ratio ~1:1 for the same queries. With that, an average ratio of ~0.8:1 of tumor suppressor to oncogenes may be assumed as a random global control.

This ratio markedly differs from the ratio observed in our data. Tumor suppressor–oncogene ratio in the hubs associated with neuropathy network was 2:9 (APC and TGFB1 as tumor suppressors, ERK1/2, AKT, MYC, FOS, AP-1, BCL2L1, HSP70, HSP90, RELA being pro-growth and oncogenic, while the MAPK3, MAPK8, and MAPK14 were marked as having context dependent dual functions). Except APC, none of the hits in this *T* range was associated with “tumor suppressor” label, while AKT, MYC, FOS, AP-1, BCL2L1, RELA were denoted as “oncogenes.”

We further tested if this oncogene association is limited to our data or is more general. A PubMed query ((Alzheimer's or Alzheimer or neuropathy or neuropathic or neuro-degeneration or neurodegeneration or neurodegenerative or dementia)) was further delimited by the keywords (("oncogene" or "oncogenes")) as well as (("tumor suppressor" or "tumor suppressors")). The ratio of 3.4:1 was observed, while a control query ((disease or disorder)) produced 1.9:1 ratio. Similar queries (neuropathy or neurodegeneration) and (dementia or "cognitive decline") produced the ratio 3.5:1 above the random ~2:1, consistent with our data.

The high-T subpopulation of network associates was segregated from the initial changed gene list (Tiers 0, 1–0, 2–1 combined) and all subpopulations underwent a similar analysis as above. Specifically, the corresponding gene lists were converted into Boolean queries and were delimited with "oncogene" and "tumor suppressor" terms. The Alzheimer's related genes were compared with a random gene sample. The random control and the initial (non-tiered) list of differentially expressed genes demonstrated comparable oncogene/tumor suppressor hit ratios of 2:1, while the population of extracted network associates produced the hit ratio of 5.7:1. For sense of perspective, the corresponding ratio for oncogene BCL2-centered network was 4.6:1 and for tumor-suppressor-centered p53-centered network was 1:2.2. Considering these ratios, the Alzheimer's disease network associates were as a group more oncogenic than the associates of BCL-2, considered to be a benchmark oncogene and this result is counterintuitive, considering the degenerative character of the disease.

In another analysis, the control query (disease or disorder) and (activation or activator) generated ~125,000 hits, while the query (disease or disorder) and (deactivate or deactivator or suppressor or repress or repressor) generated ~25,000 hits. The target query ((Alzheimer's or Alzheimer or neuropathy or neuropathic or neuro-degeneration or neurodegeneration or neurodegenerative or dementia)) and (activation or activator) produced 17,500 hits, while the query ((Alzheimer's or Alzheimer or neuropathy or neuropathic or neuro-degeneration or neurodegeneration or neurodegenerative or dementia)) and (deactivate or deactivator or suppressor or repress or repressor) produced ~1300 hits. The ratios point to neuropathies being more preferentially associated with activation processes (compare 125,000:25,000 for the control and 17,500:1300 for the neuropathies).

Thus, we conclude that an analysis of entire PubMed shows that the neuropathy-related information is more closely and paradoxically associated with oncogenes and activation than with tumor suppressors and deactivation, confirming the trend observed in our data.

4.4 A Study of Intersection for Alzheimer's Disease and Angiotensin Receptor Blocker Response Pathways

The limited number of mechanistically relevant members comprising CDM list allows aligning with the literature data covering downstream effects of Angiotensin receptor AT-1. Both the results in Table 2 of the current study and the AT-1 literature review indicate AT-1 related genes as likely to mediate the effect of ARBs (angiotensin receptor blockers) on Alzheimer's development.

Literature analysis points to significant interaction of AT-1 pathway with oncogene activation as well as with luteinizing hormone and insulin dependent pathways in neurons [32–37]. The role of oncogene modulation in response to AT1R blockers is complex, with some oncogenes being inhibited [32, 33], while some being upregulated [34]. In the cases when the ARBs exhibit neuroprotective effects via c-JUN inhibition, levels of other oncogenes remain as they were and the overall impact of oncogenic activation could still be executed through collateral routes. An example of such collateral pathway is a compensatory increase in activity of oncogenic angiotensin II receptor II (AT-2)/MAS pathway after the blockade of angiotensin II receptor I (AT-1) [38]. In mouse model, the alleviation of Alzheimer's disease was experimentally achieved by hippocampal delivery of the oncogenic fibroblast growth factor FGF2 [39]. The predominance of neuroprotective effect in oncogene stimulation by ARBs is emphasized by ARB induction of IGF1, a molecule with a powerful anabolic and pro-survival impact [37]. Thus, the connection between AT1R inhibition and general activation of neuronal oncogenes is rather prominent in the body of research literature.

The LH/FSH regulation was previously linked to Alzheimer-like degeneration in murine models [40], thus lending greater significance to stimulation of luteinizing hormone expression by angiotensin II that was previously observed in neurons [36].

Another group of entries in the higher score subnetworks 1 and 2 belongs to cytoskeleton rearrangement pathways. Cytoskeleton rearrangement related signaling that is the necessary step in vasoconstriction and vasodilation, a major short term effect of any antihypertensive drug. Angiotensin pathway is certainly involved in the pathways featured in the subnetworks 1 and 2 [41–43]. According to our data, the most abundant subnetworks of the Tier 1/subnetwork 1 are the regulators of G-protein signaling and G proteins: GNAS, GNB5, GNG3, GPRASP2, RGS4, RGS6, RGS7. The vascular remodeling and vasodilating role of GRS2, GRS3, GRS5, GRS18, and GNB3 in regulating of vascular tonicity in the context of Angiotensin receptor (AT-1) signaling was described previously in [44 48]. Vascular tone is maintained by the cytoskeleton rearrangement and the intracellular motility, thus connecting it to the protein misfolding and/or defective chaperone complex formation.

To conclude, substantial literature evidence connects the CDM derived in the current report and organized in the network of Table 2 with Angiotensin Receptor I pathway or its blockers, in the neuronal setting.

4.5 Entropic Disease Model and Pleiotropic Role of Angiotensin Receptor Blockers

Based on the observations of the current report, a pleiotropic model of early-stage Alzheimer's disease could be proposed.

From very general considerations, a rigid differentiation program and complex shape of the neuron makes it an inherently disadvantaged cell type. Thermodynamically, expression of a gene in the nucleus that is followed by long route of the delivery of resultant protein to the target site at the synaptic junction either in a folded or properly prefolded state is unfavorable due to high Boltzmann entropy loss associated with long processes. In the neurons, the travel distances may reach 0.5–1 m; the maintenance of properly folded protein requires costly coordination of its intracellular traffic with the chaperone assembly sites and migration of the chaperone-protein complex to the destination. The high Boltzmann entropy loss of the process has to be matched by a high influx of free energy in the protein traffic path, derived in sufficient stimulation of anabolic and trophic pathways. This fundamental understanding is in agreement with our findings that the pathways jointly implicated in both blood pressure control and neurodegeneration are mostly anabolic and pro-survival. When anabolic pathways become downregulated in CNS due to aging, neurotoxicity or mutation, it takes its toll on energy balance within the cell and increases the risk of misfolding. Thus, the long-term sustainability of anabolic processes in the neurons may be favored by regular antihypertensive treatments that assist cell survival.

In this balance of energy, the state of cytoskeleton organization determines the Boltzmann entropy loss. Disorganized cytoskeleton has higher initial entropy and the required intra-neuronal coordination would impose higher entropy costs. Thus, the intensity of anabolic processes is not the only factor determining energy supply for proper protein folding and trafficking. The luteinizing and follicle stimulating hormones (LH and FSH) both regulate menstrual cycle in females and spermatogenesis in males serving as upstream stimulators of androgen and estrogen production. Importantly, gonadotropins were found to be involved in the earliest stages of Alzheimer's disease and in memory-related processes in humans and in multiple murine models [49, 50]. Non-pituitary expression of FSH and its co-localization with FSH receptor and GnRH receptor in rat cerebellar cortex was shown previously [50]. Brain regions susceptible to degeneration in AD are enriched in both LH and its receptor; moreover, in animal models of AD, pharmacologic suppression of LH and FSH reduced plaque formation [51]. As both the oocytes meiosis that is triggered by LH and the process of spermatogenesis that is initiated by FSH require extensive cytoskeleton remodeling [52, 53], it is tempting to speculate that the Boltzmann entropy state of neuronal cytoskeleton may be, in part, dependent on FSH and LH stimulation, possibly through G-protein activation. Respectively, G-protein regulators form a tightly connected cluster around FSH and LH nodes (Fig. 3).

It is generally accepted that the probability of any chronic disease of old age increases in parallel with an increase in genomic entropy that degrades the complexity of epigenetic landscapes [54]. Age-dependent demethylation of the genome leads to an increase in the transcription of noncoding RNAs, while CpG-rich 5′ regions of select genes may become hypermethylated [55]. In case of neurons, the global hypomehylation and site-specific hypermethylation was found to be associated with degenerative and psychotic diseases [56, 57]. In agreement with these observations, our data point to an overall decrease in transcript expression levels in the most of the functional categories showing high enrichment coefficients by GO-MINER. In some form, the downregulation bias was traced among ~200 members of the Tiers 0–2 and ~1300 members of consistency Tier 3 and PCS groups. It is possible that this phenomenon is reflected by negative downstream changes in the stability of RNA transcripts and proteins, efficiency of translation and posttranslational modifications and, again, protein folding and trafficking. An exception to this trend is prominent upregulation of NF-kB pathway (Fig. 3, NF1C), that is involved in inflammation, cellular stress, and apoptosis.

Another chromatin remodeling associated pathway is insulin signaling (Table 2). Importantly, IGF1 pathway is implicated in both life span control and antihypertensive response. For example, a protective hormone Klotho, a competitive antagonist of IGF1 in kidney, is known to reverse degenerative nephropathies in murine models, and, as well, shown as downregulated in aging primates through chromatin methylation [58]. Interestingly, an inhibition of angiotensin II signaling by counteracting expression of IGF-II receptor is also shown to upregulate Klotho [59, 60]. Taken together, these data suggest a potential of antihypertensive agents to oppose the long-term age-related chromatin remodeling.

4.6 Literature Validation of the Entropic Model Built Based on CDM Filtered Gene List

The principle assumption of our study is that the pathways that relevant to the disease mechanism should be consistently discovered in a number of independent studies. Many pathways highlighted by our enrichment strategy were also described as experimental findings relevant to the context of early stages of Alzheimer's disease, or, in general, the process of neurodegeneration. Extreme functional enrichment for protein traffic vesicle proteins observed in the CDM dataset points to a substantial role of this mechanism in the neurodegeneration, and is likely to be an early pathogenetic event. Some experimental reports confirm impairment of protein vesicle traffic in early stages of neurodegeneration [61, 62]. The body of literature that discusses protein traffic vesicles in the context of neurodegeneration is relatively small and recent, as compared to more common and more general discussions of cytoskeleton and heat shock protein involvement in Alzheimer's disease. Hence, we may conclude that the proposed CDM building technique aids the acquisition of relatively novel mechanistic insights underrepresented in broader literature.

Another important finding of the report is abundance of oncogenes in the Alzheimer's disease interaction network built around the CDM gene list cut-off at a Tier 1 consistency. The independent literature search uncovers numerous publications describing the connection of oncogenes to improper, but possibly compensatory reactivation of cell cycle in terminally differentiated neurons that eventually leads to a cell death [63–65]. An alternative hypothesis points to the fact that patients with Alzheimer's disease have lower risk of incident cancer than general population [63, 66, 67]. One explanation to that paradox is a mitochondrial dysfunction that is both implicated in early stages of Alzheimer's disease development and impacted by the oncogene-tumor suppressor balance [65–67]. The connection between oncogene activation and bioenergy available to a neuron appears to be well described in the literature, in agreement with the conclusions of CDM-based analysis. The mechanistic support to the bioenergetic view of oncogene role over improper reactivation of cell cycle is provided by much higher score rank of the Ingenuity subnetwork 2 (Table 2) comprising mitochondrial ATPase subunits vs. subnetworks 8 and 9, comprising cell cycle components.

Additionally, the CDM gene list analysis uncovers the prominent role of follicle-stimulating hormone, luteinizing hormone, and gonadotropin in the development of early Alzheimer's. The independent literature search presents evidence of increased expression of LH in the neurons vulnerable to Alzheimer's disease [68]. In aged transgenic mice with Alzheimer-type of brain degeneration (Tg 2576), an ablation of the luteinizing hormone by a gonadotropin-releasing hormone analogue leuprolide acetate significantly attenuated cognitive decline and decreased amyloid-beta deposition as compared to placebo-treated animals [51]. Hence, the data presented in [68] and [51] and supported by CDM model indicate an involvement of FSH/LH pathway in Alzheimer's.

The gene list distilling steps and its subsequent compacting are crucial to CDM-based hypothesis generation. If these steps would be omitted, the resultant CDM would be represented by impractically large gene network, dominated by the nonspecific pathways common for many pathologies. In Alzheimer's, non-compacted gene lists are dominated by stress response and inflammatory pathways marked as the highest scoring subnetworks. Without denying the aggravating role of inflammation in Alzheimer's disease, inflammation abating approaches are unlikely to produce sustainable therapeutic results as they target relatively late stages of pathogenesis. Importantly, the distillation of the gene list into compact model (CDM) introduces an opportunity to catch a glimpse at possibly causative mechanistic alternatives that otherwise would took years to uncover through hypothesis-driven experimental studies that tend to look after overall plausibility of possible findings at the stage of the study design. While some models caution us against too stringent cutoffs for initial CDM composition [17], milder cutoffs that are combined with a cross-platform analysis appear to be a promising direction that requires further efforts.

4.7 Implications for Therapy Development in Alzheimer's Disease

Development of radical therapies for delaying and/or reversing of Alzheimer's disease is the most frustrating area of pharmaceutical research with very high failure rate for clinical trials [69, 70]. Detailed discussion of pro and contra of the current approaches in clinical trials is beyond the scope of this chapter. The entropic pleiotropic model, suggested by the Alzheimer's disease CDM described above, provides a framework for novel therapeutic approaches that may be immediately tested in clinical trials.

One of the examples is the glucose utilization deficiency commonly observed in Alzheimer's disease predisposed subjects [71–73]. This primary glucose utilization deficiency may be related to Warburg effect, and may possibly explain the link between Alzheimer's disease and cancer morbidity [63–67]. As intervention at the level of oncogenes/tumor suppressor genes in the brain may be difficult to achieve, the modulation of the hormonal master switches common for both an Alzheimer's and cancers provide an attractive option.

Among CDM-shortlisted molecular targets are insulin, estrogen, follicle-stimulating hormone, and luteinizing hormone pathways already well studied from pharmaceutical viewpoint. It is tempting to speculate that the pharmacological modification of these pathways may provide sufficient compensation for the pathological oncogene signaling that predisposes the patients both to Alzheimer and to tumorigenesis. The assessment of these interventions in clinical trials may by complicated by brain specific isoforms of the cognate hormone receptors and by endogenous production of the hormones, but the promise is apparent at least for insulin therapies [74, 75].

Moreover, age-related decrease in the level of reproductive hormones and strong transcriptomic signatures of this pathway in the Alzheimer's disease [76] produce an intuitively attractive concept of using hormone replacement therapy (HRT) for Alzheimer's prophylaxis. Indeed, menopause and andropause represent the points when the evolutional mechanisms of dying can be triggered. The initial results of HRT are promising, but also equivocal [77, 78] and the methodologies need more refinement, possibly focusing on brain isoforms of the corresponding receptors and the bioavailable small molecular modulators of these receptors capable of passing blood-brain barrier. This principle applies of FSH, LH, GNRH, ESR1 and other brain steroid hormone receptors.

Exposure to growth factors such as NGF and FGF2 may also aid in elimination of the metabolic deficiency by direct trophic activation of the neurons that will produce both stimulation and neuroprotection [39, 79, 80]. The difficulties experienced in delivering these peptide agents across blood-brain barrier (BBB) and the promise of therapy points once more to the modulation of the cognate brain receptors by low-molecular weight mimics, capable of BBB penetration. Especially promising seems combining of regenerative NGF delivery with activation of stem cells, restoring the brain tissue [79].

In both hormonal and regenerative treatments, the lasting improvement depends on sustainable disruption of the positive feedback loop that instigates the disease propagation. In Alzheimer's, these positive feedbacks would persist if the misfolded particles accumulated during the pre-intervention history would remain [81, 82]. This feature contributes to the difficulty of reversing even initial stage of Alzheimer's as opposed to its prevention that, at least from the molecular standpoint, looks more feasible. The study of the network 2 points to the broad role of cytoskeleton and traffic vesicle function in the disease origin, coupled to metabolic deficiency through high ordering requirements of normal neuronal state. The very fact that structurally dissimilar proteins such as APP and TAU were selected as causative agents suggests the possibility that other, yet unknown, proteins contribute to misfolding events. The heterocomplex misfolding may result in an array of toxic oligomers disturbing metabolically distressed neurons with diminished ability to control entropy.

In our opinion, the generic amphyphylic ligands such as methylene blue [83–85], brilliant blue G [86–88], Chicago Sky Blue 6B [89] can be considered as initial leads in design of misfolded protein disaggregants. The alteration of scaffolds and side-chains of these molecules may be an interesting avenue that may produce less toxic molecular derivatives capable of BBB penetration.

The CDM of Alzheimer's disease suggest that it is an extremely complex pathology that results from concerted deregulation of a number of metabolic pathways both in the neurons and in the supportive cells of the brain, as well as in the brain vasculature. Therefore, reducing its pathogenesis to single causative agent (i.e., misfolded protein aggregates) is a harmful oversimplification. The CDM of Alzheimer's simultaneously highlights multiple processes that contribute to metabolic deficiency of aging neurons, with further translation of this deficiency into the destabilization of cytoskeleton/vesicle network, support for the misfolded protein aggregation and subsequent progression of the pathology. Hence, the treatment and/or prevention strategies must also be combinational. We envision the prophylactic and disease modifying therapies of the future as a combination of metabolic deficiency repair, and neuronal stimulation with pharmaceutical control of the protein aggregation and an abatement of inflammatory component. The combination aspect is essential. The aggregates are known to be cleared from neurons [90] by glia, thus enabling the reversion of the disease. We envision Alzheimer's disease eventually becoming a manageable disease, not different to diabetes, some forms of cancer, chronic infection with HIV, and other ailments that in its time were considered fatal.

5 Conclusions

Here we present a novel knowledge-based algorithm that generates network clustering-validated, highly prioritized shortlists of potential targets pertinent to pathogenesis, the Compact Disease Models, or CDMs. This algorithm allowed us to generate a distilled, tiered list of Alzheimer's disease-related genes and to derive a pleiotropic, network-based model for early stages of this disease. In this model, the first degree network associates were characterized by strong predominance of oncogenes. Loss of anabolic stimulation in neurons appears to progress with age due to promoter methylation, until the available free energy in the terminally differentiated cells would cease to compensate Boltzmann entropy loss that is due to the toll of the folding and long-distance delivery of the neuronal proteins. The prophylactic, anti-Alzheimer effect of the ARBs and beta blockers suggest that they play a role at the inception steps in the development of degenerative symptoms. Consequently, understanding of the pathways opposed by these agents has a substantial value since these pathways are likely to be causative to the degenerative process. Based on this logic, protein traffic vesicles, oncogenes, gonadotropin hormones, and insulin-related pathway were identified as potential players in early Alzheimer's disease. This understanding may aid in shifting the therapeutic efforts to the reversible stages of neurodegenerative disease, when the neuronal damage is relatively mild and self-perpetuating misfolded protein oligomers are not yet formed.

Acknowledgments

The authors express their gratitude to the general support provided by College of Science, George Mason University and the Human Proteome Project Program of the Russian Academy of Medical Sciences.

Authors' Contributions
Both authors contributed to the study design, interpretation of results, and producing the manuscript. All the authors read and approved the final manuscript.

References

1. Noam Levey (2010) Soaring cost of healthcare sets a record. In: Los-Angeles Times. http://articles.latimes.com. Accessed 28 Oct 2016
2. Julie Steenhuysen (2012) A look at Alzheimer's Health Costs. http://www.huffingtonpost.com/. Accessed 28 Oct 2016
3. Feldman B, Pai M, Rivard G et al (2006) Tailored prophylaxis in severe hemophilia A: interim results from the first 5 years of the Canadian Hemophilia Primary Prophylaxis Study. J Thromb Haemost 4(6):1228–1236
4. Mayburd A, Golovchikova I, Mulshine J (2008) Successful anti-cancer drug targets able to pass FDA review demonstrate the identifiable signature distinct from the signatures of random genes and initially proposed targets. Bioinformatics 24(3):389–395
5. Hu J, Hagler A (2002) Chemoinformatics and drug discovery. Molecule 7:566–600
6. Lim HA (1997) Bioinformatics and cheminformatics in the drug discovery cycle. In: Ralf H, Thomas L, Markus L, Dietmer S (eds) Bioinformatics, Lecture notes in computer science, vol 1278. Springer, Berlin, pp 30–43
7. Sambamurti K, Jagannatha R, Pappolla M (2009) Frontiers in the pathogenesis of Alzheimer's disease. Indian J Psychiatry 51(Suppl 1): S56–S60
8. GeneCards (2012) Weitzman Institute of Science, Rehovot, Israel. http://wwwgenecardsorg. Accessed 28 Oct 2016
9. Ramsköld D, Wang E, Burge C (2009) An abundance of ubiquitously expressed genes revealed by tissue transcriptome sequence data. PLoS Comput Biol 5(12):e1000598
10. Mason R, Gunst R, Hess J (2003) Statistical design and analysis of experiments: with applications to engineering and science. Wiley series in probability and statistics - applied probability and statistics section series, 2nd edn, vol 474. John Wiley & Sons, Hoboken, NJ, p 760
11. Rhodes D, Yu J, Shanker K et al (2004) Large-scale meta-analysis of cancer microarray data identifies common transcriptional profiles of neoplastic transformation and progression. Proc Natl Acad Sci U S A 101:9309–9314
12. Xu L, Geman D, Winslow R (2007) Large-scale integration of cancer microarray data identifies a robust common cancer signature. BMC Bioinformatics 8:275
13. Tsoi L, Qin T, Slate E (2011) Consistent Differential Expression Pattern (CDEP) on microarray to identify genes related to metastatic behavior. BMC Bioinformatics 2(1):438
14. Mayburd A (2009) Expression variation: its relevance to emergence of chronic disease and to therapy. PLoS One 4(6):e5921
15. Glinsky G, Berezovska O, Glinskii A (2005) Microarray analysis identifies a death-from-cancer signature predicting therapy failure in patients with multiple types of cancer. J Clin Invest 115(6):1503–1521
16. Liu Y, Koyutürk M, Maxwell S et al (2012) Integrative analysis of common neurodegenerative diseases using gene association, interaction networks and mRNA expression data. AMIA Summits Transl Sci Proc 2012:62–71
17. Barrenas F, Chavali S, Holme P et al (2009) Network properties of complex human disease genes identified through genome-wide association studies. PLoS One 4(11):e8090
18. Ochs MF (2010) Knowledge-based data analysis comes of age. Brief Bioinform 11(1):30–39
19. Li N, Lee A, Whitmer R et al (2010) Use of angiotensin receptor blockers and risk of dementia in a predominantly male population: prospective cohort analysis. BMJ 340:b5465
20. Davies N, Kehoe P, Ben-Shlomo Y et al (2011) Associations of anti-hypertensive treatments with Alzheimer's disease, vascular dementia, and other dementias. J Alzheimers Dis 26 (4):699–708
21. Shah K, Qureshi S, Johnson M et al (2009) Does use of antihypertensive drugs affect the incidence or progression of dementia? A systematic review. Am J Geriatr Pharmacother 7 (5):250–261
22. Wagner G, Icks A, Abholz H (2012) Antihypertensive treatment and risk of dementia: a retrospective database study. Int J Clin Pharmacol Ther 50(3):195–201
23. Sun J, Feng X, Liang D (2012) Down-regulation of energy metabolism in Alzheimer's disease is a protective response of neurons to the microenvironment. J Alzheimers Dis 28(2):389–402
24. Kafri R, Dahan O, Levy J (2008) Preferential protection of protein interaction network hubs in yeast: evolved functionality of genetic redundancy. Proc Natl Acad Sci U S A 105 (4):1243–1248
25. Kitano H (2004) Biological robustness. Nat Rev Genet 5(11):826–837
26. Albert R, DasGupta B, Hegde R et al (2011) Computationally efficient measure of topological redundancy of biological and social networks. Phys Rev E Stat Nonlin Soft Matter Phys 84(3 Pt 2):036117

27. Kresse S, Szuhai K, Barragan-Polania A et al (2010) Evaluation of high-resolution microarray platforms for genomic profiling of bone tumours. BMC Res Notes 3:223
28. Chang J, Wei N, Su H et al (2012) Comparison of genomic signatures of non-small cell lung cancer recurrence between two microarray platforms. Anticancer Res 32(4):1259–1265
29. Merico D, Isserlin R, Stueker O et al (2010) Enrichment map: a network-based method for gene-set enrichment visualization and interpretation. PLoS One 5(11):e13984
30. Verdile G, Laws S, Henley D et al (2012) Associations between gonadotropins, testosterone and β amyloid in men at risk of Alzheimer's disease. Mol Psychiatry 19(1):69–75
31. Bartl J, Meyer A, Brendler S, Riederer P et al (2013) Different effects of soluble and aggregated amyloid β(42) on gene/protein expression and enzyme activity involved in insulin and APP pathways. J Neural Transm 120 (1):113–120
32. Zhang T, Fu J, Geng Z (2012) The neuroprotective effect of losartan through inhibiting AT1/ASK1/MKK4/JNK3 pathway following cerebral I/R in rat hippocampal CA1 region. CNS Neurosci Ther 18(12):981–987
33. Palkovits M, Šebeková K, Klenovics K (2013) Neuronal activation in the central nervous system of rats in the initial stage of chronic kidney disease-modulatory effects of losartan and moxonidine. PLoS One 8(6):e66543
34. Hashikawa-Hobara N, Hashikawa N, Inoue Y et al (2012) Candesartan cilexetil improves angiotensin II type 2 receptor-mediated neurite outgrowth via the PI3K-Akt pathway in fructose-induced insulin-resistant rats. Diabetes 61(4):925–932
35. Mitra A, Gao L, Zucker I (2010) Angiotensin II-induced upregulation of AT(1) receptor expression: sequential activation of NF-kappaB and Elk-1 in neurons. Am J Physiol Cell Physiol 299(3):C561–C569
36. Moreno A, Franci C (2004) Estrogen modulates the action of nitric oxide in the medial preoptic area on luteinizing hormone and prolactin secretion. Life Sci 74(16):2049–2059
37. Harada N, Shimozawa N, Okajima K (2009) AT(1) receptor blockers increase insulin-like growth factor-I production by stimulating sensory neurons in spontaneously hypertensive rats. Transl Res 154(3):142–152
38. Miyamoto N, Zhang N, Tanaka R et al (2011) Neuroprotective role of angiotensin II type 2 receptor after transient focal ischemia in mice brain. Neurosci Res 61(3):249–256
39. Kiyota T, Ingraham K, Jacobsen M (2011) FGF2 gene transfer restores hippocampal functions in mouse models of Alzheimer's disease and has therapeutic implications for neurocognitive disorders. Proc Natl Acad Sci U S A 108 (49):E1339–E1348
40. Webber K, Casadesus G, Bowen R (2007) Evidence for the role of luteinizing hormone in Alzheimer disease. Endocr Metab Immune Disord Drug Targets 7(4):300–303
41. Stroth U, Meffert S, Gallinat S et al (1998) Angiotensin II and NGF differentially influence microtubule proteins in PC12W cells: role of the AT2 receptor. Brain Res Mol Brain Res 53(1–2):187–195
42. Laflamme L, Gasparo M, Gallo J (1996) Angiotensin II induction of neurite outgrowth by AT2 receptors in NG108-15 cells. Effect counteracted by the AT1 receptors. J Biol Chem 271(37):22729–22735
43. Govindarajan G, Eble D, Lucchesi P et al (2000) Focal adhesion kinase is involved in angiotensin II-mediated protein synthesis in cultured vascular smooth muscle cells. Circ Res 87(8):710–716
44. Hercule H, Tank J, Plehm R et al (2007) Regulator of G protein signalling 2 ameliorates angiotensin II-induced hypertension in mice. Exp Physiol 92(6):1014–1022
45. Heximer S, Knutsen R, Sun X et al (2003) Hypertension and prolonged vasoconstrictor signaling in RGS2-deficient mice. J Clin Invest 111(4):445–452
46. Matsuzaki N, Nishiyama M, Song D et al (2011) Potent and selective inhibition of angiotensin AT1 receptor signaling by RGS2: roles of its N-terminal domain. Cell Signal 23 (6):1041–1049
47. Fujio Y (2010) RGS2 determines the preventive effects of ARBs against vascular remodeling: toward personalized medicine of antihypertensive therapy with ARBs. Hypertens Res 33(12):1221–1222
48. Mitchell A, Rushentsova U, Siffert W (2006) The angiotensin II receptor antagonist valsartan inhibits endothelin 1-induced vasoconstriction in the skin microcirculation in humans in vivo: influence of the G-protein beta3 subunit (GNB3) C825T polymorphism. Clin Pharmacol Ther 79(3):274–281
49. Hyde Z, Flicker L, Almeida O et al (2010) Higher luteinizing hormone is associated with poor memory recall: the health in men study. J Alzheimers Dis 19(3):943–951
50. Chu C, Zhou J, Zhao Y et al (2012) Expression of FSH and its co-localization with FSH

receptor and GnRH receptor in rat cerebellar cortex. J Mol Histol 44(1):19–26
51. Casadesus G, Atwood C, Zhu X et al (2005) Evidence for the role of gonadotropin hormones in the development of Alzheimer disease. Cell Mol Life Sci 62(3):293–298
52. Karlsson A, Maizels E, Flynn M et al (2010) Luteinizing hormone receptor-stimulated progesterone production by preovulatory granulosa cells requires protein kinase A-dependent activation/dephosphorylation of the actin dynamizing protein cofilin. Mol Endocrinol 24(9):1765–1781
53. Nicholls P, Harrison C, Walton K et al (2011) Hormonal regulation of sertoli cell microRNAs at spermiation. Endocrinology 152 (4):1670–1683
54. Pantic I, Basta-Jovanovic G, Starcevic V et al (2013) Complexity reduction of chromatin architecture in macula densa cells during mouse postnatal development. Nephrology (Carlton) 18(2):117–124
55. King G, Rosene D, Abraham C (2012) Promoter methylation and age-related downregulation of Klotho in rhesus monkey. Age (Dordr) 34(6):1405–1419
56. Klein C, Botuyan M, Wu Y (2011) Mutations in DNMT1 cause hereditary sensory neuropathy with dementia and hearing loss. Nat Genet 43(6):595–600
57. Pietrzak M, Rempala G, Nelson P (2011) Epigenetic silencing of nucleolar rRNA genes in Alzheimer's disease. PLoS One 6(7):e22585
58. Johnson A, Akman K, Calimport S et al (2012) The role of DNA methylation in aging, rejuvenation, and age-related disease. Rejuvenation Res 15(5):483–494
59. Yoon H, Ghee J, Piao S et al (2011) Angiotensin II blockade upregulates the expression of Klotho, the anti-ageing gene, in an experimental model of chronic cyclosporine nephropathy. Nephrol Dial Transplant 26(3):800–813
60. Chu C, Lo J, Hu W et al (2012) Histone acetylation is essential for ANG-II-induced IGF-IIR gene expression in H9c2 cardiomyoblast cells and pathologically hypertensive rat heart. J Cell Physiol 227(1):259–268
61. Sanchez-Varo R, Trujillo-Estrada L, Sanchez-Mejias E (2012) Abnormal accumulation of autophagic vesicles correlates with axonal and synaptic pathology in young Alzheimer's mice hippocampus. Acta Neuropathol 123(1):53–70
62. Gunawardena S, Yang G, Goldstein L (2013) Presenilin controls kinesin-1 and dynein function during APP-vesicle transport in vivo. Hum Mol Genet 22(19):3828–3843
63. Driver J, Beiser A, Au R et al (2012) Inverse association between cancer and Alzheimer's disease: results from the Framingham Heart Study. BMJ 344:e1442
64. Keeney J, Swomley A, Harris J et al (2012) Cell cycle proteins in brain in mild cognitive impairment: insights into progression to Alzheimer disease. Neurotox Res 22(3):220–230
65. Sieradzki A, Yendluri B, Palacios H et al (2011) Implication of oncogenic signaling pathways as a treatment strategy for neurodegenerative disorders-contemporary approaches. CNS Neurol Disord Drug Targets 10(2):175–183
66. Demetrius L, Simon D (2013) The inverse association of cancer and Alzheimer's: a bioenergetic mechanism. J R Soc Interface 10 (82):20130006
67. Eckert G, Renner K, Eckert S et al (2012) Mitochondrial dysfunction–a pharmacological target in Alzheimer's disease. Mol Neurobiol 46(1):136–150
68. Bowen R, Smith M, Harris P (2002) Elevated luteinizing hormone expression colocalizes with neurons vulnerable to Alzheimer's disease pathology. J Neurosci Res 70:514–518
69. Godyń J, Jończyk J, Panek D et al (2016) Therapeutic strategies for Alzheimer's disease in clinical trials. Pharmacol Rep 68 (1):127–138
70. Waite L (2015) Treatment for Alzheimer's disease: has anything changed? Aust Prescr 38 (2):60–63
71. Mosconi L, Berti V, Glodzik L, Pupi A, De Santi S, de Leon MJ (2010) Pre-clinical detection of Alzheimer's disease using FDG-PET, with or without amyloid imaging. J Alzheimers Dis 20(3):843–854
72. Caldwell C, Yao J, Brinton R (2015) Targeting the prodromal stage of Alzheimer's disease: bioenergetic and mitochondrial opportunities. Neurotherapeutics 12(1):66–80
73. Rettberg JR, Yao J, Brinton R (2014) Estrogen: a master regulator of bioenergetics systems in the brain and body. Front Neuroendocrinol 35(1):8–30
74. Maimaiti S, Anderson K, DeMoll C et al (2016) Intranasal insulin improves age-related cognitive deficits and reverses electrophysiological correlates of brain aging. J Gerontol A Biol Sci Med Sci 71(1):30–39
75. de la Monte S (2013) Intranasal insulin therapy for cognitive impairment and neurodegeneration: current state of the art. Expert Opin Drug Deliv 10(12):1699–1709
76. Winkler J, Fox H (2013) Transcriptome meta-analysis reveals a central role for sexsteroids in

the degeneration of hippocampal neurons in Alzheimer's disease. BMC Syst Biol 7:51

77. Bove R, Secor E, Chibnik L et al (2014) Age at surgical menopause influences cognitive decline and Alzheimer pathology in older women. Neurology 82(3):222–229
78. Engler-Chiurazzi E, Singh M, Simpkins J (2015) From the 90s to now: a brief historical perspective on more than two decades of estrogen neuroprotection. Brain Res 1633:96–100
79. Chen Y, Pan C, Xuan A et al (2015) Treatment efficacy of NGF nanoparticles combining neural stem cell transplantation on Alzheimer's disease model rats. Med Sci Monit 21:3608–3615
80. Tuszynski M, Yang J, Barba D (2015) Nerve growth factor gene therapy: activation of neuronal responses in Alzheimer disease. JAMA Neurol 72(10):1139–1147
81. Tatarnikova O, Orlov M, Bobkova N (2015) Beta-amyloid and tau-protein: structure, interaction, and prion-like properties. Biochemistry (Mosc) 80(13):1800–1819
82. Cohen M, Appleby B, Safar J (2016) Distinct prion-like strains of amyloid beta implicated in phenotypic diversity of Alzheimer disease. Prion 10(1):9–17
83. Hochgräfe K, Sydow A, Matenia D et al (2015) Preventive methylene blue treatment preserves cognition in mice expressing full-length pro-aggregant human Tau. Acta Neuropathol Commun 3:25
84. Paban V, Manrique C, Filali M (2014) Therapeutic and preventive effects of methylene blue on Alzheimer's disease pathology in a transgenic mouse model. Neuropharmacology 76 (Pt A):68–79
85. Cavaliere P, Torrent J, Prigent S et al (2013) Binding of methylene blue to a surface cleft inhibits the oligomerization and fibrillization of prion protein. Biochim Biophys Acta 1832 (1):20–28
86. Chen X, Hu J, Jiang L et al (2014) Brilliant Blue G improves cognition in an animal model of Alzheimer's disease and inhibits amyloid-β-induced loss of filopodia and dendrite spines in hippocampal neurons. Neuroscience 279:94–101
87. Wong H, Qi W, Choi H et al (2011) A safe, blood-brain barrier permeable triphenylmethane dye inhibits amyloid-β neurotoxicity by generating nontoxic aggregates. ACS Chem Nerosci 2(11):645–657
88. Irwin JA, Erisir A, Kwon I (2016) Oral triphenylmethane food dye analog, brilliant blue G, prevents neuronal loss in APPSwDI/NOS2−/− mouse model. Curr Alzheimer Res 13 (6):663–677
89. Risse E, Nicoll A, Taylor W, Wright D et al (2015) Identification of a compound that disrupts binding of amyloid-β to the prion protein using a novel fluorescence-based assay. J Biol Chem 290(27):17020–17028
90. Pihlaja R, Koistinaho J, Malm T et al (2008) Transplanted astrocytes internalize deposited beta-amyloid peptides in a transgenic mouse model of Alzheimer's disease. Glia 56 (2):154–163

Chapter 18

Pharmacologic Manipulation of Wnt Signaling and Cancer Stem Cells

Yann Duchartre, Yong-Mi Kim, and Michael Kahn

Abstract

Wnt (Wingless-related integration site)-signaling orchestrates self-renewal programs in normal somatic stem cells as well as in cancer stem cells. Aberrant Wnt signaling is associated with a wide variety of malignancies and diseases. Although our understanding has increased tremendously over the past decade, therapeutic targeting of the dysregulated Wnt pathway remains a challenge. Here we review recent preclinical and clinical therapeutic approaches to target the Wnt pathway.

Key words Wnt signaling, Cancer stem cells, Drug resistance, Self-renewal, Clinical trial, Somatic stem cells

1 Introduction

Drug resistance remains a major obstacle in the treatment of cancer. Cancer stem cell (CSC) or cancer-initiating cell (CIC) [1] populations share the properties of self-renewal and pluripotency with their normal somatic stem cell (SSC) counterparts. CSC appear to be the root cause of drug resistance. By definition, the self-renewal of a stem cell leads to production of one daughter cell identical to the mother cell, thereby retaining its stem cell properties. Pluripotency enables stem cells to differentiate into multiple divergent committed and specialized cell types. Understanding the similarities and differences of normal and cancer stem cells, to enable safely therapeutically targeting and eliminating CSCs, may be the key to overcome drug resistance. CSC may emerge from normal SSC after genetic alterations acquired during DNA replication, via various insults and/or from microenvironmental factors [1]. CSC and SSC are regulated by the same evolutionarily conserved signaling pathways, e.g., Notch [2], Hedgehog [3] and Wnt/β-catenin [4, 5]. Here, we review recent findings on Wnt signaling in tumorigenesis and therapeutic strategies targeting this pathway.

Tatiana V. Tatarinova and Yuri Nikolsky (eds.), *Biological Networks and Pathway Analysis*, Methods in Molecular Biology, vol. 1613, DOI 10.1007/978-1-4939-7027-8_18, © Springer Science+Business Media LLC 2017

2 Wnt Signaling Pathways

Wnt signaling is often parsed into three pathway groupings: canonical, noncanonical planar cell polarity (PCP) pathway, and noncanonical Wnt/calcium pathway. The central protein of the canonical pathway is β-catenin: whose cytoplasmic and nuclear levels are normally under very strict controls. Wnt ligand binding to Frizzled receptors as well as LRP5/6 co-receptors (low density lipoprotein receptor-related protein 5/6) initiates an intracellular signaling cascade and subsequent β-catenin nuclear translocation. In the absence of Wnt ligands, cytoplasmic β-catenin is targeted by a degradation complex composed of the tumor suppressor Adenomatous Polyposis Coli (APC), the scaffolding protein AXIN and two kinases CK1α (casein kinase 1α) and GSK-3β (glycogen synthase kinase 3β) [6] (Fig. 1a). These last two components are able to phosphorylate β-catenin on several serine and threonine residues in its N-terminus. Phosphorylated β-catenin is then recognized by β-Transducin, which is part of an ubiquitin ligase complex, leading to polyubiquitination and proteasomal degradation of β-catenin [7]. Wnt ligand binding to Frizzled receptors in association with LRP5/6 induces Dishevelled (DVL) phosphorylation, which recruits Axin, thereby deconstructing the degradation complex and achieving β-catenin stabilization and subsequent nuclear translocation. In the nucleus, in the classical canonical signaling cascade, β-catenin binds members of the TCF/LEF (T-cell Factor/ Lymphoid Enhancer Factor) family of transcription factors and recruits the transcriptional Kat3 co-activators, p300 and/or CBP (CREB-binding protein), as well as other proteins, e.g., BCl9, to transcribe Wnt target genes and engender chromatin modifications [8–11] (Fig. 1b).

The noncanonical PCP and Wnt/calcium pathways are termed "β-catenin-independent pathways" and coexist and interact with the canonical Wnt pathway. The noncanonical PCP pathway is characterized by Wnt ligand binding to Frizzled receptors and activation of small GTPases such as RhoA (Ras homolog gene family member A), RAC (Ras-related C3 botulinum toxin substrate) and Cdc42 (cell division control protein 42), via recruitment and activation of Dishvelled [12] (Fig. 2a). The PCP pathway affects the cytoskeleton and triggers the transcriptional activation of target genes responsible for cell adhesion and migration [13].

The noncanonical calcium-dependent pathway is characterized by engagement of Wnt ligands with Frizzled receptors and RYK or ROR (alternative receptors) enhancing cell migration and Wnt canonical pathway inhibition through the management of intracellular calcium flux and activation of Calmodulin kinase II (CaMK2), Jun kinase (JNK), and PKC [14] (Fig. 2b).

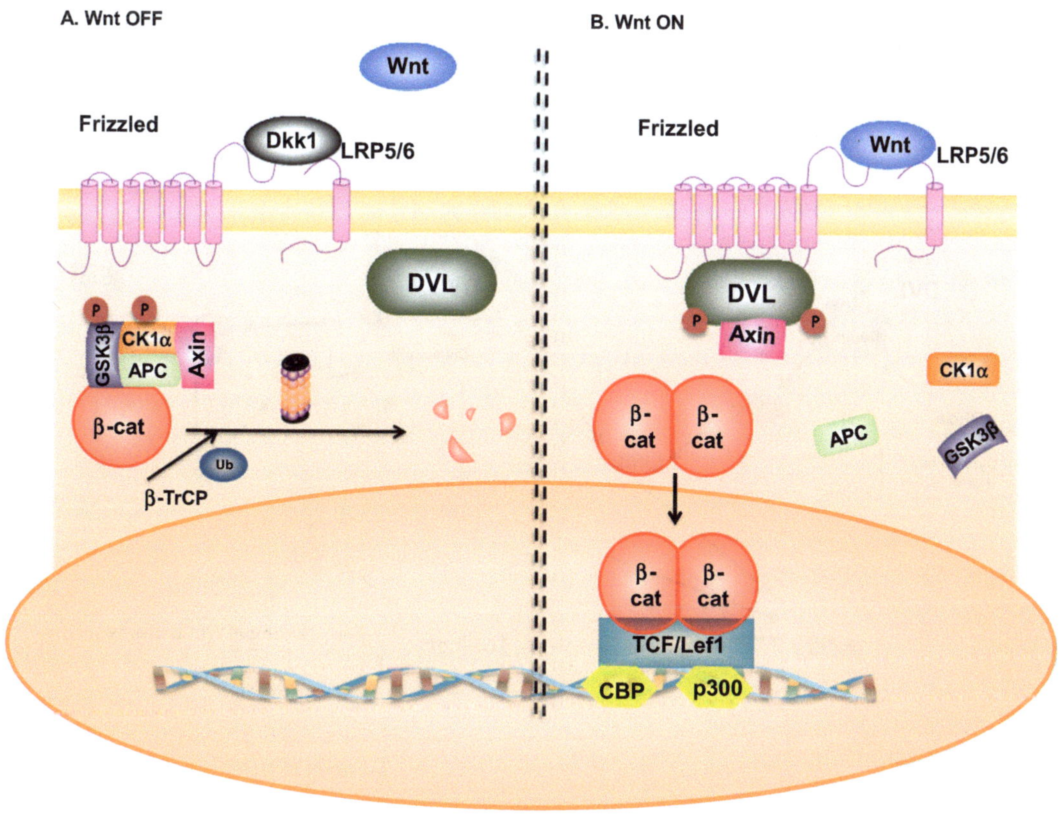

Fig. 1 (**a**) "Wnt Off." In the absence of Wnt ligands, a destruction complex composed of Axin-1 and its tumor suppressor partners Adenomtous Polyposis Coli (APC), Glycogen synthase kinase 3 beta (GSK3B), and Casein kinase 1 (CK1α) is formed. The destruction complex phosphorylates ß-catenin and targets it for proteasomal degradation, regulating the cytoplasmic level of ß-catenin. (**b**) "Wnt On." Wnt ligands bind to the Frizzled/Lrp5/6 (Low density lipoprotein receptor-related proteins 5 or 6) receptors leading to the phosphorylation of a negative regulator of the destruction complex, Dishevelled (Dvl). Dvl recruits Axin, inhibiting its interaction with other components of the destruction complex. ß-catenin is then free to accumulate in the cytoplasm and translocates to the nucleus, where it activates the transcription of Wnt target genes after association with transcription factors of the TCF/Lef family and co-activators such as CBP (cyclic AMP response element-binding protein) and p300. *Arrows* indicate activation/induction; *blunt ended lines* indicate inhibition/blockade

Even though these three Wnt pathways are separately delineated for convenience, in reality Wnt signaling involves the integration of all the three pathways [15–17].

3 The Role of Wnt Signaling in Cancer Stem Cells

Over the past decade, CSCs have been identified in multiple tumor types [18–21], including brain tumors [22], melanoma [23], breast [24], liver [25], pancreatic [26], colon cancers [27, 28], and leukemia [29, 30] and are strongly correlated with poor outcome [31, 32]. CSCs constitute a very small subset within a tumor, sustaining

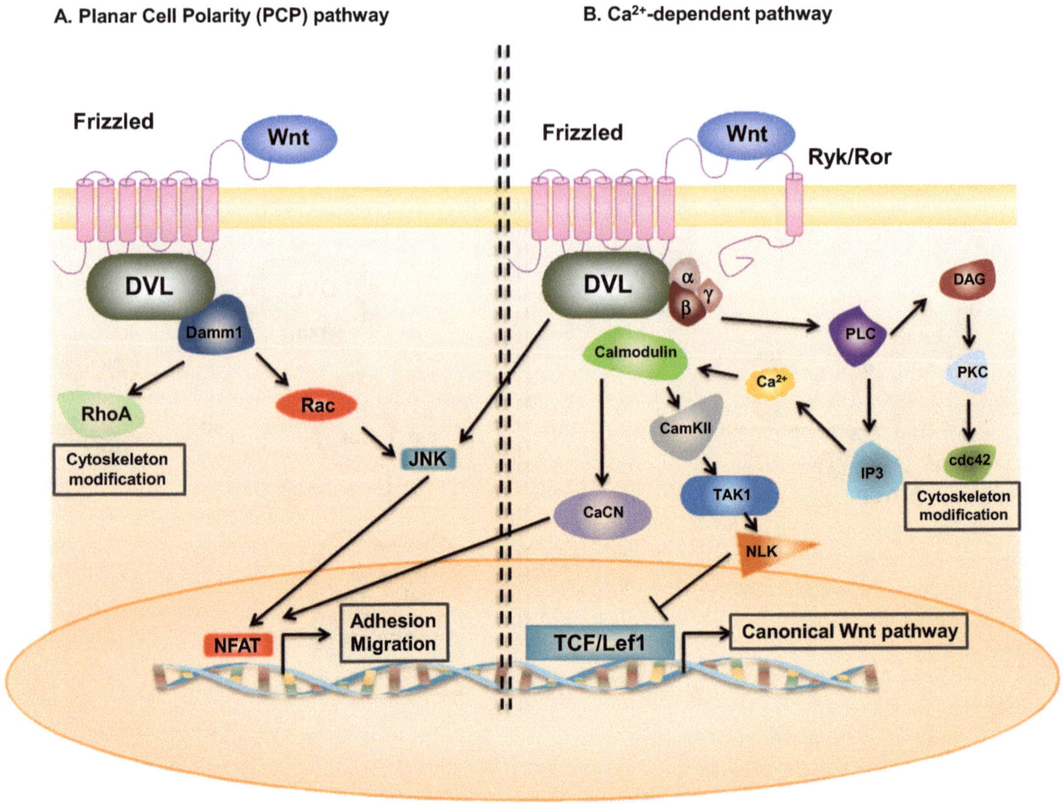

Fig. 2 (**a**) Noncanonical Wnt-signaling: Noncanonical Wnt/PCP (planar cell polarity) pathway. Wnt ligand binding to frizzled receptors leads to activation of Dishevelled (Dvl), which recruits DAAM1 (Dishevelled associated activator of morphogenesis 1), enhancing the stimulation of GTPases Rac (Ras-related C3 botulinum toxin substrate), and RHOA (Ras homolog gene family member A), leading to actin cytoskeleton rearrangement. In addition, Dvl activates Rac and finally JNK (c-Jun-N-terminal-kinase) thereby modulating cell migration. (**b**) Noncanonical Wnt/calcium pathway. Wnt ligands bind to Frizzled receptors and Ror/Ryk co-receptors, activating Dvl and trimeric G-proteins (Gα, β, γ). This leads to the generation of IP3 (inositol 1,4,5-triphosphate) and DAG2 (diacylglycerol) through PLC (Phospholipase C) activation. IP3 triggers the release of calcium ions (Ca^{2+}) from the endoplasmic reticulum activating calmodulin and subsequently CAMKII (calcium/calmodulin-dependent kinase II), TAK-1 (TGF-β activated kinase 1), and NLK (Nemo-like kinase), thereby inhibiting the canonical Wnt pathway. Moreover, calmodulin activation stimulates calcineurin and NFAT (Nuclear Factor of Activated T-cells) involved in adhesion and migration processes. This pathway activates also PKC (Protein Kinase C) and Cdc42 (cell division control protein 42), rearranging the actin cytoskeleton. *Arrows* indicate activation; *blunt ended lines* indicate inhibition/blockade

the tumor via proliferation and self-renewal [27] capabilities and telomerase expression [27]. They are also known to be more chemotherapy and radiotherapy resistant, leading to relapse and metastasis of the disease [33–36]. Resistance is also associated with their quiescent state and specific interactions with their microenvironment [37]. Therefore, targeting CSCs, specifically while sparing normal SSCs, is a critical therapeutic goal. Dysfunctional Wnt signaling has been related to the evolution of and maintenance of leukemic stem cells as well as many other different cancers.

This is not surprising given the importance of the Wnt pathway in stem cell homeostasis [38]. Examples of aberrant Wnt signaling in cancer stem cell development include the progression of chronic phase CML toward blastic crisis phase due to GSK3β mutations and β-catenin stabilization in GMP cells (granulocyte-macrophage progenitor cells) [39]. A recent study showed that despite the inhibitory effect of tyrosine kinase inhibitor (TKI) on the Wnt signaling pathway in CML stem cells, relapses occur in patients at least in part by reactivation of the Wnt pathway [40]. TKI treatment induces a downregulation of miR29 involved in CD70 promoter methylation. The overexpression of CD70 enhances the transcription of CD27, which is a known activator of the Wnt signaling pathway [41]. Wang et al. also showed that constitutive activation of the canonical Wnt pathway, via expression of a stabilized form of β-catenin, is necessary to generate AML leukemic stem cells from MLL-AF9-transduced progenitors cells [42]. This study suggests that aberrant Wnt pathway activation could give rise to leukemic stem cells (LSCs) not only from hematopoietic stem cells (HSC) but additionally from more committed progenitors. Recently, Giambra and colleagues showed, using a Wnt reporter construct expressing GFP under the TCF promoter, that minor subpopulations of bulk T-cell acute lymphoblastic leukemia (T-ALL) had highly activated Wnt/β-catenin pathway signaling and that these cells were able to transplant the disease in a limiting dilution assay [43]. Leukemic stem cells were highly enriched in the GFP^+ Wnt expressing population compared to the GFP^- (ratio of over 200-fold) population, suggesting that Wnt signaling is also required for T-ALL stem cell self-renewal. In this model, the transcriptional activation of β-catenin seems to be triggered by the transcription factor HIF1-alpha (Hypoxia-Induced Factor 1-alpha) and deletion of HIF1-alpha leads to LSC targeting [43]. Our group recently demonstrated the implication of the Wnt pathway in the self-renewal of B-cell acute lymphoblastic leukemia (B-ALL). The treatment of B-ALL cells with a small molecule that specifically binds to the N-terminal of CBP, ICG-001, inhibits the interaction between β-catenin and CBP leading to differentiation and loss of self renewal [44]. Additionally, iCRT14, a β-catenin/TCF interaction inhibitor, leads to a decrease in Wnt target gene expression, decreases the viability of ALL cell lines in combination with chemotherapy and sensitizes chemoresistant patient sample-derived ALL cells responsible for relapse [45].

In order to efficiently target CSCs, researchers initially focused on ways to identify them. Even if the normal SSCs and CSCs usually express the same cell surface markers [46], some reports have successfully characterized cancer stem-like cells in for example breast cancer based upon specific marker sets (expression of $CD44^{high}CD24^{low}$) [24]. Interestingly, both CD44 and CD24 are direct Wnt target genes [47–50]. CD44 acts as a positive regulator of the Wnt pathway by affecting LRP6 localization and activity [49–51]. The Wnt signaling pathway also appears to play an important role in another hallmark of cancer stem cells and metastasis, i.e., the epithelial-to-mesenchymal transition (EMT) [52–54]. The downregulation of E-Cadherin (usually tightly associated with β-catenin in normal epithelium) triggers the nuclear translocation of β-catenin and activation of canonical Wnt signaling [55]. The gene slug, a marker gene of EMT, also induces nuclear translocation of β-catenin [56]. Moreover, twist and slug, strong activators of EMT are both putative β-catenin targets [57]. Furthermore, a number of Wnt/β-catenin targets genes have been associated with invasion, migration, and metastasis (*S100A4*, *fibronectin*, *L1CAM*, *CD44*, *MMP7*, *uPAR*, etc.) [58]. Wnt signaling may also play an important role in the resistance of cancer stem cells to chemotherapy. The promoter sequence of the multidrug resistance gene ABCB1/MDR-1 contains several TCF binding elements triggering its transcription in colorectal cancer [59]. Fang et al. have recently shown that an inhibitor of the β-catenin–TCF4 interaction (LF3) induces strong inhibition of Wnt pathway gene expression involved in cell cycle and metastasis in a colon cancer cells [60]. Interestingly, this inhibitor, similar to the CBP/catenin antagonist ICG-001, also blocks the self-renewal capacity of colon and head and neck cancer stem cells in vitro and decreases the growth and induces differentiation of colon cancer cells in vivo. The inhibition of another interaction involving β-catenin (β-catenin/CBP interaction), using the small molecule ICG-001 decreases the expression of Survivin/BIRC5, which is an inhibitor of apoptosis and a target of CBP, leading to eradication of drug resistant ALL cells in vitro and prolonged survival of ALL engrafted mice [44]. Similar results have been obtained using ICG-001 with CML LSC [61]. Wnt signaling has also been linked to hematopoietic CSC which seem to be dependent on this pathway [42, 62]. In CML, Wnt pathway deregulation favors the progression of disease to more advanced phases [63]. The deregulation of Wnt signaling can also occur at the epigenetic level. For example, the promoters of several Wnt pathway inhibitors (i.e., SFRP, DKK, and WIF-1) were found to be hypermethylated in both ALL and AML, correlating negatively with the survival of these patients [64, 65].

4 Preclinical and Clinical Wnt Inhibitors

After decades of research and discovery on the Wnt signaling pathway, few molecules are considered to be truly specific for targeting the Wnt pathway, and to date none has been approved by the US Food and Drug Administration (FDA). We summarize here nonspecific and specific Wnt-inhibitors that are in preclinical and clinical use (Table 1; Fig. 3).

Table 1
Wnt inhibitors clinically approved

Clinical	Disease	Mechanism	Ref
NSAID (Aspirin, Celecoxib)		PGE2 generated via COX suppresses β-catenin degradation	[66–69]
Retinoids	APML	Unclear	[99]
Vitamin D	Colorectal and breast cancers	Unclear	[70]
Pyrvinium pamoate	Lung cancer, colon cancer	Unclear: Wnt signaling inhibition via CK1α activation or GSK3 activation	[72]
Sulindac		Dishevelled inhibition	[73]

This table presents the nonspecific Wnt inhibitors that are already clinically approved. The mechanism of action of these drugs (when it is known) involves the inhibition of different intracellular proteins implicated in the Wnt signaling pathway (β-catenin, CK1α, GSK3β, and Dvl)

4.1 Nonspecific Wnt Inhibitors

Nonsteroidal anti-inflammatory drugs (NSAIDS, used for treatment of pain, fever) and vitamin derivatives that target nuclear receptors have demonstrated interesting anticancer effects [66, 67] and particularly in Wnt-dependent cancers, e.g., colorectal cancer [68, 69]. Cyclooxygenases (COX1 and 2) metabolize arachidonic acid into prostaglandins (PG) that, via their G-protein Coupled Receptors, can lead to β-catenin stabilization and activation of canonical Wnt signaling [81–83]. The inhibition of COX by NSAIDS (aspirin, sulindac or specific COX2 inhibitors like celecoxib) suppresses the synthesis of prostaglandins and thereby inhibits Wnt signaling. These compounds, especially celecoxib, also showed COX-independent anticancer effects, notably in a xenograft model of COX2-deficient tumors [84–87]. NSAIDS have the capacity to decrease the number of polyps in a mouse model of Familial Adenomatous Polyposis (FAP) mouse, where the APC gene is truncated and Wnt/β-catenin signaling constitutively actived [88, 89]. FAP patients treated for 6 months with the NSAID sulindac showed a reduction in nuclear β-catenin in polyps and a reduction in polyp formation, maybe via direct inhibition of dishevelled by sulindac [73, 90–93]. The aspirin derivative NO-ASA (NO-releasing aspirin) showed even better efficacy in reduction of polyp formation in vitro and in vivo possibly via disruption of the β-catenin/TCF complex without any observable toxicity to the normal intestine [94–96]. COX2 has also been recently implicated in imatinib resistance in a model of Chronic Myeloid Leukemia [97]. Treatment of imatinib-resistant K562 cells with celecoxib, via Wnt and MEK signaling pathway modulation, downregulates the expression of ABC transporter protein family members such as MRP1, MRP2, MRP3, ABC2, ABCA2,

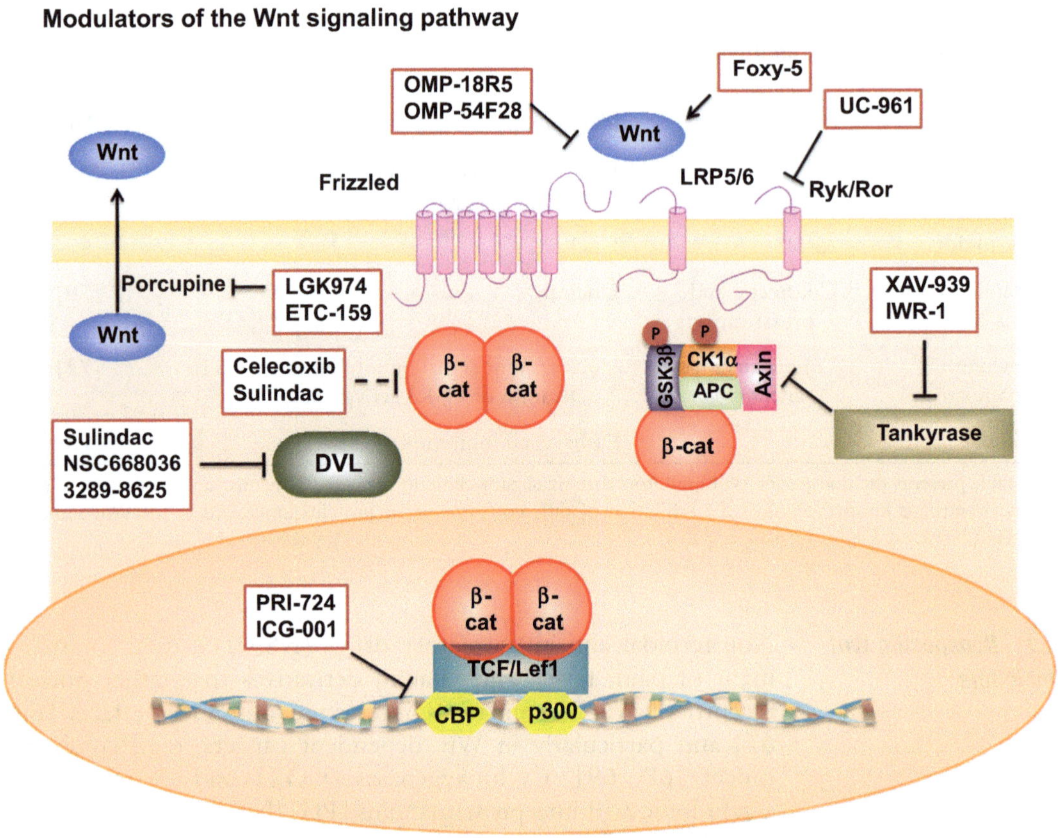

Fig. 3 Modulators of the Wnt signaling pathway. The Wnt signaling pathway can be modulated with a wide variety of drugs currently in preclinical studies, in clinical trial or already approved (*see* text and Table 2). These drugs act at different levels in the Wnt pathway: Wnt receptors (OMP-18R5, OMP-54F28, Foxy-5, and ETC-961), Wnt posttranslational modification and secretion (LGK974 and ETC-159), tankyrase inhibitors (XAV-939 and IWR-1), Dvl inhibitors (Sulindac, NSC668036 and 3289–8625), β-catenin indirect modulators (Celecoxib and Sulindac), and inhibitors of β-catenin/CBP interaction (ICG-001 and PRI-724). *Arrows* indicate activation; *blunt ended lines* indicate inhibition/blockade

and ABCG2, which are associated with drug resistance, thereby sensitizing the K562 cells to imatinib.

Retinoids, produced from vitamin A metabolism, demonstrated anticancer effects at least in part via Wnt signaling pathway inhibition [70]. 1α,25-dihydroxy-vitamin D3, the active form of vitamin D, demonstrated tumor suppressor activity, notably by formation of a transcriptional complex able to bind β-catenin and thereby enhancing the expression of E-cadherin. These effects lead to retention of β-catenin in the cytoplasm, resulting in inhibition of the Wnt pathway in both breast and colon cancers [71].

4.2 Specific Wnt Inhibitors

Besides these FDA-approved nonspecific Wnt inhibitors, several molecularly targeted agents have been developed and have entered preclinical or clinical trials. Dvl, being one of the key regulators of

Table 2
Wnt inhibitors currently in clinical trials

Clinical trials	Disease	Mechanism	Ref
OMP18R5, Vantictumab	Solid tumors	Humanized Ab against multiple Fzd receptors	[74]
OMP-54F28, Fzd8-Fc	Pancreatic, Ovarian, Hepatocellular, Colorectal, and Breast	Fc fusion protein with Fzd8, which binds all Wnt ligands	[75]
PRI-724	Solid Tumors, Colon and Pancreatic Cancer, CML and AML	Small molecule inhibitor of CBP/ catenin binding	[76]
LGK974, Porcupine inhibitor	Melanoma, Breast cancer, and Pancreatic adenocarcinoma	Wnt posttranslational acylation and palmitoylation	[77]
ETC-1922159 (ETC-159), Porcupine inhibitor	Colon, Ovarian, and Pancreas cancers	Wnt posttranslational acylation and palmitoylation	[78]
UC-961 (cirmtuzumab)	Chronic Lymphoid Leukemia	Humanized antibody against ROR1	[79]
Foxy-5	Breast, Colon, Prostate	Reduced cell migration through Fzd-5 and cytosolic calcium signal	[80]

This table summarizes the different Wnt pathway modulators with variable specificities and at different stages of development (fully described in the main text)

the Wnt canonical pathway, has been a focus of numerous studies and has engendered the development of several inhibitors. The PDZ domain of Dvl plays an essential role in Dvl-Frizzled receptor interactions and the intracellular transduction of the Wnt signal. Some inhibitors of the Dvl PDZ domain (NSC 668036, FJ9, 3289-8625—Fig. 3), discovered by in silico screening, demonstrated the ability to inhibit the Wnt pathway in vivo [98–100].

LGK974 is a porcupine (PORCN) inhibitor, which entered into a phase I clinical trial in 2011 (Novartis, NCT01351103, recruitment phase) [77]. Porcupine is a member of the membrane-bound O-acetyltransferase (MBOAT) family and is responsible for lipid modification of Wnt and subsequent Wnt secretion [101, 102]. The trial will investigate the effects of LGK974 on the Wnt signaling pathway in patients affected with Wnt-dependent cancers (pancreatic adenocarcinoma, BRAF mutant colorectal cancer) (clinicaltrials.gov).

Recently, another PORCN inhibitor, ETC-1922159 (ETC-159), developed in a collaboration between the Agency for Science, Technology and Research (A*STAR) and Duke-National

University of Singapore Graduate Medical School (Duke-NUS) entered into a phase I clinical trial in Singapore. The first patient was dosed on Jun 18, 2015. ETC-159 inhibits Wnt secretion and activity and is highly efficient preclinically in different cancers driven by Wnt signaling and notably in R-spondin translocated colorectal cancers [78].

Cucurbitacin B, a tetracyclic triterpene found in plants of the family Cucurbitaceae, has been shown to downregulate the Wnt signaling pathway, essentially via inhibition of Wnt3 and Wnt3a expression and GSK3 proteins (α and β) upregulation and activation, accelerating the degradation of the β-catenin [103]. This Wnt pathway inhibition leads to a reduction of the "stemness," angiogenic and metastatic properties of non-small cell lung cancer cells, as well as the inhibition of growth and increased apoptosis of breast cancer cells both in vitro and in vivo [103, 104].

The tankyrase inhibitors (XAV-939 and IWR-1) stabilize axin thereby inducing the degradation of the β-catenin [105] and may act as antitumor drugs also by participating in telomere shortening [106]. Wu et al. have recently shown that XAV939 has a synergistic effect on 5-fluorouracil/cisplatin-induced apoptosis of colon cancer stem cells in vitro [107]. Pyrvinium pamoate (PP) was shown to inhibit the Wnt pathway in different models of lung and colon cancers in vitro as well as in vivo [87, 108]. Even though its mechanism of action is still unclear (CK1α or GSK3 activation), this compound decreases proliferation and self-renewal of lung and colon cancer stem cells [72]. A new inhibitor of FLT3, SKLB-677, has been found to also inhibit the Wnt signaling pathway in acute myeloid leukemia (AML) cell lines and in vivo. Although the mechanism of action is not well described, this compound seems to be able to downregulate FLT3 and Wnt signaling and may improve the targeting of AML stem cells, which are responsible for AML relapse [109].

Among the few agents already in clinical trials, two were developed by Oncomed Pharmaceuticals Inc.; OMP-18R5 (Vantictumab), is a fully humanized antibody directed against minimally five different Frizzled receptors. In preclinical studies, OMP-18R5 demonstrated antiproliferative effects in various human tumors model (lung, pancreas, breast, and colon) and had synergistic effects with conventional chemotherapy [74]. The results of the Phase Ia study showed a decrease in Wnt pathway gene expression and increased expression of differentiation genes with some adverse events including fatigue, vomiting, diarrhea, constipation, nausea, and abdominal pain (ASCO, 2013). This compound is now in Phase Ib trials in combination with standard chemotherapy for solid tumors (breast, lung, and pancreas cancers).

OMP-54F28 (Oncomed Pharmaceuticals), is a recombinant fusion protein containing the extracellular ligand binding domain of human Frizzled 8 receptor fused to a human IgG1 Fc fragment

[75]. OMP-54F28 can bind native Fzd8 receptor's ligands and thereby inhibit Wnt signaling. Preclinical studies demonstrated the antitumor efficacy of OMP-54F28: reduction of tumor growth and decrease of CSC frequency as a single agent and in combination with other chemotherapeutic agents [75]. A phase I trial (NCT01608867) is currently ongoing. It is a dose escalation study in patients with advanced solid tumors. Subjects will be assessed for safety, immunogenicity, pharmacokinetics, biomarkers, and efficacy. It appears that the most common adverse events are fatigue, muscle spasms, alopecia, nausea, decreased appetite, and dysgeusia (http://www.eurekalert.org/pub_releases/2014-05/uocd-rip053014.php). Additionally, patients are followed for bone density evolution, as bone fracture was observed in one patient at the highest tested dose (20 mg/kg every 3 weeks after 6 cycles). Three Phase 1b studies have started to check the dose escalation of OMP-54F28 in ovarian (NCT02092363), pancreatic (NCT02050178), and hepatocellular (NCT02069145) cancers in combination with respective conventional chemotherapy.

Another way to modulate the extracellular part of the Wnt pathway is the inhibition of the Wnt receptor ROR1. A novel humanized antibody (UC-961, cirmtuzumab) targeting the Receptor tyrosine kinase-like Orphan Receptor 1 (ROR1), expressed by chronic lymphocytic leukemia cells (CLL), but not on normal cells, showed anticancer effects in a CLL mouse model [79]. This antibody recently entered a Phase I clinical trial to determine the safety and the effects of this antibody (NCT02222688).

The Wnt pathway, implicated in both cancer growth and drug resistance, is also highly involved in cancer cell migration and metastasis [66, 110, 111]. Recently, a hexapeptide mimicking Wnt5a named Foxy-5 has been developed and used to treat various breast cancers in vitro and in vivo [80]. The treatment of murine and human breast cancer cell lines in vitro and in vivo with Foxy-5 did not affect apoptosis or proliferation but decreased the migration and invasion of these cells and finally metastasis. A phase Ia study evaluating safety and pharmacokinetics has been completed and a phase Ib (dose escalating study) is ongoing in breast, colon, and prostate cancer patients (NCT02020291 and NCT02655952).

Wnt signaling can also be modulated very late in the pathway. Our group used a secondary structure-templated chemical library to identify ICG-001 which can efficiently modulate the Wnt pathway [112]. Despite the huge homology between the two Kat3 coactivator proteins CBP and p300, ICG-001 was shown to bind specifically to the cyclic AMP response element-binding protein (CBP) and not to the related transcriptional coactivator p300 [112, 113]. ICG-001 disrupts the ß-catenin/CBP complex and increases the proportion of ß-catenin bound to p300, leading to downregulation of survivin/BIRC5 mRNA and specific apoptosis in colon cancer cells in vitro and in vivo. Recently, Prism Pharmaceuticals developed a second-generation ß-catenin/CBP inhibitor PRI-724. In a Phase Ia safety study in colon cancer, this compound was able to decrease, in a dose-dependent manner, the expression of survivin/BIRC5 in circulating tumor cells, with an acceptable toxicity profile (ASCO, June 2013 and NCT01302405 [76]). Three patients had stable disease for 8, 10, and 12 weeks. Three Phase I/II trials are ongoing in patients with AML/CML (NCT01606579, alone or in combination with AraC or dasatinib), with advanced or metastatic pancreatic adenorcarcinoma (NCT01764477, in combination with Gemcitabine) and in patients with newly diagnosed metastatic colorectal cancer (NCT02413853, in combination with bevacizumab, leucovorin calcium, oxaliplatin, and fluorouracil). A Phase I dose escalation trial in patients with HCV-induced cirrhosis is also on going (NCT02195440).

5 Concluding Remarks

Even after more than 30 years of discovery and investigation of the Wnt signaling pathway, no therapeutic agent is available on the market that specifically and efficiently targets this pathway. Moreover, many of the potential targets like ß-catenin are also implicated in others critical functions including cell–cell adhesion, development, self-renewal [114, 115]. Clearly, precise modulation of the Wnt pathway will be necessary to balance antitumor efficacy with adverse events and will be a challenge for ongoing and future clinical trials. Despite these concerns, new regulators of the Wnt signaling cascade offer the opportunity for us to increase our comprehension of this exceedingly complex pathway and potentially for the treatment of Wnt-related diseases including cancer.

Acknowledgments

YMK was supported by NIH R01CA172896 (YMK). MK is supported by USC Norris Comprehensive Cancer Center Support Grant P30 CA014089, NIH R01CA166161, R21NS074392, R21AI105057 and NIH R01 HL112638. We apologize for the omission of any of our colleagues' work due to space limitations.

References

1. Reya T et al (2001) Stem cells, cancer, and cancer stem cells. Nature 414(6859):105–111
2. Liu J et al (2010) Notch signaling in the regulation of stem cell self-renewal and differentiation. Curr Top Dev Biol 92:367–409
3. Merchant AA, Matsui W (2010) Targeting hedgehog–a cancer stem cell pathway. Clin Cancer Res 16(12):3130–3140
4. Miki T, Yasuda S-y, Kahn M (2011) Wnt/β-catenin signaling in embryonic stem cell

self-renewal and somatic cell reprogramming. Stem Cell Rev Rep 7(4):836–846

5. Takahashi-Yanaga F, Kahn M (2010) Targeting Wnt signaling: can we safely eradicate cancer stem cells? Clin Cancer Res 16 (12):3153–3162
6. Nakamura T et al (1998) Axin, an inhibitor of the Wnt signalling pathway, interacts with beta-catenin, GSK-3beta and APC and reduces the beta-catenin level. Genes Cells 3 (6):395–403
7. Kimelman D, Xu W (2006) β-Catenin destruction complex: insights and questions from a structural perspective. Oncogene 25 (57):7482–7491
8. Moon RT (2005) Wnt/beta-catenin pathway. Sci Signal 2005(271):cm1
9. Mosimann C, Hausmann G, Basler K (2009) β-Catenin hits chromatin: regulation of Wnt target gene activation. Nat Rev Mol Cell Biol 10(4):276–286
10. Teo J-L, Kahn M (2010) The Wnt signaling pathway in cellular proliferation and differentiation: a tale of two coactivators. Adv Drug Deliv Rev 62(12):1149–1155
11. Veeman MT, Axelrod JD, Moon RT (2003) A second canon. Dev Cell 5(3):367–377
12. Lai S-L, Chien AJ, Moon RT (2009) Wnt/Fz signaling and the cytoskeleton: potential roles in tumorigenesis. Cell Res 19(5):532–545
13. Yamamoto S et al (2008) Cthrc1 selectively activates the planar cell polarity pathway of Wnt signaling by stabilizing the Wnt-receptor complex. Dev Cell 15(1):23–36
14. van Amerongen R, Nusse R (2009) Towards an integrated view of Wnt signaling in development. Development 136(19):3205–3214
15. Moon RT et al (2004) Wnt and β-catenin signalling: diseases and therapies. Nat Rev Genet 5(9):691–701
16. Thrasivoulou C, Millar M, Ahmed A (2013) Activation of intracellular calcium by multiple Wnt ligands and translocation of beta-catenin into the nucleus: a convergent model of Wnt/Ca2+ and Wnt/beta-catenin pathways. J Biol Chem 288(50):35651–35659
17. Florian MC et al (2013) A canonical to non-canonical Wnt signalling switch in haematopoietic stem-cell ageing. Nature 503 (7476):392–396
18. Chen J et al (2012) A restricted cell population propagates glioblastoma growth after chemotherapy. Nature 488(7412):522–526
19. Driessens G et al (2012) Defining the mode of tumour growth by clonal analysis. Nature 488 (7412):527–530
20. McCarthy N (2012) Cancer stem cells: tracing clones. Nat Rev Cancer 12 (9):579–579
21. Schepers AG et al (2012) Lineage tracing reveals Lgr5+ stem cell activity in mouse intestinal adenomas. Science 337 (6095):730–735
22. Singh SK et al (2004) Identification of human brain tumour initiating cells. Nature 432 (7015):396–401
23. Fang D (2005) A tumorigenic subpopulation with stem cell properties in melanomas. Cancer Res 65(20):9328–9337
24. Al-Hajj M et al (2003) Prospective identification of tumorigenic breast cancer cells. Proc Natl Acad Sci 100(7):3983–3988
25. Ma S et al (2007) Identification and characterization of tumorigenic liver cancer stem/progenitor cells. Gastroenterology 132 (7):2542–2556
26. Li C et al (2007) Identification of pancreatic cancer stem cells. Cancer Res 67(3):1030–1037
27. O'Brien CA, Kreso A, Jamieson CHM (2010) Cancer stem cells and self-renewal. Clin Cancer Res 16(12):3113–3120
28. O'Brien CA et al (2006) A human colon cancer cell capable of initiating tumour growth in immunodeficient mice. Nature 445 (7123):106–110
29. Lapidot T et al (1994) A cell initiating human acute myeloid leukaemia after transplantation into SCID mice. Nature 367(6464):645–648
30. Jamieson C, Weissman I, Passegue E (2004) Chronic versus acute myelogenous leukemia: a question of self-renewal. Cancer Cell 6 (6):531–533
31. Hussenet T et al (2010) An adult tissue-specific stem cell molecular phenotype is activated in epithelial cancer stem cells and correlated to patient outcome. Cell Cycle 9 (2):321–327
32. Wicha MS (2012) Migratory gene expression signature predicts poor patient outcome: are cancer stem cells to blame? Breast Cancer Research 14(6):114
33. Li L et al (2014) SIRT1 activation by a c-MYC oncogenic network promotes the maintenance and drug resistance of human FLT3-ITD acute myeloid leukemia stem cells. Cell Stem Cell 15(4):431–446
34. Sadarangani A et al (2015) GLI2 inhibition abrogates human leukemia stem cell dormancy. J Transl Med 13(1):98
35. Clevers H (2011) The cancer stem cell: premises, promises and challenges. Nat Med 17 (3):313–319

36. Visvader JE, Lindeman GJ (2010) Stem cells and cancer—the promise and puzzles. Mol Oncol 4(5):369–372
37. Vermeulen L et al (2010) Wnt activity defines colon cancer stem cells and is regulated by the microenvironment. Nat Cell Biol 12 (5):468–476
38. Klaus A, Birchmeier W (2008) Wnt signalling and its impact on development and cancer. Nat Rev Cancer 8(5):387–398
39. Abrahamsson AE et al (2009) Glycogen synthase kinase 3 missplicing contributes to leukemia stem cell generation. Proc Natl Acad Sci 106(10):3925–3929
40. Riether C et al (2015) Tyrosine kinase inhibitor-induced CD70 expression mediates drug resistance in leukemia stem cells by activating Wnt signaling. Sci Transl Med 7 (298):298ra119–298ra119
41. Schürch C et al (2012) CD27 signaling on chronic myelogenous leukemia stem cells activates Wnt target genes and promotes disease progression. J Clin Investig 122(2):624–638
42. Wang Y et al (2010) The Wnt/beta-catenin pathway is required for the development of leukemia stem cells in AML. Science 327 (5973):1650–1653
43. Giambra V et al (2015) Leukemia stem cells in T-ALL require active Hif1 and Wnt signaling. Blood 125(25):3917–3927
44. Gang EJ et al (2013) Small-molecule inhibition of CBP/catenin interactions eliminates drug-resistant clones in acute lymphoblastic leukemia. Oncogene 33(17):2169–2178
45. Dandekar S et al (2014) Wnt inhibition leads to improved chemosensitivity in paediatric acute lymphoblastic leukaemia. Br J Haematol 167(1):87–99
46. Klonisch T et al (2008) Cancer stem cell markers in common cancers—therapeutic implications. Trends Mol Med 14(10):450–460
47. Ahmed MAH et al (2010) CD24 is upregulated in inflammatory bowel disease and stimulates cell motility and colony formation. Inflamm Bowel Dis 16(5):795–803
48. Han J et al (2012) Small interfering RNA-mediated downregulation of beta-catenin inhibits invasion and migration of colon cancer cells in vitro. Med Sci Monit 18(7): BR273–BR280
49. Shulewitz M et al (2006) Repressor roles for TCF-4 and Sfrp1 in Wnt signaling in breast cancer. Oncogene 25(31):4361–4369
50. Wielenga VJM et al (1999) Expression of CD44 in Apc and TcfMutant mice implies regulation by the WNT pathway. Am J Pathol 154(2):515–523
51. Schmitt M et al (2014) CD44 functions in Wnt signaling by regulating LRP6 localization and activation. Cell Death Differ 22 (4):677–689
52. Ksiazkiewicz M, Markiewicz A, Zaczek AJ (2012) Epithelial-mesenchymal transition: a hallmark in metastasis formation linking circulating tumor cells and cancer stem cells. Pathobiology 79(4):195–208
53. DiMeo TA et al (2009) A novel lung metastasis signature links Wnt signaling with cancer cell self-renewal and epithelial-mesenchymal transition in basal-like breast cancer. Cancer Res 69(13):5364–5373
54. Moreno-Bueno G, Portillo F, Cano A (2008) Transcriptional regulation of cell polarity in EMT and cancer. Oncogene 27 (55):6958–6969
55. Huels DJ et al (2015) E-cadherin can limit the transforming properties of activating beta-catenin mutations. EMBO J 34 (18):2321–2333
56. Conacci-Sorrell M et al (2003) Autoregulation of E-cadherin expression by cadherin--cadherin interactions. J Cell Biol 163 (4):847–857
57. Heuberger J, Birchmeier W (2009) Interplay of cadherin-mediated cell adhesion and canonical Wnt signaling. Cold Spring Harb Perspect Biol 2(2):a002915–a002915
58. Brabletz T et al (2005) Invasion and metastasis in colorectal cancer: epithelial-mesenchymal transition, mesenchymal-epithelial transition, stem cells and β-catenin. Cells Tissues Organs 179(1–2):56–65
59. Shitashige M et al (2007) Involvement of splicing factor-1 in β-catenin/T-cell factor-4-mediated gene transactivation and pre-MRNA splicing. Gastroenterology 132 (3):1039–1054
60. Fang L et al (2015) A small-molecule antagonist of the β-catenin/TCF4 interaction blocks the self-renewal of cancer stem cells and suppresses tumorigenesis. Cancer Res 76 (4):891–901
61. Zhao Y et al (2015) CBP/catenin antagonist safely eliminates drug-resistant leukemia-initiating cells. Oncogene 35(28):3705–3717
62. Heidel FH et al (2012) Genetic and pharmacologic inhibition of β-catenin targets imatinib-resistant leukemia stem cells in CML. Cell Stem Cell 10(4):412–424
63. Radich JP et al (2006) Gene expression changes associated with progression and response in chronic myeloid leukemia. Proc Natl Acad Sci 103(8):2794–2799

64. Valencia A et al (2009) Wnt signaling pathway is epigenetically regulated by methylation of Wnt antagonists in acute myeloid leukemia. Leukemia 23(9):1658–1666
65. Roman-Gomez J et al (2007) Epigenetic regulation of Wnt-signaling pathway in acute lymphoblastic leukemia. Blood 109 (8):3462–3469
66. Dey N et al (2013) Wnt signaling in triple negative breast cancer is associated with metastasis. BMC Cancer 13(1):537
67. DuBois RN, Giardiello FM, Smalley WE (1996) Nonsteroidal anti-inflammatory drugs, eicosanoids, and colorectal cancer prevention. Gastroenterol Clin N Am 25 (4):773–791
68. Smalley WE, DuBois RN (1997) Colorectal cancer and nonsteroidal anti-inflammatory drugs. Adv Pharmacol 39:1–20
69. Thun MJ, Henley SJ, Patrono C (2002) Nonsteroidal anti-inflammatory drugs as anticancer agents: mechanistic, pharmacologic, and clinical issues. J Natl Cancer Inst 94 (4):252–266
70. Xiao JH et al (2003) Adenomatous polyposis coli (APC)-independent regulation of beta-catenin degradation via a retinoid X receptor-mediated pathway. J Biol Chem 278 (32):29954–29962
71. Pálmer HG et al (2001) Vitamin D 3 promotes the differentiation of colon carcinoma cells by the induction of E-cadherin and the inhibition of β-catenin signaling. J Cell Biol 154(2):369–388
72. Venerando A et al (2013) Pyrvinium pamoate does not activate protein kinase CK1, but promotes Akt/PKB down-regulation and GSK3 activation. Biochem J 452(1):131–137
73. Boon EMJ et al (2004) Sulindac targets nuclear β-catenin accumulation and Wnt signalling in adenomas of patients with familial adenomatous polyposis and in human colorectal cancer cell lines. Br J Cancer 90(1):224–229
74. Gurney A et al (2012) Wnt pathway inhibition via the targeting of Frizzled receptors results in decreased growth and tumorigenicity of human tumors. Proc Natl Acad Sci 109 (29):11717–11722
75. Le PN, McDermott JD, Jimeno A (2015) Targeting the Wnt pathway in human cancers: therapeutic targeting with a focus on OMP-54F28. Pharmacol Ther 146:1–11
76. El Khoueiry A et al (2013) Abstract 2501: a phase I firt-in human study of PRI-724 in patients (pts) with advanced solid tumors. J Clin Oncol 31(15_Supplement):2501
77. Liu J et al (2013) Targeting Wnt-driven cancer through the inhibition of porcupine by LGK974. Proc Natl Acad Sci 110 (50):20224–20229
78. Madan B et al (2015) Wnt addiction of genetically defined cancers reversed by PORCN inhibition. Oncogene 35(17):2197–2207
79. Choi MY et al (2015) Pre-clinical specificity and safety of UC-961, a first-in-class monoclonal antibody targeting ROR1. Clin Lymphoma Myeloma Leuk 15:S167–S169
80. Safholm A et al (2008) The Wnt-5a-derived hexapeptide foxy-5 inhibits breast cancer metastasis in vivo by targeting cell motility. Clin Cancer Res 14(20):6556–6563
81. Brudvik KW et al (2011) Protein kinase a antagonist inhibits β-catenin nuclear translocation, c-Myc and COX-2 expression and tumor promotion in ApcMin/+ mice. Mol Cancer 10(1):149
82. Castellone MD (2005) Prostaglandin E2 promotes colon cancer cell growth through a Gs-axin-beta-catenin signaling axis. Science 310 (5753):1504–1510
83. Jansen SR et al (2014) Prostaglandin E2 promotes MYCN non-amplified neuroblastoma cell survivalvia β-catenin stabilization. J Cell Mol Med 19(1):210–226
84. Grosch S et al (2001) COX-2 independent induction of cell cycle arrest and apoptosis in colon cancer cells by the selective COX-2-inhibitor celecoxib. FASEB J 15 (14):2742–2744
85. Maier TJ (2005) Targeting the beta-catenin/APC pathway: a novel mechanism to explain the cyclooxygenase-2-independent anticarcinogenic effects of celecoxib in human colon carcinoma cells. FASEB J 19(10):1353–1355
86. Yamazaki R et al (2002) Selective cyclooxygenase-2 inhibitors show a differential ability to inhibit proliferation and induce apoptosis of colon adenocarcinoma cells. FEBS Lett 531(2):278–284
87. Zhang Z, Lai G-H, Sirica AE (2004) Celecoxib-induced apoptosis in rat cholangiocarcinoma cells mediated by Akt inactivation and Bax translocation. Hepatology 39 (4):1028–1037
88. Steinbach G et al (2000) The effect of celecoxib, a cyclooxygenase 2 inhibitor, in familial adenomatous polyposis. N Engl J Med 342 (26):1946–1952
89. Yang K (2003) Regional response leading to tumorigenesis after sulindac in small and large intestine of mice with Apc mutations. Carcinogenesis 24(3):605–611

90. Baron JA et al (2003) A randomized trial of aspirin to prevent colorectal adenomas. N Engl J Med 348(10):891–899
91. Lee H-J et al (2009) Sulindac inhibits canonical Wnt signaling by blocking the PDZ domain of the protein dishevelled. Angew Chem Int Ed 48(35):6448–6452
92. Phillips RKS (2002) A randomised, double blind, placebo controlled study of celecoxib, a selective cyclooxygenase 2 inhibitor, on duodenal polyposis in familial adenomatous polyposis. Gut 50(6):857–860
93. Sandler RS et al (2003) A randomized trial of aspirin to prevent colorectal adenomas in patients with previous colorectal cancer. N Engl J Med 348(10):883–890
94. Nath N et al (2003) Nitric oxide-donating aspirin inhibits beta-catenin/T cell factor (TCF) signaling in SW480 colon cancer cells by disrupting the nuclear beta-catenin-TCF association. Proc Natl Acad Sci 100 (22):12584–12589
95. Williams JL et al (2001) NO-NSAIDs alter cell kinetics in human colon cancer cell lines more efficiently than traditional NSAIDs. Gastroenterology 120(5):A166
96. Williams JL et al (2004) NO-donating aspirin inhibits intestinal carcinogenesis in Min (APCMin/+) mice. Biochem Biophys Res Commun 313(3):784–788
97. Dharmapuri G et al (2015) Celecoxib sensitizes imatinib-resistant K562 cells to imatinib by inhibiting MRP1-5, ABCA2 and ABCG2 transporters via Wnt and Ras signaling pathways. Leuk Res 39(7):696–701
98. Fujii N et al (2007) An antagonist of dishevelled protein-protein interaction suppresses beta-catenin-dependent tumor cell growth. Cancer Res 67(2):573–579
99. Grandy D et al (2009) Discovery and characterization of a small molecule inhibitor of the PDZ domain of dishevelled. J Biol Chem 284 (24):16256–16263
100. Shan J et al (2005) Identification of a specific inhibitor of the dishevelled PDZ domain †. Biochemistry 44(47):15495–15503
101. Kadowaki T et al (1996) The segment polarity gene porcupine encodes a putative multitransmembrane protein involved in wingless processing. Genes Dev 10(24):3116–3128
102. Rios-Esteves J, Resh MD (2013) Stearoyl CoA desaturase is required to produce active, lipid-modified Wnt proteins. Cell Rep 4 (6):1072–1081
103. Shukla S et al (2016) Cucurbitacin B inhibits the stemness and metastatic abilities of NSCLC via downregulation of canonical Wnt/β-catenin signaling axis. Sci Rep 6:21860
104. Dakeng S et al (2011) Inhibition of Wnt signaling by cucurbitacin B in breast cancer cells: reduction of Wnt-associated proteins and reduced translocation of galectin-3-mediated β-catenin to the nucleus. J Cell Biochem 113 (1):49–60
105. Huang S-MA et al (2009) Tankyrase inhibition stabilizes axin and antagonizes Wnt signalling. Nature 461(7264):614–620
106. Kulak O et al (2015) Disruption of Wnt/β-catenin signaling and telomeric shortening are inextricable consequences of tankyrase inhibition in human cells. Mol Cell Biol 35 (14):2425–2435
107. Wu X et al (2016) Tankyrase 1 inhibitior XAV939 increases chemosensitivity in colon cancer cell lines via inhibition of the Wnt signaling pathway. Int J Oncol 48 (4):1333–1340
108. Thorne CA et al (2010) Small-molecule inhibition of Wnt signaling through activation of casein kinase 1α. Nat Chem Biol 6 (11):829–836
109. Ma S et al (2015) SKLB-677, an FLT3 and Wnt/β-catenin signaling inhibitor, displays potent activity in models of FLT3-driven AML. Sci Rep 5:15646
110. Chen Y et al (2011) Regulation of breast cancer-induced bone lesions by beta-catenin protein signaling. J Biol Chem 286 (49):42575–42584
111. Pacheco-Pinedo EC et al (2011) Wnt/β-catenin signaling accelerates mouse lung tumorigenesis by imposing an embryonic distal progenitor phenotype on lung epithelium. J Clin Investig 121(5):1935–1945
112. Emami EH et al (2004) A small molecule inhibitor of beta-catenin/CREB-binding protein transcription [corrected]. Proc Natl Acad Sci U S A 101:12682–12687. Proc Natl Acad Sci U S A 101(47):16707–16707
113. McMillan M, Kahn M (2005) Investigating Wnt signaling: a chemogenomic safari. Drug Discov Today 10(21):1467–1474
114. Brembeck FH, Rosário M, Birchmeier W (2006) Balancing cell adhesion and Wnt signaling, the key role of β-catenin. Curr Opin Genet Dev 16(1):51–59
115. Sawa H (2012) Control of cell polarity and asymmetric division in C. elegans. Curr Top Dev Biol 101:55–76

Chapter 19

Functional Network Disruptions in Schizophrenia

Irina Rish and Guillermo A. Cecchi

Abstract

It has been long recognized that schizophrenia, unlike certain other mental disorders, appears to be delocalized, i.e., difficult to attribute to a dysfunction of a few specific brain areas, and may be better understood as a disruption of brain's emergent network properties. In this chapter, we focus on topological properties of functional brain networks obtained from fMRI data, and demonstrate that some of those properties can be used as discriminative features of schizophrenia in multivariate predictive setting. While the prior work on schizophrenia networks has been primarily focused on discovering statistically significant differences in network properties, this work extends the prior art by exploring the generalization (prediction) ability of network models for schizophrenia, which is not necessarily captured by such significance tests. Moreover, we show that significant disruption of the topological and spatial structure of functional MRI networks in schizophrenia (a) cannot be explained by a disruption to area-based task-dependent responses, i.e., indeed relates to the emergent properties, (b) is global in nature, affecting most dramatically long-distance correlations, and (c) can be leveraged to achieve high classification accuracy (93%) when discriminating between schizophrenic vs. control subjects based just on a single fMRI experiment using a simple auditory task.

Key words Schizophrenia, Functional magnetic resonance imaging (fMRI), Functional networks, Multivariate predictive modeling, Classification, Predictive features

1 Introduction

Attributing schizophrenia to abnormal interactions among different brain areas has a long history in psychiatric research. It is often referred to as the "disconnection hypothesis" [1, 2], and can be traced back to the early research on schizophrenia: in 1906, Wernicke [3] was the first one to postulate that anatomical disruption of association fiber tracts is at the roots of psychosis; in fact, the term schizophrenia was introduced by Bleuler [4] in 1911, and was meant to describe the separation (splitting) of different mental functions. The failure to identify specific areas, as well as the controversy over which localized mechanisms are responsible for the symptoms associated with schizophrenia, have led us among others [5–7] to hypothesize that this disease may be better understood as a

Tatiana V. Tatarinova and Yuri Nikolsky (eds.), *Biological Networks and Pathway Analysis*, Methods in Molecular Biology, vol. 1613, DOI 10.1007/978-1-4939-7027-8_19, © Springer Science+Business Media LLC 2017

disruption of the emergent, collective properties of normal brain states, which can be better captured by functional networks [8], based on inter-voxel correlation strength, as opposed (or limited) to activation failures localized to specific, task-dependent areas.

However, while most of the previous work mainly focused on mass-univariate statistical hypothesis testing, investigating differences between the functional (and anatomical) networks of schizophrenic patients versus healthy subjects, we focus herein on multivariate predictive models, and investigate discriminative ability of various features (i.e., "statistical biomarkers") derived from functional networks. Unlike hypothesis testing that reveals statistically significant differences between two groups of subject (e.g., schizophrenic and control) on a given, fixed dataset, predictive framework evaluates the generalization ability of models built using the features of interest, i.e., the ability to predict whether a previously unseen subject is schizophrenic or not. Note that discriminative tasks are typically more challenging than significance testing, i.e., presence of significant (low p-value) features in fMRI data does not always imply accurate classification [9], and statistically significant variables are not necessarily the best predictors [10]. Thus, a combination of both evaluation criteria provides a better characterization of candidate features in terms of their relevance to the disease. Moreover, predictive modeling has potential applications in practical settings, such as, for example, early diagnosis of schizophrenia based on imaging data.

Herein, we considered diverse topological features of the functional brain networks obtained from functional magnetic resonance imaging (fMRI) data collected for both schizophrenic and control subjects performing a simple auditory task in the scanner [11]. In Functional Magnetic Resonance Imaging (fMRI), a MR scanner noninvasively records a subject's blood-oxygenation-level dependent (BOLD) signal, known to be correlated with neural activity, as a subject performs a task of interest (e.g., viewing a picture or reading a sentence). Such scans produce a sequence of 3D images, where each image typically has on the order of 10,000–100,000 subvolumes, or voxels, and the sequence typically contains a few hundreds of time points, or TRs (time repetitions). Standard fMRI analysis approaches, such as the General Linear Model (GLM), examine mass-univariate relationships between each voxel and the stimulus in order to build the so-called statistical parametric maps that associate each voxel with some statistics that reflects its relationship to the stimulus. Commonly used activation maps depict the "activity" level of each voxel determined by the linear correlation of its time course with the stimulus.

Our goal was to both explore the disruptions of functional connectivity due to schizophrenia and to assess whether the functional connectivity changes in schizophrenia can be simply explained by alterations in area-specific, task-dependent voxel activations. We observed that functional network features reveal highly statistically significant differences between the schizophrenic and control groups; moreover, statistically significant subsets of certain network features, such as voxel degrees (the number of voxel's neighbors in a network), are quite stable over varying data subsets. In contrast, voxel activation show much weaker group differences as well as stability, which suggests that network disruptions are not necessarily explained by local task-based activation patterns. Moreover, most of the network features, and especially pairwise voxel correlations (edge weights) and voxel degrees, allow for quite accurate classification, as opposed to voxel activation features: degree features achieve up to 86% classification accuracy (with 50% baseline) using Markov Random Field (MRF) classifier, and even more remarkable 93% accuracy is obtained by linear Support Vector Machines (SVM) using just a dozen of the most-discriminative correlation features. It is interesting to note that the traditional approaches based on a direct comparison of the correlation at the level of relevant regions of interest (ROIs) or using a functional parcellation technique, presented in [9], did not reveal any statistically significant differences between the groups. Indeed, a more data-driven approach that exploits properties of voxel-level networks appears to be necessary in order to achieve high discriminative power.

The material presented in this chapter is based on our prior work from [9, 12, 13].

2 Materials and Methods

2.1 Experimental Paradigm

The dataset in this study was previously acquired according to the methodology described in [11], and involved a group of 15 schizophrenic subjects (nine women) fulfilling DSM-IV-R criteria for schizophrenia with daily auditory hallucinations for at least 3 months despite well-conducted treatment. Their mean ± S.D. age was 34 ± 10 years (i.e., 22–49 years range), and the duration of illness was 12 ± 10$ years (3–28 years range). All schizophrenic patients were treated with antipsychotic drugs (mean ± S.D. – 425 ± 604 mg) chlorpromazine equivalent/day [14]. Four subjects were discarded because of acquisition issues, leaving us with 11 subjects that were approximately matched for gender and age by the control group of 11 healthy subjects. All subjects were submitted to the same experimental paradigm involving language. The task is based on auditory stimuli; subjects listen to emotionally neutral sentences either in native (French) or foreign language. Average length (3.5 s mean) or pitch of both kinds of sentences is normalized. In order to catch attention of subjects, each trial begins with a short (200 ms) auditory tone, followed by the actual sentence. The subject's attention is asserted through a simple validation task: after each played sentences, a short pause of 750 ms is followed by a 500 ms two-syllable auditory cue, which belongs to the previous sentence or not, to which the subject must answer to by yes (the cue is part of the previous sentence) or no with push-buttons, when the language of the sentence was his own. A full fMRI run contains 96 trials, with 32 sentences in French (native), 32 sentences in foreign languages, and 32 silence interval controls.

Data were acquired on a 1.5 T Signa (General Electric). For each subject, two fMRI runs are acquired (T2-weighted EPI), each of which consisted of 420-scans (from which the first four are discarded to eliminate T1 effect), with a repetition time (TR) of 2.0 s, for a total length of 14 min per run. Data were spatially realigned and warped into the MNI template and smoothed (FWHM of 5 mm) using SPM5 (www.fil.ucl.ac.uk); also, standard SPM5 motion correction was performed with the SPM5 realignment preprocessing. For each volume of the time-series, the process estimates a six degree-of-freedom movement relative to the first volume. These estimated parameters are combined to warping parameters (obtained by nonlinear deformation on an EPI template) to get the final, spatially normalized and realigned time-series. Finally, a universal mask was computed as the minimal intersection of thresholded EPI mean volumes across the entire dataset. This mask was then applied to all subjects.

Note that the schizophrenia patients studied here have been selected for their prominent, persistent, and pharmaco-resistant auditory hallucinations [11] which might have increased their clinical homogeneity, but they are not representative of all schizophrenia patients, only of a subgroup.

In summary, our dataset contained the total of 44 samples (there were two samples per subject, corresponding to the two runs), where each sample corresponds to a subject/run combination, and is associated with roughly 50,000 voxels × 420 TRs × 2 runs, i.e., more than 40,000,000 voxels/variables. In the subsequent sections, among other methods, we discuss feature-extraction approaches that reduce the dimensionality of the data prior to learning a predictive model.

2.2 Methods

We focus here on data-driven approaches based on various features extracted from the fMRI data, such as standard activation maps and a set of topological features derived from functional networks. Note that a model-driven approaches based on prior knowledge about the regions of interest (ROI) that are believed to be relevant to schizophrenia, was originally explored in [9], but did not reveal any statistically significant differences between the groups.

2.2.1 Feature Extraction

Activation Maps

To find out whether local task-dependent linear activations alone could possibly explain the differences between the schizophrenic and normal brains, we used as a baseline set of features based on the standard voxel activation maps, computed using General Linear Model (GLM). The GLM analysis described here is a standard component of the Statistical Parametric Mapping (SPM) toolkit.

Given the time-series for stimulus $s(t)$ (e.g., $s = 1$ if the stimulus/event is present, and $s = 0$ otherwise), and the BOLD signal intensity time-series $v_i(t)$ for voxel i, GLM is simply a linear regression $v_i(t) = \beta_i{}^* x(t) + b_i + \varepsilon$, where $x(t)$ the regressor corresponding to the stimulus convolved with the hemodynamic response function (HRF) in order to account for delay between the voxel activation and change in the BOLD signal, ε is noise, b_i is the baseline (mean intensity), and β_i coefficient is the amplitude that serves as an activation score (note that β_i coefficient is simply the correlation between $v_i(t)$ and $x(t)$ when both are normalized and centered prior to fitting the model). Given multiple trials, multiple estimates of β_i are obtained and a statistical test (e.g., t-test) is performed for the mean of β_i against the null-hypothesis that it comes from Gaussian noise distribution with zero mean and fixed noise (the level of noise for BOLD signal is assumed to be known here).

In case of multiple stimuli, the GLM model uses a vector of regressors $\mathbf{x(t)}$ and obtains the vector of the corresponding coefficients $\boldsymbol{\beta}$. For example, in our studies, the following stimuli/events were considered: "FrenchNative," "Foreign," and "Silence," together with several additional regressors, such as some low-frequencies trends and the movement parameters (additional 1-only column is added to account for the mean of the signal, as above—a standard step in linear regression with the unnormalized data). Once the GLM is fit, we focus on β_i coefficients obtained for the above three stimuli, and the corresponding three activation maps. Next, several activation contrast maps were computed by subtracting some maps from the others (hoping that such differences, or contrasts, may provide additional information). The following activation contrast maps were computed: activation contrast 1: "FrenchNative–Silence," activation contrast 2: "FrenchNative–-Foreign," activation contrast 3: "Silence–FrenchNative," activation contrast 4. "Foreign–FrenchNative" (note that maps 2 and 4 are just negations of the maps 1 and 3, respectively), activation contrast 5: "Foreign–Silence"; also, the following three contrast maps are simply the difference of the corresponding β_i coefficient (activation) and the mean (b_i); activation contrast 6: "FrenchNative," activation contrast 7: "Foreign," activation contrast 8: "Silence." For each of those maps, t-values are computed at each voxel (with a null-hypothesis corresponding to zero-mean Gaussian). The resulting t-value maps were used herein, rather than just the "raw" activation maps (i.e., β coefficient maps); to simplify the terminology, we just refer to them as "activation" or "activation contrast" maps.

The above activation contrast maps (that we will further refer to as simply activation maps) were computed for each subject and for each run. The activation values of each voxel were subsequently used as features in the classification task. We also computed a global feature, mean-activation (denoted mean-t-val), by taking the mean absolute value of the voxel's t-statistics.

Network Features

In order to continue investigating possible disruptions of global brain functioning associated with schizophrenia, we decided to explore lower-level (as compared to ROI-level) functional brain networks [8] constructed at the voxel level, as follows: More specifically, we computed ***voxel-level functional networks***, as follows: [1] pairwise Pearson correlation coefficients were computed among all pairs of time-series $(v_i(t), v_j(t))$, where $v_i(t)$ corresponds to the BOLD signal of i-th voxel; [2] an edge between a pair of voxels is included in the network if the correlation between the corresponding voxel's BOLD signals exceeds a specified threshold (herein, we used the same threshold of c(Pearson) = 0.7 for all voxel pairs; we tried a few other threshold levels, such as 0.8 and 0.9, and the results were similar; however, we did not perform an exhaustive evaluation of the full range of this parameter due to high computational cost of such experiment).

For each subject, and each run, a separate functional network was constructed. Next, we measured a number of its global topological features, including:

- ***The mean degree***, i.e., the number of links for each node (corresponding to a voxel), averaged over the entire network.
- ***The mean geodesic distance***, i.e., the minimal number of links needed to reach any to from any other node, averaged over the entire network.
- ***The mean clustering coefficient***, i.e., the fraction of triangulations formed by a node with its first neighbors relative to all possible triangulations, averaged over the entire network.
- ***The giant component***, i.e., the size (number of nodes) of the largest connected subgraph in the network.
- ***The giant component ratio***, i.e., the ratio of the giant component size to the size of the network.
- ***The total number*** of links in the network.

Besides global topological features, we also computed a series of voxel-level network features, based on topological properties of an individual voxel in functional network; the following types of features were used:

- ***(Full) degree***: the value assigned to each voxel is the total number of links in the corresponding network node.

- *Long-distance degree*: the number of links making nonlocal connections (i.e., links between the given voxel and the voxels that are five or more voxels apart from it).
- *Interhemispheric degree*: only links reaching across the brain hemispheres are considered when computing each voxel's degree.
- *Strength*: node strength is the sum of weights of links connected to the node. In our study, the full correlation matrix was used as a weighted adjacency matrix, where each pairwise correlation corresponds to the link weight; thus, for each voxel, its strength is the sum of its correlations with the other voxels.
- *Absolute strength*: same as above, but the link weights are replaced by their absolute values.
- *Positive strength*: same as node strength, but only positive link weights are considered.
- *Clustering coefficient* of a node is the fraction of triangles around a node, i.e., the fraction of node's neighbors that are neighbors of each other; herein, we first computed a functional networks by applying a threshold of 0.7 to the absolute values of the pairwise correlations, and then used the resulting graph to compute the clustering coefficients for each node/voxel.
- *Local efficiency*: the local efficiency is the global efficiency computed on node neighborhoods, and is related to the clustering coefficient. The *global efficiency* is the average inverse shortest path length in the network, that is $1/<1/d_n>$, where d_n is the shortest path for node n, such that for disconnected nodes $d_n = \infty$, i.e., $1/1/d_n = 0$.
- *Edge weights*: finally, we simply used as features a *randomly selected subset of 200,000 pairwise correlations* out of $53{,}000 \times 53{,}000$ entries of the correlation matrix (the location of pairs were randomly selected once, and then same locations used to derive features for all subjects); the rationale behind random sampling from the correlation matrix was to reduce the computational complexity of working with the full set of correlations, which would exceed 2800 million features. Nevertheless, subsequent feature ranking procedure was able to select a highly discriminative subset of correlation features, which would only improve if the feature ranking was allowed to continue running over the rest of the correlation matrix. Note that we also tried other sets of randomly selected 200,000 voxels and obtained similar results to those presented herein. Clearly, the results may vary if we keep selecting other random sets of voxels that may not include the top most informative voxel pairs discovered in our analysis. However, the point of our analysis is to show that it is possible to find predictive features among pairwise

correlations, and that our results demonstrate only a lower bound on a potentially even better predictive performance of correlation features.

For each of the above feature types, except the edge weights, we call the corresponding feature sets "feature map," since each voxel is associated with its own feature value, e.g., (full) degree maps, strength maps, and so on. These maps were utilized for further analysis of statistical significance of group differences, including *t*-test and several classification approaches, described below.

Note that, for each sample, we also computed spatially normalized activation and degree maps, dividing the corresponding maps by their maximal value taken over all voxels in the given map. As it turned out, normalization affected both statistical testing and classification results presented below. We mainly focus on normalized activation and degree maps (full, long-distance, and interhemispheric), since they yield better classification results. In case of hypothesis testing, however, unnormalized (raw) activations maps, unlike the degree maps, happened to outperform their normalized counterparts, and thus both sets of results were presented.

2.2.2 Classification Approaches

We focused on discriminating between the schizophrenic and normal subjects only, that resulted into well-balanced dataset containing 2×11 positive (schizophrenic) and 2×11 negative (healthy) samples (since there were two runs per each subject), with 50% baseline prediction accuracy.

Classifiers

First, standard off-the-shelf methods such as Gaussian Naïve Bayes (GNB) and Support Vector Machines (SVM) were used in order to compare the discriminative power of different sets of features described above. We used standard SVM implementation with linear kernel and default parameters, available from the LIBSVM library. For GNB, we used our own MATLAB implementation.

Moreover, we decided to further investigate our hypothesis that interactions among voxels contain highly discriminative information, and compare those linear classifiers against probabilistic graphical models that explicitly model such interactions. Specifically, we learn a classifier based on a sparse Gaussian Markov Random Field (MRF) model [15], which leads to a convex problem with unique optimal solution, and can be solved efficiently; herein, we used the COVSEL procedure [15]. The weight on the l_1-regularization penalty serves as a tuning parameter of the classifier, allowing to control the sparsity of the model, as described below.

Sparse Gaussian MRF classifier. Let $X = \{X_1,...,X_p\}$ be a set of p random variables (e.g., voxels), and let $G = (V, E)$ be an undirected graphical model (Markov Network, or MRF) representing conditional independence structure of the joint distribution $P(X)$. The set of vertices $V = \{1,...,p\}$ is in the one-to-one correspondence with the set X. The set of edges E contains the edge (i, j) if and only if X_i is conditionally dependent on X_j given all remaining variables; lack of edge between X_i and X_j means that the two variables are conditionally independent given all remaining variables. Let $\mathbf{x} = (x_1,...,x_p)$ denote a random assignment to X. We will assume a multivariate Gaussian probability density $p(x) = (2\pi)^{-p/2} \det(C)^{1/2} e^{-1/2\ x^t Cx}$, where C is the inverse covariance matrix (also called the precision matrix), and the variables are normalized to have zero mean. Let $\mathbf{x}_1,...,\mathbf{x}_n$ be a set of n i.i.d. samples from this distribution, and let $S = (1/n)\Sigma\ \mathbf{x}_i^T\mathbf{x}_i$ denote the empirical covariance matrix. Missing edges in the above graphical model correspond to zero entries in the inverse covariance matrix C, and thus the problem of learning the structure for the above probabilistic graphical model is equivalent to the problem of learning the zero-pattern of the inverse-covariance matrix. Note that the inverse of the empirical covariance matrix, even if it exists, does not typically contain exact zeros. Therefore, an explicit sparsity constraint is usually added to the estimation process. A popular approach is to use l_1-norm regularization that is known to promote sparse solutions, while still allowing (unlike non convex l_q-norm regularization with $0 < q < 1$) for efficient optimization. From the Bayesian point of view, this is equivalent to assuming that the parameters (entries) C_{ij} of the inverse covariance matrix C are independent random variables following the Laplace distributions $p(C_{ij}) = (\lambda_{ij}/2) \exp(-\lambda_{ij}|C_{ij}|)$ with equal scale parameters $\lambda_{ij} = \lambda$. Our objective is to find the maximum-likelihood parameters in C, subject to the Laplace prior, which yields the standard optimization problem over positive definite matrices C (denoted $C \succ 0$), frequently considered in the sparse Gaussian MRF learning literature (see, e.g., [15]):

$$\min_{C \succ 0} \ln \det(C) - tr(SC) - \lambda||C||_1,$$

where $\det(A)$ and $\mathrm{tr}(A)$ denote the determinant and the trace (the sum of the diagonal elements) of a matrix A, respectively, S the empirical covariance of the data. For the classification task, we estimate on the training data the Gaussian conditional density $p(\mathbf{x}|y)$ (i.e., the inverse covariance matrix parameters) for each class $Y = \{0, 1\}$ (schizophrenic vs. control), and then choose the most-likely class label for each unlabeled test sample $\mathbf{x}$.

Variable Selection

Note that each sample is associated with roughly 50,000 voxels × 420 TRs × 2 runs, i.e., more than 40,000,000 voxels/variables. Thus, some kind of dimensionality reduction and/or feature extraction appears to be necessary prior to learning a predictive model. Extracting degree maps and activation maps reduced dimensionality by collapsing the data along the time dimension.

Moreover, we used variable selection as an additional preprocessing step before applying a particular classifier, in order to [1] further reduce the computational complexity of classification (especially for sparse MRF, which, unlike GNB and SVM, could not be directly applied to 50,000 variables), [2] reduce noise, and [3] identify relatively small predictive subsets of voxels. We applied a simple filter-based approach, selecting a subset of top-ranked voxels, where the ranking criterion used p-values resulting from the paired t-test, with the null-hypothesis being that the voxel values corresponding to schizophrenic and non-schizophrenic subjects came from distributions with equal means. The variables were ranked in the ascending order of their p-values (lower p-values correspond to higher confidence in between-group differences), and classification results on top k voxels will be presented for a range of k values. Clearly, in order to avoid biased estimate of generalization error, variable selection was performed separately on each cross-validation training dataset; failure to do so, i.e., variable selection on the full dataset, would produce overly optimistic results with nearly perfect accuracy (e.g., 95% accuracy using GNB on just 100 top t-test voxels).

Evaluation via Cross-Validation

Since there were two samples corresponding to two runs per each subject, another source of overly optimistic bias that we had to avoid was possible inclusion of the samples for the same subject in both training and test datasets—for example, if using the standard leave-one-out cross-validation approach. Instead, we used leave-one-subject-out cross-validation, where each of the 22-folds on the 44-sample dataset (11 schizophrenic and 11 control samples, two runs each) would set aside as a test set the two samples for a particular subject.

3 Results

Empirical results are consistent with our hypothesis that schizophrenia disrupts the normal structure of functional networks in a way that is not derived from alterations in the activation; moreover, they demonstrate that topological properties are highly predictive, consistently outperforming predictions based on activations.

3.1 Voxel-Level Statistical Analysis

In order to find out whether various features exhibit statistically significant differences across the two groups, we performed two-sample t-test for each feature x_i from the corresponding feature vector $\mathbf{x} = (x_1,\ldots,x_n)$ of a particular type (activations, degrees, etc.); herein, n is the number of voxels for voxel-level features, and $n = 200{,}000$ for the weight features (pairwise correlations). Clearly, when the number of statistical tests is very large (i.e., n here is exceeding 50,000), a correction for multiple comparisons is necessary, since low p-values indicating statistically significant differences given one test may just occur due to pure chance when many such tests are performed. A commonly used Bonferroni correction is overly conservative in brain imaging analysis since it assumes test independence, while there are obviously strong correlations across the voxel-level features. A more appropriate type of correction that is now frequently used in fMRI analysis is the False Discovery Rate (FDR) method, designed to control the expected proportion} of incorrectly rejected null hypotheses, or "false discoveries." In general, FDR is less conservative than the familywise error rate (FWER) methods (including the Bonferroni correction), since it does not guarantee there are no false positives, but rather that there are only a few of them. For example, FDR with threshold 0.05 guarantees no more than 5% of false positives. Herein, we include the results for both FDR and Bonferroni corrections (see columns 5 and 6 of the Table 1, respectively). However, our discussion is mainly based on FDR results, while Bonferroni results are mentioned purely for completeness sake, to demonstrate that some of the statistical differences we observed are so strong that they survived even an overly strict Bonferroni correction.

Our main observation is that *the network features show much stronger statistical differences between the schizophrenic* vs. *non-schizophrenic groups than the activation features.* Figure 1 shows the results of two-sample t-test analysis for all voxel-level features, and the corresponding FDR threshold at $\alpha = 0.05$ level. Panel (a) shows a direct comparison between the best activation features (dashed lines) and three (spatially normalized) degree maps: full, long-distance, and interhemispheric. In all degree maps, on the order of 1000 voxels survive FDR correction (i.e., have their p-values below the black line corresponding to the FDR threshold), while only a handful (less than ten) of activation voxels do. The other measured graph features, including clustering and local efficiency, have less statistical power than degrees (i.e., have p-values closer to the FDR threshold), but yet outperform activation maps by almost two orders of magnitude, as shown in panel (b). A full list showing the number of surviving voxels for each map is shown in Table 1. (Note that for the activation maps, the results for both normalized and unnormalized maps are shown, since unnormalized ones performed better in hypothesis testing. In classification study presented next, the situation was reversed, i.e., normalized activations predicted better than unnormalized, thus, we always included the best possible results achieved by activations. In case of degree maps, we always used only their normalized versions, which performed best in both hypothesis testing and classification scenarios).

Finally, randomly selected pairwise correlations, as shown in Panel (c), behave similarly to degrees, with an order of 10,000 correlations surviving the FDR test, i.e., an order of magnitude

Table 1
Detailed *t*-test results for all activation and network-based features. Each column shows the number of voxels that satisfy a given constraint, such as having *p*-value below the specified threshold or surviving the FDR or Bonferroni correction *with the significance level* $\alpha = 0.05$ (the number of voxels common with the full degree maps is shown in parenthesis for unnormalized linear activation maps)

Map	$p < 0.01$	$p < 0.001$	$p < 0.0001$	FDR	Bonferroni	N
Norm, full degrees	2583	1046	448	1033	50	53,750
Norm, long-dist. Deg.	2335	972	398	924	43	53,737
Norm, inter-hem. Deg	1448	677	258	508	18	51,373
Activation 1 (3)	1799 (341)	317 (76)	52(9)	7(2)	0	53,456
activation 2 (4)	805 (27)	112 (0)	15(0)	0(0)	0	53,456
Activation 5	1356 (306)	262 (69)	63 (10)	0(0)	0	53,456
Activation 6	1481 (152)	303 (14)	55(1)	2(0)	1	53,456
Activation 7	1294 (130)	163 (13)	20(1)	0(0)	0	53,456
Activation 8	2369 (97)	467 (1)	53(0)	0(0)	0	53,456
Norm, activation 1 (3)	885	108	15	0	0	53,456
Norm, activation 2 (4)	688	95	13	0	0	53,456
Norm, activation 5	647	58	8	0	0	53,456
Norm, activation 6	1357	245	37	0	0	53,456
Norm, activation 7	1019	123	10	0	0	53,456
Norm, activation 8	1511	236	30	1	1	53,456
Corr subset(200 K)	23,573	6437	1718	12,240	37	199,998
Strength	10,917	2197	393	11,294	6	53,750
Absolute strength	6721	1053	154	971	0	53,750
Positive strength	8938	1594	277	5724	2	53,750
Clustering coef.	3812	955	240	789	4	53,750
Local efficiency	4142	1076	286	1077	4	53,750

The last column shows the total number of voxels N with non-zero values in the corresponding map (recall that Bonferroni correction filters out the voxels with $p > \alpha/N$). Note that for the activation maps, the results for both normalized and unnormalized maps are shown, since unnormalized ones performed better in hypothesis testing

more than for degrees. (Note, however, that the total number of correlation features (200,000) is also much larger than the number of degree features (about 50,000), i.e., voxels; therefore, the results for correlations are not directly comparable to those for degrees and other voxel features, and thus plotted in a separate panel).

The spatial localization of the network maps is shown in Fig. 2, representing the voxels surviving correction for (a) (normalized)

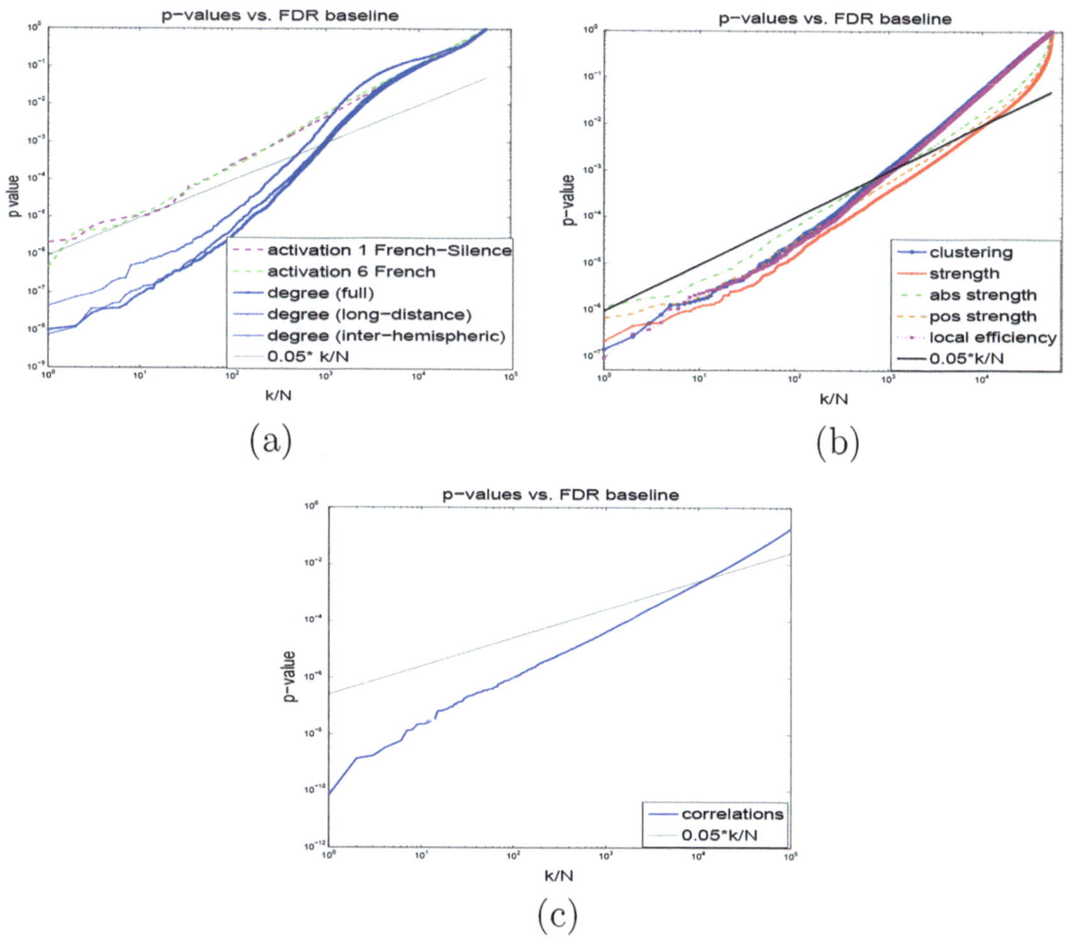

Fig. 1 Two-sample *t*-test results for different features: *p*-values vs. FDR threshold. (**a**) Activations vs. normalized degrees; (**b**) clustering coefficients, strength, absolute strength, positive strength, and local efficiency of each voxel; (**c**) 200,000 randomly selected pairwise correlations. The null hypothesis for each feature assumes no difference between the schizophrenic vs. normal groups. *p*-values of the features are sorted in ascending order and plotted vs. FDR baseline; FDR test select voxels with $p < \alpha \cdot k/N$, α—false-positive rate, *k*—the index of a *p*-value in the sorted sequence, *N*—the total number of voxels. Note that graph-based features yield a large number of highly significant (*very low*) *p*-values, staying far below the FDR cutoff line, while only a few voxels survive FDR in case of (*unnormalized*) activation maps in panel (**a**): 7 and 2 voxels in activation maps 1 (contrast "FrenchNative − Silence") and 6 ("FrenchNative"), respectively, while the rest of the activation maps do not survive the FDR correction at all

degree maps, (b) strength (red-yellow), absolute strength (blue light blue), and positive strength (black-white), (c) clustering coefficient and local efficiency maps. Normalized degrees (a) show the most spatially coherent organization, with contiguous bilateral clusters in auditory/temporal areas, prominently BA 22 and BA 21. Note also that the degree of the normal population is *higher* than the patient population. Strength-related features (b) have less bilateral symmetry and are also less spatially coherent, while clustering (c) is even more scattered.

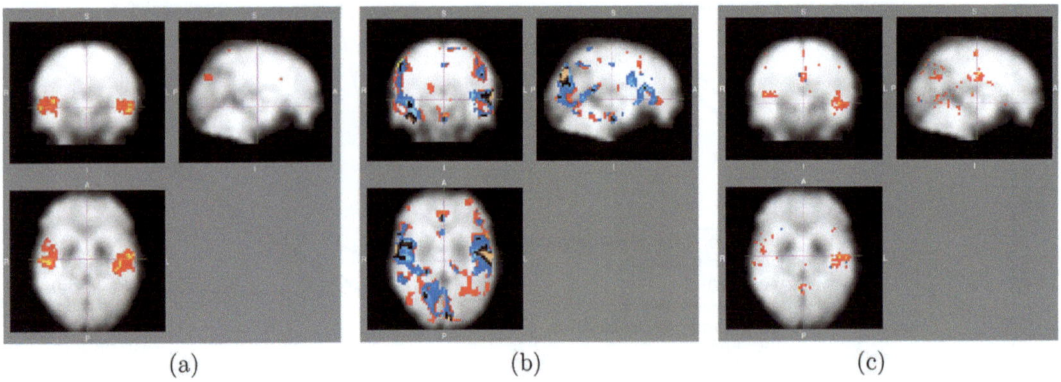

Fig. 2 Two-sample *t*-test results for different features: voxels surviving FDR correction. (**a**) Normalized degree maps; (**b**) strength (*red-yellow*), absolute strength (*blue-light blue*), and positive strength (*black-white*); (**c**) clustering coefficient and local efficiency maps. Here the null hypothesis at each voxel assumes no difference between the schizophrenic vs. normal groups. Colored areas denotes low *p*-values passing FDR correction at $\alpha = 0.05$ level (i.e., 5% false-positive rate). Note that the mean (*normalized*) degree at highlighted voxels was always (*significantly*) *higher* for controls than for schizophrenics. Coordinates of the center of the image: (**a**) and (**c**) $X = 26$, $Y = 30$, $Z = 16$, (**b**) $X = 26$, $Y = 30$, $Z = 18$

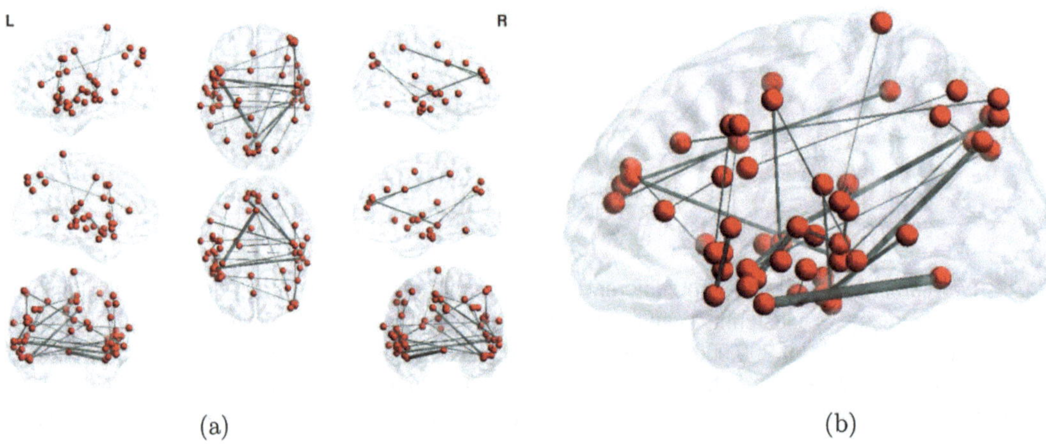

Fig. 3 Thirty top-ranked (*lowest-p-value*) edges (all surviving Bonferroni correction) out of 200,000 pairwise correlation features, computed on the full dataset. (**a**) All views and (**b**) enlarged sagittal view. Edge density is proportional to their absolute value

The network in Fig. 3 visualizes the top 30 most significantly different edges selected out of 200,000 edge features, or pairwise correlations (the total number of such features surviving FDR correction was 12,240, as shown in Table 1 and visualized in Fig. 1c. Figure 4 shows a stable subset of nine edges common to all top-30 ranked edges, over all cross-validation subsets, making it a highly robust representation. Note that unlike the degree maps, this network includes areas other than BA 22 and BA 21, prominently left precentral gyrus BA 44 (Broca's area), right middle

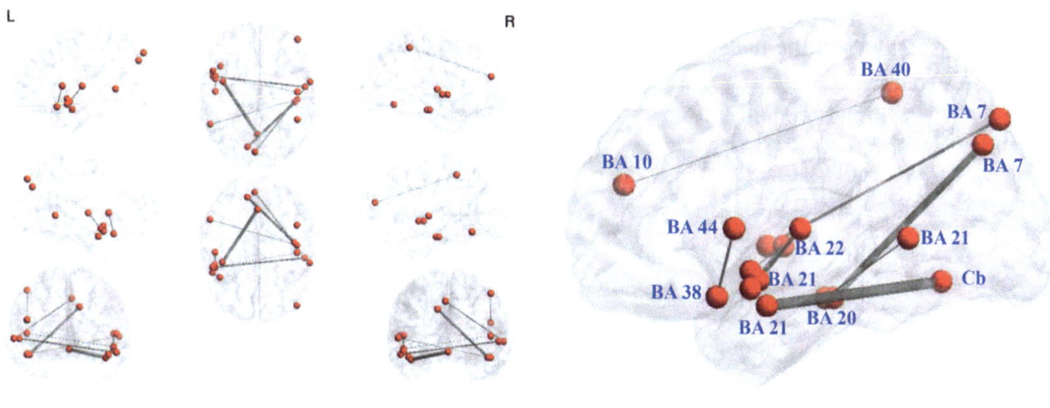

(a) (b)

Fig. 4 Nine stable edges common to all subsets of 30 top-ranked (*lowest-pvalue*) edges that survived Bonferroni correction, over 22 different cross-validation folds (*leave-subject-out data subsets*). (**a**) All views and (**b**) enlarged sagittal view. Edge density is proportional to their absolute value. The network includes several areas not picked up by the degree maps, i.e., other than BA 22 and BA 21, mainly the cerebellum (*declive*) and the occipital cortex (BA 19)

frontal gyrus BA 10, medial precuneus BA 7, and the declive of the cerebellum. A complete list of the nodes is presented in Table 2, while area-to-area functional connections determined by the nine most stable links are shown in Table 3. Note that most links span both hemispheres, and that there are no local, intra-area links, even though we introduced no voxel clustering.

Our observations suggest that (a) the differences in the collective behavior cannot be explained by differences in the linear task-related response, and that (b) topology of voxel-interaction networks is more informative than task-related activations, suggesting an abnormal degree distribution for schizophrenic patients that appear to lack hubs in auditory cortex, i.e., have significantly lower (normalized) voxel degrees in that area than the normal group, possibly due to a more even spread of degrees in schizophrenic vs. normal networks. Note that, as discussed earlier, ROI- and parcellation-level network topologies do not seem to retain information present in voxel-level networks (*see* ref. 9 for more detail), apparently due to averaging the signal over ROIs or parcels.

We also evaluate the stability of all features with respect to selecting a subset of top ranked voxels over different subsets of data. For each value of *k*, stability of the top k-ranked feature subset is defined as a fraction of features in common over all cross-validation data subsets (recall that there are 22 of them). Namely, given a fixed value of *k*, for each data subset, we rank the features by their *p*-values computed on that particular subset, choose the top k of them, and then compute the intersection over all 22 of those top-k feature subsets. The number of features common to all

Table 2
Areas corresponding to the nodes on the nine most stable links

Hemis.	Broad Anatomy	Brodmann	X	y	z
R	Temporal Fusiform Gyrus	20	45	−24	−18
R	Temporal Fusiform Gyrus	20	48	−21	−18
L	Middle Temporal Gyrus	21	−42	0	−21
L	Middle Temporal Gyrus	21	−54	6	−15
L	Middle Temporal Gyrus	21	−51	2	−12
L	Middle Temporal Gyrus	21	−57	−51	3
L	Superior Temporal Gyrus	38	−45	18	−18
L	Superior Temporal Gyrus	38	−51	6	−9
R	Superior Temporal Gyrus	22	57	−6	0
R	Superior Temporal Gyrus	22	63	0	0
R	Superior Temporal Gyrus	22	48	−12	6
L	Superior Temporal Gyrus	22	−51	−12	6
L	Precentral Gyrus	44	−54	12	6
R	Middle Frontal Gyrus	10	48	51	21
L	Medial Precuneus	7	−12	−78	36
L	Medial Precuneus	7	−3	−84	45
R	Inferior Parietal Lobe	40	48	−45	54
–	Declive	Cb	0	−63	−12

Table 3
Area-to-area functional connections determined by the nine most stable links

Left BA 21	↔	**Cb**
Right BA 20	↔	left BA 7
Right BA 20	↔	left BA 21
Left BA 38	↔	left BA 44
Left BA 21	↔	right BA 22
Left BA 38	↔	right BA 22
Right BA 22	↔	medial BA 7
Right BA 10	↔	right BA 40

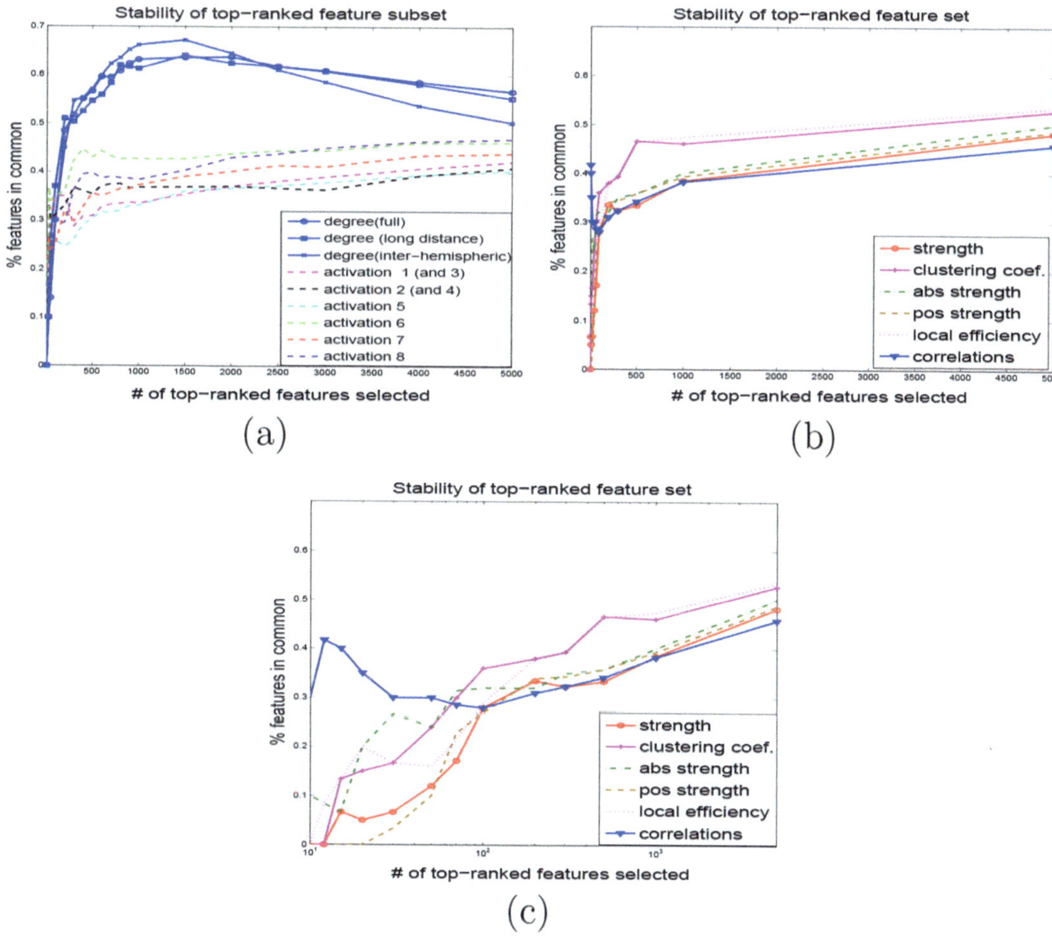

Fig. 5 Stability of feature subset selection over cross-validation (CV) folds. Stability is measured as the percent of voxels in common among the subsets of *k* top variables selected at all CV folds: (**a**) activations and degrees; (**b**, **c**) edge weights (*correlations*), clustering coefficients, strength, absolute strength, positive strength, and local efficiency: (**b**) linear scale on x-axis, (**c**) log-scale on x-axis (focusing on small number of features selected)

subsets (i.e., the size of their intersection), divided by *k*, gives us a measure of feature stability. Interestingly, network-based features, such as degrees (full, long distance, or interhemispheric) demonstrate much higher stability than activation features, as well as other network-based features. Figure 5a shows that degree maps have up to almost 70% top-ranked voxels in common over different training data sets when using the leave-one-subject out cross-validation, while activation maps have below 50% voxels in common between different selected subsets. This property of degree vs. activation features is particularly important for interpretability of predictive modeling. Stability of the other network-based features is shown in Fig. 5b, c, where the Fig. 5c shows the same results as Fig. 5b, but using logarithmic scale instead of linear, in order to focus on the regimes when only a small number of features is selected. While the overall stability of the remaining network features does not reach the high values of the degree features, it is still interesting to note that the pairwise correlations appear to be the most stable of the remaining network features when the number of selected features is relatively small, e.g., below 100.

3.2 Interhemispheric Degree Distributions

As suggested by the predominance of interhemispheric edges in the set of most significantly different pairwise correlations (Table 3), a closer look at the degree distributions reveals that a large percentage of the differential connectivity appears to be due to long-distance, interhemispheric links. Figure 6a compares the probability of finding a link in the networks as a function of the Euclidean distance between the nodes (in millimeters), for schizophrenic (red) versus control (blue) subjects. The bars correspond to one standard deviation, drawn on the top only, to avoid clutter in the figure, and the lines correspond to power-law fits for the intermediate distances (i.e., between 10 and 150 mm). The fit is $P = aD^k$, with $k = -1.46$ for schizophrenics, and $k = -1.15$ for controls. We see that for this distance range, schizophrenics have reduced connectivity, i.e., lower link probabilities than controls. Figure 6b compares the fraction of interhemispheric connections over all connections, for schizophrenic (red) versus normal (blue) groups. For each subject, a unique value was computed dividing the number of links spanning both hemispheres by the total number of links. The figure represents the normalized histogram of this interhemispheric link density for each group. The schizophrenic group shows a significant bias towards low relative interhemispheric connectivity. A t-test analysis of the distributions indicates that differences are statistically significant ($p = 0.025$). Moreover, it is evident that a major contributor to the high degree difference discussed before is the presence of a large number of interhemispheric connections in the normal group, which is absent in schizophrenic group. Furthermore, we selected a bilateral region of interest (ROI) corresponding to left and right Brodmann Area 22 (roughly, the clusters in Fig. 1a), *such that the linear activation for these ROI's was not significantly different between the groups*, even in the uncorrected case. For each subject, the connection strength between the left and right ROIs was computed as the fraction of ROI-to-ROI links over all links. Figure 6c shows the normalized histogram over subjects for this connectivity measure. Clearly, the normal group displays higher ROI-to-ROI connectivity, which is significantly disrupted in the schizophrenic group ($p = 3.7 \times 10^{-7}$). This provides a strong indication that the group differences in connectivity cannot be explained by differences in local activation.

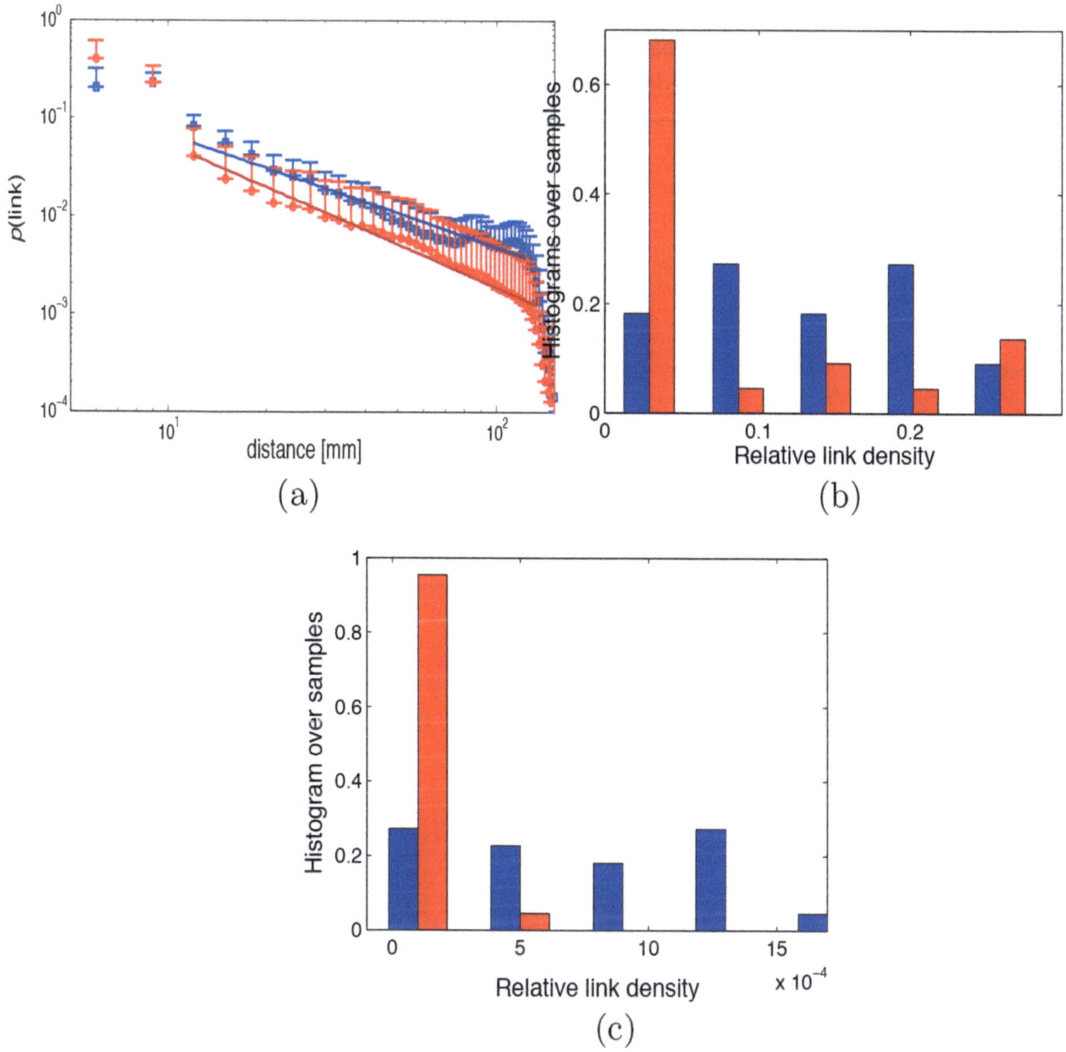

Fig. 6 Functional connectivity disruption in schizophrenic subjects vs. controls. (**a**) Probability of finding a network link as a function of the Euclidean distance between the nodes (*in millimeters*): schizophrenics (*red*) show reduced connectivity than controls (*blue*) for distances in the middle range (10–150 mm). (**b**) Disruption of *global* interhemispheric connectivity. For each subject, we compute the fraction of links spanning both hemispheres over the total number of links, and plot a normalized histogram over all subjects in each group (normal—blue, schizophrenic—red). (**c**) Disruption of *task-dependent* interhemispheric connectivity between specific ROIs (Brodmann Area 22 selected bilaterally). The ROIs were defined by a 9 mm radius ball centered at ($x = -42$, $y = -24$, $z = 3$) and ($x = 42$, $y = -24$, $z = 3$). For each subject, we compute the fraction of links connecting the bilateral ROIs over all links, and show a histogram of this connectivity measure over all subjects in each group. The histograms are similarly normalized

3.3 Global Features

For each global feature we computed its mean for each group and p-value produced by the t-test, as well as the classification accuracies using our classifiers. While mean activation (we used map 8, the best performer for SVM on the full set of voxels—*see* Table 4b) had a relatively low p-value of 5.5×10^{-4}, as compared to less

Table 4
Classification errors using (a) global features and (b) activation and degree maps; results for the SVM classifier applied to the complete set of voxels (i.e., without voxel subset selection)

(a)			
Feature	**GNB**	**SVM**	**MRF(0.01)**
Degree (D)	27.50%	27.50%	27.50%
Clustering coeff. (C)	30.00%	42.50%	45.00%
Geodesic dist. (G)	67.50%	45.00%	45.00%
Mean activation *(A)*	40.00%	45%	72.50%
D + A	27.50%	27.50%	32.50%
C + A	27.50%	45.00%	55.00%
G + A	45.00%	45.00%	72.50%
G + D + C	37.50%	27.50%	27.50%
G + D + C + A	30.00%	27.50%	32.50%
(b)			
Feature	**Err**	**FP**	**FN**
Correlations (53750)	**14%**	**14%**	**14%**
Degree (full)	**16%**	**27%**	**5%**
Degree (long-distance)	**21%**	**32%**	**9%**
Degree (inter-hemis)	32%	46%	18%
Clustering	23%	32%	14%
Local efficiency	23%	32%	14%
Strength	23%	23%	23%
Abs strength	34%	41%	27%
Pos strength	25%	32%	18%
Activation 1 (and 3)	54%	29%	82%
Activation 2 (and 4)	50%	55%	45%
Activation 5	43%	18%	68%
Activation 6	36%	27%	46%
Activation 7	32%	18%	46%
Activation 8	30%	23%	37%

For each feature, we show the average error, as well as the fraction of false positives (FP) and false negatives (FN) significant $p = 5.3 \times 10^{-2}$, for *mean-degree*, the predictive power of the latter, alone or in combination with some other features, was the best among global features reaching 27.5% error in schizophrenic vs. normal classification (Table 4a), while mean activation yielded more than 40% error with all classifiers In general, low *p*-values not necessarily imply low generalization error, as the results with other global features show. This is not particularly surprising, especially when the data violate Gaussian assumption of the *t*-test as it is in our case.

3.4 Classification Using Activations vs. Network Features

While mean-degree indicates the presence of discriminative information in voxel degrees, its generalization ability, though the best among global features and their combinations, is relatively poor. However, voxel-level network features turned out to be very informative about schizophrenia, often outperforming activation features by far. Table 4b shows the results of classification by SVM using all voxel-level network features of each type. Herein, all voxels and their corresponding features were used, without any subset selection; for correlation features, defined on pairs of voxels, we just used same number of features as in all other cases, i.e., the top 53,750 correlations out of 200,000, since 53,730 is the number of voxels used in the other features. Note that the top-performing network features are correlations (14% error) and (full) degree maps (16% error), greatly outperforming all activation maps that yield above 30% error for even the best-performing activation map 8.

Next, in Fig. 7, we compare the predictive power of different features using all three classifiers: Support Vector Machines (SVM), Gaussian Naive Bayes (GNB) and sparse Gaussian Markov Random

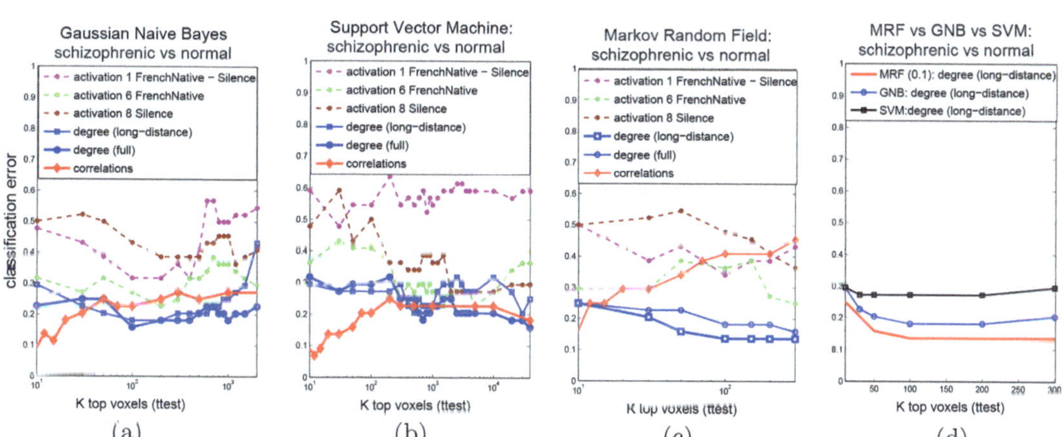

Fig. 7 Classification results: degree vs. activation features. Three classifiers, Gaussian Naive Bayes (GNB) in panel (**a**), SVM in panel (**b**) and sparse MRF in panel (**c**) are compared on two types of features, degrees and activation contrasts; (**d**) all three classifiers compared on long-distance degree maps (best-performing for MRF)

Field (MRF), on the subsets of k top-ranked voxels, for a variety of k values. For sparse MRF, we experimented with a variety of λ values, ranging from 0.0001 to 10, and present the best results; while cross-validation could possibly identify even better-performing values of λ, it was omitted here due to its high computational cost (also, using the fixed values listed above we already achieved quite high predictive accuracy as described later). We used the best-performing activation map 8, as well as maps 1 and 6 (that survived FDR); map 6 was also outperforming other activation maps in low-voxel regime. Also, to avoid clutter, we only plot the results for the three best-performing network features: full and long-distance degree maps, and pairwise correlations. We can see that:

- *Network features outperform activation maps*, for all classifiers we used, and for practically any value of k, the number of features selected. The differences are particularly noticeable when the number of selected voxels is relatively low. The most significant differences are observed for SVM in low-voxel (approx. <500) regime: using just a dozen of most-predictive pairwise correlations achieves a remarkable 7% error while the activation maps yield 30% and larger errors. Also, both pairwise correlations and degrees noticeably outperform activations on the full set of features (far right of the x-axis). Moreover, degree features demonstrate excellent performance with MRF classifiers: they achieve quite low error of 14% with only 100 most significant voxels, while even the best activation map 6 requires more than 200–300 to get just below 30% error; the other activation maps perform much worse, often above 30–40% error, or even just at the chance level.
- *Full and long-distance degree maps perform quite similarly*, with long-distance map achieving the best result (14% error) using MRFs.
- Among the activation maps only, while the map 8 ("Silence") outperforms others on the full set of voxels using SVM, its behavior in low-voxel regime is quite poor (always above 30–35% error); instead, map 6 ("FrenchNative") achieves best performance among activation maps in this regime. (We also observed that performing normalization really helped activation maps, since otherwise their performance could get much worse, especially with MRFs).
- *MRF classifiers significantly outperform SVM and GNB with degree features*, possibly due to their ability to capture inter-voxel relationships that are highly discriminative between the two classes (*see* Fig. 7d). However, with the correlation features the situation is reversed, and the overall best results (7% error) is achieved using SVM with just a dozen of top-ranked correlations.

4 Summary and Discussion

Recent advances in neuroimaging have provided researchers with tools for studying not just anatomical but also functional connectivity and its disruption in schizophrenia. The "disconnection syndrome" article by [1] was among the first ones to point out abnormalities in functional connectivity using PET imaging data (*see* also [16]). (More recently, the "dysconnection" term was suggested [2] in order to better capture the fact that schizophrenia is associated with a broader range of network dysfunctions besides just missing connections.) The paper studied functional connectivity captured by temporal correlations among different brain areas during a linguistic task, using principal component analysis (PCA) decomposition of the functional connectivity (covariance) matrix. Analysis of spatial components ("eigenimages") revealed that "profound negative prefronto-superior temporal functional interactions associated with intrinsic word generation" was strongly present in healthy subjects, but practically absent in schizophrenic patients; vice versa, positive prefronto-left temporal correlations were present in schizophrenic group but in the normal group, suggesting a reversal of prefrontotemporal integrations, attributed to "failure of prefrontal cortex to suppress activity in the temporal lobes (or vice versa)."

More recently, several studies demonstrated altered patterns in default-mode networks of schizophrenia, e.g., altered temporal frequency and spatial location of the default mode networks [5], and other patterns of aberrant connectivity [17, 18]. Also, multiple recent studies [7, 19] focused on graph-theoretic analysis of *functional connectivity networks* [8] in schizophrenia, demonstrating, for example, that in schizophrenia patients "the small-world topological properties are significantly altered in many brain regions in the prefrontal, parietal and temporal lobes" [7]. There is also continuing work exploring abnormalities in anatomical networks in schizophrenia [6, 20, 21].

In general, the importance of modeling brain connectivity and interactions became widely recognized in the recent neuroimaging literature beyond schizophrenia research ([22–24] give just a few examples). However, practical applications of such approaches such as dynamic causal modeling [22], dynamic Bays nets [23], or structural equations [24] are often limited to interactions among a relatively small number of known brain regions believed to be relevant to the task or phenomenon of interest. As discussed below, such approach can be sometimes disadvantageous, while a more data-driven, *voxel-level functional networks* analysis can achieve better results.

In this chapter, we discuss an approach to constructing predictive features based on functional network topology, and applied it to predictive modeling of schizophrenia. We demonstrated that [1] specific *topological properties of functional networks yield highly accurate classifiers of schizophrenia* and [2] *functional network differences cannot be attributed to alteration of local activation patterns*, a hypothesis that was not ruled out by the results of [6, 7] and similar work. In other words, our observations strongly support the hypothesis that schizophrenia is indeed a *network disease*, associated with the disruption of global, emergent brain properties.

Specifically, we demonstrated that topological properties of (voxel-level) functional brain networks are highly informative about the disease, unlike localized, task-related voxel activations, that were greatly outperformed by network-based features in both hypothesis testing and predictive settings. We also showed that it is highly important to use functional networks at the proper level: in our study, discriminative information present in voxel-level networks was apparently lost (perhaps due to averaging over large groups of voxels) at both regions-of-interest (ROI) and functional parcellation levels; the latter did not reveal any statistically significant differences between the schizophrenic and control groups. Unlike most traditional studies of schizophrenia networks based solely on hypothesis testing approach (e.g., [6, 7, 21]), we also employed *predictive modeling* techniques in order to evaluate how well the models built using network vs. local features would *generalize* to previously unseen subjects. Using generalization power, besides statistical significance, provides a complimentary (and often a more accurate) measure of disease-related information contained in a particular type of features, such as network properties or local activations. Moreover, predictive models have potential applications in clinical setting, e.g., for early diagnosis of schizophrenia based on abnormal patterns in imaging data. (Note, however, that multiple studies on a variety of subjects and experimental conditions may be necessary to come up with a robust predictive model).

In summary, our observations suggest that *voxel-level functional networks may contain significant amounts of information discriminative about schizophrenia, which may not be otherwise available in voxel activations or ROI-level networks.* Note, however, that the schizophrenic population studied here has been selected for their prominent, persistent, and pharmaco-resistant auditory hallucinations [11], which might have increased its clinical homogeneity and reduced its value as representative of the full spectrum of the disease. The experimental protocol may also restrict the applicability of our approach to generic cases. The areas more evidently involved in the discriminative networks, BA 22 and BA 21, are involved in language processing and are known to alter their activity in schizophrenics [25], and to display genetic and anatomical anomalies [26]. The direct analysis of pairwise correlations (as opposed to the voxel-centric degree maps) identifies anomalies in functional connectivity with Broca's area, the cerebellum and, interestingly, the frontal lobe (BA 10), in loose agreement with previous findings regarding disrupted frontotemporal connectivity associated with auditory hallucinations [27]. However, the analysis of correlations as a function of (Euclidean) distance provides a more nuanced perspective, as it shows weaker long-distance and stronger short-distance correlations for the patient population. This suggests a global reorganization of functional connections, and is further evidence of the emergent nature of the disruptions introduced by the disease. In the context of this finding, the identification of specifically affected areas, or area-to-area links, may be less relevant for the purpose of understanding functional alterations.

Note that the hypothesis of an emergent signature for schizophrenia does not necessarily reject the possibility of localized activation differences with respect to the normal population, for specific tasks or conditions. The finding that long-range functional connections are differentially affected, as demonstrated by the paucity of interhemispheric links and the weakness of long-distance correlations, may still be interpreted in terms of localized changes. Our findings may follow from subtle, undetectable changes (by fMRI at least) in the local activation of a handful of areas, that get amplified by the effect of the large number of links that are pooled when network features are computed, and bear no relationship to disruptions in the effective connectivity of the network (determined, for instance, by the lack or excess of specific neurotransmitters). The fact is, however, that there is no such thing as a completely "local" activation in the brain, since the driving input to most areas of the central nervous system is provided by the activity of other areas. In this sense, the hypothesis can be reformulated to imply that the disease is concomitant with a much stronger disruption of emergent than of local features.

While our conclusions may not necessarily apply to the schizophrenic population in general, we believe that our approach transcends the specific details of the particular population and experimental protocol we studied, and can guide future investigations of schizophrenia and other complex psychiatric diseases that can be better understood as network dysfunctions. Directions for further research include exploration of network abnormalities in other schizophrenia studies that involve different groups of patients and different tasks, as well as better characterization of connections involved in the predictive discrimination.

References

1. Friston K, Frith C (1995) Schizophrenia: a disconnection syndrome? Clin Neurosci 3 (2):89–97
2. Stephan K, Friston K, Frith C (2009) Dysconnection in schizophrenia: from abnormal synaptic plasticity to failures of self-monitoring. Schizophr Bull 35:509–527. doi:10.1093/schbul/sbn176
3. Wernicke C (1906) Grundrisse der psychiatrie. Verlag von Georg Thieme, Leipzig
4. Bleuler E (1911) Dementia praecox or the group of schizophrenias. International Universities Press, New York, NY
5. Garrity A, Pearlson GD, McKiernan K, Lloyd D, Kiehl K et al (2007) Aberrant "default mode" functional connectivity in

schizophrenia. Am J Psychiatry 164:450–457. doi:10.1176/appi.ajp.164.3.450

6. Bassett D, Bullmore E, Verchinski B, Mattay V, Weinberger D et al (2008) Hierarchical organization of human cortical networks in health and schizophrenia. J Neurosci 28(37):9239–9248. doi:10.1523/jneurosci.1929-08.2008
7. Liu Y, Liang M, Zhou Y, He Y, Hao Y et al (2008) Disrupted small-world networks in schizophrenia. Brain 131:945–961. doi:10.1093/brain/awn018
8. Eguiluz V, Chialvo D, Cecchi G, Baliki M, Apkarian A (2005) Scale-free functional brain networks. Phys Rev Lett 94:018102
9. Rish I, Cecchi G, Thyreau B, Thirion B, Plaze M, Paillere-Martinot ML, Martelli C, Martinot JL, Poline JB (2013) Schizophrenia as a network disease: disruption of emergent brain function in patients with auditory hallucinations. PLoS One 8(1):e50625. Public Library of Science
10. Lo A, Chernoff H, Zheng T, Lo SH (2015) Why signifcant variables arent automatically good predictors. Proc Natl Acad Sci U S A 112(45):13892–13897
11. Plaze M, Bartrs-Faz D, Martinot J, Januel D, Bellivier F et al (2006) Left superior temporal gyrus activation during sentence perception negatively correlates with auditory hallucination severity in schizophrenia patients. Schizophr Res 87(1–3):109–115. doi:10.1016/j.schres.2006.05.005
12. Cecchi G, Rish I, Thyreau B, Thirion B, Plaze M, et al. (2009) Discriminiative network models of schizophrenia. In: Proc. of NIPS-09
13. Rish I, Cecchi GA, Heuton K (2012) Schizophrenia classification using fMRI-based functional net-work features. In: Proc. of SPIE Medical Imaging 2012
14. Woods S (2003) Chlorpromazine equivalent doses for the newer atypical antipsychotics. J Clin Psychiatry 64:663–667
15. Banerjee O, El Ghaoui L, d'Aspremont A (2008) Model selection through sparse maximum likelihood estimation for multivariate gaussian or binary data. J Mach Learn Res 9:485–516
16. Meyer-Lindenberg A, Poline JB, Kohn P, Holt J, Egan M et al (2001) Evidence for abnormal cor-tical functional connectivity duringworking memory in schizophrenia. Am J Psychiatry 158(11):1809–1817
17. Zhou Y, Liang M, Tian L, Wang K, Hao Y et al (2007) Functional disintegration in paranoid schizophrenia using resting-state fMRI. Schizophr Res 97:194–205. doi:10.1016/j.schres.2007.05.029
18. Bluhm R, Miller J, Lanius R, Osuch E, Boksman K et al (2007) Spontaneous low-frequencyuctu-ations in the BOLD signal in schizophrenic patients: anomalies in the default network. Schizophr Bull 33:1004–1012. doi:10.1093/schbul/sbm052
19. Micheloyannis S, Pachou E, Stam C, Breakspear M, Bitsios P et al (2006) Small-world networks and disturbed functional connectivity in schizophrenia. Schizophr Res 87:60–66. doi:10.1016/j.schres.2006.06.028
20. Whitfield-Gabrieli S, Thermenos H, Milanovic S, Tsuang M, Faraone S et al (2009) Hyperactivity and hyperconnectivity of the default network in schizophrenia and in first-degree relatives of persons with schizophrenia. Proc Natl Acad Sci U S A 106:1279–1284. doi:10.1073/pnas.0809141106
21. Sui QYJ, Rachakonda S, He H, Gruner W, Pearlson G et al (2011) Altered topological properties of functional network connectivity in schizophrenia during resting state: a small-world brain network study. PLoS One 6: e25423. doi:10.1371/journal.pone.0025423
22. Friston K, Harrison L, Penny W (2003) Dynamic causal modelling. NeuroImage 19 (4):1273–1302. doi:10.1016/s1053-8119(03)00202-7
23. Zhang L, Samaras D, Alia-Klein N, Volkow N, Goldstein R (2006) Modeling neuronal interactivity using dynamic bayesian networks. In: Advances in neural information processing systems 18. MIT Press, Cambridge MA, pp 1593–1600
24. Storkey AJ, Simonotto E, Whalley H, Lawrie S, Murray L et al (2007) Learning structural equa-tion models for fMRI. In: Advances in neural information processing systems 19. MIT Press, Cambridge MA, pp 1329–1336
25. Kircher T, Oh T, Brammer M, McGuire P (2005) Neural correlates of syntax production in schizophrenia. Br J Psychiatry 186:209–214. doi:10.1192/bjp.186.3.209
26. Benedetti F, Poletti S, Radaelli D, Bernasconi A, Cavallaro R et al (2010) Temporal lobe grey matter volume in schizophrenia is associated with a genetic polymorphism inuencing glycogen synthase kinase 3-β activity. Genes Brain Behav 9:365–371. doi:10.1111/j.1601-183x.2010.00566.x
27. Lawrie S, Buechel C, Whalley H, Frith C, Friston K et al (2002) Reduced frontotemporal func-tional connectivity in schizophrenia associated with auditory hallucinations. Biol Psychiatry 51:10081011. doi:10.1016/s0006-3223(02)01316-1

INDEX

Tatiana V. Tatarinova and Yuri Nikolsky (eds.), *Biological Networks and Pathway Analysis*, Methods in Molecular Biology, vol. 1613, DOI 10.1007/978-1-4939-7027-8, © Springer Science+Business Media LLC 2017

I

K

M

N

O

P

W